Temporality

Culture in the Flow of Human Experience

A Volume in
Advances in Cultural Psychology:
Constructing Human Development

Series Editor
Jaan Valsiner, *Aalborg University*

Advances in Cultural Psychology: Constructing Human Development

Jaan Valsiner, Series Editor

Temporality: Culture in the Flow of Human Experience (2015)
edited by Lívia Mathias Simão,
Danilo Silva Guimarães, and Jaan Valsiner

Making Our Ideas Clear: Pragmatism in Psychoanalysis (2015)
edited by Philip Rosenbaum

Biographical Ruptures and Their Repair: Cultural Transitions in Development (2014)
by Amrei C. Joerchel and Gerhard Benetka

Culture and Political Psychology: A Societal Perspective (2014)
by Thalia Magioglou

Fooling Around: Creative Learning Pathways (2014)
edited by Lene Tanggaard

Cultural Psychology of Human Values (2012)
by Jaan Valsiner and Angela Uchoa Branco

Lives and Relationships: Culture in Transitions Between Social Roles (2013)
edited by Yasuhiro Omi,
Lilian Patricia Rodriguez, and María Claudia Peralta-Gómez

Dialogical Approaches to Trust in Communication (2013)
edited by Per Linell and Ivana Markova

Crossing Boundaries: Intercontextual Dynamics Between Family and School (2013)
by Giuseppina Marsico, Koji Komatsu, and Antonio Iannaccone

Cross-Cultural Psychology: Why Culture Matters (2013)
by Krum Krumov and Knud S. Larsen

Interplays Between Dialogical Learning and Dialogical Self (2013)
edited by M. Beatrice Ligorio and Margarida César

Dialogic Formations: Investigations into the Origins and Development of the Dialogical Self (2013)
edited by Marie-Cécile Bertau,
Miguel M. Gonçalves, and Peter T. F. Raggatt

Cultural Dynamics of Women's Lives (2012)
edited by Ana Cecília S. Bastos, Kristiina Uriko, and Jaan Valsiner

Culture and Social Change: Transforming Society through the Power of Ideas (2012)
by Brady Wagoner, Eric Jensen, and Julian A. Oldmeadow

Cultural Psychology of Human Values (2012)
by Jaan Valsiner and Angela Uchoa Branco

Researcher Race: Social Constructions in the Research Process (2012)
by Lauren Mizock and Debra Harkins

Cultural Psychology and Psychoanalysis: Pathways to Synthesis (2011)
edited by Sergio Salvatore and Tania Zittoun

Apprentice in a Changing Trade (2011)
edited by Jean-François Perret, Anne-Nelly Perret-Clermont, Danièle Golay Schilter, Claude Kaiser, and Luc-Olivier Pochon

Constructing Patriotism: Teaching History and Memories in Global Worlds (2011)
by Mario Carretero

Methodological Thinking in Psychology: 60 Years Gone Astray? (2010)
edited by Aaro Toomela and Jaan Valsiner

Living in Poverty: Developmental Poetics of Cultural Realities (2010)
edited by Ana Cecília S. Bastos and Elaine P. Rabinovich

Relating to Environments: A New Look at Umwelt (2009)
edited by Rosemarie Sokol Chang

Rethinking Language, Mind, and World Dialogically (2009)
by Per Linell

Innovating Genesis: Microgenesis and the Constructive Mind in Action (2008)
edited by Emily Abbey and Rainer Diriwächter

Trust and Distrust: Sociocultural Perspectives (2007)
edited by Ivana Markova and Alex Gillespie

Discovering Cultural Psychology: A Profile and Selected Readings of Ernest E. Boesch (2007)
by Walter J. Lonner and Susanna A. Hayes

Semiotic Rotations: Modes of Meanings in Cultural Worlds (2007)
edited by SunHee Kim Gertz, Jaan Valsiner, and Jean-Paul Breaux

Otherness in Question: Development of the Self (2007)
edited by Lívia Mathias Simão and Jaan Valsiner

Becoming Other: From Social Interaction to Self-Reflection (2006)
edited by Alex Gillespie

Transitions: Symbolic Resources in Development (2006)
by Tania Zittoun

Temporality

Culture in the Flow of Human Experience

edited by

Lívia Mathias Simão

Danilo Silva Guimarães

Institute of Psychology, University of São Paulo

and

Jaan Valsiner

Aalborg University

IAP

INFORMATION AGE PUBLISHING, INC.

Charlotte, NC • www.infoagepub.com

Library of Congress Cataloging-in-Publication Data

Temporality : culture in the flow of human experience / edited by Lmvia
Mathias Simco , Danilo Silva Guimarces, Institute of Psychology, University
of Sco Paulo, and Jaan Valsiner, Aalborg University.
pages cm. -- (Advances in cultural psychology)
ISBN 978-1-62396-967-7 (paperback) -- ISBN 978-1-62396-968-4 (hardcover) --
ISBN 978-1-62396-969-1 (ebook) 1. Culture--Psychological aspects.
2. Time--Philosophy. I. Simco, Lmvia Mathias, editor.
HM621.T45 2015
306--dc23

2014048360

Printed and bound by CPI Group (UK) Ltd, Croydon, CR0 4YY

CONTENTS

PART III: DEVELOPING WOMEN AS GLOBAL LEADERS

INTRODUCTION

TIME—NOT ALWAYS THE SAME

Lívia Mathias Simão

This book comes as part of a broader project the first editor is developing in collaboration with the other two, aiming critically to articulate some central philosophical issues with semiotic-cultural psychology and related areas in its frontier.

As such, the project—of which the present book is part—concerns a metatheoretical perspective aiming to bring new theoretical-methodological and ethical reflections to semiotic-cultural psychology. This metatheoretical perspective has been called semiotic-cultural constructivism (Simão, 2005, 2010; Simão & Valsiner, 2007), as it implies Science as one particular kind amongst cultural constructions in human questioning for meaningful I-Other-World relationships.

This questioning is a subjective process addressed to human development, from which I-Other-World symbolic relationships are constantly unfolding. These symbolic relationships form the sociocultural milieu. In science, this Other is also (although not only) the so-called object of research; in human sciences, as in psychology and its frontiers, this Other is the research participant. In reciprocity, the researcher is also the other for the persons we study (Simão, 1989, 1992, 2007, 2010).

In such processes of knowledge construction, each of them actively transforms the other's expressions, induced and allowed by the "research design," trying to integrate the other's message in her own personal

Temporality: Culture in the Flow of Human Experience, pp. xi–xiv

affective-cognitive basis, which will also be transformed in this process. Everyday and scientific knowledge are part of this basis. In a few words, this is a process that concerns and embraces a dynamic symbolic interplay among the involved persons, allowing them a momentary understanding of the world in the face of their expectances, hopes, and concerns about it. This process happens also inherently in scientific knowledge construction, unfolding in new proposals about the I-Other-World relationships, as those present in the chapters of this book.

Similar to the previous milestone in this effort—our book (Simão & Valsiner, 2007) published in this same series—in the present volume, we invited an international cast of authors to bring their perspectives about a possible dialogue between a central philosophical issue and the core subject of their respective research domains. This time the philosophical issue is about time.

Time is a notion regarding central aspects of living, then its philosophical relevance, as well as for other areas of knowledge. Different conceptions about time brought and still bring different conceptions about I-Other-World relationships; they are in the origins of different ways of conceiving and trying to understand our human affairs, plans, hopes, and fears. Consequently, they are also in the origins of different social and professional practices concerning us.

The notion of temporality, to which this volume addresses, is certainly amongst the most complex in Western philosophy. Without purposes of exhausting it in this introductory chapter, it can be worthwhile to anticipate here some of its main features.

Talk about *temporality* implicitly implies the making of a distinction between *time conceived as an objective dimension of the world,* which is independent of the subject who perceives it, and *time conceived as a subjective experience,* which is not independent of the subject who lives within that objective time.

Although the roots of this distinction can be traced back to Augustine (354–430), and closer to our times to William James (1842–1910) and Henri Bergson (1859–1941) (see, for instance, Simão, Baerveldt, Klempe, Tateo, and Vedeler's chapters in this volume), the notion of temporality has its roots in the phenomenological analysis of human experience of time brought by the Moravian philosopher Edmund Husserl (1859–1938) and deeply developed by the French philosopher Maurice Merleau-Ponty (1908–1961). Important to us for the present discussion is that they both highlight that people experience time in at least two simultaneous ways: *we feel ourselves here and now* and *we feel that time passes.*

Time gives rise to temporality, as it is part of the subjective processes of the personal I-World (Other) construction. This process involves synthesis of already previously emerged and established meanings that constrain the current I and the Other relations-in-the making (Valsiner, 1998, 2007). In such a way, the past never ceases being; yet it never recurs in the same form as it did before. The idea of time as a sequential series of "befores" and "afters" as if they were nondependent on personal experience is then put aside.

As we move from time to temporality, we have, then, a shift in the notional character of time as human experience; instead of belonging to the objective dimension of the world, temporality is an inner experience that constitutes our I-World (Other) relationships. The main implication of this shift for our discussion is that the I-World relationship becomes intertwined with temporality.

In such a way, one of the most valuable aspects of the present book is to bring different ways of dealing with the issue of time in semiotic-cultural psychology. This is the proposal of our project that we bring to the reader in this book.

Implied in our project is the invitation to the reader to join us in being captivated by issues such as those concerning the boundaries of temporality and personal imagination, dwelling on one of the most important human potentialities for dealing with the future in its uncertainty, which generates anguish and hope as well. Also, those about the relationship among time, disquieting experiences and negotiation of meanings, as in Klempe, Marsico, Abbey, and Pontes & Bastos' contributions to this book.

When thinking about time, we can also take the perspective of asking if the distinction between objective and subjective time makes any difference for a cultural-psychological approach to time, as discussed by Cornejo and Olivares. But objective and subjective time addresses us also to other issues, as those encompassed in the temporal perception from the point of view of the psychophysics and neuroscience, as discussed by Cravo and Haddad. Afterwards, all these touches to one's conception time of life and in life, development, posed by Vedeler in his almost monographic chapter in the book.

At the same time, we need to take into account that dealing with time in abstract terms avoids the uncomfortable complexities of ontological time by paying the price of avoiding the subject of one of her most human path marks, being temporarily in her world with others, as discussed by Ellis and Stam. In such a way, to approach the issue in diverse and contextual I-Other-World situations in which human life is challenged also becomes a challenge for researchers, as those faced and reported by Coelho, Freda et al., Laskovski & Simão, and Watzlavik.

Besides, time is culturally contextualized, as discussed by Chaudhary and Guimarães in their respective chapters, and in touch with human expression about human life, as in the arts, as approached by Sampaio & Simão as well as Tardivo.

As posed at the beginning of this Introduction, the semiotic-cultural constructivism in psychology is a *perspective*. Perspectives are constructed in the present frame, represented by reflections, research based, and communicated in the aforementioned chapters of this book. Nevertheless, perspectives also point to what is still not, but may be, or may be not. Regarding the issues about the differentiation between time and temporality, what about for psychology? In the present book, Baerveldt, Tateo, and Simão's chapters go in this direction by discussing and elaborating proposals, respectively, about the time for genetic psychology, time and generalization in psychology, and the hermeneutic notion of tradition as timing for the perspective of the semiotic-cultural constructivism in psychology.

REFERENCES

Simão, L. M. (1989). Interação pesquisador-sujeito: a perspectiva da ação social na construção de conhecimento (Researcher-subject interaction: the actional perspective in construction of knowledge). *Ciência e Cultura, 12,* 1195–1202.

Simão, L. M. (1992). Interação verbal e construção de conhecimento: aspectos do diálogo pesquisador-sujeito (Verbal Interaction and knowledge construction: aspects of the researcher-subject dialogue). *Psicologia: Teoria e Pesquisa, 8*(2), 219–229.

Simão, L. M. (2005). Bildung, culture and self: A possible dialogue with Gadamer, Boesch and Valsiner? *Theory & Psychology, 15,* 549–574.

Simão, L. M. (2007). We researchers: Unquiet people disquieting others. A commentary on Duarte & Gonçalves' "Negotiating motherhood: A dialogical approach." *International Journal for Dialogical Science, 2*(1), 277–285.

Simão, L. M. (2010). *Ensaios dialógicos: compartilhamento e diferença nas relações eu-outro (Dialogical essays: sharing and difference in I-other relationships).* São Paulo, Brazil: HUCITEC.

Simão, L. M., & Valsiner, J. (Eds.). (2007). *Otherness in question: Labyrinths of the self.* Charlotte, NC: Information Age.

Valsiner, J. (1998). *The guided mind: A sociogenetic approach to personality.* Cambridge, MA: Harvard University Press.

Valsiner, J. (2007). *Comparative study of human cultural development. Culture in minds and Societies.* New Delhi, India: Sage.

PART I

TEMPORALITY AND ITS BOUNDARIES

CHAPTER 1

TEMPORALITY AND THE NECESSITY OF CULTURE IN PSYCHOLOGY

Sven Hroar Klempe

Temporality must be regarded as an ambiguous term. On the one hand, it refers to accuracy in an exemplary way. This is the chronometric time, which is divided into exact equal parts: years, hours, minutes, and seconds. This corresponds to space, to which we also have metrical terms to express exactness. In this sense, both time and space are the "universal forms of knowledge," as Schopenhauer expressed it (Schopenhauer, 1954, p. 5). Yet on the other hand time and space are also to be regarded as "the conditions of multiplicity" (Schopenhauer, 1954, p. 5). In this respect, time and space are rather concrete and refer to particularity and diversity.

However, in experimental psychology, time and space do not necessarily belong to the same category in this respect. In dissociation research, for example, the events are primary related to spatial operations, whereas temporal dissociative events are hard to measure because this requires a comparison between two separate accurate measurable timelines, which probably do not exist (Müsseler & Prinz, 2002). Whereas we can observe how a body adjusts itself to a world perceived through a pair of reverse glasses, by which the world is given as mirrored, it is harder and almost

Temporality: Culture in the Flow of Human Experience, pp. 3–22

impossible to measure the experienced starting point of a dominating sound that lasts for a while. In the experiment of complexity (*Complicationsversuche*) from 1902, Wundt demonstrated that if a person pays attention to the second hand on a watch, and an unannounced tone starts sounding unexpectedly, the sound is experienced as having a prior entry than it actually has (Müsseler & Prinz, 2002). However it is hard, or maybe even impossible, to measure when the experienced starting point is. The experienced time is hardly observable, and in fact it would probably appear as diverse as life itself. This aspect of time opens up for what the Danish philosopher Søren Kierkegaard called "the foaming multifariousness of life," which is the formulation he applies in an attempt at defining psychology as a science (Kierkegaard, 1980, p. 23). According to Kierkegaard, temporality is the most important term for expressing the actual life in terms of existence, and it is intimately related to the core aspects of psychology as a science. Yet this implies that temporality primarily refers to multiplicity, particularity, and the more or less chaotic aspects of life. In this respect, temporality stands in stark contrast to unity, order, and generality. However, this opens up a fundamental and a contradictory conflict embedded in the term itself, which probably makes it one of the most fascinating and challenging terms to pursue intellectually, not least to achieve a deeper understanding of what psychology is supposed to be about.

Psychology therefore appears to be the core field in which this contradiction is revealed. This is why the crisis in this science is so obviously admitted (Bühler, 1926; Vygotsky, 1997). Simultaneously with these statements about crisis in psychology, in 1929, a meeting between the two intellectual giants from the early 20th century, Martin Heidegger and Ernst Cassirer, took place in Davos (Cohen, 2006; Gordon, 2012). Both are more or less to be regarded as neo-Kantians, but their conclusions about the intellectual heritage from Kant were not only different but went in two opposite and irreconcilable directions. They both also have had tremendous impact on some parts of the psychology even today. This is primarily related to the understanding of the relationship between subjectivity and culture. Yet these two aspects are by both placed in a broader perspective, which can be summarized in questioning to what extent psychology and ontology can be combined. In other words, there is a certain distinction between those two, which might be crucial to take into account in the achievement of understanding the relationship between psychology as a science and culture. Yet the key to understanding the differences is primarily embedded in temporality.Thus in this chapter, I will pursue this term with the aim of finding out in what perspective cultural psychology can be understood. Examining some crucial aspects of time when it comes to its relationship to subjectivity, objectivity, metaphysics, and culture will delineate the pathway to follow in this investigation.

The Terms Are Ambiguous

The ambiguity is probably also revealed in an apparently confusing use of the two parallel terms "time" and "temporality." On the one hand, they may refer to the same in the sense that both are primarily referring to the fact that duration is the condition to which human life is bounded. Thus, time cannot exist without temporality and vice versa. On the other hand, the term "time" seems to be a more paramount term in the sense that it may refer to a variety of meanings. When we talk about the chronometric aspect of duration, it is probably more natural to apply the term "time" instead of "temporality," exposed through the use of words in questions like, What time is it? However "time" may also refer to the actual life, demonstrated through the derived form '"timely," which refers to the limited experienced time. Moreover, the term may also refer to obligations that have to be done in the future in the sense that they have to be done in time. This perspective can be applied to acts in retrospect as well, in the sense that they were done in time. In other words, this term refers to almost all the aspects of duration, whether it is the actual experienced moment, the scientific accuracy of moments, or the actual lifespan we are living in. This is how the term will be applied in this context as well. "Time" will be used as a general term that includes all aspects of duration, whereas "temporality" will rather refer to the actual experiences of time in life.

The Absurdness of a Timeline

One of the sharpest analyses of time was made by the very early Christian philosopher Augustine (354–430 CE). He extrapolated some of the most salient aspects of time, which do not make time less ambiguous but more understandable. He sorted out some aspects, which make the notion of time quite clear and highly recognizable. On the one hand, he highlighted the importance of the moment by stating, "The drops of time are precious with me" (Augustine, 1999, p. 11.2.2.). Without saying it explicitly, time is first of all given as subjective experience. This is the basic principle for the whole analysis of time he made, and it is primarily related to some certain interests and perspectives. Thus, the experienced moments are filled with interests. That is why we all recognize the moments, not only as precious but very often even as great as they can be. If we refer to a great experienced event, we are talking about a "moment" and not a longer time span. Thus, the moment includes the whole event experienced as complete without taking into regard the duration of it. However, the moment does not have to be related to good experiences solely. An upsetting event is realized in the very moment it shows up, and afterwards it may stay as a nagging sore for a long time. Whether the moments are great or cruel, they always appear as subjective and quite private experiences. Sometimes they are understood by others, but definitely not always.

Another important aspect of time Augustine (1999) highlights is the illusory aspect of it. Time is apparently a ubiquitous event, but a more meticulous investigation reveals that it is not exactly like that. Time is experienced as a line consisting of three parts: past, present, and future, as illustrated in Figure 1.1.

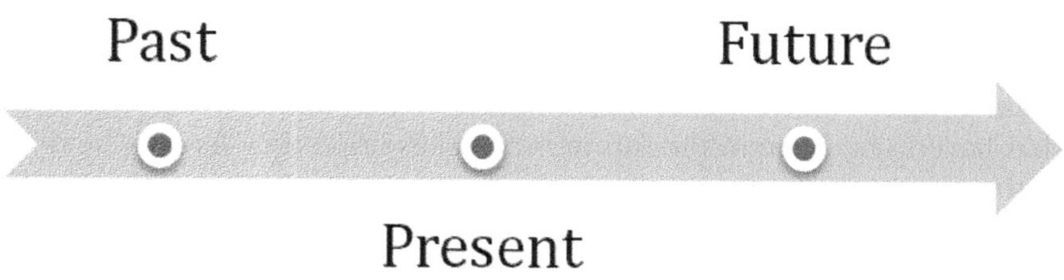

Figure 1.1. Linear timeline.

Yet Augustine even talks about these as three different times (p. 11.17.22). However his main concern is that he realized that two of them do not actually exist. The past is by definition what once existed, but does not exist anymore; and the future is by definition what has not yet come to be and therefore does not exist either. This also makes the linear understanding of time quite impossible. The timeline presupposes a row of points in which just one of them is the present and all the other points, which are an infinite abundance of points, do not exist. And if they do not exist, there cannot be any lines either. Geometrically, a line can be defined in terms of having a direction given by its starting and ending points. A point, on the other hand, can be defined in terms of having no direction, which is the same as including all directions (see Figure 1.2).

This implies that a point can geometrically never be a line. According to Augustine (1999), therefore, the only thing that exists is the point of present, which can never be presented as if it was a point in a geometric line. This supports the dissociation research mentioned above, which makes a clear distinction between time and space. In this respect, the whole metaphor of regarding time in terms of a line is quite simply absurd. Moreover, if what we have left is just a point, then time cannot exist either, because a point is rather spatial and cannot represent time at all, so "What is time?" is the desperate question Augustine ends up with: "Who can readily and briefly explain this? Who can even in thought comprehend it, so as to utter a word about it?" (p. 11.14.17). Well, Augustine gives the answer himself by saying, "Times are made by the alterations of things" (p. 12.8.8). In other words, time is not a formless and abstract entity, but the actual experiences of life in terms of movements and changes.

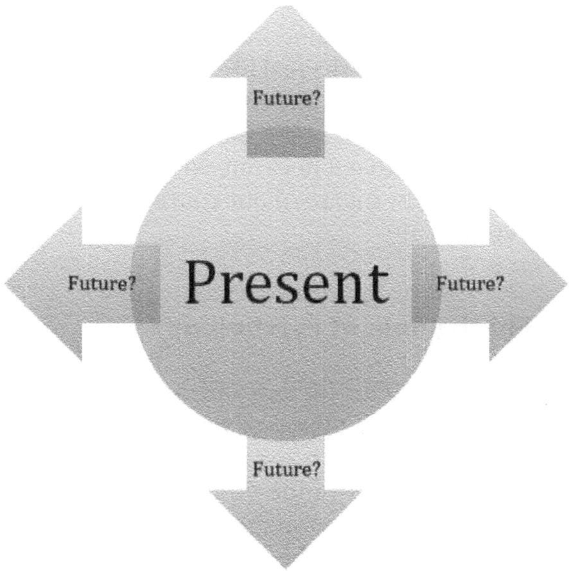

Figure 1.2. Nondirectional point in time.

The Role of Eternity According to Augustine

Since time does not exist as an abstract entity, we rather experience what appears as chronometric time because of changes and transitions. This is the subjective and private perception of time solely related to the actual life, which is exactly what we call temporality. Thus, temporality is not about the chronometric time but rather the earthly life each one of us is living. And Augustine (1999) applies this term when he talks about the life we are living here, which he relates to the "temporal nature" (p. 13.21.31) man is bounded to. However, in this analysis of time presented here, there are some black holes left standing. If it is so that chronometric time does not exist in actual lives, how can we then have watches and even let them control and regulate our daily lives? The answer Augustine presents is by referring to eternity. This is a critical point in Augustine's reasoning, and it even forms the basis for his whole analysis of time. Augustine is a Christian mystic, and the point for him is to underline the differences between the miserable earthly life and the divine happiness, and this is exactly what he does by subverting the role of time in actual life. Temporality versus eternity may illustrate exactly this, and therefore he concludes that temporality "may suffice that eternity be clearly seen, being understood by the things that are made" (p. 13.21.31). In other words, we acquire a conception of eternity by living in temporality, and the contrast between the two aspects of life generates a notion of a chronometric time. Although neither past nor future exists in the real life, we think about what happened and not least what will happen in the future, which he calls "forethinking [in] present" (p. 11.18.23). This is the most crucial aspect of Augustine's analysis of time, specifically that it does not exist in the real experienced life but solely in our thinking by means of our conception of "what will come," which points toward eternity.

Kierkegaard and the Role of Psychology

Although Augustine (1999) focuses on the subjective experience of time, he actually ends up with a kind of neglecting of its existence, both in actual life and in heaven, so to speak. However, the most interesting aspect of Augustine's analysis of time though, is how he emphasizes the distinction between something unpredictable and even chaotic in actual life and stability and order in thinking. This is of course the heritage he had from Plato, so his contribution was rather related to an elaboration of the subjective experiences of the actual life, of which Plato did not pay too much attention. The one that really makes these aspects intriguing, and in addition focuses on psychology, is Søren Kierkegaard. On the one hand, he is following up Augustine's project in making a distinction between the lived life and thinking. Yet so many thoughts have been thought during the 1,500 years after Augustine, that Kierkegaard ended up with another conclusion, specifically that *truth* is first of all related to subjectivity. However, to end up with this conclusion, he had to focus on the science that really dealt with subjectivity, and this was psychology. This is the big difference between Augustine and Kierkegaard. Augustine does not mention psychology at all, whereas for Kierkegaard, psychology is the science that reveals and focuses on actual, real life. Moreover, Kierkegaard uses psychology as a device for disclosing the fundamental distinction between subjectivity and objectivity. This is why Kierkegaard appears to be so actual—quite simply because our notion of knowledge even today is apparently related to objectivity, whereas the foundation for it is given through subjective impressions. In the premodern area, there was a tendency to overlook the distinction between objectivity and subjectivity in terms of just ignoring subjectivity. Yet we can trace the same tendency today, not by ignoring subjectivity, but rather by ignoring the distinction between subjectivity and objectivity. For some other reasons, but quite parallel, the French sociologist Bruno Latour concluded, "We have never been modern" (Latour, 1993), which probably is still true in very many respects.

However, there is certainly a change coming up after Descartes, and one of the most crucial is probably also one of the most ignored, which is the introduction of psychology as an independent science. This is exactly what Kierkegaard deliberately highlights, though without too much support from other contemporaneous scholars. This lack of support has had tremendous consequences for the understanding of psychology today. According to Kierkegaard (1980), psychology is the science of subjectivity, and in this respect he is relying on the contemporaneous author Karl Rosenkrantz, who published in 1837 a popular introduction to psychology with the informative subtitle: "The Science of the Subjective Spirit" (*Die Wissenschaft vom subjectiven Geist*) (Rosenkrantz, 1863). Yet immediately before, Kant had disqualified psychology as a science because of its lack of objectivity, and Hegel had effectively ruined it by absorbing all subjectivity into the philosophy of the absolute spirit. Yet the fact that Rosenkrantz also was one of the most committed disciples of Hegel illustrates very well what kind of mind games that existed, and psychology is probably standing in the middle of it still. Nevertheless, Kierkegaard, Kant, and Hegel represented a crossroad that points in three different directions: One is emphasizing objectivity as the criterion, the other is emphasizing subjectivity, and the third is emphasizing both, in the sense that they cannot be mixed up. Kierkegaard represents the third alternative, and this implies that his focus on subjectivity is conditional, in the sense that it is the negation of objectivity and vice versa.

Kierkegaard's interest was primarily related to religious ponderings, which dominated his whole life, even from the very beginning. However religious ponderings and theology are actually two different things in the sense that the former is about subjective experiences of religiosity, and the latter is about objective religious doctrines. Thus, all his writings are about the conflict between those two spheres, which actually appear as irreconcilable, albeit they are reciprocally dependant on each other. So in his investigations of objective doctrines in Christianity, Kierkegaard applies theology, but when he investigates his own religious ponderings, he applies psychology. In other words, to investigate the real life psychology is in his perspective a *sine qua non*. What he then discovers through his psychological investigations is that we, in addition to living the real foaming and multifarious life, also have a notion of their fundamental contrasts, namely, order, stability, and harmony. This contrast is first of all related to our conception of time in the sense that our real lives are given in temporality, whereas we, according to Kierkegaard, always regard it in the perspective of eternity, or *sub specie aeternitatis*, which is one of the most precious phrases of Kierkegaard. In this sense, Kierkegaard is following up the tradition from Augustine, but he is also to be regarded as a representative for German idealism, but slightly different from both Kant and Hegel.

Kant and the Transcendental Understanding of Time

Now we are back where we started, namely, with the inherent duality of time. Yet the understanding has hopefully come a step further. Instead of dealing with an either/or, we have, according to Kierkegaard, to include both ways. So the crucial question is, In what way? However, the point of departure has to be subjectivity. This was Kierkegaard's stance, and this is also the main content of the transition from premodernity to modernity, which Immanuel Kant described as a Copernican revolution and Edmund Husserl explained as the "Cartesian overthrow … of an absolute grounding of science" (Husserl, 1999, p. 7). Objectivity was no longer to be regarded as the basis for knowledge, which rather had to be located in the subject. So if both subjectivity and objectivity should be included as the basis for knowledge, the problem was not first of all related to subjectivity, but objectivity. Kierkegaard did not deal with that problem very much. He just referred to metaphysics and dogmatic as given and indisputable sciences. He respected them, but he did not discuss very much their content. Kant on the other hand, had the opposite perspective. He became concerned about the content of metaphysics because the basis for it was not there anymore. So Kant's critical philosophy was first of all a restorative project of the same objectivity that metaphysics had provided, but lost its basis for.

Kant's investigation of time therefore is illustrative for this restoration project. The perspective he applies in his investigation on time is first of all about finding the "pure form of sensuous intuition in general" (Kant, 1999, pp. A20, B34–35). This is what he calls "transcendental aesthetic," which is about all the principles of sensibility that exist a apriori (Kant, 1999, pp. A20, B34). Time is exactly this, because, he says, "Time is not an empirical conception. For neither coexistence nor succession would be perceived by us, if the representation of time did not exist as a foundation given a priori" (Kant, 1999, pp. A30, B46). This is what Augustine pointed at too; specifically that time does not exist in real life, but we do have a certain notion of time anyway. In this respect, Kant makes a clear distinction between psychology and philosophy. Psychology is about the immediate and actual sensations and the notions they create in the mind. In Kant's investigation, the aim is rather to investigate principles and conditions for the notions we may have without including sensations or their sources. So when he is focusing on time, he ends up with the pure idea of time, which is sequential and given in a past, present, and future. He ends up with the same conclusion as Augustine, but he is focusing on the counterpart. Whereas Augustine highlighted the actual experienced time, Kant is rather focusing on the ideal aspects of time. Yet both come up with the same result, namely, that past, present, and future only exist in an ideal understanding of time, which is different from the actual time, in addition to the fact that it is almost impossible to experience actuality without placing the experiences in either the past, present, or the future. In this respect, time, combined with temporality, defines very much the area in which actuality and ideality meet and become more or less united.

Due to Kant's (1999) terminology, "aesthetics" has nothing to do with "theories of art," but rather about "sensation," which is the original meaning of the Greek word"aisthesis." Kant even complains about what he calls a German tendency to associate "aesthetics" with theories of art, which he rather relates to the capacity of making judgments, which was the content of his third critique. However, instead of applying the term "aesthetics," Kant could have applied the term "empirical psychology," because this was the term that referred to "sensation" at that time. Yet Kant did not apply that term, and he could not either, because he had banned empirical psychology from a discussion on theoretical foundations for science. This was also a strategic move in an attempt at avoiding ending up with psychologism, that is, that all kinds of knowledge could be reduced to psychological expectations based on habits, like Hume and other British empiricists had done. Thus, not only empirical psychology had to be avoided, but he also introduced the term "transcendental," which refers to what he calls the "supersensible sphere" (p. B6). The transcendental goes beyond psychology in the sense that it is not referring to the actual object but rather our conditions for acknowledging the actual object. In this sense, the transcendental represents the ideality of knowledge.

This perspective was continued by German idealism, and it was in a sense restored by the phenomenology of Edmund Husserl. In his lectures on time, he is quite clear about the distinction between the subjective experiences of time and what we normally call an objective time. His point is that these two cannot be mixed up: "One cannot discover the least trace of Objective time through phenomenological analysis" (Husserl, 1964, p. 24). This implies that the experienced now "is not a point of Objective time" (Husserl, 1964, p. 24). According to Husserl (1964), they do not exist in the same sphere. This is not just a characteristic of time, but of all objects out there. In this perspective, he underlines the irreconcilable gap between the process of sensing and what is sensed. As long as this gap exists, all talk about sensing an object is to be regarded as a transition, not only from the inner to the outer experience, but even more from those two types of experiences to something out there—the object per se—of which we in principle do not have a clue. This is why we have to talk about the objects in general, and time especially, as transcendental. They are, in principle, out of reach, but the mysterious aspect of this is that we still have an idea of what they are about, especially when it comes to the existence of objective time. So the fundamental question therefore would immediately be, how come?

The answer to this question is first of all given through phenomenological analysis, in which detecting the essential is the ultimate goal. However, the essential is highly related to ideality in the sense that the essential refers to the objective or intersubjective aspects of the phenomena. This implies that there is a double communication when Husserl combines phenomenology with the attempt at grasping the essential. Phenomena are given through the particular or the immediate experience of the subject. In this sense, phenomenology has much in common with psychology. However, he wants to make a clear distinction between the two areas in the sense that grasping the essential in the phenomenon brings phenomenology out of the psychological sphere because psychology is not about ideality, which rather the essential actually is about. This is why Husserl tries to avoid psychologism, which is explicitly declared and elaborated in his early work *Logical Investigations* (Husserl, 1970). However, this is followed up in his investigation of time in the sense that in this investigation, he endeavors first of all to uncover the objective aspects in something that is apparently given as pure subjectivity. That is why the terms "retention" and "protention" appear to be so important in his analysis. They transform the immediate subjective experience of the moment into a sequential and linear understanding of time, which includes the past and the future. This reveals that time is not subjective at all but rather transcendental and objective and must be understood in terms of ideality.

In this sense, there are some similarities between Kierkegaard, Kant, and Husserl. All three make a clear distinction between objectivity and subjectivity. Moreover, they all regard empirical psychology as being primarily associated with subjectivity, and they have in common the suggestion that time is one of the most interesting areas in which the relationship and distinction between subjectivity and objectivity can be detected. However, there is one important distinction between Kierkegaard, on the one side, and Kant and Husserl on the other. This concerns the interest and focus they have on the particular and subjectivity. Kant and Husserl take subjectivity as a point of departure, and they are both able to build up a transcendental understanding of objectivity, whereas Kierkegaard focuses on the opposite, namely, the immediate subjective impression. This is why Husserl quite categorically concludes, "We are indifferent to the question of the empirical genesis" (Husserl, 1964, p. 28). This refers to what he calls "psychological apperception," which is about how the mind includes new empirical impressions. This was exactly what empirical psychology was supposed to be about from Wolff to Husserl, and Husserl does the same as Kant, specifically excluding psychology from theory of knowledge, although they both are deeply influenced by empirical psychology in the sense that it was empirical psychology that introduced subjectivity to metaphysics and the foundation for scientific thinking.

Heidegger and Metaphysics

The combination of subjectivity and metaphysics has always been regarded as problematic in the sense that metaphysics is about objectivity. Yet time has been an important part of metaphysics (Baumgarten, 2004), and in this sense, metaphysics has been challenged by subjective issues, not only with regard to psychology but also in terms of dealing with time. Although Kant threw out psychology, not only from metaphysics but also from the foundation he made for scientific knowledge, he never threw out time or subjectivity as such. In this respect, Martin Heidegger is regarded as a neo-Kantian, not least because he restores the role of metaphysics and the role of time. On the one hand, Heidegger is following up Husserl by building upon his phenomenology. Yet on the other hand, he contradicts Husserl completely in the sense that subjective existence precedes essence, to apply the Sartrian slogan. When Heidegger gave his inauguration lecture in Göttingen in 1929, some said he was astonishingly focusing on metaphysics (Heidegger, 1994). In this respect, he had some similarities with Kierkegaard, not least by concluding that if man is regarded from a metaphysical perspective, its substance must be nothing else but subjectivity. The essence of subjectivity is therefore an "ecstatic-existential being," and this of course violates the framework of metaphysics (Heidegger, 1994, p. 71).

This is also the main message in *Being and Time*. One of the most important sentences in this thesis is, according to Heidegger, "*The 'essence' of Da-sein lies in its existence*" (Heidegger, 1996 p. 40, emphasis in original). However, only man has existence, says Heidegger (1994, p. 70); the cliff, the tree, the horse, the angel, yes, even God are there, but they do not exist. All objects are there, but they do not exist in the sense that they do not reflect upon their own subjectivity like a human being does. On this background, objectivity does not exist as an entity outside subjectivity, but "all objectivity as such is subjectivity" (p. 69). This is why time becomes so important in *Being and Time*. Just to be is not the most salient trait of a human being's life. This is rather characterized by living in a history dominated by movements and changes. In this respect, time is the most important phenomenon to grasp because it combines this fundamental subjectivity with the most stable and apparently objective foundation of life itself, namely, chronometric time, which is just running and running apparently without taking note of the fact that individual lives are lived and they come and go.

On this background, there are several similarities between Heidegger and Kierkegaard. They are eagerly restoring the role of metaphysics, emphasizing subjectivity, and they regard time as a mediator between the two. Thus, Heidegger has from the very beginning been regarded as being highly inspired by Kierkegaard (Gordon, 2012). Yet there is one important difference that is crucial in this context. According to Kierkegaard, psychology is the most vital subject to uncover the aspects of the lived life, whereas for Heidegger, it is rather the opposite. According to Heidegger, psychology is just a positive science, like many other sciences that do not include "'being' as a matter of question" (Heidegger, 1996, p. 46). So according to Heidegger, psychology lacks a proper ontological foundation for its knowledge. This is the objection Heidegger addresses to Kierkegaard as well. Heidegger admits that Kierkegaard is outstanding by explicitly grasping "the problem of existence as existentiell in a penetrating way" (Heidegger, 1996, p. 407), but "in an ontological regard he is completely under the influence of Hegel and his view of ancient philosophy" (Heidegger, 1996, p. 407). This quotation exemplifies two important aspects that highlight the differences between Heidegger and Kierkegaard. Yet these quotations are from Heidegger's notes and not from the thesis *Being and Time*. He never mentions Kierkegaard in his thesis, but the notes reveal that he had read him. To what extent he had read him well is in fact an open question. The differences between those two concern first of all the role of ontology. Heidegger is completely right when he says that Kierkegaard is referring to an ancient conception of ontology. Heidegger is, on the other hand, trying to develop another and more modern ontology. This is one important difference between the two. The other is the role of psychology in this respect. Heidegger's ontology attempts to achieve an ontology that comprises all sciences, also psychology. However, as we have seen, psychology is put aside because it does not fit his ontology. In this respect, Kierkegaard goes in the opposite direction. He is not interested in revising ontology or other sciences. He is, rather, interested in taking the full consequences of how they have been understood, and in this respect, he sees that ontology, as a part of metaphysics, stands in diametrical opposition to psychology, quite simply because ontology and metaphysics are about objective knowledge, whereas psychology is about subjectivity. Kierkegaard does not want to mix this up, and this is why he criticizes Hegel, because Hegel mixes these things up, especially by talking about logic as an explanation of the historical development. According to Kierkegaard, logic is a stable science beyond time, and it cannot be mixed up with typical timely areas like history, and Hegel mixes these things up by explaining historical development in terms of logic. According to Kierkegaard, logic is beyond time, and it cannot be blended with typical timely areas like history. In this sense, Hegel ends up in a quite wacky situation, which is so eloquently put into words by Kierkegaard:

> If anyone would take the trouble to collect and put together all the strange pixies and goblins who like busy clerks bring about movement in Hegelian logic, ... a later age would perhaps be surprised to see that what are regarded as discarded witticisms once played an important role in logic, ... which made Hegel's logic something of a miracle and gave logical thought feet to move on, without anyone's being able to observe them. (1980, p. 12)

Thus, it is hard to say that Kierkegaard is "completely under the influence of Hegel," as Heidegger stated (1996, p. 407). It is rather a question if Heidegger does justice to Kierkegaard by emphasizing this perspective on his relationship to Hegel and in addition claims that Kierkegaard's analysis of anxiety is made in a "theological context" (p. 405). The subtitle of Kierkegaard's book (1980) rather highlights a psychological context, which Heidegger mentions, but put in quotation marks, as if this is to be regarded as a more vicarious context and rather does not exist actually. There are probably reasons to conclude with the opposite, namely, that anxiety is to be understood as a psychological state, and that psychology is the main field Kierkegaard is focusing on. In this respect, Heidegger is right when he says that Kierkegaard "defines the Moment with the help of the now and eternity" (Heidegger, 1996 p. 413). This is the main difference between Heidegger and Kierkegaard, specifically that Kierkegaard refers to the contradicting references of the "now" and "eternity." The "now" is the existential experience of life, which neither theology nor any other objective sciences like metaphysics can say anything meaningful about; whereas the objective sciences can say something about the eternity, of which the "now" at the same time gives us a notion. In this sense, psychology is the subject that makes us aware of eternity and objectivity, but this does not imply that there are no distinctions between psychology and the objective sciences. Heidegger calls Kierkegaard's conception of time "vulgar" (1996, p. 412), but Heidegger precludes completely the role of psychology and makes it superfluous, not exactly in the same way as Kant or Hegel did, but at least with the same result.

Heidegger and the Transcendental "Now"

The role of our conception of eternity is critical. This was one of the core issues addressed in the dispute between Heidegger and Cassirer in Davos in spring 1929. Heidegger quite simply denied the existence of eternity. However, this is not exactly how he formulated his answer to this question. Rather, he focused on the illusory aspects of eternity in the sense that the eternal is just "possible on the grounds of an inner transcendence of time itself" (Gordon, 2012 p. 171). Yet the only transcendence Heidegger accepts is the transcendence of being (Heidegger, 1996). His understanding of this is quite illuminating when he adds, "The transcendence of the being of Da-sein is a distinctive one since in it lies the possibility and necessity of the most radical individuation" (Heidegger 1996, p. 34). In other words, only being can form the basis for transcendental knowledge, but being is at the same time to be in the moment, which stands in opposition to the abstract and generic aspects of the term "time." Thus, time is only real as long as it is experienced as being, which is in the moment of the now. This is in fact not very different from what Augustine, Kant, and Husserl actually said. Rather, the difference is related to the conclusion, namely, that instead of trying to seriously take into account the fact that we have a concept and a notion of eternity that contradicts the now, Heidegger rather reduces the eternal to a kind of hypergeneralized NOW.

In this sense, the strategy would be to deconstruct that kind of illusory conception of time and rather focus on the authentic manifestation of it, namely, the now. In this respect, the time is not only the now, but also the then, which implies that time, according to Heidegger, is the main gate in an attempt at understanding the actual existence, which is given in terms of the particular. Yet, as long as the particular forms the basis of ontology, and this is what he is saying when he states that individuation is transcendental, Heidegger is very much repeating Fichte and Hegel in the sense that subjectivity is replacing objectivity, which also made philosophy replace psychology.

Cassirer and the Role of Culture

Ernst Cassirer represents the opposite perspective. Time is not regarded as just being derived from an experienced being as an *a posteriori* entity, but rather as "created by thought itself a priori" (Gordon, 2012, p. 147). This reveals Cassirer as a neo-Kantian, but this does not imply that Cassirer is just reproducing Kant's project. The main difference lies in the understanding of science, which Kant referred to as pure, whereas Cassirer refers to it as changing and historical. This is the critical aspect of Cassirer's perspective; specifically that something apparently stable appears as changing over time. This is also true when it comes to science. Even mathematics is not stable in a longer time span. Infinitesimal calculus and functions are both modern forms in mathematics. The crucial question then is to find the point or an area that may guarantee an apparent stability yet at the same time opens up for changes. According to Cassirer, the answer to this is culture. However, culture might be so many different things that it is of course hard to define. Epistemologically, Cassirer applies psychology as having a mediational role in an attempt at uniting object and objectivity on the one hand and subject and subjectivity on the other; yet by itself this does not make him in fact very different from Kant, Hegel, and Heidegger. The big difference is that he *admits* that psychology is the main entrance to subjectivity. This is the role of the whole Chapter 3 in *The Philosophy of Enlightenment*; specifically to prove historically how psychology introduced subjectivity as a factor in modern thinking and the result of it (Cassirer, 1968). The chapter is entitled "Psychology and Epistemology" and was written immediately after the Davos meeting, and was first published in 1932. In this chapter, he quite simply states, "Psychology is designated as the foundation of epistemology" (Cassirer, 1968, p. 93).The reason for this is that the explanation of the human mind "is only to be found in its evolution" (Cassirer, 1968, p. 93). In other words, the human mind is changing, although it appears as stable and universal, and psychology is telling us how this can be so.

To bring in psychology as a core term in defining the basis for epistemology does not make things simpler, because psychology is too broad and can refer to almost everything. However, in a historical context, the term is more specified in the sense that it primarily refers to perception, which was the content of empirical psychology in the 18th century. Yet perception is not just a sensational process but it is also a concept that combines sensations with representations and ideas. In other words, psychology is first of all about how the mind mirrors the outer world by means of sensation. This is why Cassirer calls the Aristotelian statement, "Nothing is in the intellect which was not first in the sense" the "psychological axiom" (Cassirer, 1968, p. 99). Yet this is not just an argument for empiricism, but in Cassirer's understanding, rather the opposite. "When the minds become a mirror of reality, it is and remains a living mirror of the universe, and it is not simply a sum total of mere images but a whole composed of formative forces" (Cassirer, 1968, p. 124). These "formative forces" are exactly what German idealism historically has been exploring, and Cassirer is following up this line, but by ending up with different sorts of solutions. Thus, the purpose of Cassirer's chapter about psychology is to demonstrate that empirical psychology is explicitly used as a device for demonstrating the role of subjectivity in epistemology. In this respect, he is not only following up Kant, but also Kierkegaard and Husserl in avoiding ending up in psychologism."The formative forces" are not empirically given but rather the principles that form and structure the impressions we get from sensation. What these forces might be may change over time, and therefore they are to be found on such an abstract level that they hardly can be conceived directly, but rather given through human expressions, which are summarized in the term "culture."

According to Cassirer, culture is defined in terms of communication. One may easily apply Clifford Geertz' classical definition of culture "as a symbolic system" in this respect (Geertz, 1973, p. 17). Yet it is in fact more correct to say that this conception of culture can be traced back to Cassirer and especially his 3-volume work on *The Philosophy of Symbolic Forms* (1982), which was completed in 1929. In this work, the epistemological foundation for both mythical and scientific thinking is placed in language. The first volume therefore focuses on language, whereas the two others treat myths and science. There is a close connection between those three pillars, and this is expressed in a later small work on Culture:

> What we conceive as the "meaning" of the world we will meet all over where we direct us towards the super-individual, the general, what is valid for everyone, instead of closing ourselves into our own world of conceptions. And this option to and necessity of breaking through the barriers of individuality is no other places so obvious and certain as in the phenomenon of language. (Cassirer, 1994, p. 33)

Language, in other words, is transforming the individual conceptions from the particular to the general in terms of being a "formative force." The forms represent the *a priori* understanding of the world, and they are materialized in symbols, which get their highest form of expression in language. This understanding forms also Cassirers understanding of time, which both follows up and contrasts Heidegger's conception of time.

The third volume of *The Philosophy of Symbolic Forms* was completed immediately after the Davos meeting, and Cassirer makes a comment on Heidegger's conception of time in an added footnote. Although Cassirer agrees that the fact that time reveals the primordial existence of human being, we cannot relate existence to time unless we have a conception of a "time-form" (*Zeit-Form*) (Cassirer, 1982, p. 190). Time, in other words, represents the immediate existence manifested in the now, which is the finite aspect of life, but also our general and supra-individual understanding, which includes a conception infinity, which at least exists in our use of language.

Thus, the crucial aspect of course is how language is conceptualized. In this respect, Cassirer is referring to Humboldt as a kind of mixture of Kant and Herder, which makes language to be regarded as a function. "Language is quite simply not a product, but rather a continuous process, which constantly renews itself, and as this process proceeds, it gradually reveals for man the contours of the world in which he or she is living" (Cassirer, 1994, p. 34). In this respect, Cassirer is referring to the child's development of language, which he says follows exactly this pattern in the sense that the child's conception of the world goes from the undifferentiated to the differentiated. Although Cassirer does not mention the psychologist Heinz Werner, this is exactly his thesis about the development of language in the individual (Werner, 1957). Werner and Cassirer knew each other very well and were even colleagues in Germany before both of them had to flee the country. However, this understanding of the development of language is not just related to the childhood but for the individual's whole life span as well. This is, in other words, how language is to be understood, not as a stable system but something that changes due to the individual's needs to express him or herself in a process where the conversation is a question about being understood and accepted by other people. Thus, language expresses both community and individuality at the same time. This requires some conventions and a common framework, but also individual space for nuances that differentiate and highlight personal characteristics.

Wilhelm Wundt also regarded language as the main gate to the understanding of the human mind. "It follows, then, that psychological analysis of the most general mental products, such as language, mythological ideas, and laws of custom, is to be regarded as an aid to the understanding of all the more complicated psychical processes" (Wundt, 1902, p. 11.) This quotation reveals, on the one hand, the sort of elementarism Wundt aimed to follow in his psychological research, but on the other hand, when it comes to the higher mental functions, they are supposed to be found primarily in such ephemeral forms as language, myths, and norms, and those three form the pillars on which Wundt's folk-psychology was based. In this respect, his lectures on the gesture are of certain interest. They formed a basis for how to understand language, but they also tell us something about how the aspect of time is to be understood. The gesture has in common with language that it is an abstraction from the actual event to which it refers. It also has in common with language that the meaning of a gesture is established, not primarily by any connections to the reference, but rather by being in opposition to other gestures. To get any meaning out of gestures, the distinction between them appears as more important than the similarities they may have to their references. What gestural communication does not have in common with language is the same level of abstraction. "Gestural communication reports events exactly in the order in which they happen" (Wundt, 1973, p. 125). Therefore, "the time sequence in gestures is a replication of the temporal passage of the events themselves" (Wundt, 1973, p. 125). Thus, there is a gradual transition from gestural communication to language, and this is exactly why language is embedded with this double meaning or even ambiguity when it comes to the understanding of time. The particularity in temporal being, which Heidegger highlighted, is primarily related to the gestural aspects of language, whereas language's ability to abstract from the actual being results in a differentiation between past, present, and future. How these distinctions are expressed is subordinated to the fact that language actually is able to make these distinctions and by this, give us exactly the ambiguous understanding of time.

CONCLUSIONS

The ambiguity of time referred to in the beginning of this discussion is now ready for a revisit. And the question is of course if the term has been less ambiguous by now. As a matter of fact, we have to conclude that it has not. Rather, what we have found is what this ambiguity actually is about. It is about the fact that we live in actual reality, in which time is given as nothing less than a point, which is the actual now. However, this is just one aspect of our conception of time. We cannot avoid the need for placing the point of the now in the context of a past and a future, which are, on the other hand, not perceived as points but rather as a line that has a certain direction. What we actually do is that we mix up two geometrical phenomena, specifically, a point and a line, which are geometrically very close, but still two different entities. Yet this tendency to turn time into geometrical terms is also a fallacy in the sense that geometry is about space, and we found that time is different from space, both when it comes to the subjective experience and the understanding of it. So what this ends up with is that time represents two different realities at the same time: the lived and the thought. Time demonstrates therefore that lived and thought realities are completely different spheres, but equal in their ontological status. This implies that both realities have to be taken into account, and the fundamental question has always been where to find them, both as separate entities and in a natural combination. The answer to this question is "culture." This is the reality in which we both live and think.

Culture is the answer to a situation when metaphysics has fallen apart, and psychology is the science that makes "metaphysics run aground" (Kierkegaard, 1980, p. 18), but our conception of time also gives us a glimpse of some supra-individual aspects of thinking. In this respect, it is demonstrated two trenches into which one may easily fall. One is to neglect the subjective perspective, which is the premodern solution, which only focuses on the objective aspects of time and scientific knowledge. And if this is primarily related to a premodern perspective, we may conclude with Latour (1993) that at least some of us probably never have been modern at all. The other trench is to neglect the supra-individual aspects of time and other objective notions like logic and mathematics, which apparently are based on something detached from pure subjectivity.

To avoid falling into one of the two trenches it has been focused very much on Kierkegaard and Cassirer. Yet Kierkegaard does not actually present any final solution to this problem in the sense that he cannot tell on what basis objectivity is founded. He is just focusing on the fact that psychology, through its emphasis on subjectivity, is forcing us to get an idea of objectivity through the notion of its negation, namely, eternity. Thus, his contribution is first of all to maintain the distinction between subjectivity and objectivity, and not least to emphasize the role of psychology as the science of subjectivity. Cassirer also points at psychology as inevitable in an attempt at finding a solution to the problem of knowledge when metaphysics has been stranded. His solution is to refer to culture in terms of communication. The communicational process in culture is first of all characterized by self-constitutive systems that appear as stable and universal, but continuously go through changes and processes of development. The double meaning of time in terms of being both universal and individual is easy to explain through the self-constituting system of language. In this sense, there is a gradual transition from the particular to the general, also when it comes to the conception of time. Time refers, on the one hand to the immediate experienced "now," but also to the general. In this respect, our conceptions of the immediate life and the general aspects of thinking are both provided by culture. Yet this implies also that culture appears to be an inevitable source for a psychological investigation of time.

REFERENCES

Augustine (1999). *The confessions of St. Augustine* (E. B. Pusey, Trans.). New York, NY: Modern Library.

Baumgarten, A. G. (2004). *Metaphysik* (G. F. Meier, trans.). Jena, Germany: Dietrich Scheglmann Reprints.

Bühler, K. (1926). Die Krise der Psychologie.*Kant-Studien. Philosophische Zeitschrift, 31*, 455–526

Cassirer, E. (1968). *The philosophy of enlightenment.* Princeton, NJ :Princeton University Press.

Cassirer, E. (1982). *Philosophie der Symbolischen Formen, Dritter Teil: Phänemenologie der Erkenntniss.* Darmstadt, Germany: Wissenschaftliche Buchgesellschaft.

Cassirer, E. (1994). *Kulturvitenskapenes logikk*. Oslo, Norway: Pax.

Cohen, R. A. (2006). Introduction. In E. Levinas (Ed.), *Humanism of the other.* Urbana: University of Illinois Press

Geertz, C. (1973). *The interpretation of culture.* New York, NY: Basic.

Gordon, P. E. (2012). *Continental divide. Heidegger, Cassirer, Davos.* Cambridge, MA: Harvard University Press.

Heidegger, M. (1994). *Hvad er metafysik?* Fredriksberg, Denmark: Det lille forlag.

Heidegger, M. (1996). *Being and time: A translation of Sein und Zeit.* Albany: State University of New York Press.

Husserl, E. (1964). *The phenomenology of internal time-consciousness*. Bloomington:Indiana University Press.
Husserl, E. (1970). *Logical investigations*. London, UK: Routledge & Kegan Paul
Husserl, E. (1999). *Cartesian meditations: An introduction to phenomenology*. Dordrecht, The Netherlands:Kluwer.
Kant, I. (1999). *Critique of pure reason* (J. M. D. Meiklejohn, trans.). Kila, MT: Kessinger.
Kierkegaard, S. (1980). *The concept of anxiety: A simple psychologically orienting deliberation on the dogmatic issue of hereditary sin* (R. Thomte & A. B. Anderson. trans.). Princeton, NJ: Princeton University Press.
Latour, B. (1993). *We have never been modern*. New York, NY: Harvester Wheatsheaf.
Müsseler, J., & Prinz, W. (2002). *Allgemeine Psychologie*. Heidelberg, Germany: Spektrum Akademischer Verlag.
Rosenkrantz, K. (1863). *Psychologie oder Die Wissenschaft vom subjectiven Geist*. Königsberg, Germany: Verlag der Gebrüder Bornträger.
Schopenhauer, A. (1954). *The world as will and idea* (Vol. 1, 7th ed.). London, UK: Kegan Paul, Trench, Trübner.
Vygotsky, L .S. (1997). *The collected works of L. S. Vygotsky, Vol. 3: Problems of the theory and history of psychology* (R. W. Rieber & J. Wollock, Eds.). New York, NY: Plenum.
Werner, H. (1957). *Comparative psychology of mental development* (Rev. ed.).New York, NY: International Universities Press.
Wundt, W. M. (1902). *Outlines of psychology* (C. H. Judd, trans., 2nd Rev. English ed.). Leipzig, Germany: Engelmann.
Wundt, W. M. (1973). *The language of gestures*. The Hague, The Netherlands: Mouton.

CHAPTER 2

DEVELOPING WITH TIME

Defining a Temporal Mereotopology

Giuseppina Marsico

Each of us knows the story of Columbus. What we in Europe have learned about this very important historical event is that a man started his journey in 1492, leaving from the Spanish coast, and reaching, many months later, another coast, which was supposed to be the rich and alluring India. We knew a lot also about what happened after he arrived with his men in the new land and all the historical implications started from this event, which transformed, almost completely, the known geography and history of the world.

However, rather little is known about what happened during the trip. Let us imagine for a while that *we* are wearing Columbus' captain's uniform, and are in command of his ship. What did he see during his trip? What was in front of his boat, what was behind it? Day after day, week after week, month after month he navigated the ocean, moving away from Spain into the uncertain waters that had not been tested before. Behind him was a concrete borderline—the coast of the Iberian peninsula. Ahead of him, there was no new border in sight. It was the unknown, it had never been seen and, when Columbus was watching ahead, it was not visible anywhere,

Temporality: Culture in the Flow of Human Experience, pp. 23–40

yet. Or, at least, he had not crossed it so far. He definitely moved from a familiar territory to a new foreign land, losing more and more of his points of reference while standing in a strange "between-ness."

Columbus took a lot of time in reaching America, coping with several problems such as the fatigue of the crew, the ordinary demanding life in an narrow space of a ship, the changes in weather conditions, and so on. Yet the horizon was the same every day—just a visible line on the ocean surface. That line moved with him, never to be reached. Columbus did not have any clear idea—only a vague ones—of what he would find in the place "over there" toward which he was moving and with very few instruments for facing any kind of possible risks or unexpected events there. Like all of us in our lives, he was an adventurer.

Colombus' adventure seem to me like the adolescent's growing task, who is asked to deal with the uncertainty of a trip where he has to go from a well-known space (childhood) to unfamiliar terrain (the *not yet* region of adulthood) while he has to build up his own orienting tools (see Figure 2.1).

Figure 2.1. A boy trying to walk in a straight line (see http://www.123rf.com/photo_11414707).

Of course, as in the case of Columbus, the devolvement of any moment in our life's trajectories is not a *solo voyage* (after all, Columbus wasn't alone either), but the analogy of his journey becomes relevant, especially with respect to the crossing boundaries phenomenon and his temporal features.

Moreover, boundary (in its inherent ambivalence) is a very intriguing issue in itself; it simultaneously belongs to two parts while being defined by neither one entirely. Thus, boundaries look different, when established from two sides: European discovery means, at the same time, the South American occupation.

Development as Guided: Framing the Trajectories of Living

The idea of life's trajectory implies the notion of *being on the move*, where the movement is defined as a goal-oriented process occurring within spatiotemporal coordinates.

This basic assumption leads to the idea of human development as an open-systemic process that guides the organism's actions toward unknown immediate future states in relation to the environment (Valsiner, 2008). Striving for uncertainty is an unavoidable characteristic of the human being who constantly deals with crossing the boundaries between different social worlds. The sociocultural locomotion the person makes (or in other words, his/her life trajectory) takes place in culturally organized life contexts within the insurmountable "boundaries" of irreversible time.

Any institutional organization of person-environment relations takes place in time. If on one side, a person is recognized to actively interact and reconstruct his/her own immediate environment. On the other side, this interaction is culturally guided along the life course. The cultural framing of human development set up by institutions is the *core issue* here. Different social institutions such as schools, church, local communities, and many others can promote, on the basis of their goal orientation, various ways of social suggestion. They could prescribe how specific social settings have to be structured, what kind of social discourses are acceptable or not about some target objects, and also what people are expected to do a certain points in their life. The institutions frame our social life in contiguous segments that are distributed over time. They establish temporally bounded stages whose entrances and exits are rather rigidly fixed. For instance, it is expected to access college at a certain time and not another, and it is expected to be out (after getting a degree) at certain point and not another.

Only a short delay is allowed along this path. So one may term this phenomenon as an "institutional setting up of temporal borders" along life's course, which strongly guides human development, prescribing, consequently, what is normal, abnormal, or deviant.

Vygotsky and *The Problem of Age*

A look back in the history of psychology reveals a valuable attempt made by Vygotsky in 1934 to deal with the problem of dividing developmental trajectories into periods. In his chapter in *The Problem of Age* (1998),[1] Vygotsky placed the schemes proposed by the developmental psychology of his time into three groups which had in common the isolation of one single trait or psychological criterion or, even, pedagogical principle as an arbitrary basis for dividing age levels. Since development is a self-reorganizing process in dialogue with the environment, these schemes forgot the simple evidence that the single trait changes continuously its value and its role with the transition from one period to another. Thus, sexual maturation becomes primary during puberty but has less significance earlier. Yet, even in this case, one assumes a dialectical understanding of developmental process as a progressive appearance of better-organized new psychological formations which didn't exist before. However, something is still missing. What is missing? The basic explanation of the dynamic transition from one stage to another over life's course.

From a temporal point of view, the changes in age levels can happen gradually and smoothly over a long-term molecular process (the *stable age* period, which usually takes several years) or they may occur suddenly and stormily through discontinuities, ruptures, shifts, and alterations, which take place in a very short time (the *development crises*). The paradoxical feature of these crises deals with the temporality issue. While they seem to occur abruptly and concentrated in some months (or 1or 2 years), they create a sort of temporary suspension of the progressive developmental flux. Breaking down the continuity in constructing new developmental structures along the lifetime, these critical periods open up to apparently floating phases where what the person loses (interests, activities, and relations) is more that he/she gains. The temporal paradoxicality of crises (while suddenly emerging, they make up a time slowness) is one of the reasons why many authors exclude them from a general theoretical model of developmental trajectory, considering them a "deviation" from the normal path.

The ambivalence of temporal features of the critical age level actually challenges the culturally oriented framing of the life course provided by some institutions, first and foremost, the school. Educational settings fix a quite firm and subsequent levels of training, defining in such a way a so-called normal course of education from which children (and later adolescents) at the critical period, could easily drop out because of the decrease of interest in what has been earlier acquired, experiencing some internal and external troubles at the same time. This kind of transition (Zittoun, 2006) presents a great inner-individual variability, depending on the internal life (sometimes connected with disorienting experiences or conflicts) and the external conditions that strongly influence the characteristics the crises concretely assume. In addition, the boundaries that set the beginning and the termination of the crisis from the contiguous age periods are not well defined, creating a fuzzy intermediate area (Figure 2.2).

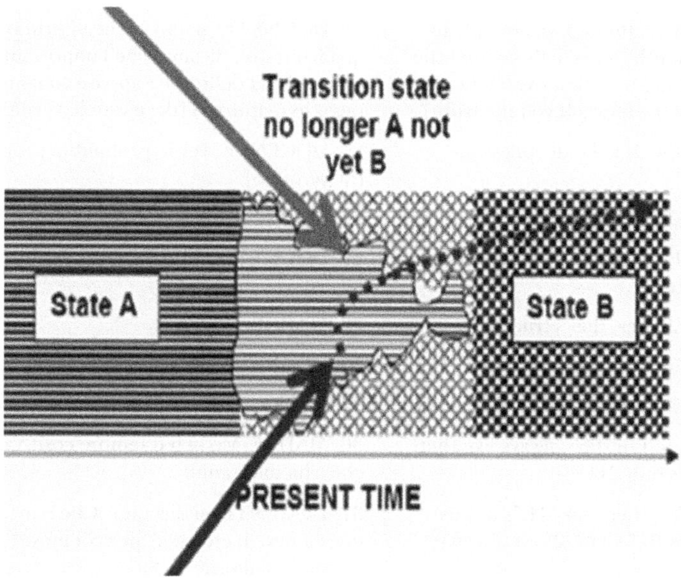

Figure 2.2. Transition state (after Valsiner, 2011).

The floating "temporal borders" of the transition state where a person is no longer A, but not yet B, is well depicted in the following excerpt, where parents and teacher, during a school-family meeting,[2] are talking about the very intermittent behaviors of the 13-year-old boy who seems to go "to and fro." The boy appears to them sometimes very mature and other times totally childish. The boy himself feels this oscillation at the point that he represents his own actual condition (the pre-adolescence phase) like a shrimp (see Table 2.1).

This excerpt, describing the ambivalence of the passage from childhood to adolescence, is a good example of how development sometimes puts in question the social timing and challenges the rigidity of the temporal border framed by institutions. Along the path of guided development, the school (as well as other educational settings) shows its limits in dealing with neither speeding up nor slowing down. The institutionalized segmentation in bounded age levels over time doesn't keep up the inherent temporal parodoxicality of the critical periods which become unsolved disquieting objects to treat for both educational practices and theoretical analysis. This call for an abstract model is able to create a set of ideas for understanding qualitative changes in developing a system.

Table 2.1. Floating Temporal Borders

English Version:	*Italian Version:*
1. TEACHER: Although we are only in the first 4-months term, let's say that the important thing is ... that overall he seems a mature boy. I [must tell you the truth]	11. DOCENTE: siamo anche al primo quadrimestre, diciamo che l'importante che ... il più delle volte appare come un ragazzo maturo. vi [devo dire la verità]
23. TEACHER: For the deepness [of thinking]	23. DOCENTE: Per la profondità proprio [di pensiero]
24. MOTHER: [Yes]	24. MADRE: [Si]
25. TEACHER: Yes, but at other times, on the other hand ...	25. DOCENTE: Si, poi in altri momenti invece ...
26. MOTHER: Yes, that's true.	26. MADRE: Si, è vero.
27. TEACHER: He seems a child to me.	27. DOCENTE: Mi sembra un bambino.
28. MOTHER: yes that's true, that's true	28.MADRE: Si è vero, è vero,
36. MOTHER: OK, he's always like that! OK, well he's only 10!	36. MADRE: Va, beh è sempre così! Va, beh che c'ha dieci anni!
37. TEACHER: Let's say "He's in this phase of growth," so he shows alternate phases.	37. DOCENTE: si diciamo "Che è in questa fase di crescita," per cui mostra queste fasi alterne
38. MOTHER: [()]	38. MADRE: [()]
39. TEACHER: [yes, yes, yes, yes]	39. DOCENTE: [si, si, si,si]
40. FATHER: He needs to leave childhood behind.	40. PADRE: Si dovrebbe lasciare alle spalle un poco di infanzia.
41. TEACHER: Eh! Eh! He made a poster in the classroom on adolescence and I asked him, I said, "Why it [have you drawn a shrimp?]"	41. DOCENTE: Eh! Eh! Ha fatto un cartellone in classe sull adolescenza e io gli ho chiesto dico ma "Perché [rappresenta un gambero?]"
50. TEACHER: That is what he said later on actually.	50. DOCENTE: Poi me l'ha raccontato cosìeffettivamente.
51. MOTHER: ()	51. MADRE: ()
52. TEACHER: Let's say when he wants and he's 100% mature and at other times ...	52. DOCENTE: Diciamo quando vuole e ... è maturo al 100% altre volte poi
53. FATHER: He is like that at home too.	53. PADRE: Lo fa anche a casa.
54. MOTHER: At home he's like that.	54. MADRE: Pure a casa è così.

Toward a Temporal Mereotopology

The institutional setting up of temporal borders in the life's course is based on arbitrary criterion which become the guideline for establishing a *diagnosis* of development (Vygotsky, 1998).

In applying the terminology and the conceptual ontological framework provided by mereotopology, which is a mathematical part of contemporary philosophy dealing with the part-whole relationships (Smith, 1997; Smith & Varzi, 2000; Varzi, 1997, 1998), one may say that this case is a quite evident example of *fiat boundaries* (artificial and conventional demarcations resulting by a human's decision), which are transformed in *bona fide* boundaries (natural and objective differentiations).[3]

The question of qualitative forms of borders is very central in mereotopology as it is directly involved in the representation and organization of the surrounding world. It faces, from an ontological point of view, the part-whole issue and therefore the question of the relationship between a border and the thing it bounds. Mereotopology rises from the ambitious attempt to provide a unified framework of the way we represent space, the objects that occupy it, and the relationships between them. It consists of a combination of *topology*—the discipline which deals with the qualitative aspects of geometric structures—and the "theory of parts and the whole" (or *mereology*), whose Aristotelian roots have been systematized by Brentano (1981). It has yet to establish its relations with psychology.

Assuming a subjective criterion (i.e., the level of training and education), the developmental trajectory is divided into segments delimited by temporal *fiat* boundaries (i.e., the preschool age, the primary school age, etc.), which are later more rigidly conceptualized as a *bona fide* boundaries and used as a marks in a fixed temporal stepwise construction.

What is lost here is the flowing stream of development as well as the ambivalent richness of transitional phases where, despite their chaotic features, the novelty could emerge and the transformation of the actual psychological structure takes place.

As a consequence, the developmental process is conceptualized more on the basis of a static ontology of being rather than on an ontogenetically perspective of becoming.

This is a dramatic result for developmental psychology and in itself a contradiction in terms since the development, *de facto,* deals with the unfolding of novel processes (Baldwin, 1906). Thus, the phenomena of emergence, becoming, and transformations should be the very core of any developmental sciences:

> In all issues pertaining to development and education the reliance on ontological discourse is misfitting. We are not interested in the ways of being

> of a child of age X or school grade P. Instead, we are looking at the child of age X as s/he is moving towards becoming a child of age X+1 (or of school grade P+1) and—in the case of education—how that "what-is-becoming" might be directed towards a specific meaningful (for the directing agent) outcome state. Thus, educational and developmental theories need to break with the ontological conceptualizations that may be sufficient for the non-developmental sciences. (Valsiner, 2008, p. 132)

Development is a "disquieting object" for psychology at-large, very much focused on the ontology of psychological functions in an axiomatically time-free perspective; and even more for the analytic philosophical framework provided by the mereotopology which is interested in the existence of some static entities at a given time (Marsico et al., 2013). In dealing with qualitative time-based changes, instead, development is antithetical to any ontological system. It is founded upon the *epistemology of becoming* and is bounded into *irreversibility of the time* (Valsiner & van der Veer, 2013).

The theoretical focus on the phenomena of "becoming" implies the centrality of time and of transitional processes. The reliance of time needs to go together with the idea of directionality: any developing person moves from a previous settled—already known—state to an area of indeterminacy (a *not yet* region), crossing a series of boundaries in this space in between (Marsico, 2011, 2013). While meterotopology asks very complex questions about the qualities of the forms leading the focus on boundary notions in the part-whole relations, it fails in dealing with both the movement between parts within the whole and the time-dependent change of development. Although these limits have been recognized in the mereotopological model (Smith & Brogaard, 2000), they are not solvable by merely reformulating the ontological stance into an ontogenetic one. What is needed is a *temporal mereotopology*, able to provide a formal set of ideas for understanding the qualitative transformations occurring in any developing system over time. This general science of transforming forms should provide conceptual tools able to explain the role of the borders (very central in mereotopology) in their dynamic movements[4] toward a teleological direction in irreversible time (the very core of any developmental perspective).

One Step Further: The Temporal Border Zone

Some recent attempts to conceptualize the interplay between "border" and "time" have been made by Tateo (this volume) and Abbey (this volume). In his "Time-as-a-context" model, Tateo presents a multilevel idea of time as a context for any psychological phenomena. It seems to contain *in nuce* the basic assumptions from which it is possible to start laying the foundations of a *temporal mereotopology*. By reframing Bronfenbrenners' ecological model

(1979), time is depicted as a multilayered concentric construction moving from the individual psychological experience of temporality (microtime) to an intersubjective experience of time (mesotime), where the communicativeness and sharing processes are made possible among people, to a wider societal time (esotime), which includes, for instance, the social timing, the institutional framing of a life's course, to a historical and phylogenetic time regarding the evolution of human species (macrotime). What is more valuable in this model, for the sake of initial definition of a *temporal meretopology*, is the focusing on the border between different time levels, which are considered to be interrelated in a mutual shaping relationship that is constitutive of the entire temporality experience. Expanding the Time-as-a context model, one may say that the semipermeability of the borders and their inherent ambivalence (they divide while unifying bounded contiguous regions) allows the emergence of an intermediate area, a *Temporal Border Zone* on each and every external frontier between two near temporal circles where not only the mutual shaping takes place, but also the coordination and negotiation of different time levels (Figure 2.3).

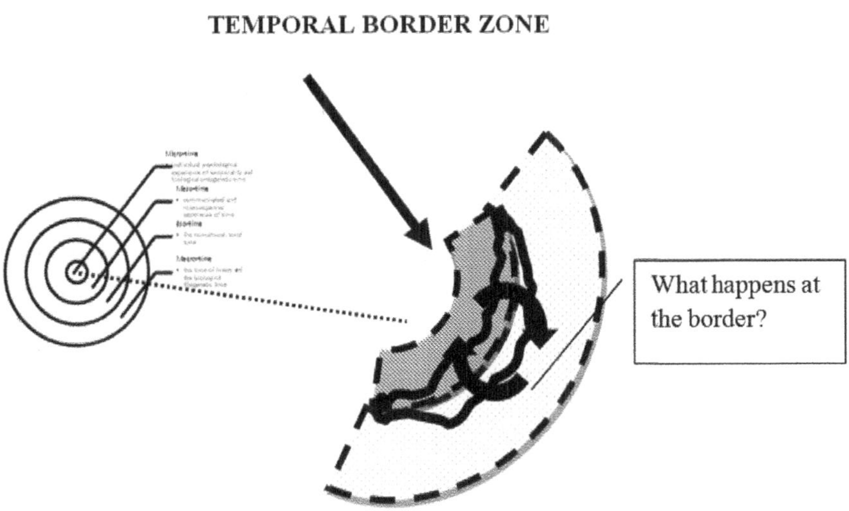

Figure 2.3. Temporal Border Zone (Tateo, this volume).

In addition, what happen in any Temporal Border Zone between two specific time levels (i.e., the microtime and the mesotime) is always influenced by all the other temporal dimensions (the esotime and macrotime), as they are coexistent but with different weights, depending of the life's circumstances (Simão, 2012). Thus, each person experiences simultaneously the presence of several time levels—from the very inner temporal dimension to a distal ones (the macrotime)—that are actively involved in the Temporal Border Zone.

Consider, for instance, a 23-year-old student, still an undergraduate.. One may suppose he "feels late," with a sense of discomfort. Where does this inner temporal experience of "feeling in delay" come from? Following the above argumentations, it is one of the possible outcomes of the modulation of different temporal levels in a Temporal Border Zone between his microtime and the meso-, eso-, and macrotime. In this case, the intrapsychological experience of "feeling late" is due to a temporary desynchronization of personal time (as a student) with the institutional timeframe (provided by university) and the wider sociocultural timing—historically set—which values young age as a right life-stage for learning.

Although this expanded version of the Time-as-a-context model, which includes the Temporal Border Zone, seems the most likely answer the Tateo's question (this volume) about what happens on the border between time levels, it doesn't look sufficiently dynamic for lack of reference to the notion of irreversible time that implies the inherent impossibility of "returning" to a previous state back in time.

Since the human being grows with time (not in time) (Valsiner, 2000), the irreversibility of time is both the axiomatic core of any developmental discourse and very central for the foundation of a *temporal mereotopology*.

As Abbey (this volume) points out, the irreversibility of time comes together with the idea that novelty is a main feature of human life and therefore the future—as well as the present and the past—plays a central role in the temporality of human experiences. A developing process links the past with the future in the present, making the boundary between present (including elements from the past) and the future the real space for "becoming." This perspective on time meets with the Bergson's notion of duration (*durée*):

> For our duration is not merely one instant replacing another; if it were, there would never be anything but the present—no prolonging of the past into the actual, no evolution, no concrete duration. Duration is the continuous progress of the past which gnaws into the future and which swells as it advances. (Bergson, 1911, pp. 6–7)

This view on time is that of irreversible time, which entails the continuity of the change from the infinite past toward the infinite future. According to Abbey (this volume), a living being seems to experience his relation with the world on the border between immediate present—an infinitely small time that links the vanishing past—and the approaching unrevealed future. The Boundary Zone between here-and-now and the unknown future she depicts (Abbey, this volume), could be seen as a complementary conceptual tool in defining the basic assumptions for a *temporal meretopology*. By inserting Abbey's notion of a boundary between present and future into the expanded version of Tateo's Time-as-a-context model, a wider and detailed understanding of the relation between "time" and "border" in a *temporal mereotopology* perspective is reached (Figure 2.4).

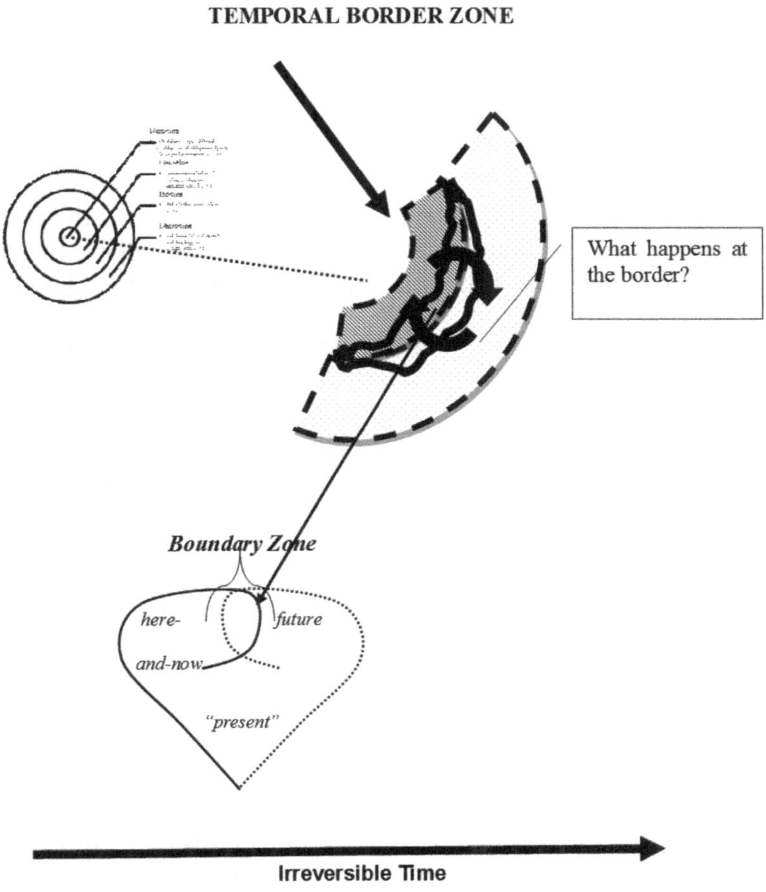

Figure 2.4. Temporal Border Zone (Abbey, this volume; Tateo, this volume).

As argued above, the Temporal Border Zone in the expanded version of the Time-as-a-context model is the place where the mutual shaping between contiguous temporal levels and the coordination among all of them occurs. Nevertheless, these processes happen in the irreversible time and the temporal coordinates (here-and-now, present, unknown future) provided by Abbey and are allowed to zoom into the Temporal Border

Zone looking at the very axiomatic principles for any human temporal experiences. Acting as a basic process in each and every Temporal Border Zone between different time circles, Abbey's notion itself gains a wider significance by its inclusion in the multilayered structure of Tateo's model. The next step in the direction of a *temporal mereotopology* is trying to conceptualize, in a more abstract way, how this double contemporaneity—mico-, meso-, eso-, and macrotime levels, on one side, and past-present-future coordinates, on other side—co-generate the temporality of human being.

Expanding TEM Model: The Point of Bifurcation Is Not a Point

The rudimentalry attempt above depicted to account for the relation between "time" and "borders" in a *temporal mereotopology* perspective could be combined with TEM—Trajectory Equifinality Model (Sato, Hidaka, & Fukuda, 2009; Sato & Valsiner, 2010)—whose main goal, to conceptualize the variability of life's trajectory taking into the account the quality of the whole in relation to its parts, fits with the mereotopological discurses.

The TEM model is not merely a track of a retrospectively reconstructed life's course, but it is a new way in time-dependent theorizing to keep together the real components (the actualized trajectory) and the imaginary components (the future set of possibilities) of a developing system in its progression. Past and future, possible and actual—and their intertwined relations—are here considered as conditionally permeable boundaries. TEM has a major implication for the developmental psychology since it supports the understanding of the psychological phenomena in "becoming." As an analytic tool, it captures the future multiple possibilities in the actualized past, through present actions, making evident, in such a way, the time-dependency of developmental processes taking place always "in-between" (Marsico, 2011) the organism and the environment and the past and future. In this perspective, the "present moment" in which alternative options appear is conceptualized as a Point of Bifurcation—BFP (Sato et al., 2009, p. 229). The developing person constantly deals with choice points in his trajectory, where new "under construction" actions are close to the already existing ones. But the different options available at a given time are neither disconnected nor mutually exclusive possibilities. They instead form a set of potential trajectories projected into the future. The Point of Bifurcation, thus, appears as the birthplace of a certain number of new possibilities (not all the possibilities), a sort of "window of potential options," that emerges under the pressure of the irreversible time flow, the past course of the events, and the social guidance (Sato & Valsiner, 2010). One may say that by merely visually speaking, the point of bifurcation seems too narrow for this complex phenomenon. Expanding the Point of Bifurcation in a wider *Field of Future Extended Duration* (FFED)seems more suitable to better grasping situations in which a constellation of processes happens simultaneously but not instantaneously taking, in itself, an unreliable period of time (Figure 2.5).

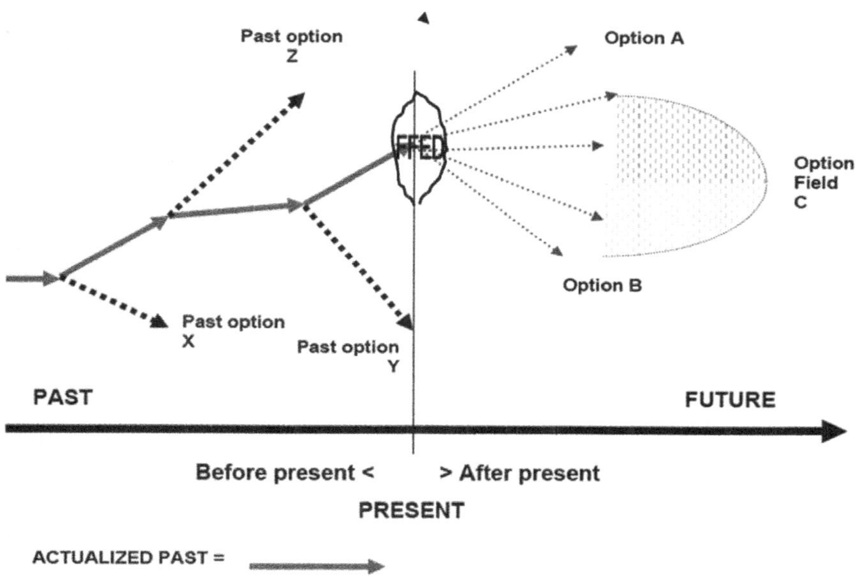

Figure 2.5. Field of Future Extended Duration (FFED) in TEM model (Valsiner, 2011).

A microview of a Point of Bifurcation reveals that it is more than a "*very point*" (Sato & Valsiner, 2010, p. 90). Rather, it needs to be described as space (*Field of Future Extended Duration*) with its own time of *durèe* (Bergson, 1911), where the "window of selected potential options" is constructed and where, out of multiple possibilities, one trajectory rises.

If one looks microgenetically at these interconnected processes, it becomes evident that they are the synthesis of a set of constraints (the actualized sequence of past options, the social guidance, all the realm of constructed meanings, etc.) and of the uniqueness of the subject with his personal orientations, autonomous actions, and intentionality. According to Zittoun (2012), the inherent tension produced by all these elements is negotiated by the person who keeps creating, in such a way, his unique life trajectory. By doing that, people have constantly to deal with the coordination of different time levels (Tateo, this volume) at the boundary between present and future (Abbey, this volume). Therefore, what has been above depicted in the elaborated version of the Time-as-a-context model (see Figure 2.4) could be seen as a basic microgenetic temporal process taking place, over a certain period of time, in the Field of Future Extended Duration. For this reason the FFED is more a field-like notion—rather than a point-like one—of future trajectories in their emergence in the current moment. Its temporal extension allows the articulation of the coexistence of multiple concentric time levels with the process of generating multilinear possibilities toward the future. Zooming into the Field of Future Extended Duration, one may describe how the multilayered time perspective (and its mutual shaping) intersects the emerging trajectories and then how, still in this time in between, it contributes to making distinctions among them as a process to overcome the uncertainty of human existence.

According to Sato (2006, p. 326), no one can adopt multiple options at the same time, and even if the trajectories are depicted and perceived as different, with their own complex architecture (Zittoun, 2012), a person cannot choose totally freely any possible option. The selection of the future trajectory happens within an already bounded "window of possibilities," on which other complex phenomena intervene, as in the case of the coordination between some social constraints (social guidance, institutional timeframing) and personal orientations. Thus, the not-realized options still remain as traces along the irreversible timeflow until the next Field of Future Extended Duration emerges. There, they could be incorporated and reactivated in the present moment as elements of the past (Abbey, this volume; Valsiner, 2011). With respect to this, Bergson wrote, "We choose in reality without ceasing; without ceasing, also, we abandon many things. The route we pursue in time is strewn with the remains of all that we began to be, of all that we might have become" (Bergson, 1911, p. 111).

How extended over time should the FFED be, or, in other words, where are its temporal borders? This question addresses, once again, a mereotopological issue. FFED needs to be described as a temporal experience of "between-ness" in a lived (and irreversible) time. Assuming the fluidity of irreversible time, it could be quite impossible to set rigid contours to this Field. Nevertheless, a sense of "appropriate dimension" of its temporal extension seems to appear where the coordination within different time levels fails or a desynchronizations emerges. Coming back to the example of the 23-year-old student, still an undergraduate and limiting attention only to his life as a student, one could depict his present moment as a Field of Future Extended Duration, where a set of future possibilities are emerging (i.e., university, nonuniversity, job, nonjob, etc.) and among them only one will become a realized trajectory in the future. In defining how long this process should be, a very central role is played by the sociocultural temporal parameters (for instance, the social timing at esotime level in Tateo's model), which establish, even if a general terms, the temporal borders (*fiat boundaries* in a mereotopological words) of the FFED. Overcoming this invisible fiat boundary means to create a "syncopation" in the rhythm of the social timing, even judgeable as a pathologic lifestage in case it is prolonged with some effects in the subjective temporal experience. Here the social guidance in framing life's trajectories is quite evident as well as its connection with institutionally guided development.

Pathway to Synthesis for a Rudimentary Foundation of a Temporal Mereotopology

Amplifying the Point of Bifurcation in the TEM model with a Field of Future Extended Duration notion, it is possible to place, in a wider and abstract way, the process of coordination of different time levels at the boundary of present and future (as in the expanded version of Tateo's model and Abbey's proposal) on the landscape of human temporality.

As argued, in the FFED, several emerging options are co-present while the person, in making distinctions among them, also experiences the coexistence of different time levels he must negotiate. The coordination among them affects the process of creating a trajectory of further development. Yet, in order to define a rudimentary basis for a *temporal mereotopology*, a formalization of the qualities of the FFED is needed, whose fiat temporal boundaries are artificial but not, for this reason, less binding. By analyzing the way in which the double contemporaneity (one along the axis of the concentric mutually shaped time circles, through the Temporal Border Zone, and the other along the axis of irreversible flow of time—past-present-future) acts over the emerging set of potentialities, it would be possible to define a conceptual framework able to explain the qualitative transformation of previous configurations into a new one. This embryonic attempt should provide the basic temporal mereotopological rules to support the understanding of the complex phenomena of dynamic border on move toward a direction over the time that is a fundamental feature of any developmental process.

Borders and time are neither avoidable nor fixed coordinates of human life. On the contrary, they shape the humans experience in its perennial transformation. Development constantly deals with overcoming borders in a movement toward the unknown future. Columbus' trip, in this respect, seems a beautiful metaphor of this most ordinary human effort.

ACKNOWLEDGMENT

The intellectual impact of the highly caffeinated early morning discussions together with Jaan Valsiner is gratefully acknowledged. I wish to thank Livia Mathias Simão for her invitation to think about "temporality" and for all the personal and professional stimuli she and her students gave me during my stay in São Paulo in August 2012. The Kitchen Group at Clark University is thanked for the comments on earlier versions of this manuscript.

This chapter was written during a stay as a visiting professor at Clark University, November 2012–March 2013, and supported by the University of Salerno, Italy, by an International Mobility Grant (2010).

NOTES

1. This chapter was included in the book on developmental child psychology that Vygotsky was preparing during the last years of his life. It was first published in *The Collected Works of L. S. Vygotsky*, Volume 5, 1998, pp. 187–205.
2. This excerpt is a part of a wider research program that documents what happens when the parents of pre-adolescent children take part to the school's report cards delivery event. For a detailed descriptions of the results see Marsico & Iannaccone, 2012 and Iannaccone, Marsico, & Tateo, 2013.
3. As Smith and Varzi (2000) pointed out, boundaries may be natural or artificial, spatial or temporal, sharp or vague, insubstantial or substantial. The distinction between the natural boundaries (*bona fide* boundaries based on some kind of natural discontinuity) or artificial ones (*fiat boundaries* produced by human action as an outcome of negotiation and agreement) makes evident how both of them constantly regulate our daily actions, affecting our lives and guiding our plans (Marsico, Cabell, Valsiner, & Kharlamov, 2013).
4. A border is a line which demarcates and connects at the same time. The border, then, is the point of contact and not of division. It is both the place of demarcation and definition and of openness to the dialogue. According to Neuman (2003), a boundary is not a fixed line, but an *oscillatory activity* that forms the ground and figure, the being and nonbeing. "Therefore, from a phenomenological perspective the boundary is not parasitic of the entities it defines, but precedes and constitutes them through its indeterminacy" (Neuman, 2003, p. 103).

REFERENCES

Baldwin, J. M. (1906). *Thought and things: A study of the development and meaning of thought, or genetic logic (Vol. 1.): Functional logic, or genetic theory of knowledge.* London, UK: Swan Sonnenschein.

Bergson, H. (1911). *Creative evolution*. New York, NY: Henry Holt.

Brentano, F. (1981). *Philosophical investigations on space, time and the continuum.* London, UK: Croom Helm.

Bronfenbrenner, U. (1979). *The ecology of human development: Experiments by nature and design*. Cambridge, MA: Harvard University Press.

Iannaccone, A., Marsico, G., & Tateo, L. (2013). Educational self: A fruitful idea? In M. B. Ligorio & M. César (Eds.), *Interplays between dialogical learning and dialogical self* (pp. 219–252). Charlotte, NC: Information Age.

Marsico, G. (2011). The "non-cuttable" space in between: Context, boundaries and their natural fluidity. *Integrative Psychological and Behavioral Science, 45*, 185–193.

Marsico, G. (2013). Moving between the social spaces: Conditions for boundaries crossing. In G. Marsico, K. Komatsu, & A. Iannaccone (Eds.), *Crossing boundaries: Intercontextual dynamics between family and school*. Charlotte, NC: Information Age.

Marsico, G., Cabell, K. R., Valsiner, J., Kharlamov, N. A. (2013). Interobjectivity as a border: The fluid dynamics of "betweenness". In G. Sammut, P. Daanen, & F. Moghaddam (Eds.), *Understanding the self and others: Explorations in intersubjectivity and interobjectivity* (pp. 51–65). London, UK: Routledge.

Marsico, G., & Iannaccone, A., (2012). The work of schooling. In J. Valsiner (Ed.), *Oxford handbook of culture and psychology* (pp. 830–868). New York, NY: Oxford University Press.

Neuman, Y. (2003). *Process and boundaries of the mind*. New York, NY: Kluwer Acadmic/Plenum.

Sato, T. (2006). Development, change or transformation: How can psychology conceive and depict professional identify construction? *European Journal of School Psychology, 4*(2), 321–334.

Sato, T., Hidaka, T., & Fukuda, M. (2009). Depicting the dynamics of living the life: The trajectory equifinality model. In J. Valsiner, P. Molenaar, M. Lyra, & N. Chaudhary (Eds.), *Handbook of dynamic methodologies* (pp. 217–240). New York, NY: Springer.

Sato, T., & Valsiner, J. (2010). Time in life and life in time: between experiencing and accounting. *Ritsumeikan Journal of Human Sciences, 20*(1), 79–92.

Simão, L. M. (2012). The other in the self: A triadic unit. In J. Valsiner (Ed.), *Oxford handbook of culture and psychology* (pp. 403–420). New York, NY: Oxford University Press.

Smith, B. (1997). Boundaries: An essay in mereotopology. In L. H. Hahn (Ed.), *The philosophy of Roderick Chisholm* (pp. 534–561). Chicago, IL: Open Court.

Smith, B., & Brogaard, B. (2000). Quantum mereotopology. *AAAI Technical Report WS-00-08*, 25–31.

Smith, B., & Varzi, A. (2000). Fiat and bona fide boundaries. *Philosophy and Phenomenological Research, 60*, 401–420.

Valsiner, J. (2000). *Culture and human development*. London, UK: Sage.

Valsiner, J. (2008). Open intransitivity cycles in development and education: Pathways to synthesis, *European Journal of Psychology of Education, 23*(2), 131–147.

Valsiner, J. (2011). Constructing the vanishing present between the future and the past. *Infancia y Aprendizaje, 34*(2), 141–150.

Valsiner, J., & van der Veer, R. (2013). Encountering the border: Vygotsky's zona blizaishego razvitia and its implication for theory of development. In A. Yasnitsky & R. van der Veer (Eds.), *Handbook of cultural-historical psychology*. Cambridge, UK: Cambridge University Press.

Varzi, A. (1997). Boundaries, continuity, and contact. *Noûs, 31*, 26–58.

Varzi, A. (1998). Basic problems of mereotopology. In N. Guarino (Ed.), *Formal ontology in information systems* (pp. 29–38). Amsterdam, The Netherlands: IOS.

Vygotsky, L. S. (1998). The problem of age. In M. Hall & R. W. Rieber (Eds.), *The collected works of L. S. Vygotsky. Vol. 5. Child psychology* (pp. 187–205). New York, NY: Plenum.

Zittoun, T. (2006). *Transitions. Development through symbolic resources*. Coll. Advances in Cultural Psychology: Constructing Development. Greenwich, CT: Information Age.

Zittoun, T. (2012). On the emergence of the subject. *Integrative Psychological and Behavioral Science, 46*, 259–273.

CHAPTER 3

TEMPORALITY AND THE BOUNDARY BETWEEN PRESENT AND FUTURE

Emily Abbey

Time and human experience are inextricably linked with one another. As such, humans live within an irreversible flow of experience where no two experiences can be considered identical to one another. This flow of experience mirrors that of many open systems more broadly, from the functioning of a cell to the growth and transformation of solar systems. It reframes much of how we can understand human psychological functioning, especially the process of development. The impact of an individual's past experiences is often the focus of developmental inquiry. The questions posed frequently consider how an event from one's past is now influencing the person in the present. Yet accepting that no two experiences are identical, the future is an equally important aspect of the developmental process. The fact that two experiences are never the same grants axiomatically a basic asymmetry—the person knows some of what has happened, but is always developing toward a somewhat uncertain future for which they must prepare. It is here assumed that the irreversible nature of human life experience constitutes a core aspect of what is meant by temporality, and it is the aspect of temporality on which this chapter will focus. Such a

Temporality: Culture in the Flow of Human Experience, pp. 41–56

focus is warranted for it assures that the site of development is not (as is often assumed) between the past and present, but rather at the boundary between the present (incorporating elements of the past) and the future.

The boundary of the present and future is the focus of this chapter. It will be characterized as a dynamic space in which the person experiences ambivalence between what is thought to be the case in the present and what they imagine might happen in the future. Consideration will be given to outlining how ambivalence can be productive, driving the process of emergence as the person strives to overcome it. By analogy, it will be suggested that just as when one reads a poem, at the boundary of the present and future literal and imagined senses of a situation meet, with novelty emerging through their interaction. Thus, these considerations on the temporality of human lives will lead to uncovering the inherently poetic nature of everyday living.

Irreversible Time

In some areas of psychology, as well as other social sciences, it has not been unusual to consider time as something separate from human life experience. Time can be used as a means to track development, as something to order events by, but it is usually not considered to influence those events in any particular way. For instance, a researcher may chart stages of development along a timeline ranging from 0–5 years of age, but the passage of time in and of itself is but a vessel for those changes to occur within and is not understood to directly influence the nature of the process through which that transformation happens. An alternate understanding of time, and the one this chapter adopts, builds a much closer relationship between time and experience. For the purposes of what follows, time is considered to be deeply entwined with human life experience such that every moment, no matter how superficially similar to another, is nonetheless tinged with novelty.

A basis for this notion of time can be found in Bergson's writing on time, understood as duration (*durée*). As Bergson (1913) explains, the irreversibility of experience happens though a *pure duration* of the ego—where our consciousness is freed from keeping past and present experiences separate and instead is allowed to endure through time. Bergson writes,

> Pure duration is the form which the succession of our conscious states assumes when our ego lets itself *live*, when it refrains from separating its present state from its former states ... it need not be entirely absorbed in the passing sensation or idea; for then it would no longer *endure* (Bergson, 1913, p. 100; emphasis in original)

Bergson likens this sense of time to a rolling snowball which has an ever-changing shape as it moves. As the ego endures, he explains, our experiential whole grows continuously. As such, no matter how similar two actions appear to be on a superficial level, they are not experienced as completely identical to one another (Bergson, 1913). As such, the notion of irreversible time means that as the person moves through life, it is impossible to "return" to the same position one occupied in a previous time, even if it was only moments before. The notion therein highlights novelty as a foundational quality of human experience and in so doing reframes how one can think about the process of development.

Frames for Inquiry

Within irreversible time, the common practice of positioning the past as *the* driving force of development comes into question. Often in developmental inquiry, the focus is to explore how past events influence the present (see Table 3.1 below).

Table 3.1. Different Notion of Time and Corresponding Frames of Inquiry

Notion of time	*Focus of Developmental Inquiry*
Time as separate from experience	Past events ➞ Present
Time as duration	Present (incorporating elements of the past) ➞ Future events

By contrast, the notion of time as duration means that the future is also an important part of the developmental process. Because all experiences are tinged with some degree of novelty, the person is always moving toward a somewhat unknown set of circumstances for which they must prepare. In this case, the focus of inquiry falls on the boundary between the present moment and the future (see Table 3.1).

The Boundary Zone

Adopting Bergson's notion of time-as-duration, we can see that a notion of the "present" as an elongated, stable, and unchanging space is unwarranted. Rather, as the flow of experience moves along, the person can be said to symbolically act at the *boundary* of an infinitesimally small here-and-now and unknown future (Simão, 2012). In irreversible time, the sense of the present as described above is perhaps better understood as a boundary marker, which delineates what is now known (the past) and through its realization—stipulation of the known—axiomatically introducing the next experiences as part of the unknown future (Matte Blanco, 1975, 1988). As depicted below in Figure 3.1, our lives happen within what can be described as a *boundary zone* of the just barely known moment and the unknown future.

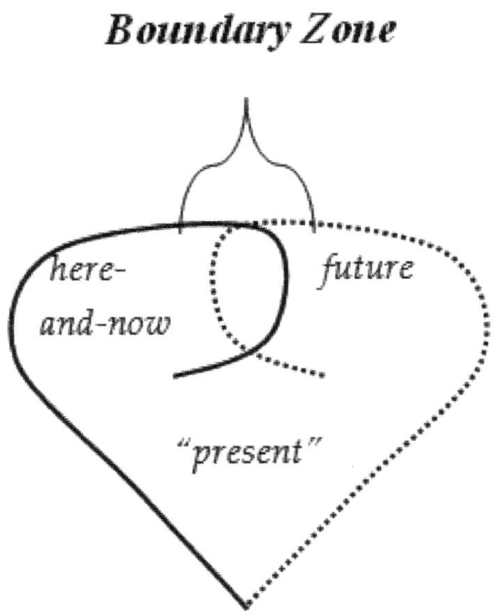

Figure 3.1. Boundary zone between here-and-now and unknown future.

Sign Use in Irreversible Time

Humans are sign making and use signs as organisms, and these symbols allow them to organize their relationship to the environment. Yet the concept of irreversible time suggests that signs serve a dual function in their use. Within the boundary zone of the present and future, the sign speaks to two things. On the one hand, the sign fulfills a representational function by communicating something about the here-and-now. On the other, signs also serve a presentational function by suggesting possibilities for what will come. In its presentational sense, the sign allows the person to pre-adapt to the future, which will necessarily be different from the present. As depicted in Figure 2 below, as the person operates in the boundary zone, each contextualization of the sign speaks to what is in the here-and-now, while also creating an open field of imagined ideas for *what could come next* (Josephs, Valsiner, & Surgan, 1999).

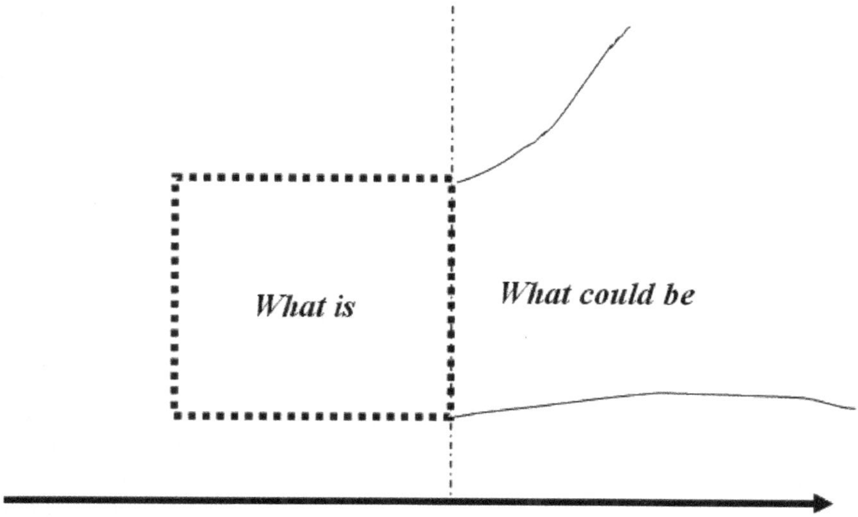

Figure 3.2. Openness of the sign to possible future meaning.

These fields of possible future meaning are often not clearly delineated. They can be thought of as open suggestions to the self which are guiding, but not determining, aspects of the meaning-making process.

As an example of how the sign functions within the boundary zone, consider the experience of hearing a noise in the distance. As one attends to it, he or she may begin to process it as "a siren." This immediate meaning allows them to make some sense of the moment. At the boundary of the present and future, emerging simultaneously with this notion of "siren," is a field of possible meanings for the future. For instance, a person who is injured may imagine "the impending arrival of aid." These imagined possibilities for what comes next guide the person, allowing them to begin to plan for the next moment and pre-adapt to uncertainty. Improvisational jazz musicians function in much this way, where what is played at a given moment is simultaneously novel and yet guides the musical stream forward, through to what it is not exactly clear.

Ambivalence of Literal and Imagined

As the person goes about the process of actualizing the future from within a field of possibilities, he or she functions within *ambivalence* (Abbey, 2012; Abbey & Valsiner, 2005). On the one hand, ideas about what is the case in the present moment lead thought and action in one direction, while on the other, ideas for what could come next suggest alternative orientations. This sense of ambivalence is not limited to the classic understanding of polar opposition, but rather, can be defined as *a tension produced by a system entailing a kernel and at least two vectors that are nonisomorphic in size and direction*. Thus, any difference between the representational and presentational sense of the sign constitutes a basis for ambivalence.

Ambivalence between the representational and presentational sense of the sign drives the process of semiotic emergence. That is, as the person experiences an imbalance between them, he or she strives to overcome this tension and in the process arrives at a new meaning. For instance, returning to the above example, a person who hears a siren and believes it may signal the arrival of aid may arrive at a new meaning, "I am going to be okay now." However, in irreversible time, from this new sense, there emerges immediately a new field of possible future meanings (e.g., "Will I be okay?") that re-creates ambivalence and continues the meaning-making process as the person tries to overcome this new tension. In general, the meaning-making process proceeds in such a "stop-and-start" manner with each new contextualization of the sign providing both a resolution to ambivalence while simultaneously creating it anew.

Emergence at the Nexus of the Literal and Imagined

As explained in the previous paragraphs, within irreversible time, the sign functions at the boundary of the present and future, offering both representational and presentational senses of the situation. These different senses of the sign create ambivalence, and the meaning-making process develops as the person attempts to overcome this imbalance. But what exactly is occurring within this boundary of the present and future? A more detailed look within the boundary between the present and future offers that the tension of representational and presentational senses of the sign not only pushes the meaning-making process forward, but more specifically, it creates the set of conditions necessary for novel meanings to emerge.

As the person contrasts the immediate moment with the field of possible suggestions for the future, he or she engages in a process of psychological distancing. This experience has been explored in aesthetic philosophy (Bullough, 1912; Cupchik & Winston, 1996). When one distances from the immediate situation, he or she psychologically moves away from the object of perception while also remaining connected to it. This metaphysical distance allows one to assume an inclusive separation from the environment (Valsiner, 1997) and allows a space of reflection within which new ideas can develop.

Bullough describes well an anecdote conveying an example of how the contrast between two alternate conceptions of a situation can lead to the emergence of a new meaning. In his introspective recollection of being on the deck of a ship surrounded by fog, he describes experiencing a sense of immediate sense of danger in conjunction with an equally powerful sense of peacefulness. Bullough writes,

> For most people it is an experience of acute unpleasantness. Apart from the physical annoyance and remoter forms of discomfort such as delays, it is apt to produce feelings of peculiar anxiety, fears of invisible dangers, strains of watching and listening for distant and unlocalised signals. The listless movements of ships and her warning calls soon tell upon the nerves of the passengers; and that special, expectant, tacit anxiety, and nervousness, always associated with this experience, make a fog the dreaded terror of the sea (all the more terrifying because of its very silence and gentleness) for the expert seafarer no less than the ignorant landsman.
>
> Nevertheless, a fog at sea can be a source of intense relish and enjoyment. Abstract from the experience of the sea fog, for the moment, its danger and practical unpleasantness, just as everyone in the enjoyment of a mountain climb disregards its physical labour and its danger (though, it is not denied, that these may incidentally enter into enjoyment and enhance it); direct the attention to the features "objectively" constituting the phenomenon—the veil surrounding you with an opaqueness as of transparent milk, blurring the outlines of things and distorting their shapes into weird grotesques ... note the curious creamy smoothness of the water, hypocritically denying as it were any sense of danger; and, above all, the strange solitude and remoteness of the world, as it can be found only on the highest mountain tops. (Bullough, 1912, pp. 88–89)

This is the telling point in Bullough's description, for he then relates how the contrast of these two different senses of the situation—one of danger and one of peace—come together and at their nexus, a new feeling of delight emerges: "*The experience may acquire, in its uncanny mingling of repose and terror, a flavor of such concentrated poignancy and delight as to contrast sharply with the blind and distempered anxiety of its other aspects*" (Bullough, 1912, pp. 88–89; emphasis added). As Bullough's narrative description makes clear, it is the simultaneity of these opposing senses—one of peace and one of extreme fear—that brings about a new experience of delight and deep meaning.

Centrality of Imagination: Poetry

The meanings described above by Bullough, and those that we attribute to any particular situation on a daily basis, are not inherently given. That is, although they may try, neither institution nor political body or individual can absolutely define the meaning of an object or event. Rather, humans are guiding themselves toward them by way of their imagined senses of a situation, which are ultimately more open to creative interpretation. One realm of meaning making in which imagination plays a similar role is that of poetry. It has been said by Ricoeur (1981) that in poems, imagination "suspends" the ordinary meaning of words, creating a space in which something new can begin to develop: "The suspension of the reference proper to ordinary descriptive language is the negative condition for the emergence of a more radical way of looking at things" (p. 240). Indeed, what humans are doing as they make meaning on a daily basis in many ways mirrors what happens when one reads a poem, for they are using imagination to distance themselves from their immediate, literal understanding of the situation, and through that distancing, finding a new idea. In such distancing, they are not losing touch with the immediate understanding entirely, rather they are approximating a new sense of it simultaneously. Such imaginative distancing is a basic phenomenon in human meaning, especially in times of change (Zittoun, 2006), and as such, the processes within the boundary between the present and future can be further elaborated through reference to poetry.

As a starting point for this elaboration, take for instance the lines, "Paper Birch tree and the poet/even in August still wrapped by winter." As a reader begins to process these lines, numerous fields of relating appear between the literal and imagined senses of these words. For example, as the reader encounters the notion of "winter," literal senses of a "time of cold weather" likely appear for some readers. That said, this is not all that may be in the mind of the reader. Alongside this literal meaning of winter, the reader's imagination may begin to offer imagined senses too, for instance, notions of "despair," "isolation," or "sadness" (see Figure 3.3).

A similar process could be described for the other concepts within the poem, such as "wrapped." "Wrapped" can be understood for both its literal meaning of "being covered by some material" while also for possible imagined meanings such as "surrounded." This process of interaction between the literal and imagined can continue in the mind of the reader, and through a relating of literal and imagined senses, there may emerge a new meaning. For instance, the reader may come to think of "wrapped by winter" as conveying a deeper, affectively laden meaning, perhaps that of "a pervasive sense of sadness or isolation." Importantly, this meaning is not coming solely from literal or imagined senses alone, but rather appears through their relating (see Figure 3.4).

The equally affectively rich sense of "delight" that emerges for Bullough (1912) in the narrative above comes about in much the same way as metaphoric meaning appears in these poetic lines. The imagined notion of "peacefulness" as he contemplates the fog allows him to distance from the sense of danger, and through overcoming the ambivalence of these different senses, Bullough arrives at a new understanding of an event.

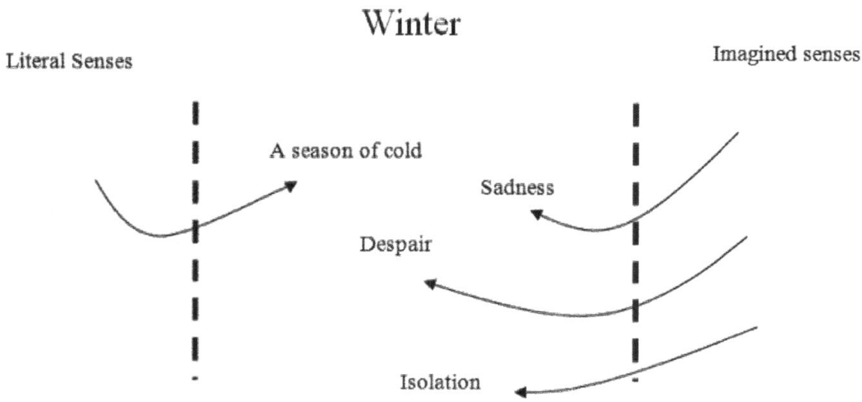

Figure 3.3. Inside the boundary of the literal and imagined in illustrative example.

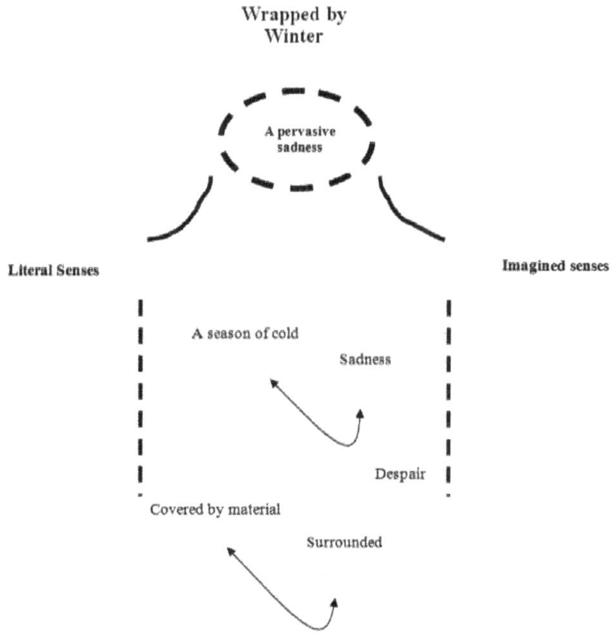

Figure 3.4. The process of affective generalization in poetry.

A Gentle Breath

Vygotsky's analysis of Bunin's story "Gentle Breath" has quite a bit in common with what is described in Bullough's narrative and the process of affective generalization in poetry. The story, "Gentle Breath," originally published by the Russian author in 1916, is used by Vygotsky (1971) to describe the aesthetic response of the reader to a short story. It is of interest to Vygotsky, for one, because its form contrasts quite centrally with its content. In so creating this tension, the author, Bunin, manages to illicit a quite unexpected response within the reader. Strictly speaking (though not as it actually unfolds on the page), "Gentle Breath" depicts the experiences of a exceptionally beautiful and equally naïve schoolgirl, Olia Meshcherskaia, who becomes entangled in a series of love affairs with older, powerful men. Ultimately, one of these relations goes sour. She is shot to death by this jealous lover on the platform in a train station. Vygotsky points out that if the story was told in a linear manner, from the moment Olia first meets one man until the time of her death by the second, it would bring the reader to a clear moment of climax, tension, and terror. And in that sense, at its essence, it is a story no less full of uncertainty and unhappiness, as well as a dark suspense, than is Bullough's description of being on the deck of a ship lost in a bank of fog. In each, it is clear that the dominant theme and feeling should be one of tragedy and fear. And yet, just as Bullough's narrative ends with a profound sense of peacefulness, Vygotsky points out in his analysis, in "Gentle Breath," the feeling the reader is left with is not at all one of suspense, terror, or tragedy. It is, in fact, the sense of a "gentle breath" and the profound feeling of the ephemeral nature of life itself.

The reader's arrival at a feeling of the ephemeral rather than the tragic is accomplished, as Vygotsky (1971) points out, through the manner in which the story is told. Just as imagination contorts lived experience in Bullough's (1912) narrative, in "Gentle Breath" the author breaks from a linear telling of the story. Bunin tells the story in a random, nonlinear order, beginning in fact with the reality of Olia Meshcherskaia's death, understood by the reader through a description of her tombstone. From there, it meanders in a convoluted path back through different experiences in her life, such that the moment of her death is but almost an afterthought, lost within the vivid description of other events and details. And in this extreme breaking from the linear—from the literal, one could say in terms of our terminology in Figure 3.4—a new meaning can arise. Specifically, the story concludes as a former teacher, herself somewhat obsessed with Olia's death as a means to transpose drama into her own dull life, sits at her graveside on a cold spring day and recalls a conversation with Olia. In this conversation, Olia is joyfully remarking on what she believes makes a woman of beauty, such as herself. Of these many attributes, one is that such a woman should possess "a gentle breath." And Olia happily remarks to the teacher that she indeed possesses just this sort of ability to exhale in a slow and soft manner. Bunin then, in the very next line, uses this idea of a gentle breath as a distancing device, similar to Bullough's imagined sense of the beauty of the fog, pulling back from Olia's description and into the scene at hand—that of a gravestone in a graveyard on a cool spring day:

> In one of my father's books ... I read a description of what a woman's beauty should be ... eyelashes black as night, a gentle glow of color, a slender figure, hands longer than usual ... a small foot, a moderately large bosom, a nice round calf, knees the pink color of a seashell, fairly high shoulders.... And most important, you know what?' "A gentle breath!" "I have it, don't I?" "Listen, how I sigh." "Its there, isnt it?"
>
> And now this gentle breath is dissipated again in this world, in this cloud-covered sky, in this cold spring wind. (Bunin, 1916, p. 12, as cited in Vygotsky, 1971).

Thus, the reader herself experiences a similar thing as Bullough (1912) himself. The tragedy of this lovely, happy young girl's life ending is distanced from through the notion of a gentle breath. And in this removal from the immediate, the break from the literal and linear, the reader arrives at a profound sense of how one life slips seamlessly, and almost unnoticed, into the ongoing flow of life. As a conclusion, this sense of a "gentle breath" is far devoid of the terror and drama of her actual shooting, as one would be left with if the story would be told in a linear fashion, and as Vygostky writes, the reader instead experiences, "an almost pathological lightness" (1971, p. 160).

Vygotsky (1971) himself is quite clear in how necessary the phenomenon of contrast and tension is for the development of these highly generalized aesthetic experiences. He points out that while many theorists would claim that the form and content should reinforce one another, in fact the opposite is true. Just as in Figure 4 above, where the *contrast* between the literal and imagined meaning of words leads to the emergence of a higher-order aesthetic experience, in his analysis of the reader's response to Bunin, Vygotsky writes how the presence of contradiction is exactly what allows the aesthetic experience to transpire:

> We seem now to have reached the point where we can say that a work of art always contains a certain amount of contradictions, a certain inner incongruity between the material and the form. We can say that the author intentionally chooses a difficult, brittle material which will resist his efforts to say what he wants to say. *The more brittle and hostile the material, the more suitable will it be for the author*. The form given by the author to the material is not intended to

> reveal the feelings hidden in the material itself. It is not intended to expose the life of a Russian schoolgirl in a provincial town in all its depth. On the contrary, its purpose is to overcome these properties, *to compel the horror to speak the language of gentle breath, and to induce the trouble, pain and turbidity if life to breathe and move like the cold spring wind.* (p. 160; emphasis added)

Temporality and the Poetic Nature of Human Meaning Making

Given what has been presented thus far, the temporality of human lives suggests an inherently poetic quality to daily meaning-making processes. Always operating at the boundary of the present and future, humans imagine possibilities as a way of guiding themselves toward the unknown. These imagined possibilities contrast with literal senses of the situation, and in overcoming the ambivalence of different ideas, new meanings emerge much as they do when one reads a poem. It is through this process that individuals can survive many of the undesirable or painful circumstances which they encounter. Imagination allows one to distance from the immediate circumstances, and perhaps in this act, even arrive at a rich, deeply felt beauty amidst what is otherwise mundane or difficult to manage.

Some examples of this life poetry can be seen in the journal entries made by Vygotsky during a 1925 summer excursion to London. The journal entries, recently translated (van der Veer & Zavershneva, 2011), document Vygotsky's voyage to London, as well as his experiences at a conference on the education of deaf individuals, where he had been sent as a representitive of the Soviet Union. Vygotsky was of ill health during much of this trip and in unfamiliar surroundings without the companionship of his young child and wife. Among other things, the journal entries, in their frankness, allow a window into how a person can use imagination to find peace and even beauty in experience despite the presence of considerable hardship. Take for instance the following entry made near the start of his journey. In it, Vygotsky has left Holland and is on the ship which will take him to England. He writes,

> 5. 25 P.M. A very strong wind. The sun. A rainbow. The enormous sea. Delight. All my sorrow and my fear sink in this enormous space. The white crests of the waves. The boat is rolling. Infinity is with me. Gokha you are forever with me. (van der Veer & Zavershneva, 2011, p. 462)

In this passage, the meaning-making process begins with a statement of what is literally the case at the moment, with Vygotsky noting the environmental conditions: "A very strong wind," "the sun," "a rainbow." Yet simultaneously, his thoughts also move to the notion of "delight," and through this distancing, there emerges a deeply meaningful sense: "All my sorrow and my fear sink in this enormous space." Just as in a poem, here the imagaination allows for an inclusive separation from the literal and creates the space for this new, affectively laden sense to emerge. This process repeats itself as the passage continues. Vygotsky returns to the immediate situation, noting the characteristics of the waves on the ocean and the motion of the boat. Yet again, simultaneous to this, he begins to distance: "Infinity is with me." And as before, in the contrast between these literal and imagined senses of the situation, a novel meaning emerges as he is drawn into thinking of his wife, fondly refered to as "Gokha": "Gokha you are forever with me." This emergent sense is, again, deeply poigant and rich with emotion (see Figure 3.5).

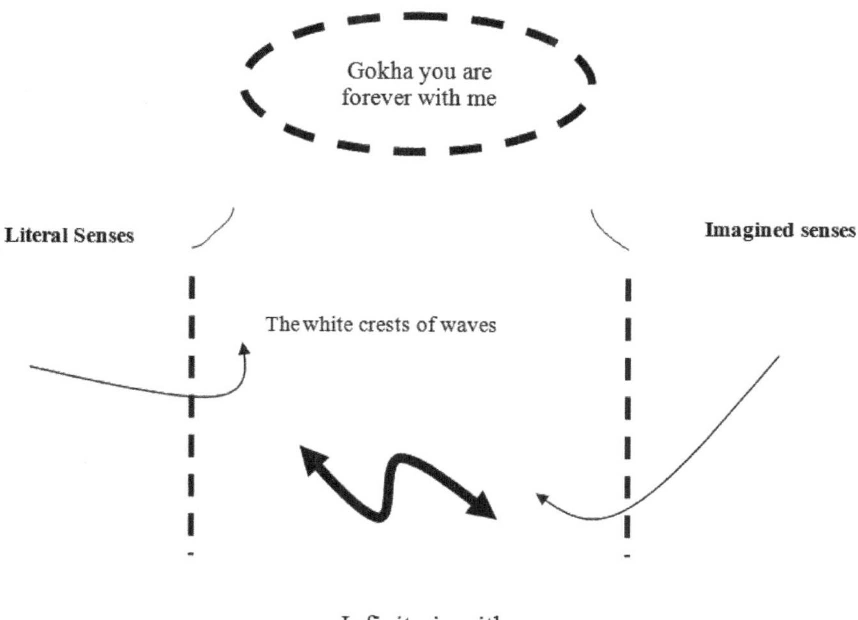

Figure 3.5. The poetic nature of daily meaning making.

Another example comes in an excerpt from Vygotsky's journal later in the trip, after he has already arrived in London and is attending the conference. He makes the following comment:

> 10. 15. A.M. In 15 min the conference will be opened. I was presented to the President and I was moved from the 4th to the 1st row. I am alone. Slightly

> cold. Calm. I am a stranger to myself—in the mirror and in feeling, but your name G. is with me and I am calm. Not without reason the ancients believed in the name as a reality (res)—a genius. In your name, Gokha my beloved wife. Let be. (van der Veer & Zavershneva, 2011, p. 464)

As above, the meaning-making process begins with Vygotsky noting aspects of the immediate environment, such as his physical location, his lack of companionship, and certain aspects of his bodily state. Also with this, imagined notions begin to enter, and he writes, "I am a stranger to myself—in the mirror and in feeling, but your name G. is with me and I am calm." In this cycle of interplay between the literal and imagined, this distancing act allows him to re-create the sense of calm he first notes. And from this more immediate sense of calm to which he has just returned, the process begins again as he imagines "Not without reason the ancients believed in the name as a reality—a Genius. In your name, Gokha my beloved wife." Here, Vygotsky's imagined senses allow him to transcend the immediate situation, while also remaining connected to it. And through overcoming this ambivalence, between a sense of calm and a contemplation of deep love for his wife, Vygotsky arrives at a new meaning: "Let be." As in previous instances, this sense is highly generalized, deeply affectively laden, and likely serves as a higher-order semiotic mediating device for future thoughts, feelings, and action. Indeed, it would seem that the generalizations humans make from the most mundane life events can lead to deeply powerful meanings that guide our immediate present and future. Just as Vygotsky (1971) notices the reader of Brunin's story becomes awash in the feeling of life's ephemeral nature through his last line, Vygotsky here becomes moved by the contrast of being a stranger with his intense love for his wife, concluding, "Let be."

CONCLUSION

Considerations of the temporality of human lives lead, for one, to the reality that our meaning-making processes are inherently poetic in their nature. As presented here through examples from poems—as well as the poetry of living—imagination serves as a means to transcend what is, and allows us to consider what could be. These latter considerations guide our experiences, and the ambivalence they create drives the meaning-making process forward as the person attempts to overcome it. At times in this process, the person may simply reach another mundane meaning closely resembling a previous sense, as in Vygotsky's return to a sense of "calm" in example just presented. Yet there is also the potential for the creation of deeply felt moments of beauty, as the examples from Vygotsky's journal entries also document well. In both of these cases, it is by looking within the boundary of the present and future that the complex dynamics of the meaning-making process begin to emerge and offer to us a view showing the beauty of this process.

As depicted ain Figure 3.6, humans continually move from the mundane daily experiences they have—such as Vygostky's move from being in the fourth to first row above—and construct higher-order meanings though contrasts they themselves create. These new meanings, arrived at through overcoming ambivalence of literal and imagined senses, guide future meaning making.

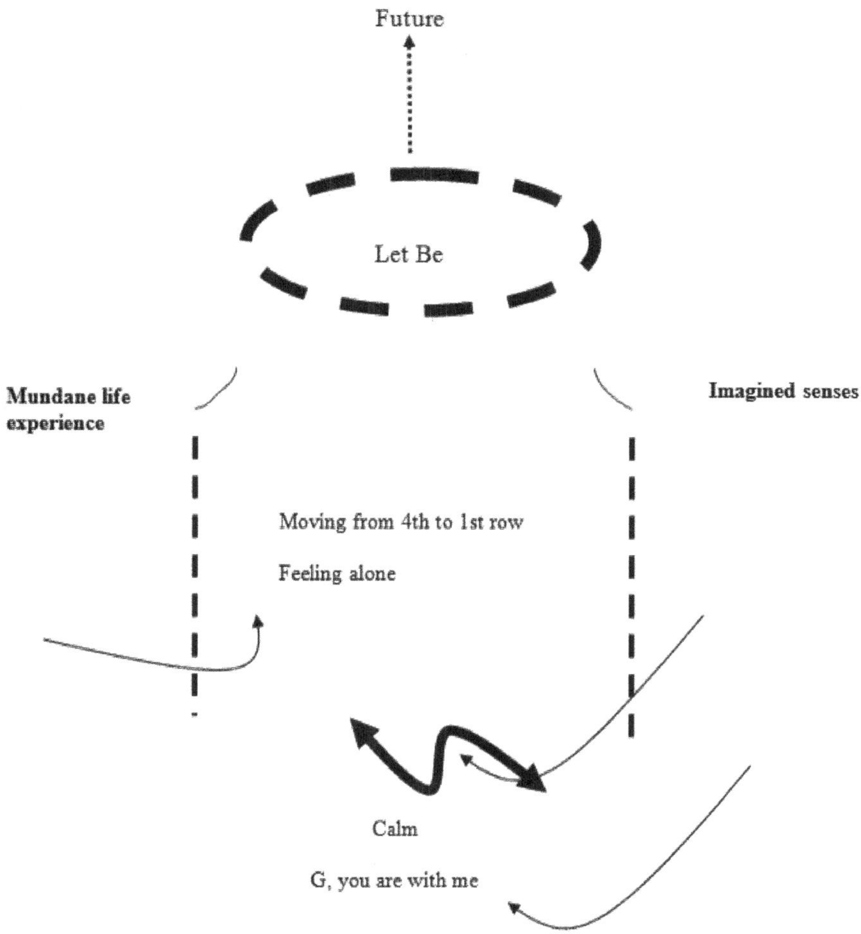

Figure 3.6. The poetic synthesis of mundane daily experience.

REFERENCES

Abbey, E. (2012). Ambivalence and its transformation. In J. Valsiner (Ed.), *Oxford handbook of culture and psychology* (pp. 989–997). New York, NY: Oxford University Press.

Abbey, E., & Valsiner, J. (2005). Poetiken des Selbst: Zwischen Ambivalenz, Bedeutung, Formlösigkeit und Wandel. *Psychologie & Gesellschaftskritik, 115*, 133–149.

Bergson, H. (1913). *Time and free will*. London, UK: George Allen.

Bullough, E. (1912). "Psychical distance" as a factor in art and an aesthetic principle. *Journal of Psychology, 5*(2), 87–118.

Cupchik, G., & Winston, A. (1996). Confluence and divergence in empirical aesthetics, philosophy, and mainstream psychology. In M. P. Friedman & E. C. Carterette (Eds.), *Handbook of perception and cognition: Cognitive ecology* (2nd ed., pp. 61–85). San Diego, CA: Academic.

Josephs, I. E., Valsiner, J., & Surgan, S. E. (1999). The process of meaning construction. In J. Brandtstätder & R. M. Lerner (Eds.), *Action and self-development* (pp. 257–282). Thousand Oaks, CA: Sage.

Matte Blanco, I. (1975). *The unconscious as infinite sets*. London, UK: Duckworth.

Matte Blanco, I. (1988). *Thinking, feeling and being*. London, UK: Routledge.

Ricoeur, P. (1981). The metaphorical process as cognition, imagination, and feeling. In M. Johnson (Ed.), *Philosophical perspectives on metaphor* (pp. 228–247). Minneapolis: University of Minnesota Press.

Simão, L. M. (2012). The other in the self: A triadic unit. In J. Valsiner (Ed.), *Oxford handbook of culture and psychology* (pp. 403–420.) New York, NY: Oxford University Press.

Valsiner, J. (1997). *Culture and the development of children's action* (2nd ed.). New York, NY: Wiley.

Van der Veer, R., & Zavershneva, E. (2011). To Moscow with love: Partial reconstruction of Vygotsky's trip to London. *Integrative Psychological and Behavioral Sciences, 45*, 458–474.

Vygotsky, L. S. (1971). *The psychology of art*. Cambridge, MA: MIT Press.

Zittoun, T. (2006). *Transitions: Development through symbolic resources*. Charlotte, NC: Information Age.

CHAPTER 4

UNACCOMPLISHED TRAJECTORIES

Shadows From the Past in the Present and Future

Vívian Volkmer Pontes and Ana Cecília Bastos

> *It's just when we feel that little one inside ... I felt like having that sensation again, of having him alive in my arms.*
>
> —Beatriz

Life trajectories can be understood as continuous movements that depart from present time, based on past- and future-oriented experiences. Uncertainty about the future, about becoming, leads to the construction of meanings in the present, based on generalizations from past experiences, enabling people to face the future with some illusion of predictability, through imagination (Sato & Valsiner, 2010). The basic human capacity that allows imagination and coordinates the past and the future in the present is the semiotic function, that is to say, the ability to use signs (Zittoun et al., 2012).

Temporality: Culture in the Flow of Human Experience, pp. 57–91

This, for example, happens when a woman discovers she is pregnant. The pregnancy represents for many women the experience of a waiting period, an event somehow (personally and socially) imagined and expected, whose outcome is supposedly the birth of a new life. However, for other women, this relatively predetermined pathway from conception to childbirth leads to an unexpected interruption, and the life cycle is reversed: the death occurs before birth. The experience of pregnancy loss leads to the experience of many other significant losses: not only the loss of the "baby," but also of a certain ideal of family, the social role of mother, a certain control over the woman's body, and even her own life. Thereby, the experience of a pregnancy loss threatens these women's sense of self, causing them to experience discontinuities in their developmental course, launching them into an unknown field, in which uncertainty over the future is acutely perceived. Uncertainty is part of our human experience, of our relationship with the immediate future. In this sense, the past is constantly being rebuilt as we face the uncertainty of the future, through the construction of several signs (Valsiner, 2007).

The present work has the purpose to investigate a seemingly contradictory reality: what happens when a woman chooses to become a mother—allegedly a choice available to her female constitution—but instead of this experience, she experiences recurrent gestational losses. In this context, gestational losses are seen as markers of transition in the life course of these women. After all, these are events that can result in changes in the course of development, to the extent that there is interruption of pregnancy—a critical period in a woman's emotional growth, which enables restructurings, modifications, and reintegration of personality (Maldonado, 2002). The experience of an unexpected event, nonnormative, can be understood as a time of transition, resulting in a qualitative reorganization, both psychological and in behavioral levels (Cowan, 1991). Moreover, it can also reflect changes in these women's meaning systems, such as those related to motherhood, femininity, and the perception of themselves (Volkmer, Covas, Franco, & Costa 2006).

Therefore, a gestational loss consists of an event that threatens the sense of self and maximizes uncertainty. From this idea, we propose to deepen the understanding of the dynamics of the self during the transition of a critical moment (Thompson et al., 2002), leading to a rupture of what was expected to happen and threatening the sense of self. The purpose of the study consists of a theoretical and empirical deepening on *how the self builds continuity when going through successive ruptures along the reproductive trajectory.* It involves understanding the dynamics of the dialogical self moving in the flow of time aiming at reorganizing and rearranging in order to stay cohesive, in spite of ruptures and recurrent discontinuities, through an incessant movement of self-differentiation, of self-transformation. These transformations make it possible to understand life within the time.

THE EXPERIENCE OF DEVELOPMENTAL RUPTURE-TRANSITION FROM THE PERSPECTIVE OF CULTURAL DEVELOPMENTAL PSYCHOLOGY

This chapter is grounded on two central theoretical assumptions about development: the irreversibility of time and the semiotic and dialogical nature of human experience. Both assumptions have been systematically explored by two major theoretical traditions: developmental science and social-cultural psychology (Zittoun et al., 2012).

Developmental science is configured as a general perspective that is oriented toward the study of developmental processes. Development is the property of open systems to undergo changes in qualitative ways, under a constant relationship with the environment, and within irreversible time. To transform implies a change of form, that is to say, some type of form becomes a new form (Sato, Hidaka, & Fukuda, 2009). Transformation is thus a concept of inclusive time (Sato & Valsiner, 2010).

In turn, the social-cultural psychology states that human development is regulated by signs. The semiotic function mediates everything from personal to cultural experiences. Humans create signs by which they organize and give meaning to their subjective worlds throughout their entire lives. It is through the lens of these theoretical perspectives that experiences of pregnancy, maternity, and gestational loss will be analyzed. It is noteworthy that pregnancy comprises an event in which the idea of irreversible time becomes more explicit, illustrating the relationship of humans with the coming to be, with the future. By knowing herself pregnant, or even before that, a woman anticipates, projects for the future, a symbolic place for this new person in her life, wondering about physical and personality characteristics of the child, and the type of dialogic and affective relationship that will be established between them—mental constructions that will gradually give a symbolic form to a future child, a future, and to a future mother. Thus, this future and potential relationship is anticipated.

The occurrence of an involuntary gestational loss thus acquires a dimension that is not limited to the present moment, the loss of the fetus itself, but reaches the imagined loss of a baby, child, and family configuration anticipated in the future. For this reason, the spontaneous gestational loss is understood in terms of rupture, an abrupt and unexpected interruption of the ongoing identity-construction process, resulting in discontinuation of the sense of self. It leads to the interruption of a process of adaptation to a new condition, that is, the social identity of being a mother, through which practices and fields of personal and collective meanings prepare changing individuals for the future.

The domain of knowledge that unites human development and culture is designated by Valsiner (2007) as Developmental Cultural Psychology.

Developmental Cultural Psychology

The notion of motherhood, as well as the notion of abortion or a childless woman, is value-laden in our sociocultural context, being socially promoted and personally internalized. According to Farinati, Rigoni, and Müller (2006),

> The constitution of the desire for parenthood is part of the symbolic chain constituted by the identity of the subject him/herself. We are born tangled in a web of maternal and paternal desires (conscious and unconscious), carrying the marks of being linked to a symbolic plot, which transcends biology, but which through it reveals our belonging to a family, a generation, a place in the world. (p. 436)

Motherhood and abortion are socially and culturally regulated situations, permeated by social cues present in human environments, which are semiotically organized or structured based on the combination of several signs that work as social guides. The power of such social situations and experiences and the hypergeneralized meanings that emerge from them guide and organize the behavior, thought, and human affections, thus demonstrating the centrality of culture within human minds (Valsiner, 2007).

According to Bruner (1997a, 1997b), culture shapes human lives and minds, gives meaning to action, and situates its underlying intentional states (beliefs, desires, and meanings) in an interpretative system. It can be defined as a set of tools with techniques and procedures that enable the individual to understand his/her world and deal with it. The mind is the creator of meanings and thus at the same time it is constituted by culture, it also constitutes culture

Valsiner (2007), in turn, assigns a dynamic and procedural nature to the operation of culture within human psychological systems, which includes both the intrapersonal domain, as for example, thinking, feeling, and acting; as well as the interpersonal domain, which involves behavior and discursive practices in relation to others. Culture can be understood as a system of semiotic mediation—mediation which is established through signs—and that includes an inherent part of the systemic organization of human higher mental functions. Thus, to Valsiner, culture is a process of internalizing and externalizing problems, personal and social world constituting mutually.

People are engaged in constantly rebuilding their intrapsychological worlds through continuous exchange of perceptual and semiotic materials with the environment. In this sense, internalization consists of the process of constructive perception and analysis of experiences or of existing semiotic materials in the social world and their synthesis, in a new form, within the intrapsychological domain (Valsiner, 2007). That is to say, it consists of constructive transformation of the meanings of the social environment to the subjective world, promoting innovation of personal meanings (Zittoun et al., 2012). In turn, externalization consists of the constructive process of analyzing existing materials in personal cultures—intrapsychologically—during the transposition of the person's internal field to the external domain (environment) and the subsequent modification of the environment as a way of a new synthesis of these materials. The results of externalization prospectively feed the internalization process in progress, inasmuch as the progress of ongoing internalization promotes the process of externalization (Valsiner, 2007).

Therefore, according to this perspective, people create signs using their story of building signs, under the guidance of other human beings, who in this venture of manufacturing signs, are collectively driven by different social institutions. The sign, according to Vygotsky (1984/2008), is a psychological instrument that is produced in the relationship and affects and transforms individuals, who start to work in the realm of the symbolic. In the author's words, the sign has the function of mediation. It is a means of internal activity, directed at the control of the individual him/herself, "the sign is internally oriented" (Vygotsky, 1984/2008, p. 55). And because it is a human production, it operates in various ways: "as a mediating element (works between, addresses), as an operator (turns objects into another thing), as a converter (transforms) of social relations into mental functions" (Smolka, 2004, p. 41). It can also be designed in its "constitutive, flexible and reflective" aspect (Smolka, 2004, p. 45).

Thus, this dual process of internalization and externalization makes each individual a singular person, although under the bases of the same cultural matrix, in that cultural messages, although similar for different individuals, are transformed and reconstructed in a necessarily unique way. In the cultural transmission of knowledge then, the individuals actively transform cultural messages, which leads to multiple courses of reconstruction of messages, enabling the emergence of novelty.

Furthermore, Valsiner (2007) also defines culture as a mechanism of psychological distancing. This means that, due to the capacity and propensity of human beings to create and use semiotic resources, humans become able to distance themselves in relation to their immediate contexts of life. Thus, the individual is both an actor who is immersed in a given context and a reflective agent who can distance him/herself from the scenario in which he/she is immersed. This reflection, which is both cognitive and affective, allows the psychological system to consider contexts of the past, imagine in future contexts, as well as take the perspective of others.

Studying such human experience, especially under circumstances in which uncertainty is fully present, requires an analysis based on theoretical models capable of taking into account development within irreversible time.

TRAJECTORY EQUIFINALITY MODEL: RUPTURE AND RECONSTRUCTIONS IN PERSONAL CULTURE OVER TIME

The trajectory equifinality model (TEM) is a structure of methodological research developed in order to analyze, examine, and describe qualitative data in terms of the diversity of developmental trajectories and life courses (Yasuda, in press). TEM was developed based on the concept of equifinality, which, according to Sato et al. (2006), was originated from Von Bertalanffy's general theory of systems in 1968 and is based on the first biological works from Hans Driesch by 1908. According to Von Bertalanffy (as cited in Sato et al., 2006), equifinality, a basic characteristic of open systems, means that the same outcome can be reached through distinct initial conditions and over different paths. Thus, the notion of equifinality suggests that certain points represent common destinations for cultural and social contexts, although individuals may follow different paths to achieve their goals.

TEM originates in the theoretical necessity of contemporary science to reconcile two central aspects in their analytical schemes: the time and the transformation of potentials into reality—elements essential for the study of human development. It is based on a systemic view of psychology—that considers human beings as open systems—and takes into account the notion of irreversible time, that is, the flow of time from an infinite past toward an infinite future (Sato et al., 2009; Valsiner, 2010). According to Sato and Valsiner (2010), time lies within biological, psychological, and social events and is inseparable from all growing systems. In this sense, the transformation, which implies a change in form, is an inclusive concept of time. It involves moving toward the future based on the network of actual and potential future trajectories.

As is clear by definition, any human development is possible only in the case of open systems, which exist due to their exchange relationships with the environment. Thus, the developmental process is potentially open to multiple pathways (multilinear process), although only one of them is followed. The translation of potential trajectories into the actual trajectories is the focus of TEM. Therefore, TEM includes in its analytical method both the real aspect (the real trajectory of development until the present moment) and the "*un*real" aspect (possible trajectories that existed in the past and are supposed to exist in the future), which includes reconstructions and imaginations; that which is not real or which is not real yet or that which will never be real. In the Figure 4.1, Valsiner presents the minimum systemic unity of analysis of TEM:

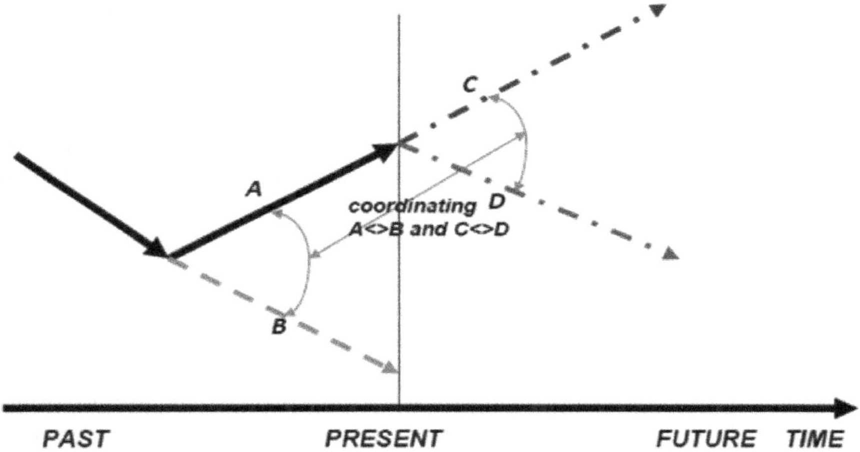

Source: Valsiner (2010).

Figure 4.1. Minimum systemic unity of analysis of TEM.

Figure 4.1 above includes three imaginary parts (B, C, D) and one real part (A), representing the predominance of the imaginary over the real aspect. Moreover, Valsiner points out that the relationships between these four elements (A<>B, C<>D) and their metarelationships ({A<>B} <> {C<>D}) constitute the unity of structural analysis (Valsiner, 2010).

Considering the issue of time, one can observe an asymmetry of the future in relation to the past: the future is full of potential courses of events, inasmuch as the past is mainly characterized by unilinearity, which sets the path for real development until the present moment (A). Thus, insofar as the future is constantly in the process of becoming the past, the movement toward the "new present" (to then become past) implies maintaining the characteristics of possible future trajectories. In this sense, the past has no valid alternatives (B), which marks the impossibility of going back—time is of an irreversible nature (Sato & Valsiner, 2010).

The TEM is thus a method of describing human life within irreversible time, after researchers have focused on important events as *Equifinality*

Points. As Sato et al. (2009) state, the term "equifinality" does not imply "sameness," a condition that is impossible in any historical system. Instead, it suggests a region of similarity in the time course of different trajectories. After establishing the equifinality point, trajectories should be drawn. Beyond the point of equifinality, turning points should also be considered. According to Sato et al. (2009), there are the *obligatory turning points* (OTP), consisting of a phase and/or event that a person inevitably experiences. There are two types of OTP: *endogenous*, which includes species-specific biological turning points, such as menopause, and *exogenous*, configured by the environment (Sato et al., 2006).

There are also turning points that have options between two or more pathways to follow, which are called *bifurcation points* (BPF). The bifurcation points are coined by Sato and Valsiner (2010) as points of transformation, where one looks toward the future and makes a choice. They are full of tension, as in these points two conflicting forces may be present: social direction and social guidance. The former, the social direction, refers to sociohistorical power, which provides tradition, social norm, and social pressure. Social guidance, provided by intimates such as family, friends and others, is the power of defense against social direction. Both forces are transmitted through everyday social exchanges in social situations. However, whereas social direction refers to the power that inhibits progress toward the equifinality point, social guidance consists of the power of promotion to reach that point (Sato et al., 2009). In this context, a person proceeds with his/her *synthesized personal orientation* (SPO) and struggles to accomplish his/her own orientation against social direction, with the support of social guidance provided by his/her intimate social relations. The synthesized personal guidance assumes that each person has his/her unique orientation to the Equifinality Point, which is an open-systemic orientation of the human being within irreversible time (Sato & Valsiner, 2010; Sato et al., 2006).

In this chapter, we are interested in the analysis of reproductive trajectories of women marked by recurrent gestational losses, understood as disruptive experiences that threaten the sense of continuity of self. That is to say, it threatens the sense of identity (of who they were, who they are, and what they will be), requiring efforts to forge a sense of continuity through transitions, as well as to build a sense of integrity or consistency among their values and actions over time (Zittoun & Grossen, 2012). The disruption in sense of self demands new subjective movements toward the reconstruction of the self.

In order to understand the mechanisms of reconfiguration of the self system—triggered in moments following the disruption in the flow of experience—an analysis of the processes of emergency and semiotic mediation, which occurred in the imaginary space of self and were expressed, was performed in some way through Beatriz's autobiographical narrative (described below).

In the process of emergency and semiotic mediation, people create and use signs as an effort to actively relate to the world, to be prepared for what is about to happen or make it happen. In each representation made by a sign, there a suggestion for the future, which therefore builds a bridge between past and present to a sense of possible future. Therefore, in a context of uncertainty, people create signs to build stability, a process which is always necessarily dynamic (Valsiner, 2005, 2007; Abbey & Valsiner, 2004).

THE CASE OF BEATRIZ: AUTOBIOGRAPHICAL NARRATIVE OF REPRODUCTIVE HISTORY

In the present chapter, we focus on a singular case—Beatriz's case—whose interview is part of a broader research project about the experience of recurrent pregnancy loss in different health contexts held in Salvador, Bahia, Brazil, between 2009 and 2012, in order to analyze how the self builds continuity when going through recurring ruptures (Pontes, 2013). At the time of the interview, Beatriz had been seeking a private clinic in order accomplish an investigation and immunological treatment after suffering recurrent miscarriages. At this point of her reproductive trajectory, she was invited by clinical psychologists to participate in the study, where a narrative interview was conducted about her reproductive history.

Beatriz has a reproductive history marked by three gestational losses: one miscarriage and two stillbirths. She was not pregnant at the moment the interviews were conducted.[1] The first pregnancy occurred in 2005, when she was 22 and had been married for four years. She reported that the couple had planned to have a child, but not at that moment, for Beatriz still considered herself very young. In spite of this, she reported that finding herself pregnant was an unexpected event, which was considered as positive:

> To me it was good news [laughs], especially finding out it was a girl ... because my biggest dream was to have a girl [laughs]. It was really good for us, we were very happy ... I didn't expect too much at that moment, but I didn't expect much at that time, but I really liked.

Besides the age issue, Beatriz reported she worried about the moment of childbirth, an event in which people say there is severe physical pain. She also worried about her ability to care for a child due to a perception relating youth to immaturity:

> I think all women, before getting pregnant, are afraid of labor pain. I was very much afraid [laughs] and wanted to become more mature ... I didn't think I was responsible enough [laughs], ... because I was young, I don't know, I don't know how to explain, but I think I wasn't prepared at that moment, you know, previously, at that moment.

Despite her fears, her pregnancy went on with no complications. However, on the seventh month of pregnancy, her blood pressure began to rise and other symptoms appeared in the form of generalized body swelling, stomach and back aches. Doctors also realized that the unborn baby was losing weight. These symptoms seemed to indicate a diagnosis of eclampsia. However, in order to perform laboratory tests to confirm this diagnosis, Beatriz had to resort to the public health service of another nearby town, because she lived in a small town in Bahia which did not have this service. During the wait for test results—about a week—the symptoms persisted.

Beatriz reported that she understood what eclampsia represented, due to previous experience with an aunt who had this diagnosis during pregnancy, resulting in the death of the baby. In spite of this information, she chose to believe in the wisdom handed down by her grandmother, who said: "When a woman is pregnant and becomes very swollen, people say she is going to have a good delivery." Besides, the body experience of feeling the baby moving was, to Beatriz, evidence that there was nothing wrong with the baby: "Since the baby is moving a lot, the baby moved a lot, 'Oh, a baby that moves is a healthy baby,' so I wasn't worried." However, when she was 30 weeks pregnant, she was doing an ultrasound examination, when it was found that the baby's heartbeat had stopped—an event that was perceived as "an unexpected death":

> When the day came for me to do the prenatal, I went and the doctor didn't hear any heartbeats anymore, then *we were in shock*, because we had everything ready, the baby's room was ready, everything was ready, I still have the baby's room decoration ready [laughs], and then *I found out that she wouldn't, she wouldn't come home anymore*, but ... *I wasn't afraid of thinking about it, I was already making plans for the next pregnancy* [laughs], I never thought about giving up, I was afraid, I was always thinking positive, so much that I got pregnant, I wanted to get pregnant soon.

As the fragment above illustrates, the narrative of her reproductive trajectory, marked by gestational losses and risks to her own life, was accompanied by an often cheerful and smiling expression, which is apparently incompatible with the contents of her report. Moreover, the experience of the first gestational loss, a meaningful departure from what was expected to happen, seems not to have been acknowledged and developed under an affective field and was automatically replaced by the planning of the next pregnancy.

Thus, a necessity to induce labor emerged, based on the diagnosis of fetal death and a severe preeclampsia. Beatriz reports that she even saw her daughter, but did not want to touch her because of her fear of dead people:

> I saw her as soon as she ... was born, they put her close to me, I didn't touch her, I'm terrified of dead people . . . When I saw the baby, I couldn't hold her, I felt like, I don't know how to explain. I couldn't hold her ... they put her by my side, I even touched her little hand, but I couldn't hold her in my arms.

The stillborn baby was buried by her husband and some close relatives. Beatriz did not attend the ceremony: "We did the burial, yes, we did all that. She was really beautiful, it looked like she was asleep. It was hard to believe ..., because everything happened so fast, it all went well ... it happened very quickly."

For Beatriz, the first gestational loss was the result of deficient healthcare assistance—she associates it with the fact that she was assisted by the public healthcare system—and the passivity of the couple regarding the events:

> Because of the town, small towns only have public healthcare assistance, which is terrible.... Due to the fact that the pregnancy had some complications, I think we got a little conformed and we were very confident, since we, he is evangelical, and used to think a lot, "God will help, everything will turn out fine." The doctors also used to say, "No, everything will work out fine," so we trusted them a lot.

Following medical recommendations, Beatriz became pregnant again one year later: "I didn't think about giving up for one second.... Previously I was assisted by the public healthcare system, from the second pregnancy on, I wanted private healthcare assistance, since I realized it was a risky pregnancy." The second pregnancy was precociously interrupted due to miscarriage, when Beatriz was approximately 6 weeks pregnant. It was an anembrionary pregnancy. Inasmuch as the fetus was retained in her womb, she had to go through a curettage procedure: "I was kind of ..., I left that experience desolate, I had no ground, just to think of the possibility of going through all that again." After the second gestational loss, the couple started to pay for a private healthcare assistance for Beatriz, so she could have access to, according to them, assistance of better "quality": "Because it happened, we are going to look for a private doctor, one who has more experience in this field."

In 2008, Beatriz became pregnant for the third time. She had prenatal care in the private healthcare assistance context, in a city located close to her home town. She reports that every 20 days she went to this city to go through prenatal examinations. However, when she was about 7 months along, some changes in the pregnancy were detected by the doctor: "When the time for a morphological ultrasound came, I arrived there and the doctor realized the baby was underweight, my blood pressure was too high, so he realized something was wrong." Beatriz was urgently referred to a private clinic in another city, Feira de Santana, in order to repeat the ultrasound examination, in a more modern facility. Despite the medical information about pregnancy complications, Beatriz reports she had brought with her a DVD in order to record images of the baby—which, once again, was a girl—but was discouraged by the physician, who reported that they could not see the baby because she was too small: "Then he told me, 'You won't be able to see anything. I won't even explain it to you because you won't understand anything.'" On the day after the exam, Beatriz was admitted to a public hospital with a diagnosis of severe eclampsia, feeling sharp pain in the stomach and back.

Despite the perception of risk to her own life, Beatriz emphasizes in her narrative that, at that moment, she thought only about the baby and her small chance of surviving:

> I myself didn't think much of me, I thought about the baby. I wondered, what are her chances of life? ... I entered the hospital thinking a lot. I only thought about the baby, I just wanted to know what her chances were. I think I forgot about myself at that moment [laughs].

Since the baby was still alive, there was an attempt on the part of the medical team that accompanied it, to prolong pregnancy, once they had managed to stabilize her blood pressure:

> Then I became confident, I trusted everything was going to work out fine. This was on Thursday, if I'm not mistaken. I felt the baby moving and everything, but on Saturday and Sunday I didn't feel her. I felt something similar to the last time, something like a mass in my belly ... on Tuesday I did the ultrasound, which confirmed she was already dead, so it was a big shock for me because I wasn't expecting such a thing. *I was hopeful ... I hoped that a miracle would happen.*

Thus, after a few days of hospitalization, the death of the baby was diagnosed. Beatriz then had to undergo an induced labor again. The termination of pregnancy was perceived as a sudden and unexpected event: "Everything went so fast, my concern was this: that everything happens really fast, on one day I'm normal, just as I talk to you, but when I get to the doctor, he realizes something is not right." Because the stillborn baby was too small, Beatriz returned to her home town 15 days after hospitalization, bringing the baby in a container with formaldehyde. Her husband buried the baby in the same place as the baby from the first pregnancy.

From this experience of gestational loss, in dialogue with health professionals, Beatriz constructed the meaning of risk to her own life:

> I went to a great risk because I had HELLP.[2] HELLP is ... risky for the kidneys, one may fall into coma, the blood pressure increases ... I was at great risk, *my life was really endangered.... It was much more serious*, they told me it was more serious than the first time.

After hospitalization, Beatriz reported she had initiated a new plan for the next pregnancy: "I came home and there I stayed making my plans, of how the next pregnancy would be." She and her husband agreed to accumulate funds until they had enough money to have access to specialized medical treatment offered by private health care: "[Preparing] especially financially, since I think I have always been psychologically prepared [laughs] because I always had positive thoughts, I never thought about giving up. I was always strong, I always thought positive." Thus, in order to save money to go through medical treatment, Beatriz saved coins. At the end of 2 years, she reports that she had gathered more than 20 kilos in coins worth one Real, totaling approximately 7,000 Reais. Thus, she concluded that the time had come to initiate treatment and then try another pregnancy.

Therefore, she sought specialized medical care 2 years after the third pregnancy loss. When performing the tests ordered by specialists, they identified some immunological changes, such as thrombophilia (blood clotting changes, with greater risk for thrombus formation, or thrombosis) and alloimmune factors (maternal immune response of the body, which does not recognize and accept the conceptus). She went through the alloimmune treatment and was released by the doctor to get pregnant in December 2010. But almost a year later, Beatriz reported she had not yet been able to conceive. For this reason, she requested psychological treatment in November 2011 (when the second interview was conducted), believing that the reason for not being able to get pregnant was the high level of anxiety. It is noteworthy that this meaning was co-constructed with health professionals and others in her surrounding social network: "I wasn't able to get pregnant yet, so I'm thinking, the doctors speak of anxiety.... Today, I need a psychologist, I have to talk, I have to let everything go."

The Self-Dynamics: Dialogical Processes Involved in Semiotic Self-Regulation

The reproductive trajectory of Beatriz is marked by miscarriages and serious risks to her own life, due to her persistence in getting pregnant. In this scenario, in which life is constantly threatened or lost, many people expressed significant opposition to her choice for biological motherhood. These voices intensified ambivalence related to decision-making regarding whether to get pregnant again or quit trying.

Despite this opposition, and with the purpose of empowering herself and giving support to the decision to get pregnant again, Beatriz used a powerful ally: biomedical discourse. After all, such discourse legitimizes the chosen path, especially in technological advancement, by promising her a few certainties in a scenario filled with uncertainties:

> We talked to the doctor. If he had said there wasn't any way, then it was OK, but when general people talk to us and say we can't, *the one to say so is the specialist*, isn't it ? *He said yes, you can get pregnant if you are in treatment* ... because general people talk too much, they keep talking on our heads. I said: "No, I'll think, *first I'll go to the doctor to know what he's going to tell me*, whether I have a chance to get pregnant or not, *then, thank goodness, he said what I wanted to hear* [laughs]."

The opposition of meaningful persons to Beatriz's persistence in trying to become pregnant again was intensified after the third loss, especially due to serious complications in her health. Then, even her husband, who always offered important emotional support to Beatriz, as well as an incentive to continue the pregnancy, also grew averse to a new gestation. His voice echoed to Beatrice as an attempt to broaden the scope of future possibilities for the realization of other paths, like the alternative path of adopting a child or being a childless couple:

> When I got home, he said, "*Beatriz, you are not getting pregnant again. I don't want it, let's be only the two of us, I don't want to run risks anymore.*" Because he thought of me first. Then I said "No, I want to try again. If the doctor said I have chances, then I will try." He responded, "Then, you know, *if it were for me, it would be only the two of us, if you wanted to adopt I would adopt.*" But I said, "No, I'll try first, let's say, not for the last time, but let's try again."... "If there is treatment available, then let's try it."

The voices of other important family members were also present to attempt to dissuade her from her purpose of continue trying to get pregnant. One of those voices is highlighted in her narrative: a cousin of her husband, who she considered like her sister. This cousin has a university degree and used the Internet to research what had happened in Beatriz's last pregnancy. Armed with information about the severity of what had happened (severe eclampsia, HELLP syndrome), she used the semiotic strategy of emphasizing the worst that could have happened in the past—renal failure, cerebral edema and, therefore, coma—to warn Beatriz of the possibility that these events could happen in the future if she insisted, again, on following that route. In doing so, she attempted to destabilize the semiotic threading built by Beatriz, which uses signs from the medical sphere to minimize ambivalence and reduce the degree of uncertainty for the future. Thus, this cousin calls into question the power and promise of certainty attributed to the biomedical discourse. However, the affective power of her voice was not sufficient to master the voice of the specialist physician:

> Alberto's [her husband] cousin even tells me she does not agree much with the idea of me getting pregnant again.... She worries a lot about me, she is his cousin, but we have a strong friendship. It feels like she is my sister.... So, she researched a lot, she worried about me.... She told me I ran a lot of risks, "You risked having kidney failure because your blood pressure was increasing too much," and she sort of explained to me what she had researched, about "the possibility of entering into coma at any moment due to the elevated blood pressure".... And she told me "All right, the doctor said you could try it again with the treatment," *but she always says that "no doctor gives 100% certainty,"* and I said, "No, but I think they are grounded on some level, so they won't mislead me into something with which I may be risking my life later, so if he is saying I can, if he is assuring me that, because I think that if I were risking his life, he would not go ahead with my treatment, isn't it so?"
>
> But then I explain, I say, "No, the doctor already told me *what had already happened will not happen again, and if it does happen, it won't be as intense, I won't run as much risk,* because they will be close to me, and will interfere so nothing serious happens, either with me or with the baby.

Strong signs (Abbey & Valsiner, 2004)—signs that overly narrow the field of meanings while at the same time offer a clear path for the person to follow at a certain moment, in this case, deriving from the medical sphere, such as the general sign "treatment"—are internalized in a particular way by Beatriz. Such signs seem to support her movement toward motherhood. Even as they constrain her field of future possibilities, including only determined trajectories, such as becoming a mother and having a child; they seem to exclude other possibilities, such as the repetition of losses and the risk to her own life (see Figure 4.2). Thus, these signs allow her to imagine the foreseeable future, with a promise to prevent or minimize damage in a subsequent pregnancy attempt, thus reducing the ambivalence. The making of distinctions, through mechanisms of selective attention and perception, and its semiotic organization, work to enable the construction of a subjective world that seems relatively stable (Valsiner, 2007). However, as Abbey and Valsiner (2004) point out, this condition is likely to rigidify the dialogical process of meaning construction, markedly neglecting the complexity of the phenomenon, being only temporarily functional.

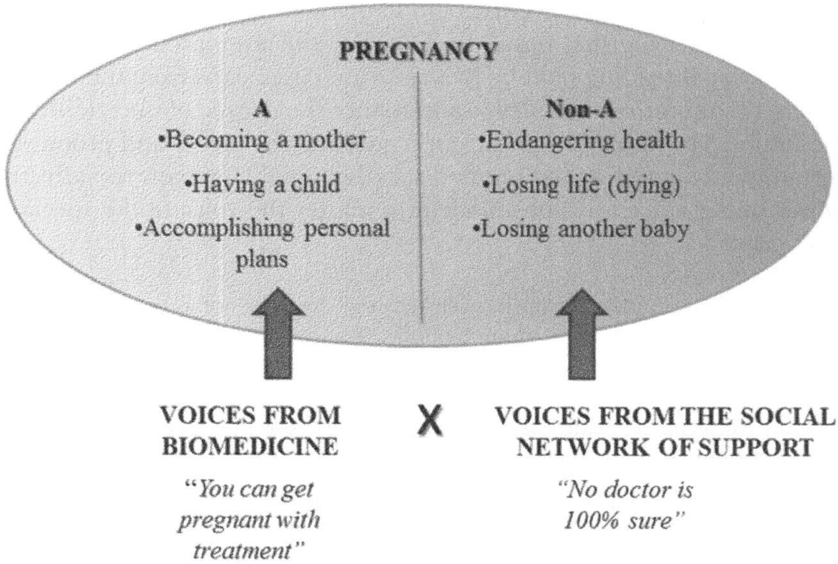

Figure 4.2. Meanings of pregnancy: Unity of opposites and their dynamic relations.

In the attempt to build a sense of security, as if getting pregnant again could constitute a trajectory whose outcome had become predictable—thanks to specialized medical treatment—a major obstacle presented itself: the fact that getting pregnant is "taking too long." The time elapsed after the attempts to get pregnant after the immunological treatment with the specialist and its lack of success increased the ambivalence in Beatriz regarding trying to get pregnant again. The network of support contributed to the expansion of the ambivalence, constructing a meaning for the difficulty in Beatriz getting pregnant, which is related to "God's will" (see Figure 4.3). Adding to the questioning of the signs from the medical sphere, in this scenario there was the introduction of another powerful and hypergeneralized sign—"God":

> So this cousin of his was telling me, "Look, Beatriz, you get pregnant so easily, why is it now taking so long? *Isn't it God showing you that you should give it up and stop it all?*" And *that stayed in my mind.*

The internalization of this sign, through a process of active construction through bidirectional interaction with social others, led Beatriz to a reflective prayer with a future-oriented focus. That is, it led her to pray in order to direct and reflect on future actions: should she become pregnant again?

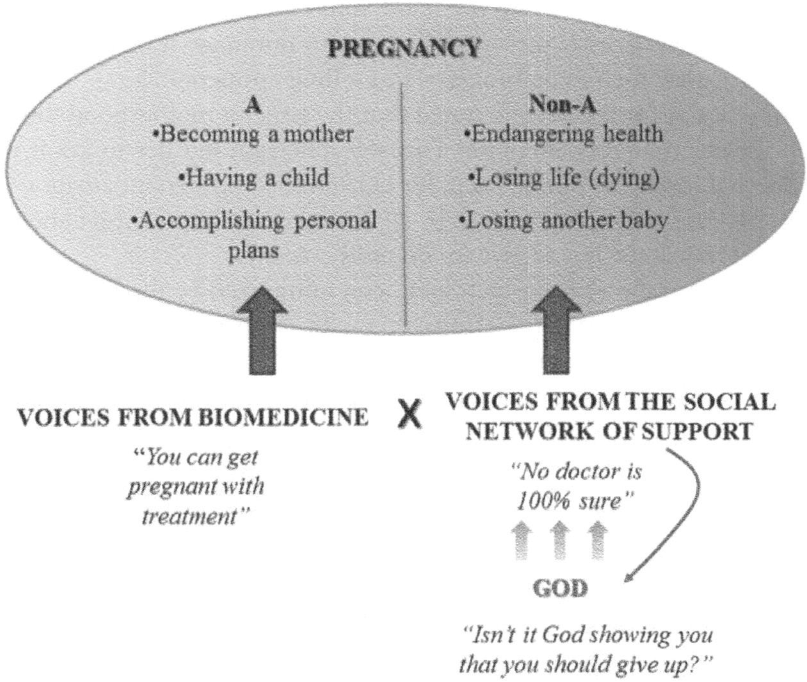

Figure 4.3. Meanings of pregnancy: Unity of opposites and the changing in its dynamics of relationships.

Thus, she established a communication with God, asking Him for answers on how she should act:

> So ... when I got home, *I talked to God*. I said, "Oh, my God, *if it is for me to give it up* ... if it is for me to give it up, *then take it out of my mind* so I can forget about it."

mention that, despite that communication *with* God can be addressed internally and individually, the communication *from* God is indirect and occurs in the form of metacommunication (Peskek, Kraus, & Diriwachter, 2008; Valsiner, 2000). This metacommunication depends on the interpretation, that is, what one might consider to be God's messages and how these messages are internalized and analyzed (Peskek et al., 2008).

Given the alleged condition of infertility, Beatriz resorted to the medical sphere in order to build some meaning to explain her difficulty conceiving. However, since there was no evidence of physical problems in her and her husband's respective reproductive tracts, there was the suggestion that the medical difficulty to become pregnant could be related to anxiety. From this suggestion, Beatriz undertook actions to minimize it by engaging in other activities, such as embroidery and selling cosmetics.

However, at one point, she noted that the issue of motherhood did not occur in her thoughts anymore. This self-reflection led her to ask if that meant that it was God's will, then, for her not to get pregnant anymore, as if God was answering her request by allowing her to simply forget about it, if it was His will for her to give up on motherhood.

Thus, the ambivalence which had been minimized by the construction of the meaning of "anxiety"—from the medical suggestion—is intensified again, promoting confusion in her thoughts. Therefore, at this moment, there is the construction of other meanings, such as the idea of being punished by God:

> So I was wondering, did I forget a little bit about pregnancy? "My God, was it because I asked you to forget?" And God was already showing me that I wasn't supposed to get pregnant? The *confusion started in my head*, "Am I supposed to stop trying, did I forget about it? ... At the same time, I thought I wasn't supposed to get pregnant, I questioned myself: "*What did I do, why is God punishing me for it*,"since I get pregnant so easily, why is it taking so long? ... and I used to question myself "*Why? Oh what did I do wrong?*" and came back to that thought, "Does not God want me to get pregnant? Am I supposed to stop trying, to give up?" [laughs] So *it made me confused*, and I don't really know what to do.

The self-reflection of Beatriz about "God's will "—supposedly superior to her own will—seems ambiguous, since Beatriz does not seem to have clarity about what direction this powerful sign seems to indicate her to follow. After all, on one hand, there is the difficulty of getting pregnant and the perception of the time period in which the issue of pregnancy was not present in her mind—events that seem to indicate God's will regarding her giving up motherhood. On the other hand, she reports that medical treatment that she has been performing, for pregnancy without risk to her life or the baby's life, has a very high cost, which worries her because she does not know until when she will be able to afford it. Nevertheless, she constructs the meaning that God is helping the couple to continue with the treatment: "But thank goodness, *God is helping*, for now everything is still normal. I think it is very likely that we won't stop [the treatment] for a while [laughs]. But I'm very confident because of it."

Therefore, Beatriz's desire to get pregnant again seems to persist in her future plans, despite the ambiguity of meanings attributed to divine will or other significant voices against this decision. This persistent movement of "swimming against the tide"—reducing the ambivalence and uncertainties related to the future—is supported by the supposed certainty and predictability "sold"—at a high cost—by the biomedical discourse:

> Look, if I had to give up, it would be because of the financial situation and not because I wouldn't want to try to become pregnant again, never, only if the doctor says "You can't, no way. If you get pregnant, you'll risk your life; you might die," then I would stop ... but no doctor has ever told me that. They always say I have chances to get pregnant, with the treatment, because Dr. Mn told me, "*Without treatment, yes, you run risks. But going through the treatment, no, if you do everything right, you are not running any risks.*" *So I trust him. I don't think I would give it up so easily.* [laughs]

Thus, some signs coming from the medical sphere were internalized by Beatriz—from research conducted on the Internet, in consultations with the medical specialist, and medical reports—which allowed the reduction of ambivalence regarding uncertainty about the future. Such signs, like the idea of "placental thrombosis," "blood incompatibility," "advanced technology," and "treatment," allowed her to construct a meaning to the earlier losses (unexpected disruptions in the trajectories of the past), as well as the imaginary construction of stability, security, and a measure of control, directed to the future:

> I have researched a lot, saw several cases. I have researched on the Internet, and *thanks to God, technology is now so advanced,* it starts to give us the opportunity to try again ... based on what I could understand from the reports, because I had *placental thrombosis, and incompatibility* ... our blood is not compatible for the fact we have the same blood, type O Positive. There are some components that do not match, so I think it leads my body to reject. The doctor explained that my body rejects, detects it as a foreign body and wants to expel it. I think this is why I had those complications and all that, and also the placental thrombosis.... And due to the fact that the blood is stuck there, you know, my blood pressure tends to increase because it's stuck, it does not circulate regularly, so the doctor told me.

According to Valsiner (2007), each interpretation is prescriptively value-laden regarding the action. Therefore, the medical information, which is internalized in a particular way by Beatriz, constrains the range of future possibilities, influencing the present moment, to the extent that they empower her decision making and subsequent action, that is, becoming pregnant again after going through medical treatment (see Figures 4.4 and 4.5).

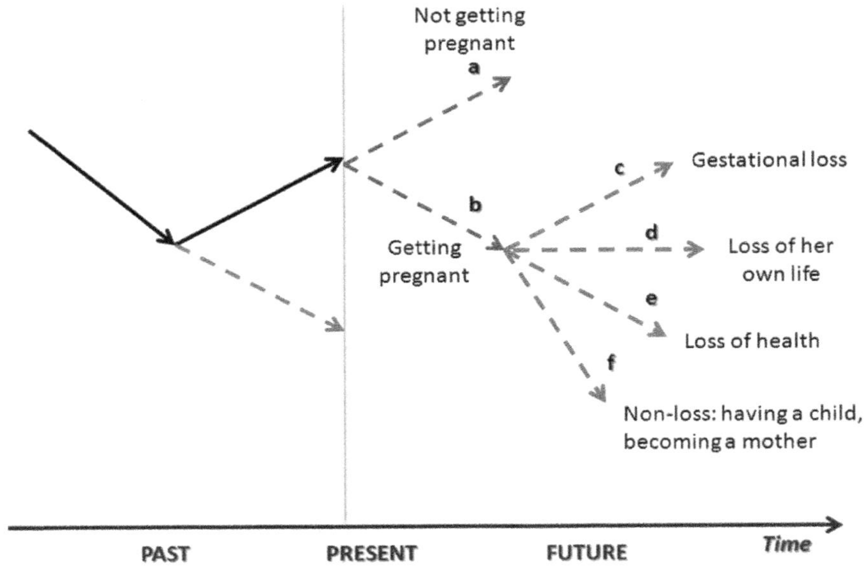

Figure 4.4. Range of future possible trajectories.

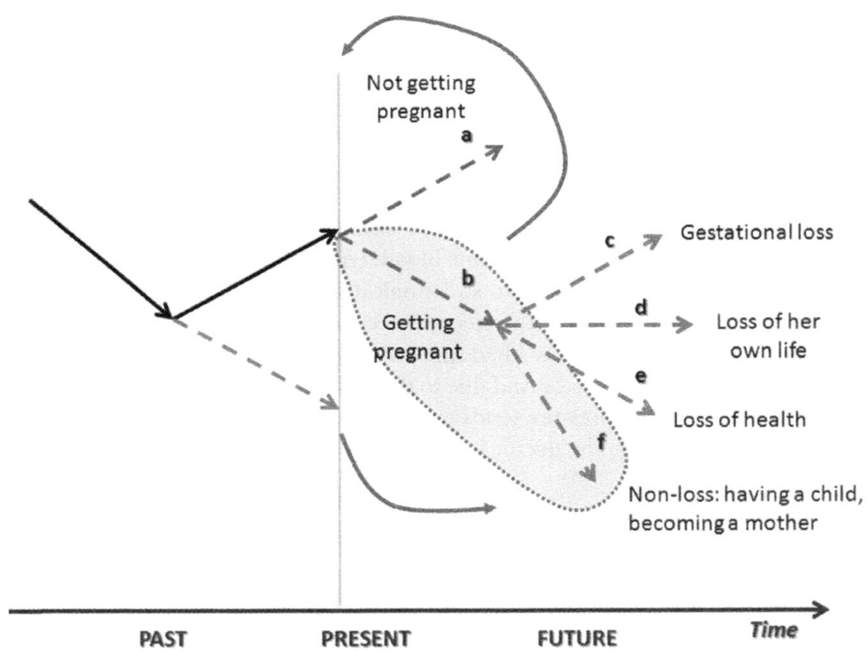

Figure 4.5. Future trajectories constrained by signs deriving from the medical sphere.

SEMIOTIC MEDIATION: STRATEGIES FOR THE CONSTRUCTION OF CONTINUITY

Unaccomplished Mourning

The process of construction of meanings grows out of the need for pre-adaptation to future environmental conditions, which, at present, are not known. The role of the other units in that process, as the physicians and the close social network, is to guide the microgenetic process. The associated meanings are built in motion. Another key aspect of this movement of construction of meaning consists of the affective processes. After all, the regulation and organization of meaning is a function of these processes. Affect is the basis upon which semiotic organizers are established and configured (Cabell & Valsiner, 2011).

The subjective experience is always built on an affective relationship with the world. People create meanings in their relations with the world, and the world itself, through their feelings, which are culturally organized through the creation and use of signs (Valsiner, 2007). This affect-laden relationship with the world is constituted as a totality—a combination of past, present, and future directions simultaneously felt as one (Cabell & Valsiner, 2011). The domain of feelings is crucial to the construction of personal cultures (Valsiner, 2007).

Thus, human affective processes are closely related to the meanings emerging from them (Valsiner, 2007). Based on this assumption, one can analyze the relationship between affective processes and meanings in the trajectory of Beatriz. Throughout her narrative is present some evidence of an attempt to minimize or deny the seriousness of experienced pregnancy complications, as well as attempts at replacing a lost baby with another one, as soon as possible, putting her own life at risk—discourse peppered with laughs calls the attention to this, as if the history narrated were comic instead of sad. These elements seem to indicate either a process of unaccomplished mourning or the lack of affective acknowledgment of the risks and losses experienced. The following report illustrates Beatriz's attempt to deny or minimize risks to pregnancy and her own life, based on the construction of meanings of a situation of imaginary normality:

> I stayed home waiting to get the exams, but then it started: high blood pressure, swelling persisted, headache and all, *just these symptoms*, and *we listen to the conversations of the elderly. My grandma used to say that when a woman is pregnant and gets very swollen, it means that she will have a good delivery*, but today it isn't like this anymore, so I don't know. I think I used to think a lot about it, and also *I didn't feel anything besides these symptoms*. So, *since the baby is moving a lot, she moved a lot, "Oh, a baby that moves is a healthy baby,"* so I wasn't worried at all. But when the day to have the prenatal exam came, I went and the doctor

> didn't hear the baby's heartbeats anymore. We were in shock, because we had everything all set up, the bedroom was ready.

The complications in a pregnancy consist of a critical situation that requires the construction of a generalized meaning that somehow reduces ambivalence toward the events, promoting some kind of action. In the example described above, Beatriz seems to deny the possibility that the symptoms she feels may represent some important gestational complication. Instead, she resorts to the bricolage of meanings and some tinkering in order to mitigate the severity of the situation. Thus, she adds to her narrative a saying of her grandmother, rescued from a different context and historical moment, whose belief explained that the swelling in pregnancy was an indication of a good delivery. Related to this meaning is the medical-commonsense belief that a baby moving in the womb is a healthy baby. With this mosaic of different representations, Beatriz manages to reduce the intrapsychological ambivalence, saying to herself that all was well. But she is not able to change the course of events, whose symptoms actually comprised a diagnosis of preeclampsia, which culminated in the death of the baby in the last trimester of pregnancy and endangered her own life. Events like these took Beatriz by surprise. She was expecting the birth of a healthy baby, an easy delivery.

Following this direction, the narrative built on the first experience of pregnancy loss— the narrative referring to the memory of her dead daughter—and her emotional reaction, laughter, also draw attention. Such a reaction seems incompatible with the content covered in the report, at least when contextualized within the discourse of our culture, which relates death to sadness. The narrated story seems to lack emotion, which may suggest that Beatriz has not yet faced the loss, which is not possible to replace: "On the first [pregnancy], I saw, I saw the baby, she was perfect, *it looked like she was asleep*, she was very cute *[laughs]*."

Beatriz reported feelings of "tranquility" when the baby's death was found, as well as during delivery, reflecting a certain inadequacy of this emotional reaction to the situation: "I was surprised with myself." Furthermore, she reports that her inclination, seeing the dead baby, was to consider it as if it were alive, simply "sleeping." She also states that after the loss, she located the lost baby in another baby (the daughter of a neighbor) whom she considered to be much similar to the daughter she had lost. All these aspects not only seem to reveal a remarkable absence of sorrow, but also a failure to acknowledge the emotional loss, as illustrated by the stories that follow:

> Look, I was surprised with myself, because I *was quite easy*, many doctors and nurses thought that, because that was my first pregnancy, it wasn't good to go

> through all these problems. I was very easy. I went through labor, everything. It was a situation in which I stayed tranquil. *It didn't even look like all that was happening to me.*
>
> Until a while ago, I, oddly enough, there was a neighbor of mine, who, as soon as the girl arrived the next day, the next week, that ... when she came in, *if you closed your eyes it would be as if you were looking at my baby*, she was very much like her. There I was, with that in my mind, you know, that her face ... resembled the girl, when I saw this girl, she reminded me of the baby.

Another important aspect is the persistence in naming her next child—whom Beatriz expects to be female—with the same name chosen for the first baby lost: "Ester would be the name of my daughter, so, in fact it remains. If I have a girl, her name will be Ester again. [laughs]."

The case of Beatriz, thus, seems to illustrate a situation wherein there is a denial of death. There is a denial of irreparable loss as something that could not be avoided in the past and cannot be repaired in the future. Similarly, there is a denial, supported mainly by the biomedical discourse and its promises of certainty grounded in technological advancement, of the possibility of such loss being experienced again in the future.

Thus, Beatriz constructs the meaning that the loss is likely to be recovered in a near future, if the couple reaches some conditions considered essential, such as being accompanied by the best medical experts, whose private services—and high cost—are inscribed in the market of a big city. Therefore, at the same time she experiences the corpse in the present moment—the death of a baby—her thoughts and feelings are already designed for the future, planning for a new pregnancy. There is the emergence of a meaning, according to which it is possible to recover her initial project in life, to recover her daughter Ester, without realizing that this reparation is impossible, given the irreversible nature of time.

Thus, examining more closely the process of meaning construction undertaken by Beatriz, other elements should be considered. After all, she reports that the circumstance of death arouses emotions which cannot be exactly expressed in words: "I don't know exactly how to explain." However, in the narrative construction, she makes a reflective effort, semiotically anchored, to constrain this negative affect in terms of emotion, which she calls "terror": "I'm terrified of people who die."

Concerning this intense negative-affect-related death, the experienced gestational losses occurred within her body without there being any concrete alternative—such as avoidance, for example—to encountering something that arouses deep aversion. However, at that moment, a certain psychological mechanism of self-protection seems to be triggered, as Beatriz performs an attempt to deny death, an already consolidated loss. And this happens in affective and semiotic terms: she acts and feels as if the baby had not died ("seemed to be asleep"), she refuses to touch her cold body and declines to participate in the funeral performed by the husband. The veracity of death is put at stake and shared by her husband, since when he saw their daughter in the coffin, he reinforced the idea that the baby seemed to be alive:

> When I saw the baby, *I couldn't hold it*. I felt like, *I don't know exactly how to explain it*, but I couldn't hold it ... put her by my side, I even touched her little hand, but I couldn't hold her in my arms ... *I wasn't even able to cry* ... that round little face ... *because she was born perfect, she didn't look deformed, not at all, she was perfect....* [In the funeral, when the husband opens the coffin] He said *she didn't even look like she was dead, it looked like she was asleep*, she wasn't swollen, she was very beautiful, he said she was perfect, he even doubted: "*Is she really dead or not?*"

The whole meaning construction involves signs of a dual nature, consisting of the core A and its immediate interdependent context non-A, forming a complex whole. Non-A is its denial or its opposite (Cabell & Valsiner, 2011; Valsiner, 2007). Thus, for example, when a woman becomes pregnant, it is possible to outline the construction of meanings on life (core A), and its counterpart (non-A), death (see Figure 6). The occurrence of a miscarriage or gestational loss reverses this dominance: death becomes the core of this complex of meanings, whereas life becomes its counterpart, the context.

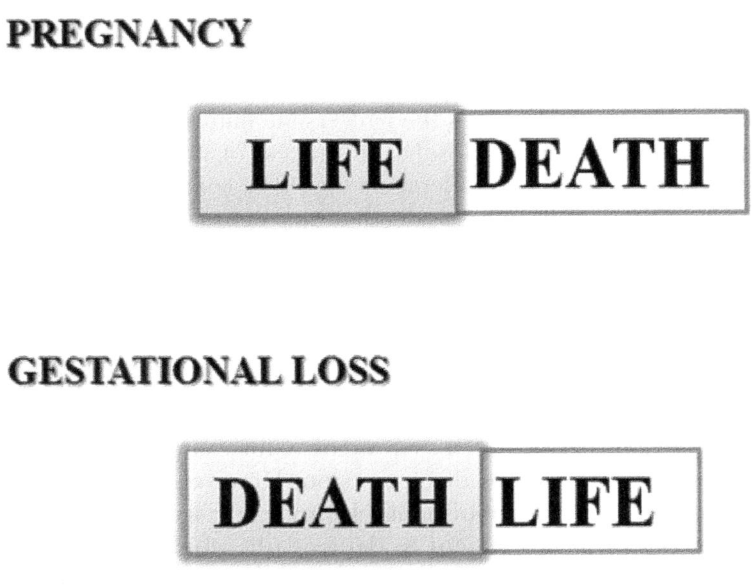

Figure 4.6. Complex of meanings during pregnancy and gestational loss.

In the analyzed case, one of the field components of meaning—death—appears temporarily on hold while Beatriz strives vigorously to generate life. She carefully plans the next pregnancy, researching the subject on the Internet, informing herself about the medical treatment and the specialist with whom she may get in contact, collecting, coin by coin, the financial resources to realize her plans. However, the condition of infertility, after her third gestational loss and after specialized medical treatment, triggers a crisis which she has vigorously tried to avoid: acknowledging the occurrence of the loss and her inability to repair it.

Therefore, the mourning process or affective-semiotic process of repairing the rupture, which until then had not been experienced in its fullness (because she had not acknowledged the finitude of the loss), seems to move, triggered by the present crisis. Accordingly, in the second interview with Beatriz, she expressed, at times, another kind of emotion—sadness—through crying. In one of those moments, she reported her mother's reaction to the loss of the last baby, reflecting on the maternal role, which for her, relates to the protection of the child. At that moment, then, she cried and addressed changes in her way of feeling, thinking, and acting as a result of acknowledging the irreversible loss:

> She [her mother] was also very, very sad, because she used to say that she wanted to go through all that, she wanted it that way, if it were possible, to feel all the pain in the world so I wouldn't feel it [laughs]. Then she was like this because she has that motherly way; she is very protecting, she doesn't want any of her children to feel any pain, so she, I might say she ..., uh ... [silence; she cries] .... I'm sorry [laughs] .... It's just that lately, it has been difficult for me ... because I always, from the first pregnancy, as I told you, I cried a lot and all, but when I came home, my husband seemed to suffer more than I was suffering, *but for me, it is as if I hadn't realized what had happened*, from one month on, one month and so on, I felt that void, all that thing, then I realized the loss I had. The same happened during the last pregnancy, when I came home, there were no signs of all that sorrow, but after a while, I started to feel it, and now.
>
> Lately, I mean, *previously, I didn't feel so sensitive*. There are moments in which I feel like this, *sometimes I cry for nothing*, something like that. I might listen to a song, especially if it's an evangelical song, then *sometimes I feel that void, that loss, that feeling of loss*, you know, *I start to cry for no apparent reason*. Even before I wasn't that way; I was pretty strong, I used to be surprised with myself, after all that I have been through and still being so strong, but lately, sometimes I haven't been able to bear it [laughs].

Among the emotional reactions triggered by the acknowledgement of loss, a high level of anxiety, crying for no apparent reason, irritability, agitation, nervousness, and anger—usually directed at her husband—are the most common ones: "I get so angry with him. Sometimes I say things I shouldn't say; sometimes I'm rude with him.... Then the anger came unannounced, because he didn't do anything.... The problem is all in my head."

According to Parkes (1998), the most common characteristics of the mourning process are acute episodes of psychic pain, accompanied by great anxiety. In addition, both anger and irritability are related to the initial phase of mourning. It is noteworthy that these were the symptoms that led Beatriz to seek psychological help. After all, there was the meaning, co-constructed with health professionals, which indicated that these symptoms, especially anxiety, could be related to the condition of infertility. So in order to placate them so that she could get pregnant again, Beatrice resorted to psychological assistance.

UNACCOMPLISHED TRAJECTORIES: SHADOWS OF THE PAST IN THE PRESENT AND FUTURE

According to Valsiner (2007), people feel prospectively, that is to say, they affectively anticipate the future, as a way to have control, create some stability in the face of the unpredictable and unprecedented next instant of experience. It is exactly this imaginative and affective prediction of the future that guides the person in the present moment. "Human experience is constantly directed toward the future, as a pre-adaptation to the uncertainty of the future" (p. 258). Human effort toward this unknown, creating novelty, is an inevitable aspect of the human being—"an act of crossing borders" (Valsiner, 2007). In this process of feeling ahead, there is a coordination with the past of two types: one that really happened and what could have happened (but did not). Similarly, there are possible future paths: potentialities and possibilities. Therefore, the cultural psychology perspective advocates an orientation of the past toward the future (Cabell & Valsiner, 2011).

In some moments of her narrative, Beatriz selectively recalls the past and, by using the semiotic function of imagination, she reflects on what might have been if the babies who had died were alive. These potential, yet unaccomplished, trajectories of the past, denominated by Bastos (2014) as shadow trajectories—acting on the present through imagination—make her reflect about an alternative present (one in which she would be mother of two children). The contrast or the tension between the real present (*what is*, meaning not being a mother, not having children) and the alternative future (*what could have been*, meaning being the mother of two children) appears in the narrative projected into the future: she is concerned with recovering as soon as possible what she should have, which she does not have yet. She worries about her age (previously she was too young, and now the perception she is growing "old"), and this tension modifies the present and future prospects, making her reflect on the paths which were not taken:

> So I'll try again to have immediately my grace, because I'm so very excited only by thinking, once I had friends who got pregnant with me ... whose children are now at the age my first one *would be*. I get more anxious just to think about it and to see them with their children, ... Now, I'm finding myself very old.... When I got pregnant for the first time, I was 22, and now I'm 28. So I wonder, *if I had had a child, he would be 6 now*.... I always see my friends. I have a friend who got pregnant at the same I did, at first, she already has a boy, mine was a girl, her boy is now 5–6 years old, the same age group, the difference is of months, and then she got pregnant, also almost on the same period I got pregnant the last time, so she also has a 2-year-old girl. So now, when I see her, I'm, I'm thinking, "Look, *I could also have had two*," so every time I see those friends of mine, those *children that were born at the same period mine were, who were supposed to be born*, I usually think, "*Why do I not have my children with me now?*" So, sometimes it worries me a little bit.

Thus, Beatriz selectively remembers the past, combining those memories with desires imagined and projected into the future. Thus, she reconstructs the past by simultaneously changing both the future perspective and promoting the emergence of a sense of urgency—expressed through anxiety—in the present moment, in order to follow the trajectory of motherhood. These are the dualities present in her narrative: the tension between *what is* (real present) and *what could have been* (potential past/shadow trajectory), as well as between *what is* and *what is not yet*, but *may become* (potential imagined future). Such dualities guide, in the present, her development toward motherhood.

THE SACREDNESS OF MOTHERHOOD: PROMOTING WAYS OF FEELING, ACTING, AND THINKING

The subjective world of people is set up as a "complex totality of the immediate experience" that is constantly and dynamically changing in the irreversible time (Valsiner, 2007, p. 256). Thus, the semiotic processes that people introduce into their own lives are articulated in order to guide and regulate this flow toward some selected direction in the future (Valsiner, 2007). In the case of Beatriz, motherhood is the direction selected for the future, therefore she acts and feels in this direction, toward the future.

Through the narrative of Beatriz, one may notice that motherhood consists of a future goal, which she had planned to achieve when she was 30 years of age. Thus, there emerged the construction of a symbolic border—established on the basis of the criterion of age—as well as the meanings "young" (under 30 years of age) and "old" (over 30) to become a mother. However, the first pregnancy, unplanned, occurred when she was 22 years old, a moment in which she considered herself too young to have a child. Therefore, after she became pregnant, she crossed the border into her own personal culture toward what she had planned. Despite the occurrence of an unplanned event, the precocious pregnancy was affectively experienced by her with some joy. However, the involuntary experience of a late gestational loss (on the last trimester of pregnancy) led to the development of a sense of urgency in relation to pregnancy, destroying the symbolic previously built age border. The emergence of that sense of urgency was mainly related to the physical sensation of fetal movements inside her own body and with the anticipation of a future contact with the living baby and recognition of the continuity of something that is hers (physical traits) in a new generation, as illustrated in the following report:

> I wanted it again, because that *sensation of feeling the baby, the sensation of having a baby in my arms, of knowing she is ours, of her having our characteristics*, then I said "Oh, no, *I cannot wait that long anymore*, so if God sent her to me ahead, then I will try again, in order to *soon have my grace*."

Thus, the bodily experience of fetal movements allowed Beatriz to feel motherhood in her own body, to be taken by this feeling, by this hypergeneralized sign, filled with affection. As if the sacredness of motherhood—the female potential to give birth, to create life, like a demigod—had taken her away, guiding her thoughts, feelings, and behaviors.

Notwithstanding, considering that meaning construction involves signs of a dual nature, the sign motherhood (core A) needs to be considered together with nonmotherhood (non-A). That is to say that, at the same time that the sign "motherhood" refers to the meaning of generating life, it also grasps the meaning of generating death—generating in Beatriz an emotion nominated by her as "terror." The feeling associated with both (life and death) happens at the border. Death is a part of life, both forming a complex, while at the same time, death is the opposite of life.

Thus, the sacredness of motherhood is a hypergeneralized affective field. A devastating and intense feeling, which Beatriz felt in her guts through fetal movements, led to the emergence of an affective state such as, "I'm feeling something that I cannot specify what it is."

> *I think that, only that experience, of feeling that being inside, that movement, I don't know, I wanted to feel that again, to have her alive in my arms*, after she was born, because I haven't been through this part [laughs]. I already *felt it*. I *enjoyed very much the sensation of having the baby move, all that, I really liked it*. Then, I said, *I miss it*, I want to try it again right now, and have a child.

According to Valsiner (2007), we can consider the construction of a hierarchy of levels of semiotic mediation of affective processes with different levels of generality. The affective phenomena "are organized at different levels, from those located near the immediate physiological processes, to those hyperabstract supergeneralized feelings of wholeness" (p. 260). At the highest level of generalization, feelings mediated by signs are undifferentiated. In this sense, "the person 'just feels' something, but cannot put that feeling into words" (p. 261). It is a hypergeneralized feeling, semiotically mediated, to which rational discourse about emotions is subordinate. One example of affective phenomenon of higher order is understood as "basic human resources of affective orientation ontogenetically internalized, however, their externalization can be observed in many aspects of human behavior.... [They are] dynamic semiotic fields" (p. 262). Inasmuch as values reach the condition of hypergeneralization, they cannot easily be verbalized. Nevertheless, people can act in a decisive way, directed by their values. Thus, becoming a mother can be considered an internalized value along the pathway of life and intensified or awakened by the experience of a pregnancy loss.

The affective fields of the higher type regulate experience in its entirety. And once the affective sign of the type field becomes hypergeneralized, it "colors" each new experience of a person. According to Valsiner (2007), "a stream of generalized sentiment takes over the intra-psychological world of this person and starts to control his/her concrete actions and to weaken any opposite efforts" (p. 262).

Thus, along the reproductive trajectory of Beatriz, there was the emergence of an overwhelming feeling: to feel another life develop and manifest within, to feel within her own body the sacredness of motherhood. This feeling seems to indicate that the affective field became undifferentiated—the result of extensive abstraction of emotions and their overgeneralization to general feelings: "to feel that being inside, that movement, *I don't know*, I wanted to feel that again." Therefore, as Figure 4.7 illustrates, the affective fields related to motherhood grew, became larger, and started to cover hypergeneralized fields. In turn, the affective fields linked to death tend to narrow toward a point-like state, exemplified by the specific category of emotion called "terror," an emotion that, although intense, became constrained, accessible to cognition and verbalization.

Therefore, the sacredness of motherhood—felt in the guts of Beatriz's own body—serves as a promoter sign, an abstract sign, which acts as a guide for the whole range of possible constructions in the future (Valsiner, 2007). But besides this powerful promoter sign, another sign emerges in operation: the repairing sign (Pontes, 2013, 2014). This is a sign constructed in the present moment, which repairs the break triggered in the past and guides the person toward the future. In the case of Beatriz, the repairing sign was the specialized medical treatment. After all, this sign provided a meaning for the pregnancy losses experienced in the past—that is, the losses occurred because Beatriz had no access to specialized medical treatment—as well as potential paths to be experienced in the future; when she has access to this treatment, miscarriages will not occur anymore, death does not dominate life, instead, life conquers death, and she will finally be able to achieve the idealized experience of motherhood, the sacredness of motherhood, which guides her toward the future.

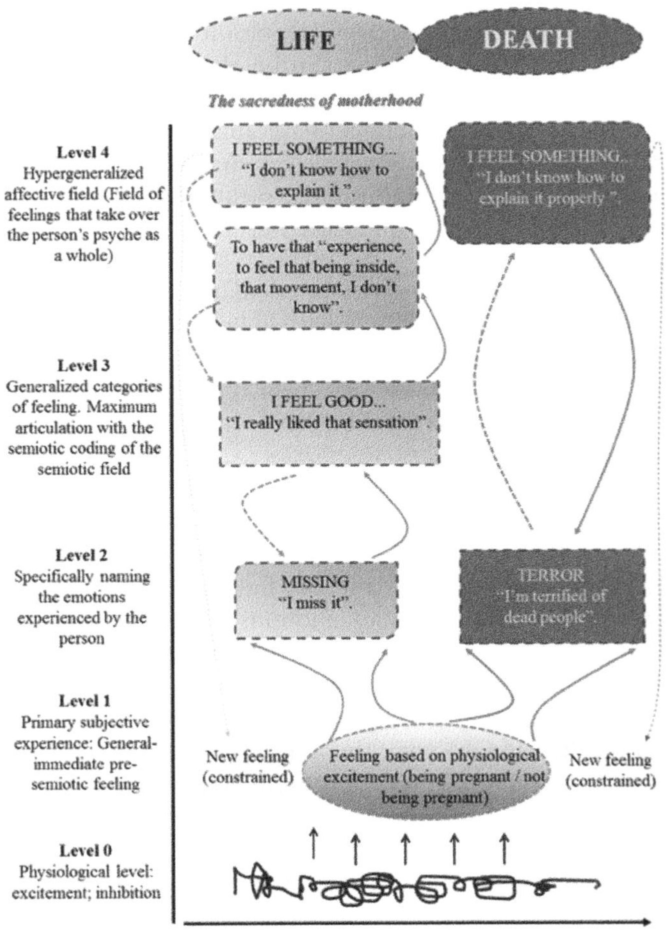

Source: Valsiner (2007).

Figure 4.7. Processes of generalization and hypergeneralization in the affective regulation of the flow of experience.

Figure 4.8 illustrates the main aspects analyzed in Beatriz's reproductive trajectory, with emphasis on the semiotic strategies for building and maintaining continuity of the position I-mother after the third pregnancy loss.

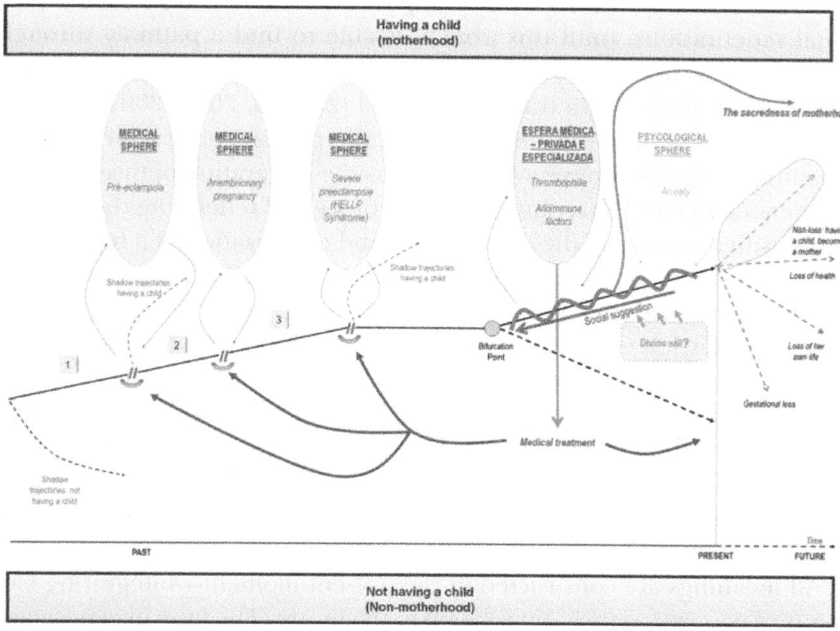

Figure 4.8. Beatriz's reproductive trajectory.

FINAL REMARKS

The experience of a miscarriage or an involuntary gestational loss is understood in the present work in terms of rupture, both in the sense of a developing pregnancy, of what was expected, and what was about to happen (becoming a mother and the birth of a child), but also in the sense of discontinuity in the development of the self (understandings and expectations that had been built about the self and the world). Thus, it promotes ruptures in various levels—from the physical body to the social role. Therefore, all meaningful framework that had been built over the course of the pregnancy is abruptly interrupted by its nonfulfillment. In other words, the expectation was violated by the happening of another trajectory: the pregnancy loss and not becoming a mother. Thus, when a system suffers an unexpected interruption through the significant modification of the environment, there emerges of a magnified uncertainty toward the future and increased ambivalence perception—events that constitute a rupture as discontinuity in the self system.

These disruptions in the flow of experience demand repositioning processes and may request further acquisitions, understandings, and personal redefinitions, until this woman is able to find a pathway through uncertainty toward a new regularity (Zittoun, 2007). This may require time and space, thus setting a transition period (Zittoun, 2004, 2007). During the transition period, it is necessary that the person construct a personal meaning for the disruptive event as well as the integration of this meaning in a general personal narrative. Also, the process of redefining the identity is intensified, involving the construction and mobilization of a representation of the self in the past and possible "selves" in the future, in a particular social-cultural context (Zittoun, 2004).

In everyday life experience, people strive continuously to build a sense for the world in which they are immersed. The construction of meanings consists of a central human psychological process through which people interact with their environment. However, such a construction is configured in an active process in which one cultivates his/her own self and environment. Thus, people relate to the world through semiotic mediation (Valsiner, 2007; Cabell, 2010).

All meanings are constructed in the present moment—integrating elements of past experience—in relation to the future. The human experience lies thus in a temporal dimension that is future-oriented. The occurrence of an involuntary gestational loss, for example, exceeds the present moment—of the loss itself—and encompasses the loss of an imagined child, of the future anticipated child. Inherent in this relationship with the future is the experience of uncertainty. After all, the future can never be fully determined at the present time. The future is uncertain and the past is constantly being rebuilt as we face the uncertainty of the future, through the construction of many signs. Thus, people experience and represent the world through signs. And through semiotic mediation, they face the future. (Abbey & Valsiner, 2004; Valsiner, 2007).

On the other hand, it is necessary to maintain a sense of being the same, a sense of continuity through change. "A crucial necessity for the dialogical self is to maintain a dynamic stability in the self" (Valsiner, 2002, p. 258). It is relative stability, since it is always related to change. Thus, when the immediate experience is disturbed, it is necessary to reconstruct the past and future in order to enable action in the present. With a new perspective on the past and future, continuity is restored.

NOTES

1. Beatriz was interviewed by the first author at two different sittings. The first interview was conducted in November of 2010, when the investigation and immunological treatment started, due to previous gestational losses in private practice. Although she was accompanied by her husband during the interview, she narrated her reproductive trajectory virtually alone. The second interview took place almost a year later, when she came to see me to talk about her difficulty to conceive. This time, she was alone, and I asked her to narrate her reproductive trajectory again.
2. Among the hypertensive disorders during pregnancy preeclampsia is defined as high blood pressure associated with proteinuria, edema, or both after 20 weeks of gestation. Preeclampsia can be mild or severe. The definitive treatment for preeclampsia or eclampsia is delivery itself. The urgency of it depends on the severity of the case. In the case of severe preeclampsia, the pregnant woman's safety must be considered first. The priority is to evaluate and stabilize the maternal condition, particularly coagulation abnormalities. HELLP syndrome, which consists of hemolysis, elevated liver enzymes, and low platelets, is a form of severe preeclampsia. The treatment is the same for severe preeclampsia, i.e., to perform delivery. The HELLP syndrome is associated with bad maternal and perinatal outcome. The perinatal fetal mortality reported in the medical literature ranges from 7.7% to 60.0%, and maternal mortality from zero to 24%. Maternal morbidity is common. Many patients with HELLP syndrome require transfusions of blood and blood components and have an increased risk of acute renal failure, pulmonary edema, ascites, cerebral edema, and hepatic rupture (Reddy & Witter, 2001).

REFERENCES

Abbey, E., & Valsiner, J. (2004). Emergence of meanings through ambivalence. *Forum: Qualitative Social Research, 6*(1). Retrieved from http://www.qualitative-research.net/index.php/fqs/article/view/515

Bastos, A. C. S. (2014). Motherhood & movement (over time): The poetic motion of motherhood meanings through the lens of lived temporality. In K. Cabell, J. Valsiner, P. Marsico, & C. Cornejo (Eds.), *Annals of cultural psychology, 1*. Charlotte, NC: Information Age.

Bruner, J. (1997a). *Atos de significação*. Porto Alegre, Brazil: Artes Médicas.

Bruner, J. (1997b). *Realidade mental, mundos possíveis*. Porto Alegre, Brazil: Artes Médicas.

Cabell, K. (2010). Mediators, regulators, and catalyzers: A context-inclusive model of trajectory development. *Psychology & Society, 3*(1), 26–41.

Cabell, K. R., & Valsiner, J. (2011). Affective-hypergeneralization: Learn from psychoanalysis. *Cultural Psychology and Psychoanalysis: Pathways to Synthesis*, 87–113.

Cowan, P. A. (1991). Individual and family life transitions: A proposal for a new definition. In P. A. Cowan & M. Hetherington (Eds.), *Family transitions*: Hillsdale, NJ: Lawrence Erlbaum.

Farinati, D. M., Rigoni, M. S., & Müller, M. C. (2006, October–December). Infertilidade: Um novo campo da psicologia da saúde. *Estudos de Psicologia, Campinas, 23*(4), 433–439. Retrieved June 29, 2012, from http://www.scielo.br/scielo.php?pid=S0103-166X2006000400011&script=sci_arttext

Maldonado, M. T. (2002). *Psicologia da Gravidez: Parto e puerpério* (16th ed). São Paulo, Brazil: Saraiva.

Parkes, C. M. (1998). *Luto: Estudos sobre a perda na vida adulta* (M. H. F. Bromberg, Trans.). São Paulo, Brazil: Summus.

Peskek, D. J., Kraus, M. L., & Diriwachter, R. (2008). Construction and internalization of prayer practices to cope with transitional life periods. *Yearbook of Idiographic Science, 1,* 135–161.

Pontes, V. V. (2013). *Construindo continuidade frente a sucessivas rupturas: estratégias semióticas de reparação dinâmica do self.* Doctoral thesis, Instituto de Psicologia, Universidade Federal da Bahia, Salvador, Brazil.

Pontes, V. P. (2014). The experience of recurrent gestational losses: Semiotic strategies of dynamic self-repairing after sequential ruptures. In K. Cabell, J. Valsiner, P. Marsico, & C. Cornejo (Eds.), *Annals of Cultural Psychology, 1*. Charlotte, NC: Information Age.

Reddy, U., & Witter, F. (2001). Distúrbios hipertensivos da gestação. In N. C. Lambrou, A. N. Morse, & E. E. Wallach (Eds.), *Manual de ginecologia e obstetrícia do Johns Hopkins*. Porto Alegre: Artmed Editora.

Sato, T., Hidaka, T., & Fukuda, M. (2009). Depicting the dynamics of living the life: The trajectory equifinality model. In J. Valsiner, P. Molenaar, M. Lyra, & N. Chaudhary (Eds.), *Dynamic process methodology in the social and developmental sciences* (pp. 217–240). New York, NY: Springer.

Sato, T., & Valsiner, J. (2010). Time in life and life in time: Between experiencing and accouting. *Ritsumeikan Journal of Human Sciences, 20,* 79–92.

Sato, T., Yasuda, Y., Kido, A., Arakawa, A., Mizoguchi, M., & Valsiner, J. (2006). Sampling reconsidered: Idiographic science and the analyses of personal life trajectories. In J. Valsiner & A. Rosa (Eds.), *Cambridge handbook of sociocultural psychology* (pp. 82–106). Cambridge, UK: Cambridge University Press.

Smolka, A. L. B. (2004). Sobre significação e sentido: uma contribuição à proposta da Rede de Significações. In M. C. Rossetti-Ferreira, K. S. Amorim, A. P. S. Silva, & A. M. A. Carvalho (Eds.), *Rede de Significações e o estudo do desenvolvimento humano*. Porto Alegre, Brazil: Artmed.

Thompson, R., Bell, R., Henderson, S., Holland, J., McGrellis, S., & Sharpe, S. (2002, May). Critical moments: Choice, chance and opportunity in young people's narratives of transition. *Sociology, 36*(2), 335–354.

Valsiner, J. (2000). *Culture and human development. An introduction*. London, UK: Sage.

Valsiner, J. (2005). Transformations and flexible forms: Where qualitative psychology begins. *Qualitative Research in Psychology, 4*(4), 39–57.

Valsiner, J. (2007). Comparative study of human cultural development. *Culture in Minds and Societies*. New Delhi, India: Sage.

Valsiner, J. (2002). Forms of dialogical relations and semiotic autoregulation within the self. *Theory & Psychology, Calgary, AB, 12*(2). Retrieved May, 17, 2008, from the http://www.psych.ucalgary.ca/thpsyc/

Valsiner, J. (2010). *Climbing the sacred mountain of knowledge: Psychology at its eternal crossroads*. Keynote address at XIV Congreso Colombiano de Psicologia, Ibagué.

Volkmer, V., Covas, I. G., Franco, A. L. S., & Costa, O. L. N. (2006). "Você se pergunta: por que não pode ter?": Aspectos psicológicos e sociais de mulheres com história de abortamento de repetição. *Revista Vivência,* 31.

Vygotsky, L. S. (1984/2008). *A formação social da mente: O desenvolvimento dos processos psicológicos superiores* (7th ed.). São Paulo, Brazil: Martins Fontes.

Yasuda, Y. (in press). *How can the diversity of human lives be expressed using TEM? Depicting the experiences and choices of infertile women unable to conceive after infertility treatment.*

Zittoun, T. (2004). Symbolic competencies for developmental transitions: the case of the choice of first names. *Culture Psychology, 10,* 131–161.

Zittoun, T. (2007). Dynamics of interiority: ruptures and transitions in self-development. In L. M. Simão & J. Valsiner (Eds.), *Otherness in question: Labyrinths of the self* (pp. 187–214). Charlotte, NC: Information Age Publishing.

Zittoun, T., & Grossen, M. (2012). Cultural elements as means of constructing the continuity of the self across various spheres of experience. In B. Ligorio & M. Cesar (Eds.), *The interplays between dialogical learning and dialogical self* (pp. 96–126). Charlotte, NC: Information Age.

Zittoun, T., Valsiner, J., Vedeler, D., Salgado, J., Gonçalves, M., & Ferring, D. (2012). *Melodies of living: Developmental science of the human life course*. New York, NY: Cambridge University Press.

PART II

LIVING TEMPORALITY

CHAPTER 5

LIVING AND OBSERVING

Two Modes of Understanding Time

Carlos Cornejo and Himmbler Olivares

Nothing, Lucilius, is ours, except time.

—Seneca, ca. 62 A.D.

While "time" usually refers to a linear sequence of moments, "temporality" denotes the *sense* of time, that is, the internal feeling of the continuous transformation of the landscape, along with all its inhabitants. Written records of units of time have been found in many ancient cultures. Typically these include regular astronomical units, such as the lunar and solar cycles, in close connection with the planning of agrarian labors. Temporality, on the other hand, is a relatively recent discovery of Western culture. Up to beginnings of the 19th century, the greater part of the intellectual milieu in Western nations followed a literal understanding of the Bible according to which God created the world and its creatures in the same shape as we observe them in the present. By the end of 18th century, it was generally believed that the Earth was about 6,000 years old and that its geomorphology was largely the same since the divine creation. Fossils

Temporality: Culture in the Flow of Human Experience, pp. 95–114

and ammonites found on mountainsides were considered evidence of the great deluge, reaffirming a nonevolutionary worldview.

The static view of the world changed radically during the 19th century. The insight that we are living in a continuously evolving planet had emerged by middle of the 18th century and rapidly expanded during the 19th century through different fields of social and natural inquiry. Otto Jespersen addressed this in a straightforward manner:

> On the whole, it must be said that it was reserved for that century [the 19th century] to apply the notion of history to other things than wars and the vicissitudes of dynasties, and thus to discover the idea of development or evolution as pervading the whole universe. (Jespersen, 1922, p. 2)

Our aim in this chapter is to sketch the evolution of this radical change by highlighting important landmarks of this movement. In the first section, we will argue that by the end of the 19th century, the last field to be reached by the temporalizing movement was precisely that of conscious experience. This happened during the genesis of modern phenomenology, immediately before the emergence of its transcendental variant, which took the temporality of consciousness back to the stable domain of the eternal entities of objective time. In the second section, we will discuss the distinction between subjective and objective time, which has been dealt with by several authors concerned with phenomenological experience. In the third and final section, we argue that this distinction is crucial to advance a cultural-psychological approach to time.

The Temporalization of the World

The idea of dynamicity is traceable to the classical era of Western culture. Hecateus of Mileto, Anaximander, and Heraclitus are often mentioned as the precursors of a historical perspective with respect to cultures, biological species, and the natural world (see Toulmin & Goodfield, 1965). This spirit can, for example, be recognized in Marcus Aurelius' *Meditations*, the epitome of the stoic philosophy of life. In this text from the 2nd century, the author encourages perceiving the wholeness of universe, realizing the minimal position that humankind has within this totality, and extracting ethical guidance from this to lead the good life:

> Think of the universal substance, of which thou hast a very small portion; and of universal time, of which a short and indivisible interval has been assigned to thee; and of that which is fixed by destiny, and how small a part of it thou art. (Marcus Aurelius, 2006/167 AD, Bk. VI, para. 24)

In some sections of the *Meditations,* time is described in universalistic objective terms, as in the previous quote. In other parts, however, there is the strong appeal to *feel* the flow of time:

> Time is like a river made up of the events which happen, and a violent stream; for as soon as a thing has been seen, it is carried away, and another comes in its place, and this will be carried away too. (Marcus Aurelius, 2006/167 AD, Bk. IV, para. 45)

Already in the Aurelian text, one can find two concepts of time intertwined: time as sequence of objective and external events and time as temporality, that is, the sense of flow captured here by the metaphor of the irreversible flow of a river.

However, it was the concept of time as a sequence of external events that proved to fit better with dominant Aristotelian cosmology. Less important than to feel the permanent movement of Nature is to describe its order. This divine Order of Nature pervades every part of the universe, from the stars and planets to the classification of living species. The zealous description of this order turns out to be the basis of the definition via *genus proximum* and *differentia specifica*. Such *logical* order presupposes fixed and immutable concepts ("essences"). Questions concerning development accordingly lacked the relevance or even the pertinence of the descriptive efforts to give an accurate depiction of the essential order. The patristic reconstruction of Aristotelian philosophy brought together the logical Order of Nature with God's plan. The mission of the rational man was precisely to decipher such a plan through God's gift of *rationality*—observing and theorizing—and through faith—praying and piously searching for revelation.

In the Aristotelian-patristic worldview, there was little space for a sense of evolutionary development of the world. Instead, the rational discovery of the divine plan for this steady, unchanging world occupied most intellectual works. Questions concerning the origins of the world were assumed to be answered by the authority of the Bible. During the Middle Ages, the Bible-based chronology of the Earth became popular, which interpreted the unit "day" in Genesis as a thousand years:

> Julius [Africanus] took from Jewish literature the idea of a "millennium"—the thousand-year-long Kingdom of the Messiah which prophecy declared would end the history of the world—and used it as a symbolic key for interpreting Old Testament chronology.... On this interpretation, the whole of history corresponded to a cosmic week, each of whose days lasted a thousand years.... (The justification for this step was found in the words of Psalm 90, verse 4, "A thousand years in thy sight are but as yesterday"). (Toulmin & Goodfield, 1965, pp. 60–61)

In this interpretation, the Earth was about 6,000 years old, considering the six days that the Creation took, according to Genesis. This is the classical ahistorical worldview, which basically remained uncontested until the 18th century. Developmental accounts, such as the aforementioned, became rather marginal in the dominant vision that assumed the world as being in essence the same as it was since its creation.

Civil History

It is not an exaggeration to say that the rapid irruption of the historical view of the world began in the field of social sciences. In *La nuova scienza* (1948/1725), Giambattista Vico discussed the revolutionary insight that social products are in permanent change, so that they come to life just after a certain development. In Vico's time, history, understood as the report of past events, was consolidated. Everybody knew that the Roman Empire had risen and fallen and similarly that every culture had a history to be told or written. The revolutionary point in Vico's perspective was his idea that social products are anchored to people's mentality from the particular culture that produces them. Expressions such as language, law, civil order, traditions, and architecture are dependent on the historical moment of the nation that created them. By analyzing the history of several nations, Vico arrived at what he called "the natural law of nations": there are three periods through which every nation shall pass: (a) the age of gods, in which people believe they live under divine government; (b) the age of heroes, who are the governors possessed of a superior nature over common people, forming an aristocratic lineage; and (c) the age of men, in which all people are recognized as equals in human nature. In each particular age, a different language is spoken: (a) the sacred or secret language, composed of rituals and mute acts; (b) the poetic language, full of comparisons, images and metaphors; (c) and the epistolary or vulgar, which served a variety of everyday uses.

A corollary of the theory of the three ages is that to really understand an alien culture, one has to make a deliberate effort of imagination—the *fantasia*—trying to re-create the mentality of the nation beyond the words of its language. Vico himself described as a key discovery that Homeric poems should not be read as a literary version of what could be written in literal language, since such poems were products of a culture that did not separate poetic and "vulgar" language:

> We find that the principle of these origins both of languages and of letters lies in the fact that the early gentile peoples, by a demonstrated necessity of nature, were poets who spoke in poetic characters. This discovery, which is

> the master key of this Science, has cost us the persistent research of almost all our literary life, because with our civilized natures we cannot at all imagine and can only understand by great toil the poetic nature of these first men. (Vico, 1948/1725, p. 19)

A further corollary of Vico's (1948/1725) revolutionary conception of society is that our conceptual framework is permanently evolving. Since our present civil order is rooted in our present worldview, we can neither assume that it was the same throughout history or that it will stay unchanged in the future. The understanding of every culture—our own included—is to be done under the principle that social life is always temporally in the process of becoming something else.

Vico's account of cultural development is manifestly at odds with the common 17th century view of the natural and social world. Vico (1948/1725) argued directly against modern Cartesian philosophy, which defends a formal and decontextualized approach to timeless Nature. The methodical advance to a correct description of Nature supposes the researcher adopts a perspective detached from a particular spatial or temporal context. A formal description of Nature captures the explanandum from any possible temporal-spatial point. Nevertheless, according to Vico, such pretension is impossible to satisfy since science is itself a social product and therefore *historically* determined. Thus, our theories on Nature can never reach the status of certainty. On the contrary, we can still obtain certain knowledge on the social world thanks to our imaginative capacity that allows us to enter into the alien culture. All we can really know is social; and this is by definition in perpetual evolution. Therefore, every piece of information concerning the natural or the social world has to be contextualized to the specific cultural-historical context in which it emerged.

Geology

The next major burst in developmental thinking took place in the field of geology. In fact, this discipline was the first natural science to become genuinely historical. The changes that took place between 1750 and 1850 are remarkable and full of consequences for other natural sciences, particularly 19th century biology and zoology. During the last decades of the 18th century, geology not only showed the irruption of a historical view of the Earth's morphology, but more importantly, it offered a new timescale that replaced the notion of the 6,000-year-old Earth of the biblical interpretation. The first attempt to overcome the scriptural time boundary is the series *Histoire Naturelle*, written between 1749 and 1788 by Georges Louis Leclerc, Comte de Buffon (see Richet, 2007; Rudwick,

2005). By combining physical theory with geological descriptions of strata and composition of rocks, he traced back the first modern estimation of the Earth's age to between 168,000 and half a million years. No matter how inaccurate this calculation was, the real importance of this pioneering work was that the timescale of the Earth's development was radically expanded. Interestingly, Buffon tried to conciliate his scientific calculations with scriptures, suggesting that the word "day" in the "six days of Creation" should be interpreted as an indefinite period of time whose length could range from 3,000 to 35,000 years. For Buffon, the 6 days of the Creation were actually 6 *epochs*, from the formation and cooling of the Earth until the appearance of men.

Another significant step in the historicity of the Earth was taken in 1765 when Jean-Étienne Guettard argued, comparing rock exemplars from different parts of Europe, that the peaks of the Massif Central in the French province of Auvergne were really extinct volcanoes (Rudwick, 2005). This interpretation was obviously at odds with the traditional timespan of 6,000 years for the Earth. It increasingly became implausible to think of these kinds of colossal geological transformations within the narrow biblical scale. The Earth, Guettard suggested, had to experience radical morphological changes to adopt the forms we presently observe. These modifications are not abrupt but so slow that we cannot perceive them and rather infer them by observing traces of geomorphologic events. From this, it follows that there is no reason to think of the present state of the Earth as stable and definitive. On the contrary, if agents of geomorphologic transformations, such as water and volcanoes, work on a timescale imperceptible from a human point of view, it is reasonable to think that Earth is modifying its forms permanently—even now. James Hutton, in his *Theory of Earth* (1788/1795), was one of the first authors to express this insight:

> Time, which measures every thing in our idea, and is often deficient to our schemes, is to nature endless and as nothing; it cannot limit that by which alone it had existence; and as the natural course of time, which to us seems infinite, cannot be bounded by any operation that may have an end, the progress of things upon this globe, that is, the course of nature, cannot be limited by time, which must proceed in a continual succession. (p. 8)

Geological mutations are so amazingly slow that they deserve to be considered outside *our* human time; our "schemas" are deficient to capture the actual duration of these processes. Water sedimentations forming rocks, volcanic eruptions, erosion by wind, and such are forces as active today as they were in primordial times. Yet they act so slowly that they are better understood in the perspective of "infinite time." Hutton (1788/1795) concludes:

> We have now got to the end of our reasoning; we have no data further to conclude immediately from that which actually is: But we have got enough; we have the satisfaction to find, that in nature there is wisdom, system, and consistency. For having, in the natural history of this earth, seen a succession of worlds, we may from this conclude that there is a system in nature; in like manner as, from seeing revolutions of the planets, it is concluded, that there is a system by which they are intended to continue those revolutions. But if the succession of worlds is established in the system of nature, it is in vain to look for any thing higher in the origin of the earth. The result, therefore, of our present enquiry is, that we find no vestige of a beginning, no prospect of an end. (pp. 66–67)

The Earth develops permanently at such rhythm that there is "no vestige of a beginning, no prospect of an end." Equally important in this quote is Hutton's observation that there is a "system of nature," suggesting that the perception of temporality requires the contemplation not of an object (be this a fossil, mountain, or the strata), but the situation of this object within the landscape along infinite time. Only in the contemplation of the wholeness do we feel the entire Earth in movement.

By 1830, when Charles Lyell published *Principles of Geology*, the discussion in the field had moved from the origins of Earth to its development and the forces that explain its morphology (Toulmin & Goodfield, 1965). Lyell gave detailed evidence to Hutton's theory of the manifold forces acting upon geomorphology. His position founded "uniformitarism," the view that the agents of geological formations cannot be reduced to mere opposition between fire (volcanoes) and water (rivers and seas), but must include many other factors acting in cooperation with or opposition to each other over extended temporal scales. The complex combination of these factors explained the different accidents of the Earth's crust. Thus, by the middle of the 19th century, geology conceived of the Earth as *being* rather than as a thing in a definite state. The Earth extends over time, and its morphology does not stop being continuously modified by factors whose effect can only be observed when we approach the Earth temporally.

Linguistics

Another salient example of the replacement of the classical worldview by a dynamic one is the recognition that most modern European languages were originally related to other geographically distant languages. In the context of the increasing number of accurate descriptions of languages from distant continents, it was unquestionably the interest in Sanskrit that prompted a major change in how languages were approached. During the second half of the 18th century, some authors pointed to peculiar similarities among Indian Sanskrit, Greek, and Latin. In 1786, Sir William Jones first posed the hypothesis that the affinities among these languages were neither accidental nor the result of borrowings on the side of the Indian speakers (Jespersen, 1922). Rather, they were the proof that these languages come from the same common source, now probably extinct. This insight developed further in a Romantic spirit in precursors of comparative linguistics, such as Friedrich von Schlegel, Jakob Grimm, and especially Franz Bopp, who is considered the father of comparative philology. In his first book in 1819, Bopp moved beyond Jones' hypothesis, to speculate on the characteristics of the original tongue that gave rise to Sanskrit, Greek, Latin, and other European languages. Given their common root, he considered the set of languages as belonging to the same "family of languages," namely, "Indo-European" (Jespersen, 1922). After Bopp, it is clear that languages evolve adopting first dialectal variations and after centuries even constituting different tongues, although similar in recognizable ways. In 1836, Wilhelm von Humboldt expressed the developmental view of language straightforwardly:

> Language, regarded in its real nature, is an enduring thing, and at every moment a *transitory* one. Even its maintenance by writing is always just an incomplete, mummy-like preservation, only needed again in attempting thereby to picture the living utterance. In itself it is no product (*Ergon*), but an activity (*Energeia*). Its true definition can therefore only be a genetic one. For it is the ever-repeated *mental labour* of making the *articulated* sound capable of expressing *thought*. (Humboldt, 1999/1836, p. 49; emphasis in original)

By the 1840s, the idea of language as an organic whole that is in permanent evolution had already been settled. Far from being a ready-made divine gift, language was seen as the expressive activity of a nation that mutates jointly with the life of the community. As a social activity, it has its origin in human interactions. Similarities among languages should also be understood developmentally. Historical thinking offers the vision of major totalities—the linguistic families—where the similarities among present languages can be better comprehended. The kinship of tongues within the same linguistic family offered an evolutionary alternative to the Tower of Babel story, wherein different languages were created from scratch in an unrelated way.

Temporality in the Phenomenological Tradition

From the end of the 19th century until the middle of the 20th, diverse authors in the phenomenological tradition advanced the study of consciousness from a dynamic point of view. The starting point of this approximation is the idea of progress or flow of experience in the consciousness. Psychic phenomena have to be understood in the course of their being manifest and not as a succession of static images or concepts. It is necessary then to establish the conditions and the organizing principles based on which it would be possible to conceive the flow of consciousness in a course of time to realize the human experience. In other words, it is from the idea of elucidating the human experience from a dynamic point of view that the phenomenon of the lived experience of time appears in the discussions of the epoch. What is sought is to understand how it is that in the subjective consciousness, each psychological state has a duration that is due exclusively to the course of the immediate experience in which the person finds him/herself and not to any type of formalization when trying to describe it. It is worth noting that understanding the phenomenon as a lived experience of time is transformed into a common denominator in the descriptive psychology and the philosophy of phenomenological orientation that emerged at the end of the 19th century. It is thus that Bergson, James, Husserl, Heidegger, Merleau-Ponty, among others, have addressed time from the point of view of lived experience.

The phenomenon of the lived experience of time comes into conflict with the concept of time that until then had greater cultural roots in the epoch, heir to a static conception of natural phenomena and of the modern tradition. In this sense, *time* is used as an *objective* notion to address geometric and/or arithmetic form within the organization of experience in the consciousness. This can occur whether based on quantification or from a sequence. In both cases, there is an effort to organize the phenomenon based on the relationships that exist between a number or a part of the sequence and another, the precision of their limits and their qualitative relationships. These problems of mechanistic approximation of the phenomenon are at the heart of the discussion between approximations of lived experience versus an objective approximation of time.

Bergson was the first author to address from a phenomenological point of view the distinction between these two modes of understanding time. In his doctoral thesis *Time and Free Will* (1910/1888), the author discussed the organization of immediate experience in the consciousness and proposed two types of organization in which time is the essential condition. The first of these is objective. This organization of experience is sequential and quantifiable. In this case, a homogeneous conception is manifested in the relationship among psychological states. The second mode of organization is subjective. This is a holistic type of organization that assumes the heterogeneity of the psychological states and the relationships among them. In terms of time, in the case of the objective organization of psychological states, according to Bergson, experience is organized within space, which implies that each of our psychological states and its corresponding time is sequentially organized. In the context of situating each of our psychological states in space, the idea of the number is introduced as an organizing element of the psychological states with the aim of understanding their mutual relationships. With the appearance of the number, symbolic representation is used. With symbolic representation, an abstract space is introduced within this organization. "Space … clearly conceived by the human intellect, enables us to use clean-cut distinctions, to count, to abstract, and perhaps also to speak" (Bergson, 1910/1888, p. 97). The introduction of the space to understand the duration of psychological states turns out to be problematic, since it leads us to put one psychological state homogeneously side by side with others (or juxtaposed to one another). Thus, immediate experience and its temporal dimension lose the qualitative substrate in the holistic organization of which each psychological state is articulated with others without precise limits, based on a delicate process of interpenetration:

> We can thus conceive of a succession without distinction, and think of it as a mutual penetration, an interconnection and organization of elements, each one of which represents the whole, and cannot be distinguished or isolated from it except by abstract thought. (Bergson, 1910/1888, p.101)

In the case of the heterogeneous conception of psychological states, in describing immediate experience as a heterogeneous and qualitative experience, each psychological state occupies its time within the experience. Subjective temporality pertains to a holistic organization of the experience in which each psychological event elapses according to its own manifestation. In this sense, any compromise by breaking down the progress of consciousness via abstraction is not pertinent. This is what Bergson understood as *pure duration*:

> Pure duration is the form that the succession of our conscious states assumes when our ego lets itself live, when it refrains from separating its present state from its former states. For this purpose it need not be entirely absorbed in the passing sensation or idea; for then, on the contrary, it would no longer endure. (Bergson, 1910/1888, p. 100)

For Bergson, time is irreducible to space, given that all the careful relationships among each aspect of our experiences are lost, making abstraction prevalent as a guide for its organization. Through the idea of *pure duration*, Bergson (1910/1888) attempted to reinstate a conception of temporality as a fundamental aspect of the flow of experience that occurs in the consciousness. To this idea of temporality is added the concept of *mobility*, which brings the same principles of qualitative organization to movement.

Bergson proposed a conception of multiplicity to take in account the flow of consciousness. In this flow, there is a time for each state or event in our lived experience, in which each moment of its passage reveals a qualitative organization that is progressively expressed. In synthesis, in Bergson these two modes of understanding time are part of two modes of understanding psychic life.

A few years later, in the early 1900s, Husserl also saw the problem of in understanding subjective time as if it were an abstract phenomenon that occurs beyond the realm of lived experience. Following a line of analysis similar to that of Bergson's, for this author, it is essential to distinguish between objective quantified time and the lived experience of time, given there is no possibility of qualitatively describing the formation of an object in the consciousness from the point of view of measured time. In this sense, Husserl shared with Bergson the idea that time should be considered within the flow of the experience of the person. However, despite these coincidences, their conceptions of objective and subjective time are radically different.

For Husserl, to understand the phenomenon in the course of the experience of the person, we have to think of a time that is adjusted to each moment of the lived experience. In this sense, Husserl proposed the existence of a temporal structure of consciousness (*structure-time-consciousness*). Through this structure, the author proposes a specific organization of the subjective experience in which the *primal impression* of what is perceived or experienced coexists simultaneously and in an organized manner along with *retention* and *protension*, defined as brief instants that retain and anticipate the experience in progress, a question that always occurs in the present time in the consciousness or the *now-phase-of-the-object*. Husserl described how the temporal structure of consciousness works—the *now-phase-of-the-object*—through successive primal impressions, retentions, and protensions: "This consciousness is in a state of constant change . . . the impressional consciousness constantly flowing, passes over into ever new retentional consciousness" (Husserl, 1991/1893–1917, p. 112). This structure of temporal organization of the consciousness is not only circumscribed to the perception of objects, but also, according to Husserl, recognizes the subjective experience, understood from the temporal point of view as a network of first impressions, retentions, and protensions.

Considering the subjective experience as a temporally organized phenomenon, Husserl (1991/1893–1917) proposed the existence of a *double intentionality* of temporal phenomena. As well as the aforementioned temporal structure of the consciousness of time (primal impression-retention-protension), there is a temporal flow of consciousness that is always present in every experience. This flow of subjective temporality is described as absolute and *atemporal* by Husserl, as it does not make reference to any lived time or to some experience within the time. Rather, it is an absolute structure that provides a backdrop to the subjective flow of experience of time, extending the domain of any lived experience.

It is important to realize that Husserl's (1991/1893–1917) analysis of the experience of time deviates from the distinction between objective time and subjective temporality that we are stressing in this chapter. Husserl's outline of a temporal structure of experience is remarkable, as is the idea that its organization arises in the course of an experiential flow. However, his analysis takes a resolute step toward transcendental idealism since Husserl seeks to define a stance of *transcendental subjectivity* in which all the conditions of subjective temporality are possible. The direction toward idealism is also appreciable when the author postulates the realization of an *epoché* or phenomenological reduction as a condition of possibility for the instantiation of the temporal structure of the consciousness. This procedure consists of suspending the "prejudices" of the lived experience such as they are, that is to say, in Husserlian terms, "bracketing the natural attitude." Hence, the structure of immediate experience that Husserl proposes is actually an ideal structure that can only be understood based on an inquiry of a metaphysical order. By introducing this a-phenomenical structure of subjective time, Husserl deviates from the Bergsonian description of the experiential dimension of time.

A good example to show the difference between Bergson's (1910/1888) descriptive phenomenology and Husserl's (1991/1893–1917) transcendental phenomenology is the use that both made of melody to show how immediate experience is organized. This resource, which was also used by William James, has a common starting point for all of these authors: melody puts in relief the subjective organization of the experience given that the harmony is always perceived as an organized totality that is irreducible to the physical qualities of the notes that make it up. This starting point takes different directions in their different phenomenological approximations to the immediate experience. In the case of Bergson, melody is used as an analogy to show the careful relationship among the parts—the phenomena or psychological states—and the whole. The multidimensional organization of psychic phenomena within the course of immediate experience was described by Bergson in the following manner, in the context of describing the relationship between past experience and immediate experience:

> Nor need it forget its former states: it is enough that, in recalling these states, it does not set them alongside its actual state as one point alongside another, but forms both the past and the present states into an organic whole, as happens when we recall the notes of a tune, melting, so to speak, into one another. (Bergson, 1910/1888, p. 100)

For Bergson, it is fundamental to distinguish the *multidimensional* organization of experience from a simply *sequential* organization of elements (for example, musical notes), given that in the latter case the idea of *space* is surreptitiously introduced and with this, a serial-unilinear mode of understanding psychological phenomena. The multidimensional organization of immediate experience, with its own temporality, also involves a very particular sensation of being imbued completely in our own experience, such as happens when we listen to a melody. The musical notes do not add up as discrete units one after the other; instead they are intermingled, fusing themselves and forming a harmonic totality that exceeds the properties of the isolated notes.

> If our conscious point A does not yet possess the idea of space—and this is the hypothesis which we have agreed to adopt—the succession of states through which it passes cannot assume for it the form of a line; but its sensations will add themselves dynamically to one another and will organize themselves, like the successive notes of a tune by which we allow ourselves to be lulled and soothed. (Bergson, 1910/1888, pp. 103–104)

Husserl made a completely different use of melody. When Husserl tried to show the structure of the internal consciousness of time, he used melody, as well as sound, as examples of what he termed *temporal objects*. With this, every "now" of the melody has remnants of the preceding sounds. The latter, without being the "now" of what is perceived, are maintained in *retention*, giving each moment of the melody its own time in the phase of the "now." In the "now, incessantly, experience is "drowned" in the unfolding of consciousness. In Husserl's descriptions of the unfolding of the structure of time in the consciousness, a tendency can be appreciated to describe the flow or succession of primary impressions and of retentions of serial form as if they could be treated sequentially: "The tone now changes into a tone-having-been; the *impressional* consciousness, passes over into even new *retentional* consciousness.... Going along the flow or with it, we have a continuous series of retentions pertaining to the beginning point" (Husserl, 1991/1893–1917, p. 31).

While Husserl is clear in indicating that there is an ongoing transformation of primary impressions and their corresponding retentions, such as occurs in a flow, it is not uncommon that the flow of temporal experience is described in serial terms. In this sense, it is important to keep in mind that for Husserl, it is his methodology of phenomenological reduction that makes possible the analysis of the internal structure of time in the consciousness. This resulting analysis tends to separate a holistic phenomenon into parts, without taking care that its parts are maintained in totality. Husserl will found the unity of experience phenomena at a transcendental level. Concordantly, Husserl proposes a reflexive mode of analysis of the temporality of consciousness, dissenting from a holistic conception such as proposed by Bergson (1910/1888).

The case of melody also serves to show other aspects that differentiate Husserl's approach to time from Bergson's (1910/1888). In Husserl, the melodic experience is no longer considered as a phenomenon in which it is possible to distinguish a subjective temporality that is specific to the consciousness. Instead of this, Husserl understands the experience of a melody as a concrete instantiation of an unlimited range of temporal possibilities in which the flow of consciousness is understood in terms of *absolute timeless consciousness*:

> Like the violin tone (thought of as something physical), the appearance of the tone has its duration; and in this duration it changes or remains constant. I can focus my attention of one phase or another of this appearance: appearance here is the immanent tone or the immanent tonal movement, apart from this signifying. But this is not the ultimate consciousness. The immanent tone becomes "constituted"; namely, together with tone now of the moment we also continuously have tone adumbrations, and in these adumbrations the extent of tone-past that belong to this now presents itself (Husserl, 1991/1893–1917, p. 112).

This quotation from Husserl expresses the distancing of the concrete experience of time—temporality such as we have understood it in this chapter, and its replacement by absolute time that exceeds it. This obviously exceeds the descriptive phenomenology of Bergson with respect to lived time and its place in immediate experience.

Regarding Husserl's (1991/1893–1917) approach to time, he shares the Bergsonian spirit of recognizing the temporal dimension of consciousness, assuming that this has a particular and unique organization. From this follows the affinity between Bergson's (1910/1888) idea of subjective, heterogeneous, and holistic temporal organization and Husserl's temporal structure of consciousness. However, this commonality vanishes as soon as Husserl moves from immediate experience as it occurs in daily life in the quest for the ideal conditions of the temporal in an absolute consciousness. By doing this, Husserl's description of the consciousness of time distances also from the temporality of time, which happens phenomenologically within the experience of the situated, concrete person. At a more general level, the direction of Husserl's thinking accentuates the schism between the transcendental phenomenology that he represented and the budding movement of descriptive phenomenology, represented by Carl Stumpf, Theodor Lipps, and later by Max Scheler.

In July 1924, Martin Heidegger presented his own reflections on time and temporality, moving away from his former advisor Husserl. In his work, Heidegger stresses clearly the distinction between objective time and subjective time. In the first case, time is measurable. If there is a "now," this is arbitrary and has no privilege over any other possible "now." Objective time is uniform and homogeneous, and hence it is measurable. Conceived objectively, also, time is a sequence of stages that stand in a relation of earlier and later to one another (Heidegger, 1992/1924). Subjective time, on the contrary, leads us to think of our very existence: "Am I myself the now and my existence time?" (Heidegger, 1992/1924, p. 5E) Heidegger undertakes the question in Augustine's *Confessions*. In Chapter XI, the Christian thinker devotes important reflections to the origins of temporality. Augustine writes,

> In you, my spirit, I measure time ... the transitory things encountered bring you into a disposition which remains, while those things disappear.... The disposition I measure in present existence, not the things that pass by in order that this disposition first arise. (Augustine, as cited in Heidegger, 1992/1924, p. 6E)

For Heidegger, these lines reveal the close relation between time and human experience. Thus, according to Heidegger, the primordial source from where temporality flows is the *Dasein*, that is, in the kind of being that we call human life. The subjective time intimately addresses to myself, or more precisely, to my *Dasein*. The concept of subjective time finds also its roots on that "entity that we each ourselves are, which each of us finds in the fundamental assertion: I am.... The assertion 'I am' is the authentic assertion of being pertaining to the Dasein of man" (Heidegger, 1992/1924, p. 6E).

Note that the connection between time and experience does not presuppose a reflective instance such as the "Ego." In a radical separation from his master Husserl, Heidegger argues that the human Dasein is not product of a reflection upon the Ego or the Self. Dasein and, by extension, temporality, are not objects of reflective contemplation, but they are to be found in the act of being and doing: "On average, the interpretation of Dasein is governed by everydayness, by what one traditionally says about Dasein and human life.... It is governed by the 'One,' by tradition" (Heidegger, 1992/1924, p. 9E) With the notorious exception of the act that we call reflection, the figure of a cogitative Ego simply does not exist: "No one is himself in everydayness" (Heidegger, 1992/1924, p. 8E). Thus, the Dasein exceeds the individual Selfhood.

For Heidegger, temporality does not reside in the Ego, but in the peculiar being of human life, that is, the Dasein. How can we characterize this new source of meaning, which is neither the Cartesian Ego nor the collective mind of the society? Let us describe some features of Dasein according to Heidegger (1992/1924). First, the Dasein deploys as *being-in-the-world*. This means that it is to be found in the use of instruments and signs as *ready-at-hand* (*Zuhandensein*) rather than in the reflection of things that are *present-at-hand* (*Vorhandensein*). Dasein is knowledge *in* the world, not knowledge *of* the world. By trying to define Dasein, we feel precisely the same Augustinian perplexity by defining "time": Dasein is that kind of knowledge, which is clear when we are not reflecting on it, but is opaque when we reflect on it. It is ready-at-hand in our skillful dealing with the world, not in our theoretical discourse about it.

A second feature of Dasein is that it is not individual. As the Augustinian *spirit*, the Dasein cannot be interpreted as being private or subjective: "It is not I myself who for the most part and on average am my Dasein, but the Others" (Heidegger, 1992/1924, p. 8E). Dasein is *being-with-one-another*, that is, being with Others. Dasein configures also a prereflexive, meaningful background in which I find myself imbued in everydayness by being-in-the-world-with-others. My skillful dealing with the world is, in normal circumstances, accessible and transparent for my fellow human beings.

However, at the same time, the Dasein can only be personal. This is an additional feature of Dasein, which Heidegger calls *specificity* (*Jeweiligkeit*): "The *specificity* of the 'I am' is constitutive for Dasein ... Dasein is therefore also *my* Dasein.... It is *in each case its own and is specific as its own*" (Heidegger, 1924/1992, 8E.; emphasis in original). The peculiar form in which the individual Self participates in the Dasein is the *specificity*. In its specificity, the Dasein presents itself as *my* Dasein.

Crucial for our purposes, Heidegger ascribes the *temporality* to the Dasein. Insofar as my Dasein is "in each case mine," "it is still something that is not yet at an end . . . when it has reached the end, it precisely no longer is" (Heidegger, 1924/1992, p. 10E). A core feature of the Dasein is that "it never authentically is what it can be" (Heidegger, 1992/1924, p. 10E). Heidegger captures the permanent movement of the Dasein with his definition "*Dasein* as human life *is primarily being possible*, the Being of the possibility of its certain yet indeterminate past" (Heidegger, 1992/1924, p. 12E; emphasis in original). The permanent movement of Dasein is a "running ahead to the past":

> The past is not some occurrence, not some incident in my Dasein. It is *its past*, not some "what" about Dasein, some event that happens to Dasein and alters it. This past is not a "what," but a "how," indeed the authentic "how" of my Dasein. (Heidegger, 1992/1924, p. 12E, emphasis in original)

The temporality of the Dasein does not also consist of the concatenation of events (that is, of a series of different "what"), but a "how." By running ahead to the "certain yet indeterminate past," my Dasein becomes futural and deploys its specificity. In brief, Dasein is not objective time, but subjective temporality.

The distinction between objective time and subjective temporality appears also years later in the work of Merleau-Ponty (2005/1945). His analysis of the phenomenon begins with searching the concrete structure of the subjective dimension, for which it is essential to take time into account. Merleau-Ponty observes that in the immediate experience (i.e., nonreflexive experience), time is indissolubly linked to things. This indissoluble relationship between the subject, what is perceived, and the world has its own time. The time that emerges in the person-world encounter cannot be understood as a simple succession or sequence. Nor should it be understood as a record already contained in some structure. According to Merleau-Ponty, time is deeply intertwined with the relationship between the person and the world. It is in fact from this relationship that temporality emerges. "Time is, therefore, not a real process, not an actual succession that I am content to record. It arises from *my* relation to things" (p. 478; emphasis in original). In this sense, Merleau-Ponty describes a *subjective dimension* of time, such that time intermingles the person, her lived experience, and her environment. As the person is permanently involved in activities in the world, time is in constant achievement. Because of its subjective and dynamic character, subjective temporality escapes the objective descriptions of lived experience:

> There can be time only if it is not completely deployed; only provided that past, present and future do not all three have their being in the same sense. It is of the essence of time to be in process of self-production, and not to be; never, that is, to be completely constituted. Constituted time, the series of possible relations in terms of before and after, is not time itself, but the ultimate recording of time, the result of its *passage*, which objective thinking always presupposes yet never manages to fasten on to (Merleau-Ponty, 1992/1924, p. 482; emphasis in original).

CONCLUSIONS

Subjective time is a recent achievement of Western philosophy. It emerged at the end of the 19th century after a gradual transformation of our worldview. Until 1750, the Western world was supposed to have been 6,000 years old, with its existing geomorphology and biosphere being basically the same as it was when created. The feeling that our world is in movement appears during the 17th century and rapidly expands through diverse fields of knowledge, from historiography and geology to linguistics and (conspicuously) biology. By the end of 19th century, the temporalizing wave reached psychology and phenomenology, turning human consciousness into a stream of thought instead of a virtual space of prefabricated representations. There is no way to understand the radical irruption of the temporality in the 19th century by the vitalistic tradition, represented for example by Bergson, if not as the end result of a long sociocultural process of dynamization of the world.

Unlike objective time, which is uniform and homogeneous, subjective time is full of qualitative nuances and variations. Objective time offers a potentially scalable sequence of events ordered by earlier-later relations. To this metric line may belong the Self or Ego, although not necessarily. When an Ego is introduced into this line, it represents present time. This Ego works in this sense as an additional metric point, separating the past from the future. The Ego also becomes a measurable metric stage uniformly disposed in the concatenation of earlier and later events. Thus, in the objective timeline, the Ego anchors the now. But note that this now has been shorn of its qualitative profoundity, with only its formal position between past and future remaining. Within the objective timeline, the now becomes another object of reflection; it is rather a representation of the now. As any object of reflection, the objective present time appears after a certain detachment from the facticity of life, in this case, a detachment from *the experiencing of time*, that is, *temporality*.

Perhaps the evaluative dimension of present time is the first historical step toward a comprehensive notion of temporality. This is present in Seneca and Marcus Aurelius' reflections on the importance of present time. Seneca, as the epigraph of this chapter eloquently shows, confers more value to present time than past or future. From this, he followed moral consequences on the importance of correctly using the available time and not misusing it. In the same vein, Marcus Aurelius wrote, "Bear in mind that every man lives only this present time, which is an indivisible point, and that all the rest of his life is either past or it is uncertain" (2006/167 AD, Bk. III, para. 10). In both cases, even though present time plays a role structurally complementary to past and future times, it is however exalted, being worthy of special attention from a moral viewpoint. To be sure, this present time is neither ours nor a general lifestream that we are part of. Present time is something that we own, the most precious possession—perhaps the only one—that we have.

During the anti-intellectualist reaction to the increasing hegemony of natural science philosophy, the recognized importance of the present time gives way to the deeper insight that the "now" is not the possession of an external, homogeneous object, but rather is part of our way of living in the world. Subjective time is not simply "inner time," that is, a sort of objective time that became internal. It is qualitatively rich, expansive, and full of sinuosities that turn its measuring to an impossible task—or even a nonpertinent one. Temporality is also the quality through which we inhabit our world. Rather than being an object to be possessed, subjective time is the intrinsic mode by which persons feel the being part of the world. This qualitative dimension flourishes to our consciousness when we feel the wholeness of the situation into which we are taking part. Hutton's (1788/1795) insight of the dynamicity of the Earth arose precisely from his observation that "there is a system in Nature."

Ex-negativo, we know from Husserl (1991/1893–1917) that subjective time cannot be identified with a transcendental structure. Even though this solution may be suggestive from a logical-formal point of view, it does not offer a complete vision of the sense of mobility implied in subjective time. As seen above, such supra-individual, absolute subjectivity reinserts an a-temporal, static stance sustaining subjective time. By doing this, Husserl reinstalls transcendental entities to subsume the recognized dynamicity of subjective time under the metagovernment of the formally approachable world of essences.

In Heidegger (1992/1924) and Merleau-Ponty (2005/1945), we capture the insight that the sense of time we are looking for does not originate from the reflective attitude that characterizes the scientific view. Instead, it is to be found in the organism-environment encounter, or in Heidegger's words, in the *facticity* of the Dasein. Subjective time is embedded in the activities in which we are engrossed: it is a mode of encountering the world, not an object of consciousness. Consequently, subjective time cannot be contrasted with abstract space, as is the case of objective time. We feel time when we feel the movement of the world, which happens when we are occupied with a task in a particular situation of the landscape. Subjective time is felt in a specific *place*, not in abstract space. Subjective time also occurs before space and objective time are distinguished; this is, for example, the sense of the Aymara expression *akapacha* (see Núñez & Cornejo, 2012).

The conceptual analysis of time by observing and formally modeling it can only capture objective time. But at the very basis of it underlies another time, the subjective one, which is pure mobility, intertwined with the living activities of people in their environments. Ultimately, subjective time can be approached by performing our life activities; or more precisely, by feeling the movement of the wholeness of organisms and environments. Our organismic life is full of synchronies and resonances of the natural and social world. Such synchronies are the consequence of the fact that we too are part of the landscape.

REFERENCES

Bergson, H. (1910). *Time and free will: Essay on the immediate data of consciousness*. Whitefish, MT: Kessinger. (Original work published 1888)

Heidegger, M. (1992). *The concept of time*. Oxford, UK: Blackwell. (Original work published 1924)

Humboldt, W. von (1999). *On language: On the diversity of human language construction and its influence on the mental development of the human species* (M. Losonsky, ed.; P. Heath, Trans.). Cambridge, UK: Cambridge University Press. (Original work published 1836)

Husserl, E. (1991). *On the phenomenology of the consciousness of internal time*. Dordrecht, The Netherlands: Kluwer Academic. (Original work published 1893–1917)

Hutton, J. (1795). *Theory of the earth*. Edinburgh: Transactions of the Royal Society of Edinburgh. Retrieved from http://www.gutenberg.net (Original work published 1788)-

Jespersen, O. (1922). *Language: Its nature, development, and origin*. London, UK: George Allen & Unwin.

Marcus Aurelius (2006). *Meditations* (M. Hammond, trans.). New York, NY: Penguin. (Original work published 167 AD)

Merleau-Ponty, M. (1962). *Phenomenology of perception*. London, UK: Taylor & Francis. (Original work published 1945)

Núñez, R. E., & Cornejo, C. (2012). Facing the sunrise: Cultural worldview underlying intrinsic-based encoding of absolute frames of reference in Aymara. *Cognitive Science*, 1–27. doi:10.1111/j.1551-6709.2012.01237.x

Richet, P. (2007). *A natural history of time*. Chicago, IL: Chicago University Press.

Rudwick, M. J. S. (2005). *Bursting the limits of time: The reconstruction of geohistory in the age of revolution*. Chicago, IL: University of Chicago Press.

Toulmin, S., & Goodfield, J. (1965). *The discovery of time*. New York, NY: Harper & Row.

Vico, G. (1948). *The new science* (3rd ed.; T. G. Bergen, & M. H. Fisch, Ed. & Trans.). Ithaca, NY: Cornell University Press. (Original work published 1725)

CHAPTER 6

WHEN IS NOW

Measuring How We Perceive Instants in Time

Andre Cravo and Hamilton Haddad

INTRODUCTION

> It takes time for the physical changes that constitute an event to impinge on the body and to modify the sensory detectors of an organ such as the retina. It takes time for the resulting electrochemical modifications to be transmitted as signals to the central nervous system. It takes time to generate a neural pattern in the brain's sensory maps. Finally, it takes time to relate the neural map of the event and the mental image arising from it to the neural map and image of the self—that is, the notion of who we are—the last and critical step without which the event will never become conscious.
>
> —Antonio Damasio (2002)

Temporality: Culture in the Flow of Human Experience, pp. 115–126

We all live in the past, literally. Every sensation we have was physically caused by an event that took place in the past and that might even have ceased to exist. The above passage written by the Portuguese neurologist Antonio Damasio (2002) elegantly defines the concept of perceptual latency: the temporal lag between the occurrence of a physical event in the world and the moment one consciously experiences it. The considerable time delay in information transmission along sensory pathways places critical challenges to our nervous systems in everyday tasks. There are, for instance, significant differences between the perceptual and actual position of moving objects. Thus, daily actions, such as catching a ball or crossing a high-traffic street, needs to be timed adequately to be successful. Moreover, the lights we see, the sounds we hear, and the objects we touch produce sensory signals that travel by distinct neural pathways—with different processing times—before they are amalgamated into a unified and coherent percept. These differential transmission and processing latencies create a mosaic of temporal patterns spreading through networks underlying conscious perception. This issue gets even more complicated when we consider that information from different parts of our body have to travel different distances to reach our central nervous system. For example, if you touch your own toe with your finger, the sensory information from your hand and foot will reach somatosensory areas in the brain with dissimilar delays, although you'll probably judge them as simultaneous.

One could argue that as long as these delays are fixed, it wouldn't be too hard to find mechanisms to cope with them. However, as we'll see in this chapter, several factors can modulate the speed of processing of sensory stimuli, making it hard to delineate how we can act upon such an apparently chaotic flow of incoming information reaching different parts of your body and brain with different delays.

As the world we perceive is delayed in relation to its flowing content, time perception depends on an active construction of a neural state that correlates with biologically relevant elements present in the environment. The physical timing of events must be translated to the timing of neuronal processes, which is then somehow represented as a particular mental timing of experiential percepts. In the present chapter, we will briefly review how temporal perception has been studied throughout the history of cognitive sciences. From the pioneering work of Helmholtz, Wundt, and Titchener, to the recent findings of psychophysics and neuroscience that tackle these challenges, we will examine some problems inherent in the act of measuring the time of perception, and discuss possible solutions as well.

Measures of Time Perception: Past and Present

Hermann von Helmholtz (1821–1894) recorded for the first time the speed of transmission of a nervous impulse (Schmidgen, 2002). At the time, the majority of the scientific community believed that nervous and neuromuscular transmissions were achieved by a "nervous principle," or an "imponderable fluid," which either had infinite speed or was too fast to be measured physically. However, using an ingenious experiment, Helmholtz showed not only that the speed of nervous transmission was finite, but also that it was surprisingly slow (see Figure 6.1). The relevance of these experiments goes beyond interests in electrophysiology, given that for the first time, an immaterial phenomenon such as nervous transmission, which was commonly considered as a manifestation of the spirit or soul, was quantified with precision using physical instruments. Importantly, it raised a key question that has been studied by psychologists and neuroscientists ever since: With the intrinsic delays in transmission, how can we reliably judge *when* something happened in the world?

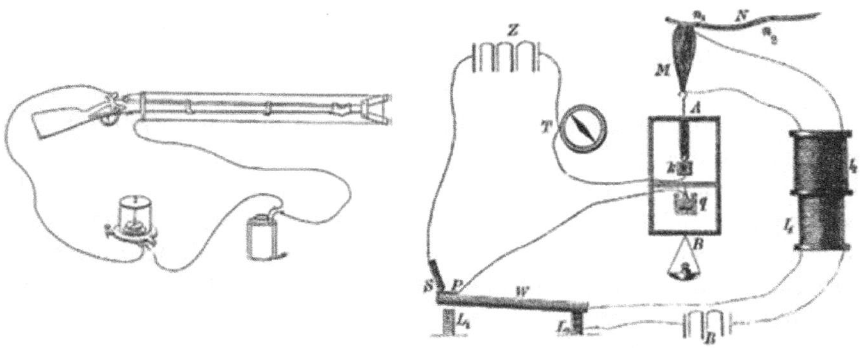

Figure 6.1. Helmholtz knew that the degree of deflection of a galvanometer pointer depends not only on the intensity of electric current but also on the time during which the pointer is exposed to this current. This fact was used to record very short time intervals; for example, to measure the speed of a bullet inside a rifle (on the left). By pulling the trigger, the bullet is fired and simultaneously closes the electrical circuit. Subsequently, the bullet cuts the wire and stops the current. The degree of deflection of the galvanometer pointer determines the time that the current was active. In order to measure the propagation speed of an impulse along a nerve, Helmholtz adapted (on the right) this method, using two interconnected circuits: a neuromuscular preparation to stimulate a frog muscle (*MN*) and another to measure the electric current with a galvanometer (*T*). The key, *SP*, closes the two circuits simultaneously, sending current to both the preparation and the galvanometer. When the muscle contracts, the device, *A*, immediately stops the current. Reading the galvanometer pointer, Helmholtz could calculate the time elapsed between stimulation and muscle contraction. The comparison between the readings when the electrode was positioned at different points of the nerve (n_1, n_2) allowed the deduction of the impulse speed (modified from Schmidgen, 2002).

Measuring the time of occurrence of physical events was never an easy task. Astronomers in the 18th century experienced better than anyone else the difficulties in measuring the very instant when an event takes place. They had to determine the precise positions of celestial events by means of the "eye and ear" method, which consisted of recording the moment when a particular star crossed the meridian marker of a telescope by counting the audible ticking of a pendulum clock. Spatial interpolation was then used to estimate the moment of transit in tenths of seconds (Baldo, Cravo, & Haddad, 2007). However, frequent discrepancies between trained astronomers produced significant measurement errors; later, this kind of inconsistency led to the formulation of a so-called *personal equation* to try to correct this difficulty (Boring, 1929).

Wilhelm Wundt (1832–1920), one of the founding fathers of experimental psychology, brought these astronomy problems into the laboratory (Schmidgen, 2003, 2005). Inspired by the eye and ear method, he designed an experimental setup, also using a pendulum apparatus, where subjects had to tell the position of a moving clock hand when a brief click was presented. Like the astronomers, Wundt observed that participants systematically misperceived the position of the moving hand when the click was presented, depending on the experimental conditions. This procedure was called "complication experiment," and it was characterized by involving the perceptual comparison of a continuous and a sudden event. Wundt suggested that his results could be explained by either differences between visual and auditory perceptual latencies or different strategies adopted by the observers; more precisely, to which of the two events, visual or auditory, the observer was paying attention during the task (Baldo et al., 2007). These observations culminated in the creation of the *law of prior entry* by Edward Titchener (1867–1927): "The object of attention comes to consciousness more quickly than the objects which we are not attending" (Boring, 1929; Titchener, 1908).

Over the years, the quest for a suitable way to measure the time elapsed between the physical occurrence of events and our perception led to the development of several behavioral and psychophysical methods. As a consequence, plentiful tasks are nowadays routinely performed in the laboratory in order to assess the temporal features of sensory perception.

One of the most common tasks is *reaction time* (RT). In a typical study, observers are required to respond by pressing a key as soon as the stimulus has been detected. It is usually assumed that the interval between the stimulus occurrence and the motor reaction depends on both sensory and motor components. Although this methodology has the advantage of being a direct measure of sensory-motor latencies, the perceptual and motor components of RT cannot be disentangled. In order to deal with purely perceptual delays, discriminative tasks are needed, such as *temporal order judgments* and the *flash-lag effect*. However, because in this kind of task subjects have to compare two stimuli, the outcome is an indirect and relative record of perceptual latencies.

In *temporal order judgments* (TOJ), participants have to report the temporal order of appearance of a pair of closely timed stimuli. When two stimuli are presented to an observer simultaneously or nearly simultaneously, the perceived order does not always correspond to the actual one. Several models about TOJ have been proposed in the last decades to account for the source of judgment errors (Allan, 1975; Cardoso-Leite & Mamassian, 2007; Jaskowski, 1991; Stelmach & Herdman, 1991; Ulrich, 1987). These models usually assume that temporal order perception depends on the arrival time of sensory information to a central comparator. In the visual system, for example, this would depend on transmission latencies of the sensory signals from the retina to this central temporal comparator, being determined by the dynamical properties of the perceptual pathways.

Similar to Wundt's complication experiment, the *flash-lag effect* (FLE) occurs when observers are asked to compare the positions of a moving object and a stationary flash presented in the visual field. In these situations, participants usually perceive the moving stimulus as being spatially advanced in relation to the positions of the flash, when in fact both stimuli happen to be physically aligned with each other in space-time (see Figure 6.2). Although this illusion was known for over 70 years, it was rediscovered and has been vigorously discussed over the past decades (Baldo & Klein, 1995; Cravo & Baldo, 2008; Mackay, 1958; Nijhawan, 1994, 2008).

Factors That Can Influence the Time of Perception

As previously mentioned, Wundt found that participants misperceived the position of the clock when the sound was presented. In his view, these results could be explained either by the fact that visual and auditory stimuli had different sensory latencies or by the different strategies adopted by participants. Specifically, he argued that the results could be influenced by which of the two stimuli the subjects were paying attention to.

In fact, the effects of attention have been widely studied with several of the methods described above. In RT experiments, for instance, it is assumed that the perceptual component depends on the properties of the sensory pathways that will determine the transmission latencies of the sensory signals. Studies have demonstrated that paying attention to where or when a stimulus will be presented can modulate this processing, thus reducing RT (Carreiro, Haddad, & Baldo, 2003, 2011; Nobre, Rohenkohl, & Stokes, 2012; Petersen & Posner, 2011). On the other hand, the motor component—the output side—is generally assumed to be less or even not affected by spatial attention. It should be noted, however, that the division between sensorial and motor pathways is far from clear (Goodale, 1998) and the problem of determining where (or when) one ends and the other starts makes arbitrary any assumption of the true place where attention is acting.

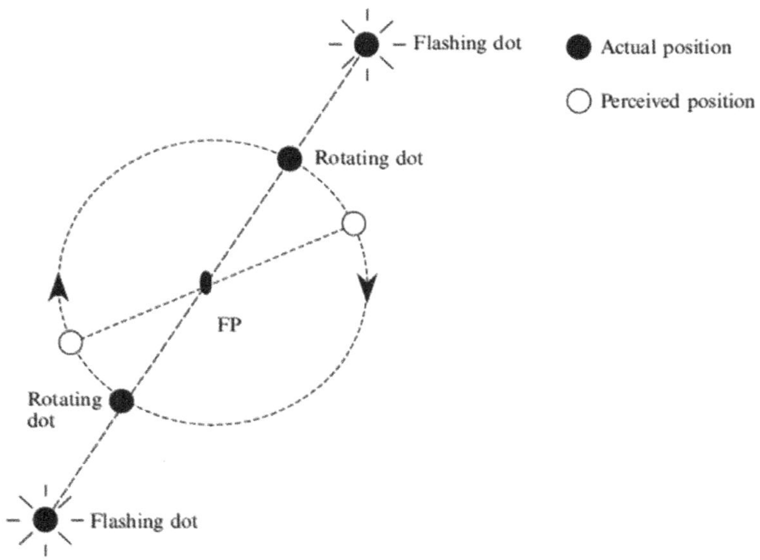

Figure 6.2. An example of the flash-lag effect (FLE). A perceptual misalignment between moving and flashing stimuli is reported by most observers: the rotating dots, moving around the fixation point (FP), are seen ahead of the flashing dots when the latter are flashed in perfect alignment with the *former set of dots*. (Modified of Baldo, Kihara, Namba, & Klein, 2002).

In TOJ tasks, attention can also alter the sensory latencies (Carver & Brown, 1997) and create an illusory asynchrony of visual information: The attended stimulus is perceived earlier that the nonattended stimulus (Casarotti, Michielin, Zorzi, & Umiltà, 2007; Haddad, Carreiro, & Baldo, 2002; Haddad, Klein, & Baldo, 1999; Stelmach & Herdman, 1991).

Several experimental results have also shown the role of attentional mechanisms influencing the delay of perceptual processing in the FLE (Baldo et al., 2002; Baldo & Namba, 2002; Chappell, Hine, Acworth, & Hardwick, 2006; Namba & Baldo, 2004; Sarich, Chappell, & Burgess, 2007). It has been proposed that the very genesis of this illusion could be in the time spent for attention to bind the flash and moving percepts into a unitary whole (Baldo & Klein, 2008). In this case, after the detection of the abrupt event (a flash or a beep), an attentional shift from the stationary object to the changing ("moving") object has to occur in order to bind them together into a unitary percept.

Apart of attention, other factors seem to be able to modulate our perception of when something happened. For example, some studies have shown that after repeated exposure to asynchronous stimuli, we start perceiving them as synchronous (Fujisaki, Shimojo, Kashino, & Nishida, 2004; Keetels & Vroomen, 2008). Importantly, it appears that the perceptual latencies of these stimuli are being modified, so that it is not just a change in our criteria of what we judge as simultaneous, but rather a change in the time we need to process the stimuli per se (Navarra, Hartcher-O'Brien, Piazza, & Spence, 2009). Recently, it has also been shown that such recalibration can also happen between our actions and their outcomes (Stekelenburg, Sugano, & Vroomen, 2011; Stetson, Cui, Montague, & Eagleman, 2006).

Curiously, motor actions seem to be able to modulate the time of perception of their outcomes even without extensive training. When participants are asked to indicate the perceived time of a beep evoked by their action, they tend to judge it as shifted earlier in time toward the action that caused it. This effect, originally called *intentional binding*, has been the subject of large discussion of its underlying mechanisms (Buehner & Humphreys, 2009; Cravo, Claessens, & Baldo, 2009, 2011; Haggard, Clark, & Kalogeras, 2002; Waszak, Cardoso-Leite, & Hughes, 2011). As in the case of temporal recalibration, it has also been shown that the perceptual latencies of the stimuli we cause are being modulated (Cravo et al., 2011).

But what would be the mechanisms underlying this effect? In other words, is there a special mechanism that modulates the perceived consequences of our actions, or is this modulation the result of a more general anticipation mechanism? A large body of evidence supports motor theories of perception, which essentially claims that perceptual contents depend not only on the relationship among inputs, but on input-output relationships as well (Crapse & Sommer, 2008; Wolpert, 1997). One could suppose the existence of an internal loop that anticipates the sensory changes caused by motor commands. This is the basis of the feed-forward models of motor control. In 1950, Roger Sperry proposed that when a motor command is sent to the muscles, an efferent copy, also known as "corollary discharge," is sent to the sensory areas of the brain (Sperry, 1950). This efferent copy would facilitate comparisons between expected outcomes of our actions and actual outcomes (Crapse & Sommer, 2008). Moreover, it would be a key factor to do online corrections of our motor commands (Wolpert, Diedrichsen, & Flanagan, 2011). Although these models were originally proposed to explain motor performance, they could help to explain effects such as intentional binding. In this view, an efferent copy sent by an internal predictive model could reduce the perceptual latencies of expected outcomes.

However, other possibilities for intentional binding might be suggested. As we discussed previously, attention is long known to have an important influence on the perception of an event's occurrence time, modulating its perceptual latency. When one causes a stimulus, its temporal predictability can be considered as the highest possible (Eagleman & Holcombe, 2002), shifting the perception of the caused event to an earlier instant in time. In this view, preparation of a movement would include the allocation of attentional resources to both the expected sensory consequences and the time of execution of the movement itself. Notice that here, action would not have a specific mechanism to modulate perception, but rather take advantage of our capacity to anticipate moments of presentation of relevant events. If this is true, than a similar effect as the intentional binding should also occur even in the absence of action. In other words, any pair of events that predict each other should also be approximated in time. The evidence in this sense is still contradictory, and while some have found that any pair of events, which are causally related, also are perceived as closer in time (Buehner, 2012; Buehner & Humphreys, 2009), others have found that action itself is necessary (Cravo et al., 2009).

As mentioned, the feed-forward models explain one of the possible mechanisms of the effect of action on perception. These models, at least in their original form, were proposed to explain processes involved in motor preparation/execution allowing fast online corrections. But could other kind of predictions use similar mechanisms? Recent accounts have suggested that in fact that might be the case. In one recent view of perception, named predictive-coding, our sensory system is conceived as making constant predictions of future events and comparing these to actual outcomes (Friston, 2005, 2010). Importantly, in this view, what is expected is attenuated and needs no further processing, while differences between the expected and the perceived are amplified throughout the system. Possibly, as has been recently proposed, a similar anatomical structure is involved in both feed-forward models for actions and for general expectations. In other words, what we normally think of as structures for our motor system can be thought of as a general mechanism of predictions (Schubotz, 2007)

CONCLUSIONS

Suppose someone asks you what time is it *now*. To give this information, you need to look at the clock, which takes time, and then the very instant we intended to grasp has already passed. This example captures the essence of the problem inherent in subjectively determining *when* something happened in the word. The substantial time delay in information transmission along sensory pathways plus the differential processing latencies among the stimuli create a mixture of temporal patterns spreading through networks underlying conscious perception.

How can we deal with such latencies is still a remarkable challenge posed to the researchers who try to study them. In this chapter, we have explored some of the problems faced by those who tried to study time perception over the last two centuries. As we have shown, the problem arises not only from the difficulty in measuring exactly when one has perceived a stimulus, but also due to the flexible nature of the inherent delays in sensory processing. Though not discussed in this chapter, factors such as attention and action, which strongly modulate the perception of time, also crucially take part in the intersubjectivity of time. How we live together and interact using a same "time" remains a mystery to be solved.

ACKNOWLEDGMENTS

The authors would like to thank Prof. Marcus Vinicius C. Baldo for his helpful comments not only in this chapter, but during the several years we have worked under his supervision.

REFERENCES

Allan, L. (1975). The relationship between judgments of successiveness and judgments of order. *Perception & Psychophysics, 18*(1), 29–36.

Baldo, M. V. C., Cravo, A. M., & Haddad, H. (2007). The time of perception and the other way around. *The Spanish Journal of Psychology, 10*(2), 258–265. CiteSeer.

Baldo, M. V. C., Kihara, A. H., Namba, J., & Klein, S. A. (2002). Evidence for an attentional component of the perceptual misalignment between moving and flashing stimuli. *Perception, 31*(1), 17–30.

Baldo, M. V. C., & Klein, S. A. (1995). Extrapolation or attention shift? *Nature, 378*(6557), 565.

Baldo, M. V. C., & Klein, S. A. (2008, April). Shifting attention to the flash-lag effect. *Behavioral and Brain Sciences, 31*(2), 198–199.

Baldo, M. V. C., & Namba, J. (2002). The attentional modulation of the flash-lag effect. *Brazilian Journal of Medical and Biological Research, 35*(8), 969–972.

Boring, E. G. (1929). *A history of experimental psychology*. New York, NY: Appleton-Century.

Buehner, M. (2012). Understanding the past, predicting the future: Causation, not intentional action, is the root of temporal binding. *Psychological Science, 23*, 1490–1497.

Buehner, M., & Humphreys, G. (2009). Causal binding of actions to their effects. *Psychological Science, 20*(10), 1221–1228.

Cardoso-Leite, P., & Mamassian, P. (2007). Temporal order judgment and simple reaction times: Evidence for a common processing system. *Journal of Vision, 7*(6), 1–14.

Carreiro, L. R. R., Haddad, H., & Baldo, M. V. C. (2003). The modulation of simple reaction time by the spatial probability of a visual stimulus. *Brazilian Journal of Medical and Biological Research, 36*(7), 907–911.

Carreiro, L. R. R., Haddad, H., & Baldo, M. V. C. (2011). Effects of intensity and positional predictability of a visual stimulus on simple reaction time. *Neuroscience Letters, 487*(3), 345–349.

Carver, R. A., & Brown, V. (1997). Effects of amount of attention allocated to the location of visual stimulus pairs on perception of simultaneity. *Perception and Psychophysics, 59*(4), 534–542.

Casarotti, M., Michielin, M., Zorzi, M., & Umiltà, C. (2007). Temporal order judgment reveals how number magnitude affects visuospatial attention. *Cognition, 102*(1), 101–117

Chappell, M., Hine, T. J., Acworth, C., & Hardwick, D. R. (2006). Attention capture by the flash-lag flash. *Vision Research, 46*(19), 3205–3213.

Crapse, T. B., & Sommer, M. A. (2008). Corollary discharge across the animal kingdom. *Nature Reviews Neuroscience, 9*(8), 587–600

Cravo, A. M., & Baldo, M. V. C. (2008). A psychophysical and computational analysis of the spatio-temporal mechanisms underlying the flash-lag effect. *Perception, 37*(12), 1850–1866.

Cravo, A. M., Claessens, P. M. E., & Baldo, M. V. C. (2009). Voluntary action and causality in temporal binding. *Experimental Brain Research, 199*(1), 95–99.

Cravo, A. M., Claessens, P. M. E., & Baldo, M. V. C. (2011). The relation between action, predictability and temporal contiguity in temporal binding. *Acta Psychologica, 136*, 157–166.

Damasio, A. R. (2002). Remembering when. *Scientific American, 287*(3), 66–73.

Eagleman, D. M., & Holcombe, A. O. (2002). Causality and the perception of time. *Trends in Cognitive Sciences, 6*(8), 323–325.

Friston, K. (2005). A theory of cortical responses. *Philosophical Transactions of the Royal Society of London. Series B, Biological Sciences, 360*(1456), 815–836.

Friston, K. (2010). The free-energy principle: A unified brain theory? *Nature Reviews Neuroscience, 11*(2), 127–138.

Fujisaki, W., Shimojo, S., Kashino, M., & Nishida, S. (2004). Recalibration of audiovisual simultaneity. *Nature Neuroscience, 7*(7), 773–778.

Goodale, M. A. (1998). Visuomotor control: Where does vision end and action begin? *Current Biology, 8*(14), 489–491.

Haddad, H., Carreiro, L. R. R., & Baldo, M. V. C. (2002). Modulation of the perception of temporal order by attentional and pre-attentional factors. *Brazilian Journal of Medical and Biological Research, 35*, 979–983.

Haddad, H., Jr., Klein, S., & Baldo, M. V. C. (1999). The contribution of attentional and pre-attentional mechanisms to the perception of temporal order. *Neuronal Bases and Psychological Aspects of Consciousness*. Singapore: World Scientific.

Haggard, P., Clark, S., & Kalogeras, J. (2002). Voluntary action and conscious awareness. *Nature Neuroscience, 5*(4), 382–385.

Jaskowski, P. (1991). Two-stage model for order discrimination. *Perception & Psychophysics, 50,* 76–82.

Keetels, M., & Vroomen, J. (2008). Temporal recalibration to tactile-visual asynchronous stimuli. *Neuroscience Letters, 430*(2), 130–134.

Mackay, D. (1958). Perceptual stability of a stroboscopically lit visual field containing self-luminous objects. *Nature, 181,* 507–508.

Namba, J., & Baldo, M. V. C. (2004). The modulation of the flash-lag effect by voluntary attention. *Perception, 33*(5), 621–631.

Navarra, J., Hartcher-O'Brien, J., Piazza, E., & Spence, C. (2009). Adaptation to audiovisual asynchrony modulates the speeded detection of sound. *Proceedings of the National Academy of Sciences, 106*(23), 9169–9173.

Nijhawan, R. (1994). Motion extrapolation in catching. *Nature, 370,* 256–257.

Nijhawan, R. (2008). Visual prediction: Psychophysics and neurophysiology of compensation for time delays. *The Behavioral and Brain Sciences, 31*(2), 179–98; discussion 198–239.

Nobre, A. C., Rohenkohl, G., & Stokes, M. (2012). Nervous anticipation: Top-down biasing across space and time. In M. I. Posner (Ed.), *Cognitive neuroscience of attention* (2nd ed., pp. 159–186). New York, NY: Guilford.

Petersen, S. E., & Posner, M. I. (2011, March). The attention system of the human brain: 20 years after. *Annual Review of Neuroscience, 61,* 450–456.

Sarich, D., Chappell, M., & Burgess, C. (2007). Dividing attention in the flash-lag illusion. *Vision Research, 47*(4), 544–547.

Schmidgen, H. (2002). Of frogs and men: The origins of psychophysiological time experiments, 1850–1865. *Endeavour, 26*(4), 142–148.

Schmidgen, H. (2003). Time and noise: The stable surroundings of reaction experiments, 1860–1890. *Studies in History and Philosophy of Science, Part C: Studies in History and Philosophy of Biological and Biomedical Sciences, 34*(2), 237–375.

Schmidgen, H. (2005). Physics, ballistics, and psychology: A history of the chronoscope in/as context 1845–1890. *History of Psychology, 8*(1), 46–78.

Schubotz, R. I. (2007). Prediction of external events with our motor system: Towards a new framework. *Trends in Cognitive Sciences, 11*(5), 211–218.

Sperry, R. (1950, December). Neural basis of the spontaneous optokinetic response produced by visual inversion. *Journal of Comparative and Physiological Psychology, 43*(6),482–489.

Stekelenburg, J. J., Sugano, Y., & Vroomen, J. (2011). Neural correlates of motor-sensory temporal recalibration. *Brain Research, 1397,* 46–54.

Stelmach, L., & Herdman, C. (1991). Directed attention and perception of temporal order. *Journal of Experimental Psychology. Human Perception and Performance, 17*(2), 539–550.

Stetson, C., Cui, X., Montague, P. R., & Eagleman, D. M. (2006). Motor-sensory recalibration leads to an illusory reversal of action and sensation. *Neuron, 51*(5), 651–659.

Titchener, E. B. (1908). *Lectures on the elementary psychology of felling and attention.* New York, NY: Macmillian.

Ulrich, R. (1987). Threshold models of temporal-order judgments evaluated by a ternary response task. *Perception & Psychophysics, 42*(3), 224–239.

Waszak, F., Cardoso-Leite, P., & Hughes, G. (2012). Action effect anticipation: Neurophysiological basis and functional consequences. *Neuroscience and Biobehavioral Reviews, 36*(2), 943–959.

Wolpert, D. M. (1997). Computational approaches to motor control. *Trends in Cognitive Sciences, 1*(6), 209–216.

Wolpert, D. M., Diedrichsen, J., & Flanagan, J. R. (2011). Principles of sensorimotor learning. *Nature Reviews Neuroscience, 12*, 739–751.

PART III

TEMPORALITY AND LIFETIME

CHAPTER 7

DURATION AND EXPERIENCE

The Temporality of Development

Dankert Vedeler

ABSTRACT

It seems evident that development takes place over time. Yet most theories and empirical research on development does not take the time dimension into account. This chapter tries to analyze what temporality assumptions a developmental theory should make. This is done through studying a number of philosophers for whom temporality was important and who present different ideas on what time is and how it may be analyzed. Some of these ideas are then associated with modern scientific thinking, notably Gibsonian Ecological psychology and Dynamic Systems Approaches, which provide basic concepts for the temporality assumptions of a developmental theory for the life sciences. A particular tribute is paid to Kierkegaard, who, with his conceptions of repetition, *Augenblick* (moment), and exception, more than any other offers a key to the understanding of stability and change in the life course of an individual organism. The chapter concludes with an extended and generalized understanding of Dynamic Systems Approaches, proposed as a theoretical basis for conceptualizing the temporality of development, akin to the scope outlined by Ken Richardson (2008).

Temporality: Culture in the Flow of Human Experience, pp. 129–214

OVERVIEW

From Charles Peirce, I draw the fundamental idea of the continuity of time, although I disagree with his mathematical analysis of time. From William James, the idea of time as a stream ("stream of thought") has been important. He furthermore agrees to a large extent with Henri Bergson, who analyzed time in terms of duration. Notably, both considered an intellectual conceptualization of temporal events to be deleterious to the proper understanding of their true nature. A further analysis of Bergson's concept of duration leads to the conclusion, also in accord with James, that time should be analyzed in "durational units," although this was not explicit in Bergson's philosophy. The question then arose of how durational units could be delimited. My conclusion is that such units should be defined in terms of a subject's activity.

This leads to Martin Heidegger, for whom being in itself *is* time. More important, though, is his analysis of the relationship between man and the world during ongoing action, which he describes as *Sorge*, care. This is based on our experience of the world on a human scale. Most of Heidegger's analysis is rejected, however; although he is credited with the fundamental insight that an analysis of time should be based on man's experiential relationship to the world.

Kierkegaard, on the other hand, stands out as more important, while antedating all other philosophers discussed here (except the classical Greek school of the Eleates and the inevitable Kant.). His analysis of the *repetition* and the "moment" (*Augenblick*) as a basis for human experience is important for my conclusion on the temporality of development, as it helps to establish a relationship between durational units and experience.

Experience is here to be understood, as in regular English, in two senses: On the one hand, we have the singular experience that we have at any particular moment. This is experience defined as a durational unit, and that can be remembered. For this kind of experience, I will use the German word *Erlebnis*. On the other hand, we have *accumulated* experiences, that is, through single and unique experiences we acquire general knowledge through similarities among the different particular experiences. This is an understanding we may have from Kierkegaard's analysis of the repetition. For this kind of experience I will use the German word *Erfahrung*.

All of the above philosophers had more far-reaching ambitions with their time-related analyses, ambitions we have refrained from discussing here in order to concentrate on the temporal relation between man and the world, a relation that I propose to be developmental.

I also refrain from discussing the far-reaching ambitions in Maurice Merleau-Ponty's discussion of time. Instead, I let his discussion of the relationship between time and subjectivity lead us to the conclusion of this chapter.

I will thus give less weight to his idea on temporality, preferring his analyses of concrete human action that Hubert Dreyfus has pointed out. These analyses point to basic and central features of concrete physical action that depend on physical properties of the human body that constrain human action. Thus, there are very concrete properties of the body that delimit what may be considered an action. These properties, I argue, therefore also delimit durational units and what is to be considered as an experience. Repeated, such action will accumulate similar *Erfahrungen*, leading to generalization and differentiation of actions in a process that may be linked to neurological changes, which, like the actions themselves, and their combination, may be described in the Dynamical Systems terminology of self-organization, attractor basins, and emergence.

INTRODUCTION

I will start this chapter by pointing to the disparity between what happens and what we describe as happening. We cannot describe what is happening without simplifying and abstracting, disregarding details and transformations involved in what is happening. By that, we also conceptualize things, events, and series of events into recognizable and describable objects and patterns of events that we may specify by name and describe. In this process, much of the details and complexity of the real happening is lost. This disparity was the main point of the French philosopher Henri Bergson at the turn of the 20th century, when he contrasted our intellectual understanding of the world to a particular kind of intuition, namely, what he considered to be "the immediate data of consciousness."

The regular, intellectual data of consciousness (Bergson, 1888/2001) are based on a spatial conception of division and identification of units, and the intuition he proposed as an alternative was supposed to give a richer and more close-to-reality understanding of the word, in spite of being more subjective. "Subjective" should not here be understood as arbitrary or distorted by personal earlier experiences, prejudices, and such: "Remarquons,[1] en effet, que nous appelons subjectif ce qui paraît entièrement et adéquatement connu, objectif ce qui est connu de telle manière qu'une multitude toujours croissante d'impressions nouvelles pourrait être substituée à l'idée que nous en avons actuellement." (Bergson, 1888/2001, p. 62/pp. 83–84).[2] Thus, the subjective is an experience that is completed, directly understood and adequate, while what is objective is accessed by a multitude of impressions that complement and even replace each other.

Both what is subjective and what is objective are facts of consciousness relating us to the world, however, in different ways. We would expect subjective experiences thus to be homogeneous, objective experiences being heterogeneous, multifaceted. With Bergson, it is the other way around: Objective experiences are homogenous, because they are considered as distributed in space, one beside the other, even if deployed in time in case events succeeding each other in time are placed in a row such that time is represented in a spatial metaphor. That also means that the events are represented *simultaneously*, such that we can go back and forth among them and find out relationships between events; for example, causal relationships. That also implies that such relationships may be reversible and that time also, thus represented, is reversible.

Bergson's view here corresponds with Newton's and the common physical conception of time and space, where time and space could be considered as a sausage (Wyller, 2001) and where time could be represented as the length of the sausage, while successive slices of the sausage represent space at different moments in time. When two events happen simultaneously, they are on the same slice—and only on one particular slice of the sausage. The problem, for Bergson, with such a spatial metaphor is that time is given a spatial representation (the slices) and that, *in the metaphor*, different moments of time are represented as simultaneous, as you may (and probably will) think of the slices as one beside the other. According to Bergson, this is a distorted representation of time.

In the history of philosophy, it is generally considered that matter and mind are essentially different (with Spinoza as the most outspoken exception). For example, Descartes and Kant put great effort in explaining the difference. This is also the main concern of Bergson. Whether you call it mind, spirit, or consciousness, it cannot be reduced to matter. Bergson's explanation of the difference was that matter is what is extended in space and mind is what is "extended" in time. You see? It seems that I'm unable to explain this to you, the reader, without using a spatial concept—"extended." Bergson's ambition was to describe and explain time without recourse to spatial metaphors.

Bergson, however, was not the only one to challenge our everyday understanding of time. He shared with William James and Charles S. Peirce the concern for the continuity of time. However, while James and Peirce were pragmatic realists, Bergson shared with philosophers like Kierkegaard and Heidegger the concern also for some kind of truth or genuineness beyond our everyday experiences. Still, none of these five philosophers were content with the unreachable transcendence of Kant's thing-in-itself. They wanted to find an immanent[3] truth of the mind or the spirit, which we as humans could attain, bypassing everyday experiences, distorted by our intelligence.

The particular issue of immanent truth is, however, not central to my argument here. I will instead consider the fact that time was *the* essential aspect of human existence both for Kierkegaard and for Heidegger, as it was for Peirce and James. I will conclude by judging them all four, together with Bergson, to be insufficient for understanding the temporality of development, which is the ultimate aim of the epistemological issue I raised in the first paragraph of this introduction. That is, I want to address this issue in terms of the temporality of our experiences. How we describe what happens depends on our experiences with what happens. These experiences are not the same in the beginning of life and at mature age, even, for example, when a child and a grown-up experience the same sequence of events. This is a disturbing fact, as we demand objectivity and truthfulness of our experiences. Bergson and James are right in their criticism of an overly intellectual understanding of experience and knowledge. Merleau-Ponty, the sixth philosopher I introduce to this discussion, has been a source of inspiration for alternative understandings of the relationship between what happens and our descriptions of it, in his emphasizing the bodily origin of this relationship. However, his analyses fall short of giving an account of what makes the body a source of experience and understanding. Concepts from Ecological Psychology, Dynamical Systems Approaches, and Probabilistic Epigenesis are suggested as tools to find answers to these questions. However, the temporal character of the processes these concepts describe needs to be emphasized in order for development to be described as change processes, rather than stages—change processes that neither of these approaches has fully integrated into their theories. It is suggested that some of the above discussed philosophers may help temporalizing development, even if they don't pay attention to development.

THE PHILOSOPHERS

Bergson, James, and Peirce: The Eleates

Let us start with the continuity of time assumed by Bergson, James, and Peirce. All three discuss time in terms of the temporality of mind (consciousness, thought, experience, etc.) as opposed to temporality in a physical sense, the temporality of matter (cf. here above, the physicist conception of time). When we think of the temporality of the mind, we have generally the idea that time is always divided into past, present, and future.[4]

Bergson makes no reference to Peirce. Their ways of explaining continuity, though, are rather similar. Namely, both make use of the idea that time is continuous because different experiences following each other merge. For Peirce, consciousness consists of ideas, which have three "elements" to them (Peirce, 1892, p. 549): (a) it is a *feeling*, (b) it may affect other ideas, and (c) may "bring along other ideas with it," that is, it may merge with other (specific) ideas to generate more general ideas. He writes, for example, "... instantaneous feelings flow together into a continuum of feeling, which has in a modified degree the peculiar vivacity of feeling and has gained generality" (Peirce, 1892, p. 555). This latter property of ideas is also his strongest argument for the continuity of subjective consciousness. In spite of this argument for the continuity of time, Peirce insists on dividing time into intervals, where the present becomes squeezed in between the past and the future. The continuity between past and present is assured by a present idea being impacted by a past idea, such that, at each interval in time (thus having a certain duration), the past idea stretches half into the interval, while the other half belongs to the future (Peirce, 1892, p. 546). This sounds arbitrary, but Peirce considers it tenable, because the intervals of time are not finite, they are not final units of time, but infinitely divisible into smaller and smaller intervals. It is not the practical division that matters, but the principle: there is no end to the partitioning of intervals into smaller intervals. This principle assures the continuity of time.

This reminds of the arguments of the Eleatic school of philosophy in ancient Greece, with arguments against the possibility of movement, for example, the argument of the Eleatic philosopher Zeno that Achilles, in running a race against a tortoise that was allowed a head start, would never catch up with and overrun the tortoise. The argument is based on measuring the distance between Achilles and the tortoise at the moment Achilles reaches the starting position of the tortoise and then measuring when he reaches the position of the tortoise at the preceding point of measurement, and so on, *ad infinitum*. We then find that this distance will be smaller and smaller, as Achilles runs faster than the tortoise. But he will never catch up when we measure this way because we just find infinitely smaller and smaller distances between the two. While there still seems to be no final and indisputable logico-mathematical solution to this paradoxical argument, Peirce and James, as well as Bergson, held different solutions to the paradox. It is particularly interesting to compare Peirce's solution to James's and the latter's solution to Bergson's.

Peirce bases his argument on the distinction between infinite and finite series of terms, such as 1/2, 1/4, 1/8, 1/16, and so on. In the case of Zeno's Achilles-tortoise example, the series is *infinite*. However, Zeno poses a *finite* series and thus confuses finite and infinite series (at least to my understanding of Peirce's argument). The important thing to remember is that Peirce poses infinitesimal divisions of time intervals as the criterion for continuity: There is no end to the division. However,

> All the arguments of Zeno depend on supposing that a continuum has ultimate parts. But a continuum is precisely that, every part of which has parts, in the same sense. Hence, he makes out his contradictions only by making a self-contradictory supposition. (Peirce, 1869, p. 200)

Another way of saying it, Peirce points out, is that Zeno attributes durations to points in time (as opposed to intervals), which, by definition, are of no length and indivisible, and thus finite in numbers (however large that number might be!).

James's (1916, ch. XI) solution goes in the opposite direction. Instead of arguing with infinitesimal time intervals, he proposes a "holistic" argument. It does not make sense to define a whole in terms of its parts. The running of Achilles has a whole to it, which is different in speed, compared to the running of the tortoise.

Bergson (1888/2001, p. 53/pp. 112–113), in my view, takes this view a step even further, in contrasting time and space. Zeno's argument is vicious by confounding the spatial variable of distance with the temporal variable of movement. Space may be infinitely decomposed; time may not. This stance is the opposite of Peirce's stance. While ignoring Peirce's writings, there is no doubt that Bergson would have considered Peirce's (1892) analysis a typical case of spatialization of time. In the Achilles-tortoise race, what should be compared is not the distances traversed, but the steps taken by Achilles and the tortoise. The steps can be counted, but each step is a nondecomposable unit, and the steps of Achilles are very different from the steps of the tortoise. Then comes an argument that is the kernel of Bergson's philosophy, namely, that *we should distinguish between the unity of something or an event when we think of it and the same thing or event as erected into an thing of thought*. The latter means making a thing out of an event, and that is a fallacy of the intellectual mind.[5]

In subjective consciousness, on the other hand, things and events are *lived, in duration, in present experience*; that means our experiences before they have become fixated, through language, through representations, through our *intellection*.[7] In *pure* duration, that is, a duration not contaminated by space, our states of consciousness would not make a succession, properly speaking, like a chain, because that would necessarily be a spatial chain, not a temporal chain. In a temporal deployment of states of consciousness, these will be interlaced such that they cannot be distinguished one from the other (Bergson, 1888/2001, p. 56/p. 122). They are contemporaneous, that is, in spite of being different from each other (heterogeneous) they are part of the present. If discarded from the present, they simply disappear into the past, to never return.

William James presented a similar idea just a few years earlier (James, 1884), when he contrasted the "stream of consciousness" to the cognitive organization of consciousness into concepts. This contrast is very similar to Bergson's main point. While Peirce does not make that contrast, his assuming continuity from a division in intervals, be they infinitesimal, is, in my view, fallacious, even under the assumption that each interval contains both the past and the future in equal halves. These halves are to be considered, under Peirce's argument, as separate and thus as intervals in and by themselves.

Both Bergson and Peirce think of duration as elements of conscious experience. While Bergson approaches this conscious experience as, in principle, unmentionable, Peirce wants to give it a rigorous mathematical definition. Peirce ends up in some kind of *reductio ad absurdum* (i.e., delimiting any "smallest" interval of time, it still is not the smallest, as it can in turn be divided into any number of even smaller intervals. That is the meaning of infinitesimal: there is no end to the division of intervals of time). Bergson, on the other hand, ends up in the paradox of mentioning, that is, writing about, the unmentionable.

In my view, the Peircian approach leads to an elaborate cognitivism (e.g., Zittoun et al., 2013, Ch. 4). The Bergsonian approach, on the other hand, is stuck in introspection. Bergson wanted to show that there was a different way of conceiving the mind's relation to the world other than the juxtaposition of events and things in space as defined and described by language, notably the language of natural sciences. For him, a more immediate[7] relation to the world should be possible if we avoided conceptualizing our experiences. That was also the gist of James's argument against the "scientific" psychology of his time. But how can we make meaning out of our lives when we abandon our intellectual capacity to elevate ourselves above other animals' more direct and unthinking relationships with the world? Peirce was wholeheartedly intellectual, while Bergson and James somehow wanted it both ways.

Kierkegaard and Heidegger

To answer that question, we turn to Kierkegaard and Heidegger. The question of temporality is best approached in Kierkegaard's book *Repetition* (Constantius, 1843; Kierkegaard, 2009). Although Kierkegaard by far antedated Heidegger, I prefer to say some words on Heidegger first. Many others, for example Heidegger's mentor, Husserl, would also have a say on temporality. I leave them out here for the sake of coming to what I consider to be the crucial point, namely, to elucidate temporality as a way of understanding the relationship between what is happening in the world and our attempt to grasp it. Heidegger dismisses both Bergson "(Heidegger, 1927/2010, pp. 338–339, note 1/p. 323, note 30) as understanding time in a "vulgar" sense: Bergson equating time to space, and Kierkegaard contrasting it to eternity.

I consider that to be a superficial reading of both. Let us focus on the fundamental question addressed in Heidegger's (1927/2010) *Being and Time*: what it means to be a human person. Heidegger discards all attempts to approach humanity through theological, empirical, ego-focused, anthropological, and such, means, which all have in common to disregard the subjective existence, the "I am" character of the human being (he also dismisses Descartes's *cogito*). As mentioned, Bergson and Kierkegaard are no exceptions, and most certainly, neither would be Peirce and James, not even mentioned by Heidegger. "Meaning-of-life" is neither an option; it should rather be *being* of life. This sounds like a rather empty issue. For Heidegger, it is *the* central issue.

Like other things, we are in the world. However, the issue of subjective existence is our human being-in-the-world (*In-der-Welt-Sein*). It is not a *cogito*, neither a "situatedness" in the world. Rather, Heidegger wants us to have a basic understanding of the conditions for our existence as human beings.

So, what does "the world" mean? For Heidegger, it is not about the world as an object. On the contrary, it is our *relationship* to the world that makes it being the world. The world is at the scale of humanity, and it is what we assume to be the source of our experiences. On the other hand, it is our experiences that make it a world for us. That is because we care (*Besorgen*) about it. We experience the world because it matters to us. That is the essence of *Dasein,* being here/there. However, Heidegger wants to address the *a priori* condition for *Dasein*, namely, *Sein*. I here reduce this question to a developmental question about the (*a posteriori*) conditions for the emergence of subjectivity.

What is interesting about Heidegger's view of things and events in the world is that he more explicitly describes them in terms of our relationships to them. They may be "ready-at-hand" (or "ready-to-hand"), in German, *zuhanden,*[8] but also *griffbereit,* a neologism in German language, however seemingly not used by Heidegger but of later date, that condenses the full sense of ready-at-hand, that is, "ready-to-be-grasped" (I cherish this word, as it fits, as hand in glove, into the conclusion that I intend to make from this discussion.)[9] *Zuhanden* should be distinguished from *vorhanden,* "present-at-hand." While *zuhanden* is about an affordance of a thing that

I care for because it is useful to me, like a tool, *vorhanden* stands for the detached, disinterested contemplation of the object, even if I care for it, in case it is broken and has lost its affordance. Its graspability is then of no use for me, or it is lost.

While Bergson conducts a principled discussion about subjective time as the basis for our relationship with the world, and Peirce and James supply with the emotional dimension that is not central to Bergson's argument, their analyses are all about consciousness. All three have in common that they investigate consciousness, something that is quite peripheral to Heidegger. He often puts the word within quotes and considers it to be present-at-hand in the same way as the things of the world when you are not engaged in them, when you don't care about them (Heidegger, 1927/2010, p. 203/p. 196):[10] "Der Terminus 'Dasein' meint sowohl das Vorhandensein des Bewußtseins wie das Vorhandensein der Dinge." From this quote we can also draw the basic conclusion that our relation with the world is primary, even overriding the subject/object distinction, which is, as to say, derived from our fundamental Being-in-the-world (*In-der-Welt-Sein*), that is, the basic meaning of *Dasein* (being there/here).

In his vocabulary, Heidegger also has *Wiedeholung*, repetition. To start with, the word is used on a metalevel, that is, when he writes about his use of words and explains his theses, for example, "this analysis need to be repeated." On page 339/324 (Heidegger, 1927/2010), however, the word is introduced as a theoretic concept: "Das eigentliche Gewesensein nennen wir die *Wiederholung*." *Gewesensein* means "being-in-the-past." It is important to understand that this does not mean that you live in nostalgia, always longing for what has been. The key concept is not repetition, but *eigentlich,* which means authentic in the generic sense, that is, something that is coming from and is about *myself.* This makes the concept of repetition relatively unproblematic (as compared to most others of Heidegger's key concepts). Heidegger has the merit of putting our relation to the world in focus, as opposed to dwelling in introspection, which has a long tradition, from Augustinus to Bergson, passing by Descartes (of course) but also Peirce and James.

Besides this, for me, crucial point, Heidegger does not much to elucidate the temporal relation to the world that is brought up by Bergson. I should just mention, to be fair, that temporality *is* central in Heidegger's philosophy (you have it in the title of his main book). Being (*Sein*) *is* time. And he defines time in the traditional subjective sense as past-present-future, while at the same time strongly distancing himself from the "vulgar" understanding of these words (Heidegger, 1927/2010, p. 325/p. 311). *Dasein* is the *unity* of these three temporal modi,[11] with emphasis on the future, namely, Death (i.e., the future end of your life) and thus the end of you, your being-in-the world. On he same page he writes that future (*Zukunft*) does not mean a now (*Jetz*) to come, but a *Kunft*, a word that is not part of any German language dictionary, except for quite a few combinations, one of which is *Zukunft*. The word is etymologically the noun form corresponding to the English verb "come." Therefore, "coming" as a noun would perhaps be a good translation.[12] To give an idea of what it is about, I cite a whole paragraph in German:

> Wenn zum Sein des Daseins das eigentliche bzw.[13] uneigentliche Sein zum Tode gehört, dann ist dieses nur möglich als zukünftiges in dem jetzt angezeigten und noch näher zu bestimmenden Sinn. "Zukunft" meint hier nicht ein Jetzt, das, noch nicht »wirklich« geworden, einmal erst sein wird, sondern die Kunft, in der das Dasein in seinem eigensten Seinkönnen auf sich zukommt. Das Vorlaufen macht das Dasein eigentlich zukünftig, so zwar, daß das Vorlaufen selbst nur möglich ist, sofern das Dasein als seiendes überhaupt schon immer auf sich zukommt, das heißt in seinem Sein überhaupt zukünftig ist. (p. 325/p. 311)

The conclusion I draw from this is that Heidegger actually thinks of *Dasein*, that is, our being in the world, as *becoming*, rather than as *being*. Has he then captured the dynamics of temporality and avoided the spatial staticity of time that Bergson attacks? Yes and no! What is then the difference between *Kunft* and *Zukunft*? As I understand, it is about the difference between ready-at-hand and present-at-hand. I understand the future (*Zukunft*), in the sense of "the now that has not yet become real" to be present-at-hand, while the future, in the sense of "the *Kunft* where the *Dasein* comes to its own capacity to be (*Seinkönnen*)" is ready-at-hand. So, as I see it, Heidegger ends up in the same dilemma as Bergson, preconizing (recommending) intuition rather than intellection,[14] while being dependent on intellection to make this recommendation. Heidegger wants the future to be ready-at-hand, but cannot promote this except by making it present-at-hand.

I think the solution of this dilemma is to be found in liberating temporality from the past-present-future schema.[15] Being-in-the-world for Heidegger implies *action*. Our caring implies action, and the future also is defined in terms of action. Which action, though, would make a difference, and here Heidegger turns to Kierkegaard, from whom he borrows the concept of *Augenblick*, as used, in English, in the expression "in the blink of an eye." It is commonly translated as "the moment" as used in, for example, "just a moment" (e.g., Klempe, 2014a, p. 21). The Danish word is essentially the same as the German. I will use the German word here,[16] to avoid the lack of specificity in the word "moment." For Heidegger, it means not the moment in time as such, but that privileged moment where you have your situation in your hand, doing the right thing knowing that it is the right thing to do *for you*, considering that you know who you are.

That does not mean that you know who you have been and how you have come to be what you are now. To dwell in your history and to be absorbed by the situation are both unauthentic being (although perfectly normal and acceptable, there is nothing wrong about it).

Even if this is yet another expression of the refusal to accept Kant's ontological transcendence of the world (the unknowable thing-in-itself) that has haunted philosophers since then,[17] Heidegger can be credited with shifting focus away from consciousness, pioneering instead man's relation to the world, even antedating the subject-object relationship (see Vedeler, 2012). This will be carried to completion by Merleau-Ponty in his philosophy of corporeality. It seems obvious that Heidegger's *Dasein* analysis has meant a lot for Merleau-Ponty to develop this philosophy.

However, Heidegger resolutely turns the back on Bergson's philosophy and thus has nothing to say on the challenges temporality puts on our *knowledge relationship* with the world. We may, however, pick up on Heidegger's inspiration from Kierkegaard, *das Augenblick*, to see what that leads to (Heidegger, 1927/2010, p. 338, note 1/p. 322, note 3).

Wiederholung in German, *gjentagelse* in Danish, repetition in English, is the core issue for the human condition. However, for our question on knowledge, we cannot as easily draw on Kierkegaard and Heidegger as we did with James and Bergson. While Bergson has in common with Kierkegaard and Heidegger a quest for liberty and free will, the latter address the existential predicament of being human more directly, overshadowing human temporality, that somehow is taken for granted. For Heidegger, man *is* time, and past-present-future is the unity of *Dasein*, "being there." For Kierkegaard, time is related to eternity.

Except for the *Augenblick*, the temporality of our knowledge relation to the world is not an issue. Furthermore, the *Augenblick* is related to action rather than to knowledge. That is not irrelevant, though, so I will discuss that later. What remains from Heidegger is a penetrating analysis of our being-in-the-world, helpful for our understanding of this problematic relationship, and probably an analysis without which we would not have the insights brought to us by Merleau-Ponty, see here below.

There is, however, an alternative understanding of repetition, as described by Kierkegaard, which I first want to consider. The general understanding of this concept is based on Kierkegaard's focus on the existential meaning of repetition in our relation to time and eternity, much the same way as in Nietzsche and Heidegger (see also Deleuze, 1968/1994). Klempe, 2014) on the other hand, focuses on the "*in-der-Welt-sein* subjectivity"[18] of repetition in Kierkegaard's writings. I use this denomination in order to highlight a possible difference between Heidegger and Kierkegaard: Heidegger departed from mainstream philosophy based on the Kantian heritage of searching for the condition of the possibility for knowledge, in particular scientific knowledge, that is, epistemology, theory of knowledge. He proposed instead (Heidegger, 1927/2010, p. 13/p. 12) a "fundamental *ontology*" focused on the conditions for the *existence* (not knowledge) of the matters for knowledge. The only way to attain that is through an analysis of human existence, that is, *Dasein*, our human being there/here (Rouse, 2005). Namely, while Kant's followers sought for conditions for knowledge that were prior (*a priori*) to any knowledge, they inadvertently assumed a knower-known relationship as given. Before we can say anything about the validity of our knowledge, we need to know the validity of our being knowers. Unfortunately, this leads Heidegger (in a Kantian tradition) to try to find out the *a priori* conditions for *Dasein*, that is, what being (in itself, i.e., *Sein*) *is*. He never managed to do that. What remains is a very interesting analysis of *Dasein* in *Sein und Zeit*, and an elaborate and inconclusive discussion on what would be needed for liberating ourselves from our present worldview in order to have a more genuine approach to *Sein* (or *Seyn*; Heidegger, 1989/2012; cf. Vernon, 2008).

Kierkegaard, with a very similar project, although not with objective knowledge in focus, but rather with a "meaning-of-life" focus, did exactly the opposite of Heidegger. Instead of trying to find the condition for all conditions in some universal *a priori* validity, he searched for a *particular* validity, also when it comes to knowledge.

Leibniz related the distinction between the particular and the universal to a distinction between thinking and perception. This implies that the senses (vision, hearing, etc.) are at the origin of particulars, while universals could only be found by thinking. The link between the particular and the universal is perception, and that's how we gain knowledge of the world.[19] Leibniz' follower, Christian von Wolff, turned Leibiniz' philosophy in this regard into an "Experimental Psychology," where the relationships between sensation, perception, and thinking are discussed. This philosophy was important for Kierkegaard and gave him a different, more "psychological" view of experiences, which has not received the same attention as his more "existential" philosophy (Klempe, 2014a and Klempe, this volume). Corresponding to (and probably at the source of) Heidegger's distinction between unauthentic and authentic *Dasein*, Kierkegaard distinguished between an aesthetical and an ethical way of living. Most of the interest in Kierkegaard was about his ideas about the ethical way of life, while the aesthetical did not draw the same interest.[20] We have mentioned two of his concepts, which are certainly not limited to aesthetics, namely, the moment (*Augenblick*) and repetition (*Wiederholung*), also important in Heidegger's *Dasein* analysis. In Kierkegaard's treatment of these concepts, however, there are more to them than existential implications.

I will here try to explain how Kierkegaard's analysis of *Augenblick* and repetition may help us understand and solve the problem of the conflicting ways of relating to the world that James named intellection and intuition. Bergson's issue was that our understanding of time was based on spatial metaphors, which deprived time of its continuity.[21] James expressed the same concern as Bergson, distinguishing between the (intuitive) stream of consciousness and the (intellectual) cognitive organization of consciousness into concepts. Bergson proposed the concept of duration to describe a more direct relationship to the world. The problem with this is that by *describing* duration, it is itself distorted into concepts, such that it is, in principle, impossible to give an adequate description of duration. Bergson proposes that it can only be *lived in* experience, not described as such. James saw the problem of this and proposed units of duration, and some of Bergson's own elaborations of the concept of duration point in the same direction. It is not intervals of time as understood by Peirce, but meaningful units related to own perception and activity.

So our problem is, can we define units of duration and still consider them as parts of the stream of consciousness that cannot be delimited in time? The solution that I will propose later in this chapter will be to recommend a negative delimitation of units of duration rather than a positive one, that is, delimit such units in terms of what they are not rather than in terms of what they are. Kierkegaard's concepts of *Augenblick* and the exception will be very helpful in explaining this line of argument.

I choose here to make no distinction between perception and action. Perceiving a thing in the world is an act of picking up affordances[22] (Gibson, 1979) of the thing for some use I may have of it, or it may be an act of contemplation. This also corresponds to the account Heidegger gives of our relation to the world with his distinction between ready-at-hand and present-at-hand. But he does not explain the continuity of perceptions/actions that is the temporal reality of our relation to the world. His analyses are principled and atemporal. You may think of it as a continuity of caring, where your changing needs continuously change the focus of our caring, because the unity of our temporality is our present past toward our future. What happens on the detailed plane of time seems of no interest to Heidegger. The *Augenblick* is discussed only from an ontological perspective as a moment of decision, not as a moment of time. The *Wiederholung* is not recognition or " the same" in any sense, it reveals for the *Dasein* its history, that's all.

Kierkegaard, however, more carefully treats these concepts: We start with the moment (*das Augenblick*).[23] Our point of departure is the distinction between the universal and the particular. However, in light of the insights of Bergson and James, we need to ask for the universality of the particular, that is, what makes the particular something that can be identified as the same over time. This may also be expressed as starting from the distinction between the universal and the particular, and then take a step further to question the particular in the same way as the universal; namely, what makes the unity of the individual thought in the stream of thought, in the stream of perception, in the Bergsonian duration?

The *Augenblick* for Kierkegaard is what is neither past nor future, it is the pure presence (Haufniensis,1844/Kierkegaard, 1980, p. 65/p. 87), however in a rather Percian, infinitesimal sense, as the transition from past to future. It is contrasted to eternity, which, on the other hand, is the pure *extended* (full) presence, where the successions in time have been canceled. Time, thus, is defined as successions. So Kierkegaard's position on time is completely opposite to Bergson's. In the perspective of Bergson's problem of defining units of duration, I choose to understand Kierkegaard to consider the *Augenblick* as a paradox, as a concept contradicting itself, being nothing, in between the past and the future, while at the same time defining "the one," that is, the particular as part of "the many," which amounts to being the universal, although named "eternal" by Kierkegaard. I thereby deliberately disregard the theological dimension of Kierkegaard's thinking in order to focus on the "experimental" part in his ambition to link the universal, the general to the particular, the individual. The *Augenblick* instantiates the particular moment of the particular subject, and that touches upon my ambition to take a step further, *from the universality of the particular, as the identity of the same individual over time, to the temporality of that same individual, what is constant and what is changing in that same individual.*

This corresponds to Kierkegaard's concept of repetition, in German, as used by Heidegger, with *Wiederholung*.[24] Kierkegaard wrote[25] a book with the title *The Repetition*. As usual, Kierkegaard made a very concrete analysis (subtitle of the book is, in translation, *A Venture in Experimental Psychology*) of himself, both in the shape of the author, Constantius, and in the shape of a young man, who Constantius first becomes the confidant of, and later received several letters from (without any face-to-face meetings). This young man has the same predicament with a woman as Kierkegaard himself had with a young woman[26] that he first promised to marry but later turned down. The choice of pseudonym underlines the importance of the argument (in translation rendered by Klempe, 2014, p. 7) that without recollection (of memories) and repetition, "all of life is dissolved into an empty meaningless noise" (in Danish: "saa opløser hele Livet sig i en tom og indholdsløs Larmen" (Constantius,1843/Kierkegaard, 2009, p. 34/p. 19). Constantius wants to be the investigator of the stability of life. However, the reader is left in bewilderment as to the success of this project. On the one hand, he concludes that repetition is not possible, on the other (as in the quote here above) he states that it is necessary. One solution might be to routinize one's life, make sure to have everything in order and to avoid that anything unexpected happens, which Constantius did. However, his young friend insisted on sending him letters where he poured out first his torments then his relief. Namely, analyzing his own situation, he understands that the background of his unfortunate handling of his love affair was the fear of falling into a life course that is normal, predictable, conventional: marrying, having children, having a professional career and finally retire. He did not want *that kind* of repetition. He wanted to have a genuine life, without knowing what exactly that meant. He found his ideal in the Old Testament figure of Job, who lost everything he had in God's trial of him to see if he would curse the Almighty God. Job stands the test, and God rewards him for that, doubling his wealth. The young man's despair was because of the guilt he felt for the predicament he had inflicted on his fiancée in the absurd way he dismissed her. He later felt relieved, however, learning that the girl has married another man and had a good life according to her wishes. Now he could start anew, with a true life according to his ideals. No doubt the young man's choice of life is quite the same as chosen by Kierkegaard himself, and one can easily think that the ultimate purpose of the book, *The Repetition*, would simply be Kierkegaard's justification of the life he himself had chosen to live.

As explored by Kierkegaard, the word "repetition" is about *experience*, where "the same" (the general, in Danish *det Almene*) is related to the exception. These two things, the general and the exception, are dependent on each other. The exception needs something general to be an exception from, and the general corroborates its generality in contrast to its exception. From the very obvious assumption that Kierkegaard's main concern is the human condition, the general and the exception are understood as different ways of living a life. I therefore feel justified in proposing that Kierkegaard's is an investigation of human subjectivity—from a different perspective, certainly—but still comparable to James and Bergson. None of these philosophers are concerned with historicity in the sense of individual life courses. While Kierkegaard has in common with Heidegger (to mention the philosophers discussed in this chapter) to not include time as such in his analysis, in the way James and Bergson do, the individual life course takes quite a different significance for him than for the others, as he focuses uniquely on the individual life condition, although arguing in general terms.

I will therefore here step down from the philosophical categories of the *universal* and the *particular*, reformulating them as applied to humanity and human subjectivity as the *general* and the *individual*. Furthermore, I want to introduce temporality into the equation, by adding, as a third term, *individual development*. Actually, *what we learn from Kierkegaard's analysis of the repetition is about the individual life course*, although not explicitly mentioned by Kierkegaard.

From now on, my main concern will be formulated in the terms of Bergson and James: Our relation to the world is paced by the continuous flow of perception and action, including as well sensori-motor action and mental action. As already mentioned, I understand perception as a form of action, that is, not only a passive registration of information from the senses. When Bergson writes about the "immediate data of consciousness," it is this constant "stream of consciousness" that James writes about. Bergson's complaint is that we do not pay sufficiently attention to these "immediate data," instead sorting them into ready-made categories through which we understand the world. Thus, the world suffers from our prejudices, as we interpret the content of the stream of consciousness in terms of previously formed conceptions. One could say, with James's terms, that intellection stands in the way of intuition. That would actually be very close to the repetition in the negative sense of Kierkegaard, where the individual simply is subsumed under the general.

However, not adopting Peirce's view on continuity as infinitesimal divisions, we are left with the issue of what makes up durational units, if such units need to be assumed for the idea of continuity to hold. Bergson has less concern for this issue than James, who, on the other hand, ends up with a theory closer to Peirce. He is more relaxed than Bergson toward spatial representations of time. However, instead of conceiving infinitesimal durations, half of which are past and the other half future, he conceives a clock-time duration of approximately 12 seconds, including past and future, as the maximal time length where we can have a clear sense of being in the present (James, 1890, p. 613).

This is incompatible with Bergson's notion of continuity, which for him is irrevocably indivisible. He considers units of duration, however, that are defined by their *content*, rather than by time length. If you think of a set of units, the description of the set is based on similarities between the members of the set. However, units of duration are radically heterogeneous (cf. this chapter, p. 110) and cannot be considered as the same in any respect. If we endeavor to consider psychological states, we are actually picking them out of the durational flux such that we may put them side by side and count them. Actually, we are quite clever at doing exactly that. Then, however, we also dwell in intellection[27] and thereby lose the durational character of these states. The durational units of consciousness thus cannot be examined from outside, they can only be lived, experienced as such.

However, Bergson has something to say about the flow of durational units. Their main common feature is that they are interlaced in each other, as pointed out here above, such that they cannot be disentangled into units. So instead of having units of consciousness, you have unity of duration and heterogeneous multiplicity of durational contents. The fact that they have content is what makes it possible to intuit them, but they may (at least, I assume, for us humans) also be intellected, that is, fixated into mental things and named. This leads us to Kierkegaard's discussion of the general and the singular, however, with Bergson, discussed in terms of homogeneity versus heterogeneity. Would not duration, as radical heterogeneity, make life "dissolved into an empty meaningless noise," as Kierkegaard feared? I conclude, so far, that there is a radical incompatibility between Bergson and James on the one hand, and Kierkegaard on the other.

How can that incompatibility be resolved? I suggest that the problem is that Bergson, as well as Peirce, James, and Kierkegaard, discusses this issue at the level of consciousness, spirit, mind, and so on. Heidegger is the first to take the discussion down to the level of direct interaction with the world. It is our way of being-in-the-world that defines us as humans. Unfortunately, like Kierkegaard sighs for eternity, Heidegger sighs for "being-as-such," *Sein*. Only he is less successful than Kierkegaard in making something out of this transcendence.[28] While Heidegger's ultimate project seems to have failed, he at least should be credited for shifting focus, from mind to man's relationship with the world.

Merleau-Ponty: The Body and the World

Maurice Merleau-Ponty has then carried on this shift of focus throughout his whole authorship. His proposals will help us to be more concrete on our relationship with the world. They will also give us some ideas on the temporality of this relationship. What will be missing in the model inspired by Merleau-Ponty is the significance of development for this relationship. Merleau-Ponty (1945/1962), as well as all of his predecessors, mentioned here or not, fails to take a developmental perspective. They all assume humanity as a ready-made condition. Recognizing the importance of the past as such is not enough. *The trajectory also matters.* However, Merleau-Ponty's concrete analyses are such that it is easy to complement them with an account of the temporality of development. Namely, by making Bergson and James's continuous and uninterrupted flow of consciousness a continuous flow of being-in-the-world, Merleau-Ponty escapes the internal contradiction of Bergson and James as to the incompatibility of the experiential flow and our experiences as defined and delimited units. By Bergson and James, our experiences (as defined and delimited units) are always and necessarily distorted. For Merleau-Ponty, on the other hand, we have access to the experiential flow through our body as being and acting in the world. The distortion only comes in when this experiential flow is conscious. Bergson's paradox becomes obvious when he makes intuition an act of consciousness, to be contrasted with another type of acts of consciousness, which are intellections, acts of conscious intelligence.

Merleau-Ponty approaches this problem by distinguishing between what is lived and what is known, transforming the problem of the relationship between body and mind to an issue on the flow of individual events and a tissue of ideal significations: “ce sera maintenant le problème des rapports de la conscience comme flux d'événements individuels, de structures concrètes et résistantes, et de la conscience comme tissue de significations idéales” (Merleau-Ponty, 1942/1963, p. 232/p. 215). On the one hand, Merleau-Ponty poses consciousness as perception in direct contact with the ever-changing realities of the world; on the other, he poses consciousness as cognition of these realities. From this quote I feel entitiled to introduce the terms “lived consciousness” and “known consciousness,” not to be found in the original writings of Merleau-Ponty.

Thus, while Bergson is faithful to the Augustinian-Cartesian, and even Kantian and Husserlian tradition, to look for truth by introspection, Merleau-Ponty favors the physical contact with the world, where subjectivity and objectivity merge. The matter of the relationship between self and the external world is turned into a relative point of contact, either by the skin or by an incorporated instrument (e.g., the blind man's cane, the end point of which replaces the man's eyes, something that already Heidegger touches upon).

As for Bergson, liberty, the freedom of the will, is important for Merleau-Ponty. However, certainly influenced by Bergson, his conception of time mostly draws on Husserl and Heidegger. For our purpose here, namely, to question the relation between what happens in the world and our understanding of it, a separate discussion of Merleau-Ponty's conception of time and temporality does not add to or complement what is already said about the philosophers mentioned above. Merleau-Ponty seems to take the concept of duration for granted, to be the same as James's “stream of consciousness” (cf. Merleau-Ponty, 1945/1962, p. 83/p. 69: écoulement de la durée/flow of duration). In consequence, the present seems to be understood as a moment of duration (cf. pp. 93–94, note 2/pp. 78-79, note 2). In this understanding, Merleau-Ponty's central concept of ambiguity (of knowledge, pp. 97–98/p. 82) would in this context cover Bergson's challenge to solve the conflict between an experience as being a moment of duration at the same time as being erected into an object of thought.

In my view, Merleau-Ponty's discussion of temporality does not fully incorporate the achievements of his concrete analyzes of interaction with the world, drawn to my attention by a paper published by Hubert Dreyfus (Dreyfus, 2002). These analyses may suggest that Merleau-Ponty might not himself have understood the full innovative implications of his philosophy. I must admit, however, that Dreyfus's references to Merleau-Ponty are somewhat “creative,” and that the innovation to a certain extent belongs to Dreyfus rather than to Merleau-Ponty, although clearly inspired by Merleau-

Ponty. The concepts Dreyfus highlights are not central in Merleau-Ponty's philosophy, and his understanding of them is not easily substantiated from Merleau-Ponty's original texts.

The title of Dreyfus's paper, "Intelligence Without Representation," points to an important dividing line in the recent history of psychology. Traditionally, psychology has favored a representationalist view of the relationship between a subject and the world: In order to perceive, remember, and act upon the world, our interaction requires mental or brain representations[29] of things and events in the world as mediators between stimulation from the world on the one hand, and goals and actions in the world on the other. Since the appearance of Gibsonian Ecological psychology (e.g., Gibson, 1966b, 1979), there have been strong objections to this view,[34] generally characterized as cognitivist (and most consistently defended by Jerry Fodor, 1981).

Dreyfus's paper proposes an elucidation of "two central concepts in Merleau-Ponty's *Phenomenology of Perception*" (Dreyfus, 2002 p. 367). I give the word to Dreyfus:

> The *intentional arc* names the tight connection between the agent and the world, viz. that, as the agent acquires skills, those skills are "stored," not as representations in the mind, but as dispositions to respond to the solicitations of situations in the world. *Maximal grip* names the body's tendency to respond to these solicitations in such a way as to bring the current situation closer to the agent's sense of an optimal gestalt. Neither of these abilities requires mental or brain representations. Rather, simulated neural networks exhibit crucial structural features of the intentional arc. Moreover, Walter Freeman's (1991)[31] account of the brain dynamics underlying perception and action is structurally isomorphic with Merleau-Ponty's account of the way a skilled agent moves towards obtaining a maximal grip. (p. 367)

The following discussion will to a great extent rely on the central point conveyed by Dreyfus in this quote, and from there I will try to draw a conclusion that solves the Bergsonian problem of the relation between the subject and the world, as well as Kierkegaard's problem of the relationship between the universal and the particular, specified as the relation between knowledge of the general and knowledge about the individual. The solution of both problems will be based on Merleau-Ponty's concrete analyses and what Dreyfus makes of them. To put this discussion in a developmental context, however, I first need to explain the understanding of development and temporality this discussion relies on.

DEVELOPMENT

Life courses are paced by both stability and change. Both stability and change are something you experience in time. Stability means that you do the same things over and over again. It gives you security and predictability. Change means new things to learn, less security, fewer challenges, and less excitement. We want a reasonable balance between the two, but some people have too much stability in their lives, while others have too much change. Furthermore, and more importantly, we all sometimes feel too much stability (boredom), as well as sometimes too much change (over-excitement).

Epigenesis

Both stability and change are inescapably linked to time, and thus in such a way that *what has happened before matters for what comes next.* This is easy to understand. I shall make it a little bit more difficult to understand. I shall give it a particular name, *epigenesis*, for which I recommend two references, Gottlieb (2003) and van Geert (2003). Epigenesis originally means

> ... individual development is characterized by an increase in novelty and complexity of organization over time. Thus, new structural and functional properties arise sequentially over the course of development. The emergence of these structural and functional novelties is a consequence of horizontal and vertical coactions among the parts, including organism-environment coactions. (Gottlieb, 2003, p. 3)

This means that ontogenesis is the result of a very complicated interaction between genes and environment *up and down and over time* at all levels of human (or any organism) functioning, from genes to society and culture (see Gottlieb, 2003, p. 3 and Figure 7.1 below).

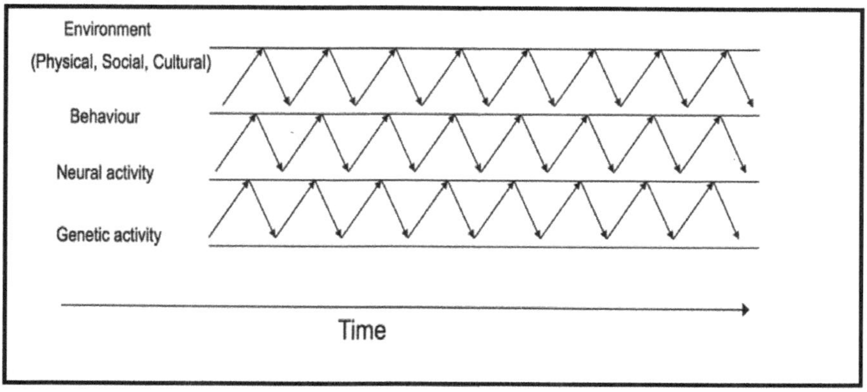

Figure 7.1. A systems view of psychobiological development (probabilistic epigenesis) as based on Gottlieb (2003, p. 11, Fig. 1.4).[32]

In Figure 7.1, Gottlieb illustrates his notion of *probabilistic epigenesis*, where genes, organism, and environment have bidirectional influences on all levels of organism structure and function. These influences are not fully deterministic and that, together with the complexity of their mutualities, made Gottlieb label his version of epigenesis as probabilistic. Thus, there is inherent uncertainty in his model of development. Furthermore, in this model, the usual nature-nurture issue becomes pointless, as influences and changes on all levels may permeate all the other levels, from genes to culture.

One implication of this model that I want to emphasize here is that, *at any time of your life, what happens next will be dependent on what has happened before in your life, up to the present moment, here and now.* Note that it is not only about "now"; "here" also matters, that is, what particular situation you are in at that particular moment. Epigenesis thus defines both your personal history and your present situation.

What does stability and change mean is this context, that is, in the context of epigenesis? It means a lot of things that cannot be disregarded but that at the same time are impossible to grasp in one single overview. Not only is it difficult to have an overview of it at one particular, as to say, frozen, moment of your life, it is further complicated by also going on, from birth to death, with incessant changes. It furthermore means stability and change on innumerable different time scales simultaneously, and, as if that was not enough, interacting with each other, as illustrated by Gottlieb's figure. The concept of temporality should thus cover both stability and change at all interlaced levels and time scales, whether in fractions of a second or over a life course. For simplicity, we restrict ourselves here to two levels: *microgenesis*, that is, the development of particular skills (e.g., as described by Dreyfus) and *ontogenesis*, the development of the individual organism.

I will not dwell on the question of whether consciousness is specifically human or not. We may just agree that we humans normally have consciousness, which gives us an awareness of our temporality. We may assume that other animals do not have this awareness; they may not think of past and future, but only live here and now. Philosophers, on the other hand, with particular interest in the human condition, have spent much ink on the question of what time means for us humans. As already mentioned, physicists who are interested in time have a very different conception of time, where there is no past, present, and future, and where even the concept of "now" is relative (Wyller, 2014). The experience of time that we humans have is something different, and that is what is discussed here.

With Merleau-Ponty, however, the experience of time is not only about consciousness but also about the body "within" which the consciousness is supposed to be found. The important shift of focus is from within to between, namely, between the body and the environment.[37] While the other philosophers mentioned in this chapter focus on the temporality of the subject and its experiences, Merleau-Ponty take one step further and focuses on the active body that is both object and subject, and that constitutes itself as subject by being a perceiver/actor in the world. Subjectivity is thus not primary, as for the other here mentioned philosophers, neither is the object, as it would be for a natural scientist. What is primary is the human body in a world and the relation between the two. If we are to take this proposal seriously, we will also have to make it the basis for our understanding of temporality, including what it means for development.

Let me try to be more concrete: To start with, we have a human organism who between birth and death traverses a life course. That's the body. It is a reference point to everything else in his or her world. It has certain features that make it natural to speak about its relations to the world in terms of ahead (from the face), behind (opposite), up and down, left and right (Todes, 2001). It is the structure of the body that, for example, makes it easier to walk forward than to walk backward and such. The world is situated relative to these features of the body. Furthermore, the body is moving in the world, such that, for example, ahead becomes behind and vice versa. As the center of all these relative properties of the world, the body becomes the subject and what is distributed around it becomes object. *But not all at one time.* The movements, the behavior, the actions of the body come one after the other in a temporal order. What it has done becomes past, what it does now is the present, and what it will do next becomes the future. This amounts to saying that the distinction between humans and (other) animals is not about a difference in kind, based, for example, on rationality, but rather a question of degree. This is my conclusion, not Merleau-Ponty's, unfortunately. I will come back to the difference between humans and animals later, including Merleau-Ponty's view on the question.

I mentioned epigenesis as a way of describing the temporality of development. For Merleau-Ponty, the subject *is* temporality (e.g., 1945/1962, p. 276/p. 239, p. 483/p. 422); the subject is not *in* time. The development of the subject should imply temporality. That sounds weird, but can be understandable from the perspective of epigenesis. In Merleau-Ponty's terms, *situatedness* means, of course, the present situation. That is the main point of Merleau-Ponty. However, situatedness is also temporal, which defines who I am, as a consequence of my history. It is also implicated in the concept of epigenesis, introduced here above.

From the perspective of the subject, what does that mean?

The Origin of Subjectivity

I have acted in the world since I was born. Whether conscious or not, these actions have been more or less successful, more or less clumsy, more or less proficient, taking place in the center or the periphery of attention. As to the delimitations of durations, I interpret Bergson as linking them to attentional actions. As duration is a matter of consciousness, it implies that it is experienced. When Merleau-Ponty distinguishes between lived and known consciousness (1942/1963, p.xx)" by When Merleau-Ponty (1942/1963, p. 232/p. 215) distinguishes between lived and known consciousness (my terms). (See also this chapter, pp. 125, 168).

The relationship between us and the world needs to be further clarified, though. We never experience duration in itself. We may experience the present content of a duration, but to experience it *as* a duration would be making it into an entity, intellectualizing it. As such, we may be able to recall a past-by duration from memory. But it may not be contemplated "live," because as soon as it has taken place, it will fade away in the past. Actually, most of the times Bergson mentions experiences, he writes about "past experiences." As to the present durations, it is enough to say that they take place in consciousness. As to past experiences, I will have more to say when I discuss the problem of the general and the singular, mentioned in the discussion of Kierkegaard's philosophy here above. Bergson (1896/2007) proposed a theory of memory where he distinguished between two kinds of memory; on the one hand, memory as habit, routine, and skill and on the other hand, "pure memories," by which we may recall a past-by duration from memory, however reified. I will not further discuss the latter concept. It is in the former, memory as habits, routines, and skills, I think, that we have the answers to our questions on the relationship between us and the world.

The point of this is that memory as habits and such is more pertinent to the body-world relationship that Merleau-Ponty focuses on, from which subjectivity and temporality will emerge. The identity relationship between subjectivity and time that seems to be assumed by Merleau-Ponty needs, however, to be explicated, and possibly restated. Let our starting point be an active body of a living organism entering the world. From the perspective of epigenesis, in Gottlieb's (2003) sense, that is a development that starts with conception. It implies, from the beginning, a certain organization of the relationship between the organism and that part of the world that constitutes its proximal environment. This organization is achieved, through epigenesis, in continuous interaction between organism and world (including the womb), as well as among the different components internal to the organism (at different levels; see Figure 7.1). The reader is here reminded that, in this process, both genes and environmental circumstances are involved in a very advanced and intricate dynamic interaction, where considerations about the relative contribution of genes and environment become pointless. Out of this organization, a subject-object relationship will eventually emerge as the internal organization of the organism more and more takes the form of "autopoiesis" (Maturana & Varela, 1980), in the sense of making the distinction between the organism and its environment more and more clear-cut, as well as more and more organized, and where the organism more and more appears as a center of activity relative to the environment.

Where does time come in with this development? While I assume that Merleau-Ponty (1945/1962) conceives a fully developed subjectivity, we would consider it obvious that the development from conception to fully developed subjectivity (whatever that is) will take time. In that sense, I suggest that early epigenesis of the human infant has an intrinsic *temporality* to it *that precedes subjectivity*.[33] I will now try to connect the development of subjectivity to a discussion on the development of intentionality. The emergence of the infant as an agent in the world will have significance for the development of both subjectivity and intentionality.

Dreyfus and Merleau-Ponty

As the infant more and more appears as a subject, the target of its activity correspondingly will appear as an object, thus laying the ground for a subject-object relationship. While other members of the species, in particular among humans, may play a prominent role in this development, I will not include in this account the very important role of the subsequent establishment of a subject-subject relationship for the further understanding of the temporality of development. I will fully agree that this is a serious caveat of this account. Merleau-Ponty pays great attention to the subject-subject relationship, and it is obvious for modern infant research that it is crucial for the development of the subject-object relationship. For our purpose here, however, that would bring us too far from the main point I try to make, namely, to give an answer to the question of what constitutes the subject-object relationship in the first place.

For the purpose of my developmental argument here, Dreyfus's (2002) use of concepts "intentional arc" and "maximal grip" plays rather different roles. I will lend less importance to the former, focusing on maximal grip, which I consider to be the relationship to the world that makes possible the intentional arc, as understood by Dreyfus. He considers the intentional arc to "capture the idea that all past experience is projected back into the world. The best representation of the world is thus the world itself" (p. 373). This is indeed a very important point also in my argument. I feel more uncertain, though, that the concept "intentional arc" gives us the best understanding of it.[34] In my view, Dreyfus draws too heavily on this concept. I don't question Dreyfus's main point, that "The best representation of the world is thus the world itself," to which Merleau-Ponty would certainly also subscribe. I think, however, that we need a more adequate concept, to make justice to Dreyfus' proposal. Merleau-Ponty often used the word from Husserl (1998), "horizon" to convey what Dreyfus implies with "intentional arc" (see, e.g., Merleau-Ponty, 1945/1962, pp. 83–85/ pp. 69–71).

As to "maximal grip" (or "maximal grasp," for that sake), this expression is nowhere to be found in the English translation of this book, neither are the equivalent French expressions *saisie* or *prise maximale* to be found in the French original. However, Merleau-Ponty's use of the noun *saisie* and the verb *saisir* as a mental act, analogous to the sensori-motor act of grasping a physical object, is very relevant for our issue here. Nonetheless, a tight reading of *Phenomenology of Perception* to a large extent supports Dreyfus's understanding of Merleau-Ponty, in spite of considerable freedom in referencing and explication of Merleau-Ponty's ideas. I here assume the same liberty *vis-à-vis* Dreyfus in order to explore the radical ideas that underlie both these two great philosophers.

More important for Merleau-Ponty is the idea of a more basic understanding of intentionality, described by him as "intentionalité opérante" (*fungierende Intentionalität*; 1945/1962, "p. xiii/p. xviii), "la motricité comme intentionnalité originale" (p. 160/p. 137)," "l'intentionnalité motrice pure" hidden behind the objective word that it constitutes (pp. 161–163, note 1/p. 138-139, note 2), and so on. This idea corresponds to my proposal on intentionality in a generic sense as "object-directedness," as opposed to the common understanding of intentionality as goal-directedness, at least in psychology (Vedeler, 1987, 1991, 1993), and to Dreyfus's (2002) proposals presented here.

In order to understand the difference between intentionality as goal-directedness and intentionality as object-directedness, we may take advantage of Dreyfus's (2002) contrasting his view with the view of his colleague at the Department of Philosophy at Berkeley, John Searle (Searle, 1981, 1983). According to Searle, intentional action needs to have a condition of satisfaction (COS) in order to be intentional. That implies a word-to-mind direction of fit; the action will bring some matter of the world in accordance with what was the purpose of the action. However, the experience we have performing the action should have a mind-to-world direction of causation. That means that our intention has caused the action leading to a change in the world. Dreyfus agrees with the first requirement, to make the world fit into the aim of the action. This however does not mean that this aim needs to be set up in advance in order to cause the action. In Dreyfus's (2002) words, the action might have a word-to-mind direction of causation (i.e., the opposite of Searle's requirement). That is, the action is not caused by the intention but by the conditions in the world. In my words, the aim is not represented by an idea of the end state to be achieved. Such an idea is what I would call *a goal*, that is, *a representation of a desired end state*. In other words, I may act intentionally without an idea of what I want to achieve. The opposite, to act with a particular goal in mind, *which may also be the case*, is what I consider to be goal-directedness, properly speaking. In my view, that is a derived form of intentionality. The action without goal, but with an aim toward an object, is the kind of "original intentionality" Merleau-Ponty (1945/1962) writes about.

According to Dreyfus (2002), this is what is implied in the reference to Merleau-Ponty's concept of maximal grip. Rather than trying to explain what Merleau-Ponty (1945/1962) originally meant, we may just as well explore Dreyfus's ideas on the issue. As an initial definition of maximal grip, he proposes "the body's tendency to refine its response so as to bring the current situation closer to an optimal gestalt." Thus, optimal gestalt is a general description of what Searle (1981, 1983) would call condition of satisfaction (COS). However, while Searle assumed a specific content to any COS, depending on the intentional content (≈ goal) of the action, Dreyfus lets the situation define the specific COS. Thus, the experience we have performing the action will not have a mind-to-world direction of causation, that the actor has changed the world according to his goal. Rather, it will have a world-to-mind direction of causation, that is, we have accommodated our action to the circumstances of the world. Thus, the COS resides in the world, not in the mind.

This raises the intricate question of what makes us perform the action in the first place? Does it make sense talking about intentional actions without goals? In trying to respond to that question, in my view, Dreyfus (2002) does not go far enough. The intentional arc is meant to cover the history of the actor, which, in turn, explains his or her wishes, desires, and such, as well as acquired skills. It takes the place of what could elsewise have been a developmental account. Nonetheless, he provides most of the ingredients we need for a developmental and nonrepresentational account of intentionality (except for its social dimension, as already mentioned). I will, however, here propose a different theoretical framework for an account that, in its essence, is the same as Dreyfus's.

Accumulation of Experiences and State of Informedness

With the epigenetic approach—inspired by Gottlieb—that I have already mentioned, development takes place on several interacting levels of the organism, but also simultaneously on innumerable different temporal scales, of which we mainly need to distinguish between the ontogenetic level (the development of the individual organism) and the microgenetic level (the development of a particular action in a particular context). These levels are implicitly present in Dreyfus's (2002) account of skill development as an illustration of the intentional arc. We need however, to give a more elaborate analysis of the relationships between them in order to get a "maximal grip" on the temporality of their development.

The key issue is how one should understand what accumulation of experiences means. This is a microgenetic issue as well as an ontogenetic one. At a microgenetic level, how does one experience things and events through the continuous flow of impressions?[35] At the ontogenetic level, once one has managed this at the microgenetic level, how does one integrate such experiences into previous experiences such that we may learn something from it? The former question should find an answer in Dreyfus's argument on the maximal grip. The latter should be understood in terms of the intentional arc.

While Dreyfus (2002) starts with the intentional arc, exemplified through an account of skill development through different stages, I prefer to start at the microgenetic level, where the particular experience emerges. We have already touched upon the relationship between them through Kierkegaard's addressing the question of the particular and the general. It will be the same issue we address here in order to grasp the temporality of development, understood in an epigenetic sense.

I would like to introduce an auxiliary concept to account for the specific developmental changes at the ontological level that use to be expressed as memory, learning, skill, and the like, as well as it accounts for short-term memory and sensory register at the microgenetic level. Sverker Runeson and Geoffrey Bingham have, in an unpublished paper (1990) proposed the concept "state of informedness" to cover the fact that earlier experiences help us understand a present situation. In this context, "memory refers to whatever acquired characteristics of the organism that make possible the (re)establishment of states of informedness in the absence of perception" (Runeson & Bingham, 1990), whether at the ontogenetic or the microgenetic level. The basis for that is not a "memory storage" but the simple fact that earlier experiences have, to varying degrees, changed the fine-tuned structure of our body, in particular, of our brain. The latter is also stated by Dreyfus (2002, p. 381), with reference to Freeman (1991) in the terms of Hebbian learning theory: "the strength of the connections between the neurons changes on the basis of experience." He also points out what we here call the epigenetic principle, that these changes take place in an organism that has a history of earlier changes, such that the initial state of the organism, together with present input, will constrain what changes will take place.

This is true, however, not only at the level of the physiology of the organism but also (at least when it comes to humans) at the level of experiences. The concept "state of informedness" covers the experiential background for present experiences, without references to stored representations of earlier experiences. The next step is then to understand *how* past experiences influences present experiences. The challenge is mentioned by Runeson and Bingham (1990) as explaining "how to prevent the masses of non-reusable situation-specific information from clogging up the system."[36] The same problem is stated by Dreyfus (2002, p. 375) as "everything is similar to everything else and different from everything else in an indefinitely large number of ways." How can we then decide which similarities between past experience and present experience are relevant for the actual situation? One simple solution would be to have a representation of that particular kind of situation to compare present experience with. We want to avoid such a solution, though, because of the problems with the assumptions of representations discussed here above.

I suggest that this is a "psychological" way of stating the issue of repetition discussed by Kierkegaard as the distinction between the general and the individual. I understand Kierkegaard's rejection of the repetition as a rejection of a *confirmatory* understanding of the relation between the general and the individual, that is, that the general is based on the accumulation of a number of similar individual experiences, and each new similar experience confirms the general. Instead, Kierkegaard proposes that, to understand the general properly, it should rather be contrasted to the *exception*: "Undtagelsen forklarer altsaa det Almene og sig selv, og naar man ret vil studere det Almene, behøver man blot at see sig om en berettiget Undtagelse; den udviser Alt langt tydeligere end det Almene selv" (Constantius, 1943, p. 150; Kierkegaard, 2009, p. 78). *Thus, the general may be understood in the light of what it is not.*[37]

This negative understanding of the relation between the general and the individual receives, however, additional significance in the light of Dreyfus's adoption of Merleau-Ponty. The whole point of Dreyfus's use of Merleau-Ponty is to provide an argument for the understanding of intentional action *without* representation of a goal. To have a representation of the goal would be a case of *confirmatory* understanding of the relation between the represented goal as something general, for example, "drinking water" as the goal for my reaching for a glass of water, and the particular—individual—action I perform at this particular moment.

What would be the case of a *negative* understanding of the purpose of my intentional action? Dreyfus (2002) also (as me here above) is careful to point out that not all intentional actions are of this kind. At least humans are also capable of goal-directed intentional actions. We may imagine a goal and act in order to achieve that goal. However, those actions are rare cases where, for example, normal skillful coping with everyday situations is not possible or fail. Normally I would not consider taking a glass of water to my mouth and drinking from it to be a goal-directed action. Rather, it could be described, in general terms, as skillful coping, or in my terms, behavioral object-directed intentionality.

I will not try to give a faithful and complete account of Dreyfus's argument. For a full account, the reader is recommended the source article, Dreyfus (2002).[38] Instead, I will only indicate some key concepts and main points and then follow Dreyfus, taking advantage of his ideas, with the same freedom he draws from Merleau-Ponty's ideas.

The reason for my comparing Dreyfus's (2002) ideas on intentions without goals with Kierkegaard's featuring the exception in the understanding of the general, is that such intentional action is guided, not by the fulfillment of a goal, but by the *discrepancy* of the action relative to a perfect action or a norm for that same action. If I'm clumsy, my reaching for the glass of water will fail, and in order to avoid that, I will continuously readjust my reaching to achieve a grasp of the glass that is as close as possible to a perfect grasp. That means I somehow have a norm for what is a perfect grasp in that particular situation. It sounds like a representation, but Dreyfus is careful to point out how come that it is not. The norm is given in the world by the glass in that particular environment and situation, it is not "in the head" of the thirsty me. The condition of satisfaction for this action is successful grasping and drinking the water. This is regulated, not by a general idea of how to grasp and drink, but by that concrete particular situation, relative to which I perform my action. *It is what it is not* (its imperfection) that guides me to make it what it should be. It implies better and better discrimination of situations, paired with better and better motor action relative to the features of the situation. *Discrimination implies sorting out what it should not be, thus implying negation in Spinoza's sense* (see Melamed, 2012, and this chapter, p. 135 and note 37, about this concept of negation). Therefore, goal-directedness does not help you improve your action, as the goal is a fixed, ready-made state. Object-directedness, on the other hand, *does* help you improve your action, as it is the quintessence of your adaptation with the world.[39]

History and Actuality

This, however, does not explain how past experiences influence present experiences. For that, we need to understand, in the first place, what makes an experience an experience. That is a huge philosophical question, of which I will not make an overview. I will approach the challenge in a way that I think could best be understood from James's concept of "stream of consciousness," without, for that matter, digging into what he thought about experiences as such. If I may summarize the philosophical literature on the subject of the relationship between past experience and present experience, there are typically two issues to consider: how past experiences influence present experiences and how present experiences become past experiences. Commonly, experiences are understood as past, accumulated experiences. Although mentioning it explicitly only a few times, Bergson rather considers the past to actually be present, that is, it is only in present time that we have memories. So the memories make the past present: "La vérité est que la mémoire ne consiste pas du tout dans une régression du présent au passé, mais au contraire dans un progrès du passé au present" (Bergson, 1888/1965). Merleau-Ponty (1945/1962, pp. 83–84/pp. 69-70) also seems to consider as well the past as the future to only be actualized in the present (Bergson, 1939/2001, p. 269/p. 129).

Let us discuss our distinction here above (cf. this chapter, p. 128), between ontogenesis and microgenesis relative to Merleau-Ponty (1945/1962). In order for us to understand experiences at the ontogenetic level, how we learn and acquire wisdom through accumulation and organization of experiences, we need to understand, in the first place, how we acquire experiences, on a microgenetic level.

I had better use my own words to explain what is at stake here: It is rather our capacity to capitalize on a *history* of earlier experiences, our epigenesis, that has led us to where we are here and now at any moment of our lives. That is what ontogenesis is about. Our challenge is to understand how experiences accumulate, how they get from being individual experiences to becoming general experiences, and notably, how they become resources for future actions. A first step in understanding that is to understand how individual experiences emerge. Merleau-Ponty addresses the issue as an issue of the permanence of the object across all the perspectives we may have on it. As already mentioned, this chapter, p. 132, he seems to rely on the horizon to knit together all the perspectives. This would be equivalent to the role Dreyfus attributes to the intentional arc. However, while linking the past and the future as horizon (Merleau-Ponty, 1945/1962, pp. 83–85/ pp. 69–71) in a duration that very much resembles James's "stream of consciousness," the present is constantly destroyed by the "flow of duration" (cf. this chapter p. 125). It makes it a weak candidate for being a unit of duration, as a basis for a particular experience.

In another passage, however, Merleau-Ponty (1945/1962, p. 484/p. 424) makes a very strong statement about the present: "Il y a du temps pour moi parce que j'ai un présent. C'est en venant au présent qu'un moment du temps acquiert l'individualité ineffaçable, le 'une fois pour toutes, ..."[40] This is impossible to understand without having a notion of unit of duration. So the *sine qua non* of present experience (individual, unique experience) is an understanding of what is a unit of duration. Certainly, we can agree that the horizon helps us to understand the unique experience. But it does not tell how it emerges. In still another passage Merleau-Ponty (1945/1962, p. 90) gives us a clue : "Si l'objet est une structure invariable, il ne l'est pas *en dépit* du changement des perspectives, mais *dans* ce changement ou à travers lui." The invariable structure of the object is there, not *in spite of* our changing perspectives, but *in* this change, or *through* it.

James J. Gibson: Invariance Through Variance

This clearly is a precursor to one of the most important—maybe *the* most important—insights of J. J. Gibson, namely that we discover *invariant* properties in the environment through the variance created by the movement *of* an ambient optic array (i.e., movement of the body and the eyes in the body), and *in* the array (i.e., changes in the perceived environment)[41] (Gibson, 1966b, ch. X, and 1979, ch. 5). Gibson does not go as far as Merleau-Ponty, though: the ambient optic array does not tell the identity of the object, what object it is. It only tells what it is *the same*, through the changes in perspectives that follow from our own movements and the movements in the environment. These movements are actually necessary for us to discover the sameness of, for example, an object.

Gibson (1966) may have restricted this argument to ecological optics. However, he considered all our sensory channels—sight, hearing, touch, smell, taste, even proprioception—as perceptual systems (cf. Gibson, 1966b). He did not go so far as to also consider them together as one coordinated perceptual system. Invariance might, however, be found in a combination of different sensory channels, for example, hearing and sight (p. 272). Furthemore, the sense should be considered "as interrelated, rather than mutually exclusive" (p. 47). This idea has been developed furthered by Stoffregen and Bardy (2001)[42] in proposing that different forms of structured energy (that may be picked up by different sensory channels) are part of a *global array* involving all sensory channels available to a particular organism. While Gibson only discusses the ambient array of light, there is nothing in his writings that contradicts Stoffregen and Bardy's claims.

The most important implication for our concern here is that invariance is perceived in time, that is, the variance through which invariance is perceived is *variance over time*. The relationships between variance and invariance in ambient arrays are *temporal* (cf. Gibson, 1966a, 1979, p. 10)

As can be understood from the quote here above (Merleau-Ponty, 1945/ 1962, p. 90)" to "(Merleau-Ponty, 1945/ 1962, p. 106/p. 90), the object is for Merleau-Ponty *given* in the perception, and *not composed* by impressions.

He does not explain that further but rather takes it for granted, relative to any particular perception. As already mentioned, Gibson (1979) was more reserved about the concept of object. He rather prefers to talk about "*medium, substances,* and *surfaces*" (p. 16). When it come to perception, he proposes that perception does not consist of projected images from the world, but of the detection of invariance in the structure of the light. We allow ourselves here to generalize this proposal to any perceptual medium that reveals how the medium is structured relative to a point of reference for which the medium is ambient, similar to the eye for which the medium *light* is a structured ambient optic array: "The eye-head-brain-body system registers the invariants in the structure of ambient light" (p. 61).

Considering Gibson's (1979) reservations regarding the concept of object, I feel confident in making no principled distinction among objects, events, and "states of affairs," although that does not help us understand what an experience is. The invariance in ecological optics is pretty well described and extended to "all ambient arrays" (p. 163) as constant physical properties of the world that remain the same in spite of changing perspectives of the array or (I would assume) changing conditions for picking up the constant physical properties in the world.

However, Gibson's use of the concept is not limited to well-defined properties of the world, but what is an invariant might be highly contextualized: "What is invariant does not emerge unequivocally except with a flux.[43] The essentials become evident in the context of changing nonessentials" (1979, p. 73). What follows directly after this quote is even more interesting:

> Consider the paradox in the following piece of folk wisdom: "The more it changes, the more it is the same thing." Wherein is it true and wherein false? If change means to become different but not to be converted into something else, the assertion is true, and the saying emphasizes the fact that whatever is invariant is more evident with change than it would be without change. If change means to become different by being converted into something else, the assertion is self-contradictory, and the paradox arises. (Gibson 1979, p. 73)

This reminds us of Kierkegaard's discussion of the exception as the criterion for the general, and we need to sort out what this means. For that, we turn to Gibson's concept of *affordance*.

Subjectivity and Objectivity of Affordances

Altogether, I take these quotes from Gibson as support for a different view of the relationship between perception and action, than the one featured by Gibson himself. Instead of seeing perception as a means for control of action (cf. Gibson, 1979, as well as 1966b; Turvey, 1977), I will follow Gibson in his strivings to overcome the inherent duality of the perception-action couple, that he expresses in his efforts to explain affordances, notably that "an affordance is neither an objective property nor a subjective property; or it is both if you like. An affordance cuts across the dichotomy of subjective-objective and helps us to understand its inadequacy" (Gibson, 1979, p 127). This quote has frequently been mentioned as confusing. I here suggest that, on the contrary, it is very clarifying. We agree that there is information on the world to be picked up by perception. This does not mean that it is "objective." As mentioned here above, for Bergson, both what is subjective and what is objective are facts of consciousness relating us to the world, however, in different ways. We would expect objective experiences thus to be more truthful than subjective experiences. For Bergson, it is the opposite: Objective experiences are "constructed" by our intelligence, while subjective experiences might be intuited, in the way that more closely correspond to actual facts of the world. The point is that the actual facts of the world, relative to our experiences of them, are *temporal* facts, not spatial (cf. our discussion of invariance, here above). Comparing Bergson to traditional cognitive science views, we could say that Bergson's "intuition" might be "intelligence without representation." Only, I concluded that Bergson finally contradicted himself in writing about intuitions, as that could not be done without representing intuitions in words, thus transforming them into concepts of intelligence. Intuitions can be lived, but not mentioned.

From the present perspective, we may see that Bergson, he also, took the subjective-objective distinction for granted, which might be at the source of his contradiction. With his "confusing" statements, Gibson challenges the assumption of the subjective-objective distinction being taken for granted. But also he contradicts himself, by, on the one hand, describing affordances as "in a sense objective, real, and physical" (Gibson, 1979, p. 129), while other passages in the same book (Gibson, 1979, p. 126; considering them as the two poles of perception) support the opposite understanding, that also Gibson takes objectivity and subjectivity for granted.

Merleau-Ponty (1945/1962, p. 203/p. 174), like Gibson, tried to overcome the subject-object dichotomy. Gibson alludes to the origins of this dichotomy in the body (cf. Gibson, 1979, pp. 121, 141), while Merleau-Ponty explicitly takes the body-world relationship as the origin of both subjectivity and objectivity. Although Merleau-Ponty does not give the issue a developmental analysis, it is implied in his statement that it is not *the idea* of a body and *the idea* of a world that is at the origin of subjectivity, it is the body and the world "in themselves" that are inseparably linked to the subject (Merleau-Ponty, 1945/1962, p. 467/p. 408). This sounds similar to Bergson's problem with describing pure duration that cannot be described, but can only be lived. However, it is easier to understand Merleau-Ponty's verbal dilemma here than Bergson's. Bergson's is a dilemma for our experience, while Merleau-Ponty's is a reference to a reality we believe exists, but which cannot be grasped without *a fundamental relationship being established between myself as a body and the world as an equally real materiality*. A developmental perspective furthermore opens for an understanding of subjectivity as emerging from such a material body-world relationship. Most evidently, this emergence is temporal. It is made possible by the epigenesis of the body as a biological organism, which gives it a history. So, if the temporality of the body-world relationship is prior to subjectivity, one wonders what it means to say that subjectivity *is* temporality (cf. this chapter, p. 129). It would be better to say that subjectivity emerges from the development of a biological organism in relation to its environment (which, of course, is temporal) .

This relationship is characterized by asymmetry, such that *the effects of the world on the body are included in the body's effects on the world*. We need a theory for fully explaining that, for which I here below will suggest Dynamic Systems Approaches. For the moment, it suffices to say that this means that in this way the body's effects on the world more and more take the form of the body's acting on the world, which in turn defines the subject-object relationship. The subject-object relationship in turn has been taken to be basis for temporality. That sounds circular. The contradiction in Merleau-Ponty lies in his discussion of temporality, as based on subjectivity, and an analysis of the body as the origin of subjectivity, which does not take time into consideration. It will be our developmental account of the relationship between body and environment that introduces *temporality at the level of the body*, as a basis for subjectivity. Neither Merleau-Ponty himself, nor Dreyfus, provide any support in this respect.

DYNAMIC SYSTEMS APPROACHES

I rely on the ideas of the Nobel laureate Ilya Prigogine (e.g., Prigogine & Stengers, 1984) to explain three central concepts that we need to use for pursuing the present line of thought. Let us start with the basic fact and the fundamental law relative to which Dynamic Systems Appproaches (DSA) should be understood, which is the *Second Law of Thermodynamics*: the universe as a whole, and most things in the universe, have a tendency toward equilibrium, which is achieved by *entropy*, that is, differences are leveled out and structure is decreased or dissolved. Think about a hot cup of coffee as a system in a cold room as environment. There is an exchange between the system and its environment. This exchange follows the Second

Law, which means that the room and the cup of coffee will end up having the same temperature.

However, there are cases that seem to defy the Second Law. For example, living organisms are proof of *increase* of structure, rather than decrease. With his theory of "dissipative structures,"[44] Prigogine (Prigogine & Stengers, 1984) proposed an explanation of these observations, such that matter, under certain circumstances, may self-organize into ordered structures, establishing an exchange relationship with its environment, for example, exchange of energy and matter in different forms. The emergence of life on earth and the development of living organisms can be considered as a paramount example of dissipative structures, something that is also reflected in the authorship of Prigogine, in spite of his own field being chemistry, not biology.

A dissipative structure defies the Second Law, by, instead of leveling out differences, increasing differences and complexity. That means that the structure literally sucks out energy from its environment in order to maintain, or even increase, its difference from the environment. Thus, it develops itself away from thermodynamic entropy. However, it tends to stabilize itself, maintaining its structure in an equilibrium *far from thermodynamic equilibrium*.

In the fields of psychology and motor behavior science, Prigogine's theory has been associated with the mathematical field of *Dynamical Systems Theory* (see, e.g., Kelso, 1995), inspiring a controversial theoretical approach that is gaining increasing influence. Esther Thelen and Linda Smith (1994), among others, have furthered these ideas under the denomination Dynamic Systems Approaches (DSA), with particular emphasis on motor development. The plural form of this denomination indicates a great variety in how to understand what is a dynamic system. A main dividing line is between a strict mathematical approach and a looser "metaphorical approach." I will use the latter to address the issues of this chapter.[45]

System and Environment

I will now use Dynamic Systems Approaches (DSA) to try out a theoretical explanation of the development of the body-environment relation as the basis for subjectivity. In the version of DSA presented here, it is up to me to define what is system and what is environment, depending on which research question I address (van Geert, 2003, p. 656).[46] Normally, one would define the organism as system and the physical and social world of the organism as the environment (cf. *autopoiesis*, as proposed by Maturana & Varela, 1980). *In our discussion, I will, from here on, change my definition of "environment" from the conventional that distinguishes between organism, animal, individual and such, and environment, to a DSA definition, where "environment" will imply everything that is not part of a system, but with which the system have an exchange relationship* (van Geert, 2003, p. 656). Practically, at our level of analysis here, this will, in most cases, mean that the organism *and* something in the world will together be considered as a system, while the rest, to which the system, as such, has an exchange relationship, is to be considered as environment. As a replacement for the usual understanding of "environment," I propose "surroundings," a term rejected by Gibson (1979, p. 43) as being too vague. I think it will do. It is organism-centered, so it is a part of our world, which contains elements that enter into exchange relationships with the organism (as a dynamic system) or that may be part of the system. In the latter case, I will speak about an "animal-slice-of-surrounding system," where "animal" may be replaced by any other organism-relevant word that is appropriate for the context.

I suggest (Vedeler, 2012) that perceiver and perceived are both system's components, with veridical perception as the vital link between the two, that is, a direct coupling between a body and some part of the world. From this, intentionality will emerge as a system's property of the body-slice-of-surroundings system. Intentionality, in the generic sense of "object-directedness" (Vedeler, 1991) will define the system as an action, from which subject (actor) and object (acted-upon) will be a description of the perception-action coupling between the organism part and the slice-of-surrounding part of the system.

This could become more specific by using Dreyfus's (2002) analysis of the maximal grip. There is a caveat, though. For Dreyfus, as well as for his target, the philosopher John Searle, the distinction between the action and the world is assumed and not considered as necessary to explain. My endeavor here is to try to explain in what consists this distinction, which amounts to explaining the subject-object distinction.

Behaving organisms interact with their surroundings as a way of surviving. Survival is enhanced by there being some optimal body-surrounding relationship. This relationship is, however, constantly under the influence of changes and stability, as well in the organism as in the surroundings. In Dynamic Systems' terms, the organism could very well be considered as the system, maintaining its optimal organization through exchange relationships with its surroundings, which in this case would be the system's environment.[47]

However, the separation between organism and world should not be taken for granted: Elements from the world (surroundings) may be part of the system. For example, when a frog has spotted a fly and puts itself in position to catch it, the fly has switched from being part of the environment to be part of the system. This would usually be considered to happen first when the fly is caught and eaten. In my argument, it happens already when the fly is spotted. The environment would then be whatever conditions in the surroundings that might influence the behavior of either the frog or the fly. Under the influence of its environment, the frog-fly system will self-organize into a catch or into an escape for the fly. From the point of view of the frog, it may be considered as an affordance to eat or an intention to eat, dependent on whether the focus is on the perception or on the action part of the behavior of the system. I elsewhere argue (Vedeler, 2012) that they are flip sides of the same relationship.

While the frog and the fly have different optimal conditions for survival, they are here seen as part of the same dynamic system, the organization of which depends on its exchange relationships with its environment, whether, for example, movements in the environment disturb the frog and/or the fly, in which case the system might dissolve, by the perception relationship being broken. If, on the other hand, the perceptual relationship persists and is integrated with motor adjustments of the frog, the system may attain an attractor basin[48] that allows for a successful catch. If and when there, in a fraction of a second, the system jumps into another attractor that is the catch itself.

The same story would go also for any skillful human action that is not based on conscious deliberation. We humans, and maybe also some primates, are capable of conscious deliberation, where we mentally manipulate representations of the world and of our relationships to the world. This I have already pointed out several times in the present chapter (see also Vedeler, 1991, 2012).The epigenesis of this skill is a story of its own that will not be told here. I mentioned here above Heidegger's distinction between two ways of relating to the world, "ready-at-hand" and "present-at-hand." I suggest that the latter is the basis for conscious deliberation, while the former is what Merleau-Ponty (1945/1962, p. xiii/p. xviii) described as "operative intentionality." It is also what Dreyfus (2002), with reference to Merleau-Ponty, calls "a more basic form of intentionality" and what he analyzes in terms of skillful action. Our ready-at-hand relation with the world is also what I propose to consider as the case where the organism makes a system together with a slice of its surroundings.

If we step down from the animal-slice-of-surroundings system to an important underlying system, namely, the brain, Dreyfus (2002), as well as Louise Barrett (2011, ch. 8) finds neuro-physiological evidence for its organization as a dynamic system in the research and theory of brain function of Walter Freeman (1991, 2000). His work very nicely provides a theoretical formulation on the functioning of a brain as a dynamic system. I'm not the right person to account for Freeman's groundbreaking neurobiological findings and insights. I will only try to extract their consequences for the questions on the temporal relationship we entertain with the world. I will furthermore exercise the same liberty in understanding toward Freeman as toward Dreyfus. Also, I will focus on a developmental and epigenetic perspective not explicit in Freeman's writing, while, of course, being vital to the understanding of ontogenesis, as well as microgenesis.

Thus, a system may go from one situation of relative stability to another, passing by situations of instability, until it, possibly, will cease to be a system and merge with its environment. In order to pursue this line of argument, we need to use three central theoretical concepts used by Dynamic Systems Approaches.

Three Concepts: Self-Organization, Emergence, and Attractor

I will here give a short explanation of these three central concepts of DSA.

Self-organization, in the context of dynamic systems, means that the relationship between the system and its environment is such that structure in the environment with which the system has an exchange relationship does not constrain or determine the structure of the system. The exchange relationship is vital to the existence of the system, without which it would simply decay and be dissolved, according to the Second Law. However, the shape of its structure is independent of structure in the environment or in the exchange relationship. The structure of the system is given only by self-organization based on the properties of the components internal to the system and their relationships.[49] However, if a component from the environment is included as a system's component, then it will be part of the organization of the system.

Attractor is the concept from Dynamic Systems Approaches that is most relevant for our present discussion. Replacing Prigogine's (Prigogine & Stengers, 1984) concept of "dissipative structure" with "dynamic system," we may rephrase our frog-fly example" (this chapter, pp. 143–144) as *a dynamic system defying the Second Law, by, instead of leveling out differences, increasing them.* Thus, dynamic systems develop themselves away from thermodynamic entropy, stabilizing themselves *far from thermodynamic equilibrium*. Such equilibria far from thermodynamic equilibrium are in Dynamic Systems theories called attractors. Attractors are preferred conditions (or "states"[50]) of the system, which give stability and low consumption of energy. Systems that are far from attractors are unstable and consume lots of energy (lots of exchange) to maintain themselves as systems. The alternative is to disintegrate, that is, return toward entropy. Under the influence of the exchange relationship with the environment, a system can switch from one attractor to another, passing by a moment of instability. This is called a *phase shift*.

Let gait be a system for the body of a horse. Horses may have several different gaits, but for the sake of illustration, let us here restrict ourselves to three of them: walk, trot, and gallop. These three gaits of horses are to be considered as examples of attractors, when you consider the horse as system and the surroundings as environment. Whether controlled by a rider, or as a member of a group of horses, the horse will have different demands on speed in different situations. Speed will thus be the "control parameter," the exchange factor between the system and its environment that makes it fall into one attractor or another. The horse naturally shifts gait dependent on the speed. While speed is the control parameter, the brain, the anatomy of the horse, and gravity as well, are related to each other in different ways under different speeds, thus constituting an "order parameter" of the system, which changes under the influence of the control parameter speed. It is an example that, even in the case where we initially defined the system as a horse, we here find that this was wrong: Gravity, as a resource from the surroundings, needs to be included in the system in order for us to account for the gait in relation to speed.[51] The weight of the legs and the momentum created by their pendular movements is taken advantage of in order to save muscular energy. Which gait will then depend on, among other things, what is the most economical use of leg movements, considering gravity and the horse anatomy (bones, muscles, joints) at different speeds. In this case, shift of gait, that is, shift to another attractor, is swift. However, if you force the horse to maintain a gait that naturally does not fit with the speed, for example, in trotting horse races, you will have a higher expenditure of energy and a more unstable attractor (that may be disorganized, or shift into gallop), at least until the body of the horse (brain-muscles, etc.) has self-organized into an extension of the trot attractor. Through training, the horse may develop a skill to trot at higher speeds, that is, the system develops toward a more advanced structure than what is initially "natural" for the horse-gravity system.[52]

The important lesson from this example is that the attractor does not "reside" in the brain. It is not an ideal state, a norm, to which the actual gait should be compared.[53] It is distributed over all the components of the system and consists in their unity as a system. Furthermore, this unity is provisional, assembled for the occasion on demand from the situation. It denotes a dynamic equilibrium within the system, that is, a highly structured equilibrium, far from, and very different from entropy, the equilibrium according the Second Law. Thus, it is not the brain that controls the movements, the movements control themselves, and the brain activity is part of these movements.

In the language of dynamic systems, such situations are called attractor basins.[54] In my description of dynamic systems, I here differ from the normal terminology of describing systems in attractor basins as being in attractor states. I prefer to avoid the word "state," as it may convey an impression of absence of change, which is never the case for a dynamic system. Only, in an attractor basin, the system is in *relative* stability. Not only is the system disturbed while in a particular attractor basin, it also may dwell in a constantly changing "landscape" of attractor basins and their opposites, peaks of instability.[55] Interaction with its environment will push it around, while it by itself searches stability. *Ultimately, I consider this restlessness of dwelling in attractor basins to be a natural basis for the temporality of a durational unit.*

This is, among other things, how the brain works. Or rather, this is a way of understanding how the brain works. Instead of trying to prove that this description is true, I prefer to make sense of the brain's role in everyday functioning in relation to other parts of the body and in relation to the world. The same description can be used to describe the functioning of any neuron or other cell in the body, as well as the functioning of the animal in its surroundings.

While describing biological cells and assemblies of cells, as well as body organs and the whole organism as dynamic systems, as well as other dynamic systems, we need also account for the relationships *between* such systems, in particular at different levels of analysis. For that, I would like to use another concept central to Dynamic Systems Approaches, namely, *emergence*. Dynamic systems in attractor basins will sometimes, or rather often, have properties that cannot be reduced to the elements or to the sum of elements in the system. Such systems will carry with them these system's properties when they themselves become elements in other systems, at a higher level of analysis. Thus, system's properties will become element properties in higher-order systems and thus be factors in the self-organization of the higher-level system. As an example, a brain system in a particular attractor basin might, as an emergent property of the system, give the organism an *experience* of a particular situation. This experience might perhaps be a feeling of satisfaction with the present situation, for example, the frog first felt well with the fine-tuning of its frog-fly system, such that it ventured a catch. Assume that it failed. In that case, this experience might in turn lead to a change in what Dreyfus (2002) would call the energy landscape, where the positioning attractor basin is slightly reorganized in view of the next opportunity, and where the frog might have improved its fly-catching skill.

Dreyfus (2002) does not mention this relationship between systems levels, while Freeman (2000) himself does, although not with the same theoretical implications as here. As I see it, it has huge implications for learning and development. Epigenesis, according to Gottlieb (2003), covers all levels of human functioning, from genes to culture. Furthermore, hierarchically organized, they influence each other (see Figure 7.1). The concept of emergence nicely accounts for this network of influences across levels of analysis.

DURATION AND EXPERIENCE

From here we may now return to the key concepts of this chapter, namely, duration and experience. Our first endeavor was to find an action-based delimitation of a unit of duration. I now propose to define the unit of duration in terms of a dynamic system's dwelling in neuronal attractor basins as described by Dreyfus (2002), with reference to Freeman,[56] *that can be experienced*, at least for humans. However, it need not be experienced, and for animals, as far as we know, experience might not the same thing as for us.

Thus, the *dwelling in an attractor basin* is here considered as an action-based unit of time that in no way corresponds to clock time. *It could be a suitable candidate to be considered as a Bergsonian durational unit*. In terms of clock time, it could last a fraction of a second, or it may last several hours, for example, if you are enjoying a whole symphony, uninterruptedly and without particular attention to any of its part. Attention is an obvious property of the system that favors experience, as a pure consequence of the organism-slice-of-surroundings coupling.

A consequence that may stir some objections is that this view on durational units would allow for slices of clock time where the system is between different attractor basins, in transition or in an unstable situation of the system-environment relationship, and this means disruption in the continuity of time, so important to all the philosophers mentioned here above. However, these moments of instability and possible transitions do not imply a standstill. On the contrary, they are more restless than the dwelling in attractor basins, as disturbances have pushed the system out of its equilibrium, and it will need a new equilibrium in order to persist as a system. So while disrupting durational continuity, they will still be highly temporal by the changes they imply for the system. Thus, they are highly relevant for the temporality of *microgenesis* in relation to clock time. As pointed out by Wheeler (2005, p. 52), biological dynamic systems have a "rich temporality" as opposed to the computational models of cognition that have an "austere temporality," that is, while in biological systems, the time it takes for such transitions from one attractor to another matters for the outcome of the transition, this is not the case for computers, whether the computer is slow or fast. In the computer, only the *steps* in the computational processes matter, while in biology, change processes take more or less time.

Thus, there is temporality in all areas of biological life. From a perspective of life sciences, Gottlieb's figure (see Figure 7.1) is universally applicable. Assumptions of emergence in dynamic systems account for the mutual influences among the levels. I have suggested that durational units were to be understood as *possibly* experienced. We are highly anthropomorphizing when we assume experiences among animals (Barrett, 2011). By linking *possible* experiences to durational units, as defined here, we have a link from action units at a behavioral level of analysis to dynamic system's situations[57] at an underlying neurological level, from which experience may emerge as a systems property of the neurological system.

History of Durations and Experiences in Epigenesis

So far, we have considered single durations and thus single experiences. From here, we have now to consider the epigenetic perspective. Freeman (1991, 2000) provides empirical evidence for changes of neuronal connections according to Hebbian learning rule, that neuronal connections are strengthened or weakened according to the "experiences"[58] of the organism. In terms of the dynamical organization of the brain, that would mean the organization of experiences into attractor basins? Freeman does not mention, but would certainly not object to the assumption that the brain of any organism has some organization already from birth, implying "innate" attractor landscapes with different attractor basins within which neonate baby-slice-of-surroundings systems with perception-action couplings may fall. Modern infant research has revealed a lot of infant inborn abilities, for example, imitation (Kugiumutzakis, 1985) to which this model could apply.[59]

However, it is not the complexity of the brain that is at stake in the first place; it is its relative flexibility, where the human brain is more flexible than the brain in most other species (Barrett, 2011). From its prenatal brain and body organization, the child develops further skills that may very well be described in terms of attractor landscapes self-organized through changes in neuronal connection strengths, according to Hebbian learning theory. Further perception-action couplings will fall into one or the other of existing attractor basins, which may in turn be modified according to new perception-action couplings, or, as often observed in video recordings of young infants, turn out as chaotic situations, which may be described as unstable child-slice-of-surroundings systems that find no appropriate attractor basin for the situation.[60] According to our definition of a durational unit, that would be a transition without duration and thus a moment of nonexperience. The transition ends, for example, if the chaotic situation of a baby turns into a stable situation of screaming and agitation, exhausting the child such that it falls asleep, again a transition to another attractor.

However, as pointed out by Freeman (2000, p. 87), but here stated in my own words, such situations may also open the system's new self-organization, thus creating new attractor basins. This is a process over time that takes advantage of the (relative) flexibility of the organism. Some species are pretty much wired-up without much change due to experience, while others are flexible, that is, capable of extensive self-reorganization of their relationships to their surroundings on the basis of experience. Chaotic situations might be fatal to nonflexible organisms, while it may provide opportunities for "creative solutions" to flexible organisms. This is illustrated by ingenious analyses of concrete examples in Barrett (2011).

The epigenesis of organisms from different species may thus have very different trajectories. They all have in common, however, that what they may be capable of and what they may do at any particular moment in their life will depend on how and in what circumstances they have developed up to that point. Thus, every organism has a history and is dependent upon its history. At any moment, this history defines the state of informedness of the organism with which it meets the world (Runeson & Bingham, 1990, cf. this chapter, p. 134).

For many species, and for humans in particular, the accumulation of experiences plays an important role for what is their state of informedness at any moment. Such accumulation, however, is particularly challenging to explain. In order for this accumulation to be ordered, and not a simple compilation of single experiences, they need to be based on similarities and differences. Kierkegaard addressed the issue in asking about the possibility for repetition and the relationship between the particular and the general. He could not, however, give any conclusive answer. Cognitive psychology has tried to handle the issue in terms of concept formation and has ended up with networks of representations that satisfy the state of the art in the field. The difficulty has been to explain how this network is established and developed in the first place. Being faithful to the basic assumptions of a representational theory, Fodor (1981) concludes that they will have to be assumed as innate, which is absurd. Connectionist networks theories have tried to understand the representations in terms of "distributed neural networks," thus, to the difference from the pure computational approach by traditional cognitive psychology, trying to relate their approach to actual biological brain structure.

Common for all these psychological explanations of concept formation is their *positive* understanding of representations. Each concept should imply certain features for experiences to be subsumed to one of them. Thus, it is a test whether an experience fulfills the criteria for a concept or not. The *non*representational approach of Dreyfus (2002) is based, *not on the similarity* of an experience with other experiences and their presumed common representation, *but on the dissimilarity of an experience, its difference* *from a criterion.* In terms of the maximal grip, it is the difference between the best grasp and the actually performed grasp. The point is that "the best grasp" is not specified (e.g., in a representation); you don't know beforehand what is the best grasp. You only know, here and now, that in this case, *it is not a good enough grasp.* It is *what it is not* that guides your action, that is, the criterion is negative, not positive. There you have the role of the exception to understand the general that Kierkegaard writes about. That is also what makes the repetition futile. You don't get anything out of an exact repetition of anything, if by chance you ever experience it. It will only be something by being different. That is also the basis of learning.

We have hitherto discussed the temporality of microgenesis in terms of the temporality within a durational unit. We now need to address the question of how durational units merge with previous durational units. As mentioned here above, past experiences may be recruited to complement the present state of informedness of the organism (Runeson & Bingham, 1990). Such experiences at a microgenetic level will, however, also merge with the whole history of experiences of the organism, along with nonexperiential changes and stabilities, up to the present experience. This is epigenesis in its essence. History is thus a resource for the organism from which skills and knowledge may be recruited when relevant for the present situation.

THE PROXIMAL PAST AND FUTURE

In Dreyfus's terminology, the past and the future, relative to the present, would be described in terms of the intentional arc, which describes "skills ... stored[61] as dispositions to respond to the solicitations of situations in the world" (Dreyfus, 2002, p. 367). His quote from Merleau-Ponty, "The life of consciousness—cognitive life, the life of desire or perceptual life—is subtended by an 'intentional arc' which projects round about us our past, our future, and our human setting" (Dreyfus, 2002 p. 373; cf. Merleau-Ponty, 1945/1962, p. 136), certainly makes the point, notably also including the future. More precise, however, would be to use Runeson and Bingham's (1990) *state of informedness* concept that links the present to the *actual* situatedness, as well as restricting the past to the *relevant* past.

The future is often, in temporality discussions, mentioned together with the past without taking into consideration an essential difference between them. The arrow of time goes from past to future. Past experiences have had an impact on the organism, while future experiences are still to come. Therefore they cannot play the same role in present duration, as defined here. One normally distinguishes between a proximal past and future, as opposed to a distal past and future, assuming clear differences between proximal and distal. The proximal past is what just has happened, the proximal future is what is imminent to happen. Peirce and James consider proximal both past and future to be part of the present, sharing each half of the present moment.

Bergson would not say exactly that. However, somehow, he says the same thing. Not quite the same, as the "time-space," that is, the duration it occupies, is not proportionally defined. Instead, he means that the proximal past and future are part and parcel of the durations they participate in, as movement and change—that defines duration—could not be described without inclusion of all three temporal *modi*—past, present, and future—or rather without merging past and future into the present.[62]

Actually, this is not very different from Heidegger's characterization of *Dasein*[63] (being-there) as caring, which is "defined" as "Sich-vorweg-schon-sein-in (der Welt) als Sein-bei (innerweltlich) begegnendem Seienden" (Heidegger, 1927/2010, p. 249/p. 240, cf. also p. 192/p. 185), in English translation: "being-ahead-of-itself-already-being-in (the world) as being-together-with beings encountered (within the world)." Neither is it different from Merleau-Ponty's understanding of the horizon:

> Chaque présent saisit de proche en proche, à travers son horizon du passé immédiat et de futur prochain, la totalité du temps possible; il surmonte ainsi la dispersion des instants, il est en position de donner son sens définitif à notre passé lui-même et de réintégrer à l'existence personnelle jusqu'à ce passé de tous les passés que les stéréotypies organiques nous font deviner à l'origine de notre être volontaire. (Merleau-Ponty, 1945/1962, p. 100/ pp. 84–85

He conceives a proximal horizon that gives continuity to the moments at the same time as it gives meaning to the past that is integrated with my existence as a person. What is common for all philosophers mentioned here is that the future carries with it possibilities that are not yet realized.

Kierkegaard: "In the Blink of an Eye," *das Augenblick*

Bergson is rather disengaged as to *what* possibilities are realized. Our inclination to anticipate the future is based on past possibilities having been actualized. However, the future is fundamentally indeterminate, as the possibilities of the future are not solely dependent on possibilities and actualities from the past. Possibilities are constantly in the making during the flow of time. So, taking an intellectual view (*sic!*), what is possible at time $t_{(t+1)}$ need not be the same as what was possible at time $t_{(t)}$. Anticipation, based on past experiences, is ubiquitous in the animal world. As such, it is part of our duration (cf. last quote from Merleau-Ponty here above), or durations, if we allow ourselves a succession of unities in multiplicity. However, the field of possibilities ahead can never be defined in advance, as the continuous changes in the flow of time will continuously change this field. Bergson labeled this "blending into the future" creativity: "dans la durée, envisagée comme une évolution créatrice, il y a création perpétuelle de possibilité et non pas seulement de réalité" (Bergson, 1934/2007, p. 12/p. 10). Thus, the immediate future is implied in the flow of the duration.

For Heidegger and Merleau-Ponty, the future *matters*. What the future offers matters for the subject; it will have consequences for its further life. This is even stronger for Kierkegaard, although he, in the *Augenblick*, sees *pure presence*, without connection to neither the past nor to the future. However, it matters for the future. Namely, the *Augenblick* connects time to eternity. We are now lured back into the far-reaching ambitions of our philosophers that we wanted to avoid, in order to concentrate on the temporal relation between man and the world. However, we need to allow an exception for Kierkegaard here because he declares that the *Augenblick* is the ambiguous moment where time and eternity touch each other which, in turn, *constitutes* temporality:

> Øieblikket er hiint Tvetydige, hvori Tiden og Evigheden berøre hinanden, og hermed er Begrebet Timelighed sat, hvor Tiden bestandig afskærer Evigheden og Evigheden bestandig gjennemtrænger Tiden. Først nu faaer hin omtalte Inddeling sin Betydning: den nærværende Tid, den forbigangne Tid, den tilkommende Tid. (Haufniensis,[64] 1844, p. 96; Kierkegaard, 1980, p. 89)

Only with temporality will the past, present, and future get their meaning. I would like to find out what that means from a psychological perspective.

If it were not for the *Augenblick*, Kierkegaard writes on the next page, eternity would appear behind as past time.[65] I see here a common ground for Kierkegaard, Heidegger, and Merleau-Ponty; that they consider the future as something that requires your action, your decision, your responsibility. This is best expressed by Heidegger as *ecstacy*[66] (Heidegger, 1927/2010, p. 329/p. 314), not in the sense of some everyday Being becoming frantic or losing oneself in emotion (e.g., orgasm), but as the unity of past, present, and future in temporality. According to Kierkegaard, the *Augenblick* catches this unity in a timeless moment and thus constitutes temporality.

Instead of seeing this as an attempt to make the transcendent immanent,[67] I would like to draw a parallel between Kierkegaard's understanding of the *Augenblick* and Bergson's concept of duration. What could be common for a moment that passes within the blink of an eye and duration, in the sense of time that lasts? Kierkegaard's reference to eternity might give a clue: In eternity there is no change. The *Augenblick* catches this glimpse of eternity. In duration, there *is* change, but change doesn't matter. I suggest another way of saying the same thing, for both the *Augenblick* and for duration. Those are moments of time in which you experience invariance *through variance*. That was Gibson's (1966a, 1966b, 1979) idea. Sounds far-fetched? It is.

However, we need to sort out some things. For Gibson, it was *properties of the world* that were the same, in spite of the flow of changing light, sound, and the like that excite our senses. Through the changes in the lower order variables in ambient array, "higher-order variables" reveal themselves to the perceiving organism as stable properties of the world. Unfortunately, Gibson was not as good at explaining how it was possible for us to extract these higher-order variables from the ambient (moving!) optic array, as he was explaining in what they consist (e.g., constant proportional ratios through changing lengths and angles on the changing retinal image.) The concept of *affordance* was an attempt to describe such higher-order variables in terms of animal-relevant properties of the world. This concept has stirred an ongoing debate on what it means and presupposes (Vedeler, 2012). I have suggested affordances to rather be considered as emergent properties of animal-perceived slices of relevant surroundings system (Vedeler, 2012). As such systems are temporal, affordances will also have to be considered as temporal, in stark contrast to what Gibson wanted to convey through the concept.

Therefore, we will have to override Gibson, borrowing his concepts and ideas, but transforming them beyond recognition in order to be comparable to Bergson's concept of duration and Kierkegaard's concept of *Augenblick*. The main point of Gibson (1966a, 1966b, 1979) was that perception takes place over time. While the properties in the world might not change, this lack of change was perceived over time, *through* change. Bergson wanted duration to mean constant and continuous change. Not a quantitative change that could be measured, but an uninterrupted qualitative change. For example, if we have a continuous change from a white surface to a black one, we will not experience this change as continuous, but as stepwise, where you attribute different shades of gray to the surface (Bergson 1888/2001, pp. 24–37/pp. 42–74). According to Bergson, that is because of our tendency to conceptualize our real durational experiences into *intellectually* conceived categories. Gibson's (1979, pp. 86f) view on shades of gray is radically different: First, normally there are no ecologically valid situations where you have a homogeneous illumination of a white surface continuously changing from strong illumination to no illumination, defined as white and black. We perceive the same shade of gray under varying illumination for the same object, dependent on other properties of the object and the illumination of its environment. Therefore, that we perceive the same shade of gray is not an intellectual classification. It is the extraction of *invariant* information from *variation* in the structure of the information in, for example, light.

This has huge consequences for our understanding of duration. Invariance is not an intellectual alienation from the stream of consciousness, but part and parcel of this stream. We may then use this insight from Gibson (1966a, 1966b, 1979) to give more substance to the idea of durational unit than we can get from Bergson (about the problem of identifying durational units in Bergson's philosophy, see this chapter, p. 123). A durational unit can be considered as a temporally extended invariance in a situation where perception is a relationship between a perceiving animal and a slice of surroundings for that animal; namely, the perception of invariance through variant structure in an array that is, obviously, extended in time.

However, while Gibson and his followers were good at explaining the relationship between variance and invariance in the ambient optic array, they could not tell anything about the skill to pick up or extract the invariance (Gibson allowed for both expressions), except that it was just a discriminatory skill. With the position that perception normally is for action, and that perception itself is action, I am here committed to trying to understand how perception of invariance is related to action. We concluded from Kierkegaard, Heidegger, and Merleau-Ponty, that the temporality of action is its "ecstacy" toward the future, that is, how the action gets beyond itself in the challenge and task that it faces. Notice here, dear reader, that I am not concerned with the subject of the action. That comes later. Neither am I concerned with the animal/human as such, to the difference from our philosophers. My only concern is the self-assembly of the animal-slice-of-surroundings system that the task has triggered. You may ask, What task? I answer, Any task!

Units of Duration as Attractors

This is what I propose to be the unit of duration: The dwelling in a particular attractor of a perception-action system, that is, an organism including some slice-of-surroundings in the system, in a particular situation, at a particular moment of time and with a particular epigenetic history, for example, a 2-year-old horse trotting for himself on a sunny day on a green meadow. The same goes for human gait, walking or running. As mentioned here above, the length in clock time of such a durational unit may vary from a fraction of a second to several hours of uninterrupted uniform activity (cf. Richardson, 2008).

The relationship between durational unit and experience, however, adds considerable complication. The human—at least as described by Dreyfus (2002), accounting for skilled action—may have his or her walking, running, biking, and lots of other skills, "automatized," that is, may let the body "run" the skill, while attention and experience is oriented elsewhere. I have earlier in this chapter assumed, temporarily, that most animals, except humans, do not have particular live and conscious experiences of their skill-related attractor situations. Nonetheless, in their epigenesis, they accumulate and refine skills, and they thereby extend their realms of attractors by the development of their brains[68] and their bodies (e.g., muscles). Our philosophers here above did not consider animal temporality. However, I take all biological processes to be temporal, in the sense that they all require time to be carried out and to develop (Wheeler, 2005). Epigenesis is a temporal process. If I define durational units as attractors, that is supposed to cover animal as well as human duration.

How then are durational units related to experiences? Dreyfus (2002) describes human skill acquisition through five stages, where in the last stage, the skill may be exercised without awareness. One particular advantage in human skill acquisition is the ability to follow instructions. You cannot instruct animals, you may reinforce them, you may train them to develop new skills from old ones, but you cannot tell them what to do in the way you instruct humans. While I could leave open whether animals have a different kind of experiences, some human experiences seem to be based on representational thinking, in which only humans excel.

Such a stance on experience needs to be related to other meanings of the word in order to delineate what we mean by the concept. It is normally taken commonsensically, also among our philosophers here above, without the need to define it. The most frequent one is the historical (or rather, the epigenetic) concept, how past experiences are important for the present situation. Anything that we have learned, we have past experiences of it in the learning process. The other is awareness of the here-and-now, the unique experience of the moment.[69] Neither of these commonsensical understandings of the word assumes that it is different for humans, as compared to other animal species. However, neither do they give any hint about what past or presents experiences "look like," what constitutes a unit of present or past experience.

One could try to delineate experiences from nonexperiences by adding awareness as a requirement. However, past experiences that one has been unaware of may still contribute to your present skill and/or to how you experience the present situation. Freud was not completely wrong: Unconscious experiences are real and will influence future actions and future experiences. So it would be easier if we stayed with durations, not bothering about experiences. But then we would have abandoned our philosophers, who all deal with human experiences. We need now to reconnect to them, hoping that they will help get an idea of what we could mean with "experience" in this context.

Peirce (1892) hardly uses the word, preferring "idea" as something in the mind. James (1884) has an interesting passage that sounds a precursor of our epigenetic stance here

> How could we, when the structure of our brain itself is continually growing different under the pressure of experience? For an identical feeling to recur, it would have to recur in an unmodified brain, which is an impossibility. The organ, after intervening states, cannot react as it did before they came. (p. 11)

It also touches upon the theme of repetition, featured by Kierkegaard. Bergson himself (1888/2001) writes about past experiences in a common-sensical way, however, hardly on present experiences. Instead, when not about past experiences, he very often means experiments. Other words are often used to denote what we would have called experiences.

Among them, the most interesting is the expression *durée vécue*, that is, "lived duration," something perceived by consciousness (Bergson, 1888/2001, p. 86/p. 193). One may comment on Bergson's conception of duration that it is irreversible and at the same moment may not appear twice. That is why the idea of durational unit, which I somehow have forced upon Bergson, is in a profound sense incompatible with his conception of duration. Namely, defining the content of durational units opens for comparing them, and for repetition, something Bergson considers also to open for the spatialization of duration. Here again we have the dilemma of Bergson's approach: Either you live the duration and you will not be able to describe it without destroying it or you fixate it, depriving it of its essence and distorting it. Experience of duration thus would leave its traces, to be retrieved after the fact from memory. That is what Bergson calls pure memories, to be distinguished from memory in the form of routines, habits, skills, and the like (see here above). While Bergson only has concern for the former as relevant for duration, we find that, on the contrary, it is the second type of memory that matters. This is also central to what we have called "state of informedness" here above.

Experience: *Erlebnis* or *Erfahrung?*

Before we look closer to this central point however, we should find out what Heidegger, Kierkegaard, and Merleau-Ponty have to say on experiences.

Most interestingly, we find with Heidegger a distinction in the German language that is not made in English. In this chapter, we have distinguished between the individual, unique present experience on the one hand, and the accumulated past experiences on the other. And we have asked ourselves, How can past experiences influence present experiences? The German distinction between *Erlebnis* and *Erfahrung* may be helpful in this respect. *Erlebnis* (what I have in my consciousness at a particular time, what I live)[70] should be distinguished from *Erfahrung*, something in my life that I have learned something from.

With Heidegger, however, the matter is not that simple. For him, *Erlebnis* amounts to triviality and carelessness. He quite obviously misses the point of the constitution of the single experience, as he also seems to take the subject-object relationship for granted (Heidegger, 1927/2010, p. 388/ p. 369). Putting more emphasis on history, and the integration of past-present-future, paradoxically, Heidegger has a higher appreciation of *Erfahrungen*. Significantly, Heidegger, has a social rather than personal perspective on human experience, that is, instead of looking for a "genuine" and original lived experience, like Bergson, he considered such experiences as doomed by our dependence on our history and our culture. That is how he understands *Erlebnisse*, in our time constrained by the technological culture in which we presently live (Vernon, 2008). Instead, he searches for εμπειρία (Latin: *experiri*), which are the classical language terms to which he links the German term *erfahren*, with the connotation of being sensorily confronted with something unexpected that evokes wonder and curiosity (cf. Heidegger, 1989/2012, pp. 161–162/pp. 125–126). In my view, that sounds like *present-at-hand*, particular for humans, as opposed to *ready-at-hand*, which is a more direct and unmediated relationship with the world (cf. this chapter, p. 115). Heidegger endorses such an understanding (Heidegger, 1989/2012, p. 444/p. 350), considering *ready-at-hand* as a technical relationship with the world.[71]

Our endeavor here, however, is to provide a theoretical basis for such original lived experiences. I will therefore stick to a more Bergsonian understanding of *Erlebnisse* and *Erfahrungen*, more aligned to the common understanding of these words.

While the approach is different, the concern is the same as with James and Bergson, namely, that we have no direct and unmediated contact with our original experiences, which are surreptitiously categorized and thereby distorted by the intellectual framework of our minds that has been forged by our culture. While Bergson—optimistically, and supported by James—sees a way to access our original experiences through intuition, Heidegger, pessimistically, considers us doomed to this intellectualization as it completely dominates the society in which we live. Ultimately, what is needed, according to Heidegger, is a catastrophic breakdown of our society, from which we may start from scratch (Vernon, 2008).

The same concern for the biasing influence of past experiences is to be found in the phenomenological movement, where bracketing of our preconceptions of the natural world[72] is a central idea, in order to get direct and unmediated access to our mental experiences. Similarly, hermeneutics is concerned with the influence of our preconceptions, based on our earlier experiences and our culture, when we interpret, for example, a biblical text, or what a patient says in psychotherapy. By the way, Heidegger does not show the same pessimism in *Sein und Zeit* regarding the analysis of *Dasein*, where human beings may be aware of the conditions of their existence in a prereflective way by considering the simple fact that we are just here and now and that we *may* reflect over our relationship to the world. So with these philosophers, we are faced with a question that is opposite to what has been our concern here. Instead of asking how our past experiences can influence present experience, we are faced with the issue how to *prevent* past experiences from influencing present experiences. That's awkward! The idea is that we don't learn from past experience, but to the contrary, they induce us to a prejudice that prevents seeing things as they are.

I save Kierkegaard for later. Let us see what Merleau-Ponty has to say! He expresses the same concern as Bergson and the others, considering science to never reveal existence in the same way as the world itself as perceived. By determining or explaining it, by acquiring knowledge about it, making myself just a parenthesis in the world, we forget consciousness as preceding knowledge (my paraphrase of Merleau-Ponty, 1945/1962, p. ix). The slogan "back to the things themselves" means coming back to the world before knowledge, the world that knowledge always talks about but which it also always alienates itself from. The "real" world is not what I think about but what I *perceive* (Merleau-Ponty, 1945/1962, p. xi/p. xvi). Rather than writing about earlier experiences, Merleau-Ponty refers to *habits* as a central concept (Merleau-Ponty, 1945/1962, p. 107/p. 91).

It Is Not the Experience as Such That Is Important, But What Is Mediated by Our Body

Thus, instead of distinguishing between past and present experiences, he distinguishes between our actual body and our habitual (or customary) body (Merleau-Ponty, 1945/1962, pp. 97–98/p. 82). The distinction is illustrated by the case of phantom limbs.[73] It is as if the habitual body takes precedence over the actual body and acts as warrant for the affordances of objects that are not presently exploitable by the actual body. It resembles the psychoanalytic concept of repression, where someone entertains a project (e.g., a love affair, career, or other achievement) that the person in question is incapable of carrying out, but also incapable of renouncing, thus wasting time and energy on something unachievable and displacing genuine presence by an impossible future.

In both cases, the past (loss of limb or failed project) invades the present and thus reveals the interdependence of biological and personal existence (particular for humanity?) that at the same time makes possible and precarious the temporal structure of our experience (Merleau-Ponty, 1945/1962, p. 100/p. 84). This structure is revealed by the temporal horizon[74] of past and future that situates the present. It surmounts the dispersion of the moments and grants the continuity of the personal existence. Our biological existence (I would say, our biological epigenesis, the history of our body) is the basis for our personal existence, while our personal existence makes our biological existence conceivable. That is from where we can understand the relationship between the psychical and the physiological. That is also from where we may understand the temporality of our personal existence.

I assume Merleau-Ponty to consider animals to be restricted to a biological existence (1945/1962, p. 103/p. 87), as he describes its existence as limited by its involvement in its present situation, while man may take his distance to it. The animal lives only in its *Umwelt* (von Uexküll & Kriszat, 1934), but man also lives in a world. What is the difference? According to Merleau-Ponty, as I understand him, the difference is between the pure duration of the animal and the (human) "experiential ecstasy[75] that make any perception a perception of something" (Merleau-Ponty, 1945/1962, pp. 85, 70).[76] While Heidegger (1927/2010, p. 329/p. 314) considers past, present, and future to be the three ecstasies of temporality, for Merleau-Ponty, as pointed out here above, past, present, and future *constitute* the temporality of human experience. Thus, the continuity of past, present, and future warrants the objectivity of our *Umwelt* (cf. von Uexküll & Kriszat, 1934),[77] thus also constituting my experience of it as the experience of an objectified world. My perspectives on this world change, but the world transgresses these perspectives by the horizons that temporalize my experiences.[95] This means that subjectivity is primary, and objectivity is derived.

I take Merleau-Ponty to mean that animals live in pure subjectivity and that there are no objects for animals, only for humans. While Bergson would reserve duration only to human consciousness, I would have suggested to him that he could find in animals—as understood by himself, as well as by Merleau-Ponty—perfect examples of pure duration, not contaminated by intellectualization or by intellections (as opposed to intuitions, see this chapter, p. 120). Namely, while completely rejecting animal consciousness and representation of an outer world, Bergson (1888/2001, p. 63/ p. 138) considers the possibility for animals to "perceive" "quality duration" (Bergson, 1888/2001, p. 58/p. 127)). That means, animals are not hit by the contradiction in Bergson's theory that carries out a discourse on duration, which by definition is beyond any intellection. Neither do they have Merleau-Ponty's problem with the ambiguity of the consciousness.

Subjectivity and Objectivity Reconsidered

Before coming to Kierkegaard, let us first recapitulate what we are trying to achieve: We started by the disparity of what happens in the world and our description of it. I blamed simplification and abstraction, and wished for the kind of intuition Bergson is considered to have achieved. The key to Bergson's solution was the understanding of time he proposed. The normal, received understanding of time, for example, as used in the natural sciences, is, according to Bergson, based on a *spatialization* of time, to which Bergson contrasted time as pure succession, where events followed each other in our experiences as the "stream of consciousness" that James wrote about. I also take the point of Bergson and Merleau-Ponty that subjectivity, in a sense, is primary and more close to truthfulness than objectivity, as represented by the natural sciences. Gibson would certainly not agree to this use of subjectivity and objectivity, but his idea of a description of the physical environment relevant for perception actually fits this idea of the primacy of subjectivity, as a condition for the possibility of objectivity. Ultimately, though, I prefer Gibson's stance, not that I would agree that the subject-object distinction is *false* (Gibson, 1979, p. 41), but rather that "the subjective and the objective are actually only poles of attention" and that "the dualism of observer and environment is unnecessary" (Gibson, 1979, p. 116). Like Gibson, I would like to be considered as a realist, that is, I believe in the existence of the physical world. So that is not the issue. The issue is about how we experience and describe this world. Gibson has a point in proposing a new and fresh approach to the concerns of James, Bergson, and the other philosophers mentioned here, by not considering a distinction between intellection and intuition, but instead finding an ecologically relevant description of the world, to the difference from the world as described by the natural sciences. Namely, these latter descriptions have not that much to do with how we experience the world. Gibson goes so far as to characterize the physical properties of the world as abstract (1979, p. 127), thus putting himself very close to Bergson.

What is missing in Gibson's (1979) account is an ecologically relevant description of how we, as living organisms, function, relative to the ecologically relevant description of the world. The question could be specified to how come we are able to discriminate higher-order variables in an optic or other ambient array. I assume that Gibson is right, and that, for example, ratios among geometrical measures in terms of retinal images may constitute higher-order variables that need not be computed, and that can specify properties in the world that may be useful for humans and other animals, what Gibson called affordances. However, the issue is not an ecological description of the properties of the world but what it means to discover these properties.

A central argument in Gibson's (1979) account of higher-order variables, with the ambient optic array as a model, is that they are revealed *through* the changes in the array, either by movements of the array itself or by movements *in* the array, that is, movements in the world, most often a combination of both. It is to experience stability through change.

We are now at a decisive point in my argument. Thanks to Gibson's (1979) insight, as described in the last sentence here above, and his insistence on the difference between a physical sciences description of the world and an ecological relevant description of the world, we have a more precise understanding of what is at stake in Bergson and James's criticism of intellectualism. In Gibson's idea of discovering (stable) higher-order variables *through* the change in lower-order variables, we have a pretty good idea of what temporality means at the level of one singular experience.

Although I, inspired by Bergson and Merleau-Ponty, have a very different view from Gibson on subjectivity and objectivity, I consider his questioning the subject-object distinction as very refreshing. With his (e.g., 1966b, 1979) and his wife's, E.J. Gibson's (e.g., 1987) singular focus on the development of discriminatory ability, they really don't offer an understanding of the origins or emergence of the distinction. None of our philosophers offer any developmental accounts that could help. Merleau-Ponty, however, is very generous in providing ample materials for such an account in his analysis of the subject-object distinction.

This is not enough, however. In Gibson's (1979) analysis of ambient arrays, the concept of experience is taken for granted. The same goes for our philosophers, except for Dreyfus (2002), who makes the connection between concrete analyses from Merleau-Ponty and Walter Freeman's use of DSA in the context of brain activity.

Attractor Landscapes of the Brain and of Perception-Action Systems

The gist of Gibson's (1979) analysis of ambient arrays is *stability through change*. In DSA, *attractors* stand for relative stability—far from thermodynamic equilibrium—in a local system that maintains a high degree of complexity and structure. When it comes to biological systems, as studied in the life sciences, such local systems may be temporary or persistent, self-assembled as demanded by contingent tasks and/or disintegrate when not needed or too much disturbed. Metaphorically, we could describe an "energy landscape"[79] of attractors as valleys, where situations of instability are peaks (Dreyfus, 2002, p. 382). As dynamic systems tend to be energy efficient, they tend to settle in the valleys rather than on the peaks, that is, in situations of stability rather than situations of instability. However, it is also possible to establish a situation of relative stability on the top of a peak, which will then have a pit on the top, with not that much high and steep walls. That means that this "attractor basin" will be more vulnerable to disturbances than a basin at the bottom of a valley. We may use as an example trotting horses in trotting horse races, where the horses are driven to run as fast as possible, without switching to a gallop. Galloping will imply disqualification, and the rider will have to slow down his horse such that it can fall into the trotting attractor again. From that example, you will also understand that this energy landscape is all but static. It is very much a dynamic landscape and thus also very much a temporal landscape. Of course, this is a metaphor for how the brain works and not a description of what it looks like.[80]

This is the image Freeman (1991, 2000) gives of the brain. From our perspective, I would say that this is a dynamic system at a lower level than the body-world level that preoccupies us here. How will then perception-action situations show up in this metaphor? That is, how will the emergent properties from brain dynamics have an impact on the self-organization of a perception-action coupling of an animal-slice-of-surrounding system? This question is at the core of Dreyfus's (2002) message.

I agree with Freeman, and Dreyfus would certainly also agree, that *mental* representations *may*, but *need not* emerge from the self-organization of neuronal patterns, at least in the human brain. That is, images, symbols, language representations, melodies playing in the head, Marcel Proust's remembering the taste of the madeleine in childhood, whatever, they *do* exist, in the sense that they may emerge as such under particular circumstances. None of us would deny that. That is also, in my view, the proper meaning of "representation," that is, "stand-in" manifestations emerging from brain activity that make absent things and gone events present. Thus, we may also represent goals, that is, future states of affairs that we want to make real. As already mentioned (see this chapter, p. 126 and p. 150), the problem with the concept of representation is that it is used in another sense, namely, as a stable brain structure that corresponds to some particular pattern of activity or state of affairs in the world, that may or may not be actualized, for example, a memory or a goal that is recruited in situations where it is needed, required, or evoked.

Freeman denounces this use of "brain representations" in the following way:

> Instead of focusing on pattern invariance and storage capacity, we began to ask how these patterns could be generated in the first place from less ordered initial conditions. What are the temporal dynamics of their development and evolution? What are their effects on the neurons to which they transmit? What kinds of structural changes in brains do they require and do they lead to? What neuromodulators do these pattern changes require?

> What principles of neural operations do these dynamical processes exemplify and instantiate? In short, we began to focus less on the outside world that is being put into the brain and more on what brains are doing. (Freeman & Skarda, 1990)

It is this latter understanding of representation that is denounced by Freeman, by Dreyfus, by Merleau-Ponty, and, last but not least, by me. Freeman, however, and also Dreyfus, somehow implicitly let the latter conception of representation in through the back door, Freeman more openly by putting great emphasis on goal-directedness (Freeman, 2000; to the difference between Merleau-Ponty, Dreyfus, and myself).[81]

Instead of Dreyfus's (2002) "energy landscape," Freeman (1991, 2000), describes "attractor landscapes," conceiving the brain as organized in terms of basins (valley bottoms) of attractors in which sensory inputs may fall, dependent on which one best suits the input. It is important *not* to understand these attractor basins as "stand-in" neuronal representations, to which sensory inputs are compared, and our greatest challenge is to avoid such an association. I'm not sure that Freeman bypasses this association, although he clearly denounces its use in brain research contexts.

This illustrates what a challenge it is to avoid a representational account of human and animal functioning relative to their surroundings. I attempt to push the issue further than Dreyfus and Freeman. Dreyfus's main point that is important in our discussion here remains, however, namely, the nonrepresentational account of ongoing action that I have described here above. This point overrides my reservations, and we may let us be inspired by Dreyfus's (2002) use of Freeman's (2000) model for brain function to convey an understanding of duration and experience based on temporality at all levels of biological functioning (actually, by extension, also at the levels of social and cultural functioning, which I will not discuss here).

The Inexistence of Attractor Landscapes

Let me first make sure that you, the reader, understand me right. The attractor landscape is *not* a metaphor for a permanent, although dynamic structure of the brain.[82] The only thing that exists in the brain is neurons, localized and connected to each other as part of brain anatomy. The dynamics that make this brain part of a living organism is the constant changes in the structure and electrochemical composition of this organ, notably the constant changes among the neuronal connections, nothing else. The attractor landscape is a metaphor for the *functioning* of relevant parts of the brain in relevant situations. That means that those parts may self-organize into dynamic systems, when required by sensori-motor or mental tasks that the organism faces. The attractor landscape is a metaphor for the possible changes and stabilities the system may go through over the time period the task situation involves the organism.

Thus, these attractors and attractor basins don't represent anything. They don't even exist. They only work as an epistemological tool for explaining the self-organization of the brain and of the perception-action system of an organism. What exist are the world and the organism. Of particular interest is the nervous system as an important component of the perception-action system of an organism where specific parts of its surroundings are included in the system.

Repetition and *Augenblicke*

Now, let me try to approach the issue from a perspective inspired by Kierkegaard. In his terms, "attractor basins" would stand for the general (the universal, if you wish). If that were to be expressed in a *brain representational* way, a sensory input, or an action, falls into one particular basin instead of another because of a "higher similarity" between the input or action and whatever is the attractor for that basin. That is, the basin would be a basket of earlier, similar experiences. Each new similar experience would then confirm the validity and applicability of the attractor basin.

However, as Kierkegaard (Constantius, 1943/Kierkegaard, 2009) sees it, new experiences would not contribute with anything. That is why he rejects the repetition as a confirmation of the general. Rather, the general must be understood in terms *what it is not.* In Gibson's terms, once we have made a discrimination that allows us to see chairs, we need no further training to identify chairs, and understand what they are good for. In spite of innumerable differences among chairs in the world, I somehow know that a chair is a chair. Gibson would have said that it is because it affords sitting. While Gibson was more concerned with the "objective" properties of the sittable thing, I think it is more important and interesting to see the sittable thing from the perspective of action, that is, in this case, sitting (cf. Vedeler, 2012). Let us assume two nearby attractor basins in the attractor landscape, one for sittability and one for the act of sitting down. I perceive a particular slice-of-surroundings as sittable, and I sit down on it, only to discover that it was not comfortable at all, to the point that I rise up again. If I had been comfortable on that particular slice-of-surroundings, it would not have been an issue, I would have taken it for granted, ready-at-hand, as Heidegger would have said. It would have been a meaningless repetition from Kierkegaard's perspective. With the "unsittable presumed chair," on the other hand, it was a difference that made a difference. It would have given me an experience that changed my attractor landscape where either a new attractor basin was formed or where the shape of two closely related attractor basins were modified.

This account addresses the question of when and how learning takes place and also when and how attractor landscapes become modified and ultimately, from Freemans perspective, how the neural connections in the brain become modified. Kierkegaard certainly was not concerned with such everyday trivialities as sittability. He had in view situations more decisive for the life course. My point is that it amounts to the same. Whether it is a banality or a life course crossroad, it is a moment of attractor landscape rearrangement that illustrates the dynamics of our temporality. Our discussion of temporality has hitherto concerned the all-too-common question of the relationship between past, present, and future. Where do we then place the moment where I'm faced with the unsittable chair? Answering that question, I again turn to Kierkegaard. I'm not sure I understand Kierkegaard right here, but taking the same libertine approach to sources as Dreyfus, that does not matter, it is my interpretation that matters. I will use his concept of *Augenblick* to characterize "the moment of truth" where a negation changes the attractor landscape.

Kierkegaard's concept of repetition will in this context mean a routine task, where the relative stability of the attractor landscape is maintained. The concept of *Augenblick,* on the other hand, implies possible failures, possible changes in the landscape, and possible new outcomes for the organism, for example, what we use to call creativity. Kierkegaard writes about the *Augenblick*: "Øieblikket betegner det Nærværende som et saadant, der intet Forbigangent og intet Tilkommende har" (Haufniensis, 1844, p. 93; Kierkegaard, 1980, p. 87). The *Augenblick* denotes the present as such, that has nothing bypassed and nothing forthcoming to it. In the *Augenblick,* we cannot rely on what we already know, and we don't know what will be the outcome. To sum up, with Kierkegaard, we have been able to connect the general and systemic idea of attractor landscape to the experiences of an individual. Also, put into the context of an attractor landscape, the routine question of the past, present, and future as the essence of temporality is being relativized by Kierkegaard's discussion of time in terms of repetition and *Augenblicke*. From this vantage point, we may be able to reformulate the temporality of development in Dynamic Systems terms that make the issue of past versus present versus future an issue derived from and designed for the discussion of the human skill in the use of mental representations (to remember the past, to imagine the future, and to contemplate the present from a *present-at-hand* (Heidegger) stance.

More important, and as a condition *sine qua non* for the possibility of human mental functioning, I propose the temporality of development to be an issue for a nonrepresentational account of the development of daily life skills to cope with our world, common for humans and other animals.

The basic question for that is an enlarged understanding of duration as a concept for understanding biological temporality and an understanding of experience as a possible phenomenon adjacent to duration, whether unique for humans or something we share with other animals.

The Temporality of Durational Units

The crucial issue for such an account has been the delimitation of durational units. The DSA gives an obvious, action-based and temporally dynamic basis for the delimitation of durational units. The dwelling of a dynamic system in an attractor basin may be long or short, but will last only as long as it is relevant for the challenge posed to the system. "Dwelling," however, does not mean "rest." An attractor may resist perturbation, but is at the same time dependent on the system's continuous exchange relationship with its environment. Furthermore, the attractor may in itself imply change, for example, cyclical change, like the movements of a pendulum or the rhythmicity typical for brain activity (Buzsáki, 2006). All this warrants the temporality within attractors as well as in the transition from one attractor to another. As the system shifts to another attractor basin, it will also shift durational unit. Through the theoretical framework of DSA and Freeman's empirical research to support it, Bergson's concept of the continuous and uninterrupted duration, as well as James's stream of consciousness, has received a temporal structure unimaginable for Bergson and James. They were using, at their best, scientific knowledge of their time. We now have at our disposal theoretical tools and empirical research that make it possible to give a much more qualified formulation of their issue.

Durational Units and *Erlebnisse*

The question of experiencing durational units is not yet addressed however. From a principled DSA stance, the issue is straightforward: if and when there are *Erlebnisse*, they are to be considered as emergent properties of perception-action systems in durational unit attractors. Having extended the issue of temporality from the level of consciousness to the level of the body, we may address the issue of the relationship between durational units and *Erlebnisse* in terms of emergence at a physiological level becoming a systems component at the level of consciousness. As *Erlebnisse* thus are subsumed under durational units, our question could be about what *Erlebnisse* ad to durational units, when they emerge.

We are experiencing what we pay attention to. I suggest that, at any moment, *Erlebnisse* are where the focus of attention is. This makes no distinction between men and animals. In usual, daily, unreflecting, ready-at-hand life, the stream of consciousness is easygoing. When we meet challenges, resistance, the unexpected, and such, we humans have the advantage to switch to a present-at-hand relationship with the world. Here, *Erlebnisse* may take another turn. Ready-at-hand life is dominated by repetition, present-at-hand implies the exception, in Kierkegaard's terminology. That makes the human race particular when it comes to innovation and creativity. Kierkegaard sighed for a present-at-hand life, considered to be more "truthful," while most of us are content with a ready-at-hand life.

The simplest understanding of consciousness would be to define it in terms of wakefulness. Both animals and humans sleep. Normally the relationship between organism and world is different during sleep than during wakefulness such that perception seems not to work during sleep. That is not completely true, though, but may work as a delimitation of *Erlebnisse*. It would allow for a ready-at-hand relationship with the world and direct perception-action couplings. In the terms proposed here above (this chapter, pp. 125, 130) and inspired by Merleau-Ponty (1942/1963, p. 232/p. 215), it would be "lived consciousness." No further citations should be needed in the next paragraph

"Known consciousness" would then correspond to a present-at-hand relationship with the world. I suggest that this relationship might involve, among other things, thinking, mental representations, and goal-directedness. I could also imagine that one could find situations that were to be considered as relationships to the world in between lived and known consciousness, such that they are not clearly delineated categories of relationship with the world. Nonetheless, the distinction is useful in understanding relationships with the world and in making sense of this relationship. When are our *Erlebnisse* ready-at-hand and when are they present-at-hand? Kierkegaard will help us answer this question. However, first we need to sort out the role *Erlebnisse* play for *Erfahrungen,* what *Erlebnisse* add to durational units in the role they play for further *Erfahrungen*.

Ontogenesis: *Erfahrungen* in the Life Course

It means, we have here taken a DSA perspective, and that implies that the impact of durational units is about epigenesis, the history of the individual and thus at the core of development. We have earlier depicted the epigenesis of an organism as consisting of interactions at many different hierarchically organized levels simultaneously and over time, at the same time up and down the hierarchy (see Figure 7.1). As mentioned (this chapter, p. 128), we here restrict ourselves to two levels: microgenesis, that is, the development of particular skills (e.g., as described by Dreyfus, 2002) and ontogenesis, the development of the individual organism. For both levels, we have adopted Freeman's (e.g., 2000) idea of a simple Hebbian rule for change of connection strengths among neurons as the neurological basis for epigenesis. This would be valid for microgenesis as well as for ontogenesis.

So far, assuming Freeman's (e.g., 2000) theory of a Hebbian rule regulated brain development,[83] we may now approach epigenesis at an ontogenetic level, that is, the life course development of an organism, embracing learning, skill acquisition, and the like. Endeavoring to give a nonrepresentational, that is, a *non-brain-representational* account of development, I propose to use simple everyday words like routine and habit[84] to describe what ontogenetic temporality is about, as well as describing the impact of durational units and *Erelbnisse* on *Erfahrungen*. We have found arguments for the temporality of the single durational unit in terms of an attractor basin. In Freeman's attractor landscape metaphor it will now be easy to extend that temporality to the whole landscape, as well as to the brain as such and by extension, to the body as such.

Considering the organism as a dynamic system in the particular sense of *autopoiesis* (Maturana & Varela, 1980), I, from the arguments here above, feel entitled to consider the basic temporality of biological functioning or, in other words, of life itself, to be the foundation of as well subjectivity as the subject-object relationship. Subjectivity, perception, action, intentionality, affordances, desires, feelings, experiences, approach-avoidance, curiosity-disinterest, evocation-habituation, and so on, all these subject-related phenomena have their basis in the temporality of physical, biological, social, and cultural activity at all imaginable levels of life functioning, as illustrated by Gottlieb's figure (see Figure 7.1).

When it comes to the impact of durational units on *Erfahrungen*, I have referred to the commonsensical notions of habit and routine. Whatever you do or perceive, it will be either familiar or unfamiliar, on a gliding scale. The familiar will fall into a very stable attractor basin, and even if there are differences from one occurrence to another, those will be absorbed by the attractor without any change or modification of it, except for possible consolidation. It might be that, if you have a conscious experience (*Erlebnis*) of it, it will feel boring. If it is useful, ready-at-hand, it will probably go unnoticed.

Only "when expectations are not met"[85] will such action draw attention to itself, with the consequence that the corresponding attractor basin, and possibly the attractor landscape around it, is reorganized for the system to solve the task it faces. Such activities in the attractor landscape will be based on changes in the nervous system, changes in connection strength among neurons, thus changing the state of informedness (see this chapter, p. 134) of the organism, and imply what use to be called learning. When such changes are integrated with the organism as such in a stable manner, the new attractor landscape may apply to new but similar situations that the organism may face.

The final question raised by Dreyfus (2002) that needs to be addressed is what similarity means, when we cannot rely on brain representations based on past experiences. This question is also critical for a DSA understanding of perception-action. How can the system find or reconstruct the "right" attractor basin for the task at hand?

Dreyfus, with reference to Merleau-Ponty, attempts an answer in terms of an inherent "prototype." Only, as he himself admits (Dreyfus, 2002, pp. 375–376), it is indeed hard to understand how such a prototype is different from a representation. Merleau-Ponty himself emphasizes that the criterion is not to be found in the mind, but in the world (1945/1962, pp. 24–33/pp. 16–25). For example, the "complete form" of an object is not composed from a collection of sense impression but is implemented simply because it is a complete form in the world.

Like Merleau-Ponty (166b, 1979), Gibson temporalizes this idea by proposing higher-order variables picked up *through* changes in the ambient array. He also adds that perception is discrimination, not composition of percept from sense impressions. From our DSA perspective, this will mean the perception (or misperception) of the object through its adherence to one attractor basin or the other. Notice, while we have defined perception as veridical, we must be open for the cases where the perceiver sees what is not there, that is, misperception. That means that in the attractor landscape of this perceiver, there is also an attractor basin for "the wrong thing" and that basin may be stronger than the basin for the "real" perception (cf. Vedeler, 2012).

The difficulty remains, however, because percepts through discrimination also need to be subsumed under general concepts in order to be understood or taken advantage of. What is it that makes up the "chairness" of all chairs? Dreyfus mentions three conditions that we may take into account. First, the innate architecture of the brain.[86] Second, the organism-surroundings interaction through the life course (my words), at all levels of physical, biological, social, and cultural functioning that I would call epigenesis and which, by the way, starts already at conception, thus incorporating Dreyfus's (2002) first point.[87] The third condition, however, is new to this discussion, namely, success. The chairness of the chair manifests itself in the fact that it satisfies my wish to sit. In fairness, to give the word to Dreyfus himself, here is how he summarizes these three points: ... these three structural body functions—brain structure, salience and order of experiences, and similarity of satisfactions ..."

Das Augenblick: "The Moment of Truth"

Here, Kierkegaard comes in. This sittability of chairs is the general. It is not specified in any way. It is not "stored" in any representation. It is not perceived by any affordance. It is just taken for granted. It needs no confirmation by being compared with a prototype. Thus, we leave Dreyfus. Only when it fails, when I cannot sit on my presumed chair, sittability becomes an issue. So rather than taking success in a positive sense, as Dreyfus does, I suggest that we take it in a *negative* sense—failure—inspired by Kierkegaard. *It is when I fail sitting on the chair that I know what a chair is.* I switch from ready-at-hand to present-at-hand, and by that, get a cognitive insight into the chairness of chairs. Not that I will store that *Erfahrung* as a brain representation, but it will imply a restructuration of my "sitting" attractor landscape, which in commonsense language would be called learning.

Kierkegaard reserves the *Augenblick* for humans who have both body and soul.[88] Time does not mean anything for nature. Only with humanity is the concept of temporality set. That means that only through the *Augenblick* is there history. That makes it also possible to speak about past, present, and future.

This somehow sounds similar to Merleau-Ponty, for whom subjectivity was the starting point for temporality. As already mentioned, I want it the other way round. This goes also *vis-à-vis* Kierkegaard: Without temporality, no *Augenblicke*. It may be that only humans experience (*erleben*) *Augenblicke*. But temporality is the essence of Dynamic Systems, in the sense of Dissipative Systems, as presented by Prigogine. The reason for that is that temporality permeates dynamic systems at all imaginable levels, whether in the life sciences or in non-life-related areas of knowledge (e.g., Prigogine & Stengers, 1984). The exchange relationship between a system and its environment, uninterrupted throughout the life course of the system,[89] all such systems will have a beginning and an end, and, through the stabilities and changes of the system (that makes it a dynamic system), the time between the beginning and the end matters for the functioning of the system, thus temporality.

Nonetheless, Kierkegaard's discussion of the *Augenblick* (Haufniensis, 1844/Kierkegaard, 1980) may be used for the elucidation of the peripeteias of attractor landscapes in dynamic systems. A free interpretation of the concept would be to consider it as the "moment of truth" that is the exception, the negation of the general, the situation where something does not fit into the prevailing attractor landscape and where then change and transition take place, for example, by a new attractor basin emerging from the reorganization of the system. Kierkegaard emphasizes the instantaneous character of the *Augenblick*, but with our DSA perspective that would also be relative. An *Augenblick* in the life course of a universe may perhaps last millions of Earth years, still being infinitely short from the perspective of the life course of the universe. So it is not clock time that matters, it is the "crisis of the moment" that changes the system.

Of course, in between such crises, the system, with its attractor landscape, will certainly not be static. As for biological systems, some are pretty "hardwired" without much changes in the overarching attractor landscape, however, nonetheless highly sophisticated, while others are flexible and capable of more or less considerable learning (Barrett, 2011). Biologically, the flexible ones mean that there is room for change in the neuronal connections in the brain, according to the Hebbian rule mentioned earlier. In terms of perception-action systems, illustrated by the attractor landscape metaphor, it means established and new attractor basins continuously modified by *Augenblicke* and consolidated by repetitions. *Augenblicke* stand for major reorganization of the attractor landscape, repetitions stand for routines and habits that reinforce and stabilize existing attractor basins. The less flexible animals will, as to say, have less *Augenblicke* and more repetitions. However, as already pointed out (this chapter, pp. 166–167), also dwelling in an attractor basin is temporal, by the fluctuations that do not disturb the attractor. Furthermore, for any animal, each task situation is new and unique, and requires adaptation from the animal. We should not forget that attractor landscapes are temporary, relative to different task situations to which temporary perception-action systems are self-organized. These systems recruit brain structures that are relevant to the tasks, where earlier experiences (*Erfahrungen*) have strengthened and/or weakened neuronal connections as described here above, inducing changes in muscle agility and strength, as well as changes in integration of neuronal, muscular coordinations and integration of neuronal and sensory systems. It is not all about the brain! Resources induced by earlier experiences are thus, through the body, recruited for task situations at hand through the state of informedness of the organism part of the actual perception-action system.

THE OBSERVATION OF TEMPORALITY IN DEVELOPMENT

I will illustrate this view of our relation with the world with two exemplified theoretical issues. Both issues will emphasize the temporality of development at all levels relevant to the examples. The first introduces two new concepts for this chapter, namely, *frame* and *co-regulation*, as proposed by Alan Fogel (1993). The second discusses a very well-known observation by Piaget (1963), where he considered the inability of a toddler to find a toy hidden at a different place than the usual. This failure has since then been called the A-not-B error. In Piaget's theory, it was a developmental stage toward object permanence, that is, having knowledge that objects exists independently of being observed or played with. This observation is interpreted from a DSA perspective (Smith, Thelen, Titzer, & McLin, 1999; Thelen & Smith, 1994; Vedeler & Garvey, 2009).

The first example illustrates the role of habits and routines in epigenesis, analyzed from a Dynamic Systems perspective. It normally concerns the interaction between two or more individuals, for example, a mother and her child. The point is that they develop a common history of interaction, where the interaction is constrained by different themes. The frame is defined by the theme of the interaction.[90] Using the conceptual framework of this chapter, I would say that the frame defines a dynamic system consisting of two persons, and possibly some thing that both persons perceive and manipulate. If the two persons have never met before and don't know each other, I would prefer initially not to consider them as *one* system, but rather as different systems, each relying on their individual history and interacting with the other person on the basis of their own subjectivity and *Erfahrungen* with other persons. On the basis of this background, different attractor landscapes will be the basis for how each of them understands the other and their common situation. Dependent on cultural similarities and differences, it will be likely that they establish some common ground on which they may have shared experiences (in the sense of *Erlebnisse*). These shared experiences will emerge over recurrent episodes of the same frame, for example, two strangers who happened to meet at a vintage car of interest to both of them. They start discussing the car, first from their different perspectives, but when they several times agree to meet again over the same car, they will have shared experiences in the frame, also when they understand these experiences differently, that is, these experiences include the same thing in the world, but will find their attractor basin in different landscapes. Over time, though, the shared experiences will make the two attractor landscapes more and more similar. Of particular interest is that they *together, by their interaction,* will have *Augenblicke* of discovery, where new aspects of the car emerge, which they both were unable to discover by themselves. This is a process that Fogel (1993) calls *co-regulation*, an important component in the history of the frame, and the basis for eventually considering the two persons in that particular situation as *one* system and as one unit of analysis.

I will also mention two examples of frame analysis, where two individuals can be considered as one system right from the beginning. The first is a mother and her child, where they actually started as one, and by conception became two, however with a common history right from the beginning. Innate propensities of the newborn child to respond in social situations will match the very basic feelings and *Erlebnisse* of the mother, such that, even if the attractor basins in interaction are radically different, they may match so well that we are entitled to see them as one system and one unit of analysis.

Another example of a frame with co-regulation, described in (Fogel, 1993, pp. 29–30) is about "wolf-circling"; two stranger wolves that approach each other, ready to attack and to respond to attack, making them perform a mutual "ritual" of circling around each other. This situation is so well defined by the physical properties of the two bodies and their perception of each other, that assuming innate preparedness for their behavior is hardly necessary. Their complete focus on each other and the tight co-regulation of their movements motivates seeing them as one system acting within a common attractor landscape. However, their co-regulation is subtle and fragile, dependent on potential hostility between them. The common attractor could, in this case, be considered to be a shallow basin on the top of an instability peak, where small disturbances would turn the ritual into a vicious circle of serious fighting deep down in a valley of the attractor landscape. Such serious fighting might also be considered as a common attractor for the two wolves as one system, with the same type of co-regulation.

This makes me think of a third example of co-regulated frame activity. Two boxers who don't know each other may still, by a very similar training history, know exactly the rules and techniques with which to respond to each other. They may have a common attractor landscape, not on the basis of a common history or similar anatomy or innate nervous wiring, but on very similar histories of training and rule knowledge. All these examples of frames and co-regulation illustrate the temporality as well on a microgenetic as well as on an ontogenetic level. At the microgenetic level, temporality is illustrated within a single occurrence of the frame by the fact that frames consist of ongoing interaction during the time the partners interact within the theme of the frame and, particularly, by the co-regulation as an ongoing process of refinement in developing the theme.[91] At an intermediate level, one can follow the development of the frame as such, with the involved partners and the particular theme, and see how the interaction changes from occurrence to occurrence, how the theme gets enriched, impoverished, changed, all dependent on the quality of the interaction and the co-regulation implied.

The other example is Piaget's (1963) observation of the A-not-B error: What makes a child between 8 and 12 months, after having gotten used to finding a toy under one particular pillow, persist in looking under the same pillow in spite of having seen the toy being hidden under another pillow? Linda Smith has studied this phenomenon from a DSA perspective. Instead of proposing an explanation in terms of lacking object concept (Piaget) or lacking memory capacity (cognitive psychology explanation), Smith and collaborators (1999) propose the DSA framework. I will reformulate her research and conclusions in our terminology here. With the play situation that functions as experiment, the child is habituated to a hide and retrieve play that includes a distance to creep, watching an adult hiding a toy under one of two pillows, always the same, waiting a while, and then creeping to retrieve the toy. This is done several times exactly in the same way. Finally, a change is introduced, namely, that the toy is hidden under the other pillow, still while the child watches. Typically, the child goes for the wrong pillow (A), the same as before, and does not try the other pillow (B) when realizing that he or she has failed to retrieve the toy.[92] An attractor landscape has been established for this situation, with routinized transitions between the different attractors for the different actions of the play.

Smith et al. (1999) introduced a change into the design, varying the time the child has to wait before being released. This proved to be a decisive change. If the child is allowed to creep to the pillows at once in the test situation, it will go for the right pillow (B). The analysis of this difference in time before retrieving is the most interesting for us here when we are discussing temporality. We may assume an attractor landscape for this game, with attractors for each component action of the game, watching the hiding, creeping, etc. We have earlier suggested that this landscape is not static but under more or less constant change, although with a relative stability. What we see in this example is a more fundamental change in the attractor landscape over time in the test case where the toy is hidden under the B pillow. As the child watches the hiding of the toy, a retrieving attractor is formed for the B pillow. However, over a short period of time this landscape changes, in favor of the old, *habitual*, landscape, which by now, leads to failed retrieval. You would say that the child forgets what he or she just saw. I cannot think of a better illustration of the temporality of a dynamic system. Except for the child being confused by the *unexpected* failure to retrieve the toy and not knowing what to do (that is, failing to see the possibility of the toy being under the B pillow) everyone, young and old, should be able to recognize oneself in the situation. We acquire skills, get use to them, and while being for them most part observant on changing conditions that require different actions, it is all too common that we let the habit rule and inattentively do the wrong thing. How often have I not taken the usual route driving home from work, forgetting that my wife has asked me to take a detour to the grocery store to buy some item, missing for dinner. Then suddenly I remember my commission, and make another, more complicated detour to buy what is needed.

Would you say that I had a plan to drive the usual route to the store just to simply switch to the home route on my way? I wouldn't say that. I had a habit (read: attractor) for the usual route that happened to overrun my planned (yes!) detour over the store route. Namely, if I don't manage to keep that plan in mind, the habitual attractor will be stronger. The organization of my thinking and behavior into the shallow store route attractor needs the "keep in mind" component to prevail. This "keep in mind" component, in turn, is an emergent property of lower-level brain organization[93] that reorganizes my habitual "return from work to home" action sequence into a weaker and more complex action sequence. Thus, planning and goal-directedness is a powerful component in the self-organization of human behavior, which, for its strength, is dependent on being consciously maintained. That it is not solely causal in explaining behavior[94] is well illustrated by the all-too-common A-not-B error. To keep things in mind is a fragile attractor for many people.[95] Lucky are those who are good at it, those who stay focused, those who have a strong attractor for it.

CONCLUSION

The strongest impetus to stability in the dynamic systems of living organisms is deep attractor basins based on routines and habits. In simpler organisms, innate, fixed action patterns will play the same role. While more flexible organisms, capable of learning, are also capable of handling new and uncommon situations, the routines and habits still have a strong impact. We thus have a biological and highly relevant illustration of Kierkegaard's issue with the repetition and the relationship between the general and the particular. This biological illustration very clearly sets the particular as the exception, the unusual, and the new. That is the sense in which the exception is the negative: There is no habitual attractor basin for the exception; it needs to reorganize the attractor landscape for the situation. By that, it will also constitute an opening to the new, to creativity.

Looking back on what has been most prominent in this chapter, it seems that microgenetic temporality has been in the foreground, and ontogenetic temporality has more or less been considered as derived. That does not make it less important. However, the discussion in this chapter had its focus on how to handle the paradox of the continuity of biological time and the conceptualization of this continuity in terms of units through which we may understand objects, events, and states of affairs in the world.

The central question has then been how to attain the temporality of the units we propose. We would need a temporally delimited, but at the same time temporally connected description of the unit that we search for. My response has been found in Bergson's contradicting himself, by half-heartedly allowing for events in an indivisible duration. My main aim in this chapter has therefore been to find a description of durational units that is temporal, that is, a description where temporality is based on a system's continuous exchange with its environment, maintaining it in the relative stability of an attractor. Thus, I consider the Dynamic System's concept of attractor to provide a temporally dependent description of the durational units I propose to be the basic units of biological time.

In order to relate those to experiences, I found it necessary to distinguish between two meanings of the English world "experience" corresponding to the two different words in German, *Erlebnis* and *Erfahrung*. The assumption is that durational units that appear as attractors in a DSA analysis of perception-actions systems are the core elements for epigenesis at a regular microgenetic level, that is, above neurophysiological and lower processes, but below the time perspective of a life course. Through epigenesis, however, these durational units are assumed to influence as well lower-level as higher-level processes and development (according to Figure 7.1) by routines and habits and by reorganization of the perception-action-related resources of the organism. As to the impact of a single *Erlebnis*, as an emergent property of a perception-action system in a durational unit attractor, beyond the durational unit it is associated with, I assume there is no general answer. It will depend on the attractor landscape and what effect the durational unit in question has on the landscape. *Erfahrungen*, on the other hand, understood as the impact of the history through which the durational units have developed, provide the state of informedness of the organism at any particular moment, including acquired skills.

I deliberately leave out *knowledge* here, as it is dependent on the human symbolic function and the mental representations associated with it. Knowledge makes a considerable difference in the flexibility and coping ability of humans that are the outcome of this particular ability to represent. I have left it out because of the confusion in the use of the word "representation" to open for an account of the *nonrepresentational* basis of biological functioning at all levels, including human knowledge. Only when this basis is recognized as nonrepresentational can a reasonable account of mental representations be attempted.

The crucial question is how we avoid making a representational account of the *generalization* of *Erfahrungen*. I have relied on the attractor landscape metaphor as a way of avoiding the idea of brain representations. While recognizing the importance of *mental* representations,[96] these are not to be considered as structural units, such as attractors, but as emergent properties of perception-action systems' self-organization in particular situations. If we take the English word "concept" as an example, we may trace the origin of the word to the Latin verb *capio*, of which the closest translation would be "grasp";[97] not in the common mental sense of "understand" but in the sensori-motor sense of taking with the hand. That the mental sense of "understand" is the most frequently used is obviously a point in this context. The word "concept" should therefore primarily be associated with the *act of understanding*, not with a representation that stands for a certain class of things in the world, or whatever. When concepts are conceived as representations for classes of things and such, they are giving such classes of things and such an ontological status they don't deserve. Each act of mentioning the name of a thing or act of using a particular thing could say to confirm the class and would thus be considered as a repetition, in Kierkegaard's sense. From the point of view of action, therefore, a concept would only be the similarity between actions. In what consists, then, this similarity?

Dreyfus (2002) addresses this question. He describes the similarity in terms of similarity between earlier experiences and present sensory input. That does not, however, avoid the problem with *what* similarity between present input and earlier experiences to deal with. The point is that earlier experiences have shaped the attractor ("energy," in Dreyfus's terms) landscape, such that a present perception-action falls into one previously formed attractor. In my view, this answer is ambiguous relative to the question of representations, when he proposes that "The attractor could be called a representation only in the very weak sense that it does incorporate past experience and leads to action on the bases of that past experience." I have stated here above that attractors reside nowhere and represent nothing. They are formed by the self-organization of brain, sense organs, and other anatomical systems coupled to a slice-of-surroundings into a perception-action system facing some particular task in a particular situation. I agree with Dreyfus, however, that "three structural body functions–brain structure, salience and order of experiences,[98] and similarity of satisfactions" constrain the organism toward a particular solution of the task. These three "body functions" are part and parcel of the constitution of the attractor landscape for the situation. In my view, the mistake is, as pointed out here above, to describe the attractor landscape as always there, as were it a net of representations for earlier experiences. I'm sure that Dreyfus does not want to say that, but he could be interpreted in that direction.

Dreyfus (2002) also uses other arguments that emphasize *differences* rather than similarities. That is in line with the point of view presented here, inspired by Kierkegaard. In terms of attractor landscape, that would mean that a particular attractor is maintained but not constituted by similar experiences, repetitions. But contrasts, negation, incompatibility of a new event, or whatever, with earlier events, may make the system reorganize into a new attractor, where, for example, one old attractor bifurcates into two attractors to better solve somewhat, but not enough, similar tasks. Having been used to opening bottles in a particular way, my attempt to open a bottle of a new type fails. I change the position of my fingers slightly until I succeed in opening such a bottle. Eventually, my skill with this new kind of bottle improves and becomes habitual. In neurophysiological terms, brain structures are consolidated by repetitions, but changed by new and different things and events. In the process of self-organization, I suggest that contrasts (Kierkegaard's negation or exception) will play a prominent role in the constitution of the attractor landscape: "This is *not* a spoon!" I have also suggested an interpretation of Kierkegaard's concept of *Augenblick* in that direction, as the "moment of truth" where something reveals itself as different from the usual.

You may ask, as Dreyfus (2002) on similarity, what makes a repetition a repetition. I would say it does by stability of routine, habit, automatisms, even boredom. You don't notice it, except by boredom, until it becomes different. Similarity is not an issue, difference is! However, as in the case with Fogel's (1993) concept of frame, repetition does not necessarily imply boredom. Repetition has its own dynamics; no situation is ever exactly the same, and small differences may contribute to "keep the attractor stable" from occurrence to occurrence without falling into boredom. The challenge for Dreyfus was the criteria for similarity; his point was that these criteria are to be found in the world rather than in representations. This is hard to grasp, but his answer in terms of satisfaction is in agreement with the view that self-organization in any life-related dynamic system is for the solution of one task or the other, the word "task" then to be understood in the broadest possible sense. It is not about intentionality, it is not about agency. It is about a relative and dynamic stability for the maintenance of the system. Therefore, the similarity is in terms of the best possible (or good enough) solution of any particular task for the system. As for our mental well-being, there is a balance between too much and too little change at all levels of life functioning. It is a balance between boredom and overexcitement that always threatens the system which, one way or the other, is at risk of disintegration. Self-organization has a tremendous potential for explanation without explaining anything at all. It is an empty explanation, except for accounting for the sheer existence of the universe as a structure rather than as entropic soup. If the universe were an entropic soup, there would be no time. Time emerges as a result of differentiation and integration, the principle for which is self-organization. This principle is the same as the self-organization of dissipative structure, in Prigogine's sense (Prigogine & Stengers, 1984). The strongest implication of this is that there is no difference between living organisms and other forms of dynamic systems, which all depend on exchange relationships with an environment.

The concept of *emergence* may be the most controversial, but also the most powerful in Dynamic Systems Approaches. I want to take full advantage of this concept. Thus, among such concepts as affordances and intentions (Vedeler, 2012) mental representations are also emergent properties of perception-action systems and not independently existing entities, neither in the brain nor in the world, for example, as words, cultural symbols or images. Those things in the world are only the material carriers (*signifiants*; cf. de Saussure, 1916/1977) of the emergent representations in the actual perception-action systems, with corresponding attractor landscapes. The act of grasping ("understand") the meaning of the *significant* is simply the symbol falling into one attractor basin or the other.[99]

Searle's (and Dreyfus's) understanding of "conditions of satisfaction" will thus here receive a different meaning than the one meant by Searle. For him, it is about some state of the world that corresponds to the intentional content of an action, thus that this state antedates the action. In the present context, condition of satisfaction would be posterior to the action. The "strongest" attractor in an attractor landscape is not necessarily the "best" one in terms of adaptation between the organism and its surroundings. For a drug addict, the injection of the drug is harmful to the person, while providing immediate satisfaction. If the same person decides to stop using drugs, and succeeds in doing so, the immediate consequence for him or her will be highly undesirable abstinence pains that need to be endured for the person to succeed in keeping away from drugs. Provided that the drug is available, the first case requires no more than the skill to inject. The second case requires striving toward a goal. Searle's model applies to the second case, not the first one.

Another aspect is the choice of the "wrong" action, in our terms, falling into an attractor that will not give any expected satisfaction. It would still be considered as a valid perception-action system with a corresponding attractor landscape. The toddler who had to wait too long for the "right" attractor to be the strongest chose to look for the toy under the A pillow, that is, under the pillow where the toy *used* to be, thus relying one earlier *habits* rather than on perceived reality. The child had his or her state of informedness changed into a maladaptive attractor being the strongest. Thus Searle's (1981, 1983) concept of satisfaction is turned from an absolute criterion for intentionality into a relative feeling emerging from the actual state of a perception-action system. As an absolute criterion for intentionality, it works for human consciously planned action. One should then not forget that most human actions are similar to animal action, not dependent on conscious planning. Humans are most of the time self-organized body-slice-of-surroundings systems in a particular situation at a particular time, and this system in turn is part of a hierarchy of systems in an epigenetic history of an ever-changing organism and its ever-changing surroundings. Thus, while looking up in the sky on a particular day of my life where things have happened and I'm in a particular mode, I look at particular cloud formations and suddenly see a face in the cloud that reminds me of a particular person. This particular situation can be analyzed as a particular perception-action system drawing, among other things, on brain resources that are innate and very strong, to self-organize into an attractor landscape where face recognition will attract the perception-action system into that particular experience.

This could be my last word. However, as the relationship between temporality and subjectivity has been central to this chapter, I will end it with a comment to a quote from Merleau-Ponty (1945/1962, p. 412): "Le temps n'est ... pas un processus réel, une succession effective que je me bornerrais à enregistrer. Il naît de *mon* rapport avec les choses" (emphasis in original). "Time is ... not a real process, an actual succession that I am content to record. It arises from *my* relation to things." While I agree with the first sentence, my conclusion will be the opposite to the second sentence: *My relation to things in the world emerges from the temporality of my life.* That is valid at all possible levels of analysis, from gene expressions to culture (see Figure 7.1). While Merleau-Ponty's proposal is a general one, my proposal is developmental.

ACKNOWLEDGMENTS

I'm grateful for the hospitality of Eric and Polemnia Amazeen and the inspiration I got from the D•PAC seminar group during my stay as Visiting Scholar at the Department of Psychology, Arizona State University, Spring 2013 and 2014, including two seminars in 2014 devoted to the discussion of this chapter.

NOTES

1. Quotes from original texts are mostly not translated. However, in most cases the content of such quotes are explicated or paraphrased in my comments to the quotes.
2. In the references to Kierkegaard, Bergson, Heidegger, and Merleau-Ponty, year and page references indicate first original edition (x)/current English edition (y) in the format 19XX/19YY, p. xx/p. yy.
3. While Kant, in his theory of knowledge, considered immanent what is possible to experience by the senses, most philosophers following him put great effort in finding a truth beyond the senses, as to say, making the transcendent immanent, for example, having knowledge of God.
4. These three concepts are of no interest for time in a physical sense. The before, the now, and the after do not exist per se in the physical world. Relativity—even relativism—is ubiquitous. You can understand neither things nor places nor displacement, except relative to other things, places, and displacements. Worse, the sausage metaphor for time, mentioned here above, is no longer valid. That is, in modern physics, it does not make sense to talk about simultaneity; time is reduced to a fourth dimension of space. So, actually, it would make no sense to speak of the temporality of matter.

5. I notice with gratitude that Bergson here mentions "unity of which one thinks" (Bergson, (1888/2001, p. 41/p. 83)) in support of my proposal on durational units here below.
6. A word that I, not Bergson, here borrow from James (cf. James, 1890, p. 478).
7. Read: immanent, as opposed to transcendent (cf. note 3).
8. I will use original language terms frequently in the text. They will then always be in *italics*.
9. See my discussion of the word "concept" in the concluding section, p. 177
10. With reference to Kant.
11. My expression.
12. We would be tempted to translate *Kunft* as "becoming." I think Heidegger would object that this would comply to the idea of a now to come, that is, something concrete and delimited. Also, the English translation would be "coming." As shown by the ensuing quote, it is not about becoming this or that (e.g., becoming a dead body). The quote points out that it is the *movement forward* that makes the *Dasein* future oriented, while this movement forward is itself only possible as the Dasein (as being) comes up to itself, in the first place, that is, that the Dasein, in its "to be," is basically a movement forward. Keeping that in mind, I will here below nevertheless use *becoming* for this implication, by contrasting it to *being*. The "this of that" part of becoming cannot be excluded from the movement forward. Only, it needs to be specified.
13. Short for *beziehungsweise*, here ≈ as well as.
14. Cf. note 6..
15. Which Heidegger does not.
16. I prefer the German word rather than the Danish, as I assume this book to have more readers familiar with German than readers familiar with Danish.
17. Bergson included! See note 3.
18. My words.
19. Cf. Merleau-Ponty's (1945/1962) insistence on the primacy of perception.
20. Klempe (this volume) has the merit of highlighting this aesthetical side of Kierkegaard's philosophy.
21. We saw with Peirce that the argument in defense of the continuity of time through a mathematical conception of infinitesimal intervals turned out to be a spatialization of time that Bergson considered so detrimental to an adequate understanding of time as a vital aspect of our relation to the word.
22. A term to be discussed here below. Cf. also affordances considered as ready-at-hand, here above, pp. 115–116.
23. As explained here above, I will continue to use the German word.
24. There is no difference in meaning between the German *Wiederholung* and the English word repetition, therefore I will continue using the English word.
25. Under the pseudonym Constantin Constantius, see reference list.
26. Regine Olsen
27. Cf. note 6.
28. Beyond reach for normal human understanding, see footnote 4.

29. Until here in this chapter, the word "representation" has denoted our cognitive capacity to mentally represent absent objects and events. Examples of such representations are mental images and language words. This understanding of the word is different from its use in Cognitive Science, where it denotes brain or neurological or even purely conceptual structures (Fodor, 1981) with the help of which we act, perceive, and understand, for example, spoken words (cf. Vedeler, in press).
30. It can be mentioned that the same issue has been debated in the science of artificial intelligence (see Brooks, 1991, having exactly the same title as Dreyfus's paper, except for the reference to Merleau-Ponty).
31. The reference (Freeman, 1991) is to be found in the reference list here.
32. Psychological level is missing: According to Gottlieb, psychological functioning must be inferred from analysis at the overt levels of behavior and the environment. Figure redrawn by myself. In it, the word "Time" replaces "Individual development."
33. In spite of evidence for a rudimentary sense of self (Gallagher & Meltzoff, 1996) in the newborn. From an epigenetic perspective, the distinction between prenatal and postnatal development becomes less relevant. Cf. also p. xx.
34. In *Phenomenology of Perception,* Merleau-Ponty only mentions the intentional arc a couple of times, described and explained as projecting "round about us our past, our future, our human setting, our physical, ideological and moral situation" (1945/1962, p. 158/p. 136, English translation chosen here, as it is referred to by Dreyfus, 2002). Thus, it constitutes the unity of the senses as well as of our motoricity and intelligence. As such, it is analyzed and understood relative to a pathological case where it is lacking. In my words, the patient in question seems to lack the temporal dimension of his experiences. He has no awareness of his past, neither of his future, and neither of other places. What he experiences is only what is here and now.
35. That also addresses our question about durational units.
36. Cf. Kierkegaard's concern for the repetition, this chapter, pp. 121–122
37. Obviously, Kierkegaard is here influenced by Hegel and his use of Spinoza's idea that (in Hegel's terms) "omnis determinatio est negatio" (cf. Melamed, 2012), "every determination is a negation."
38. The subject is also treated in other publications of his not mentioned here.
39. In Piaget's sense, that is, both assimilation and accommodation: By writing *with* the world, I imply both adaptation *of* the world and adaptation *to* the world.
40. In English translation: "Time exist for me because I have a present. It is by coming into the present that a moment of time acquires that indestructible individuality, that 'once and for all' quality."
41. Both together constituting changes of perspectives in Merleau-Ponty's terminology.
42. I wish to thank Bill Mace for drawing my attention to this chapter.
43. Notice here the word "flux," indicating the same idea as James's "stream" and Merleau-Ponty's "écoulement," translated to "flow," namely, what we

here consider as the essence of temporality. (Cf. this chapter, pp. 125 and 137).

44. "Dissipation" here denotes this exchange relationship between a system and its environment.
45. My choice of the word "dynamic" rather than "dynamical" reflects my stance relative to the dominating natural science understanding of dynamical systems as a mathematical theory, as opposed to the more metaphorical understanding reflected in the writings of, among others, Thelen (e.g.,Thelen & Smith, 1994) and Fogel (e.g., 1993): The mathematical theory has had less influence on my way of thinking about central dynamic systems concepts, such as self-organization, emergence, and attractors, than the metaphorical approach (cf. next note).
46. I need to point out that this relative understanding of what is a dynamic system is not universally recognized. That is, many proponents would claim that a dynamic system should be considered as an ontological description of reality and not simply an epistemological tool guided by a research question. For example, the mathematical (i.e., nonmetaphorical) understanding of dynamical systems is taken to be descriptions of what really happens in the world. In my view, ontological descriptions will never be completely true, only hypothetical and partial. Immanuel Kant refrained from making any knowledge statement about the "thing-in-itself," that is, knowledge about the world beyond our senses, as all knowledge is based on sensory information (for a very interesting discussion of Kant's view on sensory information, see Wyller, 2001). I agree.

    However, I take the liberty to make a metaphorical description of the ontology of a living organism (beyond our senses) to make the point that it is, in principle, unknowable as such. I propose a version of Waddington's epigenetic landscape metaphor (Waddington, 1975) that spatializes the temporal flow in terms of a marble rolling down a slope, where the topography of the slope constrains the trajectory of the marble. While Waddington used the metaphor to illustrate how biological cells become more and more specialized in a complex organism, I use the metaphor to illustrate ontological assumptions on the life course of an organism. The point is my assumption that the topography of the landscape contains all the factors, be it physical forces, chemical processes, genes, foods, climates, experiences, habits (individual or cultural), etc., etc., and their interactions that have the potential to influence any part of the life course of an organism. These factors may themselves be changing over time, as well as their interactions, such that the topography of the slope is more like an agitated sea surface rather than a stable terrestrial landscape (Zittoun et al., 2013, pp. 23ff, 34ff" to "pp. 22–24, 34–36). The complexity of this ever-changing web of influences and constraints makes it not only practically but also, in principle, unknowable as time-dependent change processes (the complexity of epigenesis, as illustrated by Figure 7.1 here above. One particularly important reason for that is the fundamental nonergodicity of the universe in general, and biological life in particular (Kauffman, 1993, 2000, also explained in Zittoun et al., 2013, pp. 24–28.

We can, on the other hand, with our senses and our cognition, make perspectival abstractions for chosen slices of the world in order to understand perceptually and cognitively delimited processes as, for example, causal relationships. In this perspective, causal models, as well as Dynamic Systems Approaches, are to be considered as epistemological tools for such an understanding.

47. It would be the case of *autopoiesis* (Maturana & Varela, 1980).
48. See note 66 here below for explanation of this concept.
49. Paul van Geert (2003, pp. 659–660) eloquently makes this point by contrasting self-organization to design (e.g., whether the development of an organism is a self-organized process or a process that is designed and determined by, for example, the genes of the organism): "Self-organizing processes use whatever possibilities are offered by both genes (and body) and environment, but those possibilities come about as a result of the ongoing dynamic process."
50. I prefer to avoid the word "state" in the context of DSA, as dynamic systems, per definition, are never at rest. More about that here below. See also note 57.
51. Thus, gravity is, in this case, not an environmental exchange factor for the system. Thanks to Beatrix Vereijken for helping me sort out this issue.
52. For a proper analysis of race trotting, a more fruitful approach would probably be to define the system as the horse-sulky-driver unit. I would guess that a trotting racehorse freely on a meadow would not trot at racing speed.
53. Cf. Dreyfus (2002).
54. The concept was used unexplained here above (this chapter, p 144), in the example of the frog catching a fly.
55. This is not the same as the epigenetic landscape metaphor referred to in footnote 46. There are however affinities between the two metaphors.
56. I only mention Dreyfus's reference to Freeman here, not Freeman's own publications, as I see an important discrepancy between the two. Freeman (2000) does not endorse Dreyfus's description (or mine) of intentionality without represented goals, and that changes pretty much the whole point of Dreyfus's analysis of maximal grip. Furthermore, Freeman makes no mention of Dreyfus, neither of Searle, although his analysis of intentionality is far closer to Searle's than to Dreyfus's.
57. A reminder: My consistent use of the word "situation" instead of the conventional "state" reflects my constant concern for not losing from view temporality at all levels of analysis. Notably, also attractor basins are temporal, not only in terms of clock time but also in terms of biological time. The duration implies relative stability through reduced but continuous changes during the duration; perturbations that are not strong enough to move the system out of the attractor basin.
58. Here we change from considering single experiences to considering accumulated experiences, see here below the section on experiences as *Erlebnisse* and experiences as *Erfahrungen*.
59. Cf. footnote 33.

60. Such chaotic situations, with no appropriate attractor available should be distinguished from "chaotic attractors," that (paradoxically) are defined as highly deterministic attractor situations that may take many different trajectories, dependent on different initial conditions, for example, weather systems.
61. With the word "store," we have a line of argument that, in my view, leads us back to a representational account.
62. We will come back on that in comparing it to Gibson's (1966b) concept of ambient array (cf. this chapter, pp, 138–139 and 154–155).
63. Although Heidegger probably would have disliked the comparison.
64. Kierkegaard's pseudonym for this book, Vigilius Haufniensis.
65. He considers that to be the case for the (classical) Greeks, who considered temporality to be sensibility *(Sandseligheden)*, because they lacked an understanding *(bestemmelse)* of the spirit. Only Christianity gives room for eternity.
66. Ecstacy, or ekstasis, comes from the Ancient Greek ἔκστασις, "to stand outside oneself." Merleau-Ponty (1945/1962, pp. 427, 430) also borrows this expression from Heidegger.
67. Cf. note 4.
68. According to the Hebb rule introduced here above, this chapter, p. 134
69. Cf. the distinction between the German words *Erfahrungen* and *Erlebnisse*, made here below (this chapter, pp. 158–158).
70. *Erleben* is a transitive verb, "live" is intransitive, does not take an object. The word is based on the German word for live, leben. With the prefix *Er*, it becomes transitive. There is no such option in English. So if you say in Continental style, "I have lived my depression," you would mean something like "I have lived through, endured, my depression."
71. While reducing in this citation present-at-hand to ready-at-hand, Heidegger seems to convey that our reflected relationship with the world is as imbued by our taken-for-granted technological culture as our simple use of things.
72. Also called epoché (ἐποχή = suspension) or phenomenological reduction.
73. Not for Merleau-Ponty, but for our epigenetic perspective, the word "habitual" may be misleading, as it implies habits acquired by past experiences. As already mentioned in note 33, the epigenetic stance on the relationship between genes and surroundings make no principled distinction between pre- and postnatal influences on the developing organism, neither between internal and external influences. Merleau-Ponty assumed the phantom limb to be based on the postnatal acquisition of a body schema, while modern infant research provides evidence that phantom limbs may be experienced also by individuals who never developed the limb in question. There is also evidence on neonatal facial imitation (see this chapter, note 39 and p. 149), implying a prenatal knowledge base for the perception-action coupling of model and own facial movements. The discussion of phantom limbs and neonatal imitation lends additional support to Stoffregen and Bardy's (2001) ideas of a global array, where different sensory channels are not distinguished (see this chapter, p. 138).. From an epigenetic perspective, the role of genes vs. surroundings (e.g., Brugger et al., 2000) becomes pointless, as birth is only one event in the epigenesis of some organisms, starting at

conception. Remains to find another common sense word than "habit" to illustrate the role of repetitions (generalizations) and negations (differentiations) that lie behind a present state of informedness.

74. What Dreyfus (2002) likes to call "intentional arc."
75. For the concept of ecstasy, see note 74.
76. My translation, different from the 1962 published English translation, p. 70, which implies a causal relationship between ecstasy and perception.
77. Depreciated by Heidegger (1927/2010, pp. 57–58/p. 58), translated "environment" in English), without mentioning von Uexküll by name, neither mentioned by Merleau-Ponty.
78. Here I must admit that I join Dreyfus in extrapolating from Merleau-Ponty's philosophy.
79. Term used by Dreyfus, not by Freeman himself, at least not in the publications I have checked. Freeman prefers "attractor landscape." I choose to use Freeman's words, not Dreyfus's.
80. It should also be distinguished from the epigenetic landscape metaphor, in it different versions, among them one proposed by myself (Zittoun et al., 2013, pp. 22–24, 34–36, cf. note 46).
81. Dreyfus hesitatingly though, by mentioning representations of goals "only in the weakest sense" (2002, p. 381).
82. Which seems implied in Freeman's theory.
83. I would prefer to say—more generally—neuronal development, embracing the whole body.
84. Well aware of their limited role in epigenesis; see note 90.
85. A way of saying I don't like it, as it puts back on the stage the brain representations I despise. How much better it is to use Kierkegaard's words, pointing out the exception!
86. The body, I would say. Even better would be to consider this as the state of informedness of the organism.
87. This is a very free paraphrase of what Dreyfus writes in his paper (2002, p. 376). Cf. Figure 1 and note 73.
88. Not to forget, they need to be Christian as well; cf. note 65.
89. Also nonbiological systems, such as, for example, weather systems, universes, or the self-organization of bromate ions in a highly acidic medium, the so-called Belousov-Zhabotinskii reaction. Cf. Thelen & Smith, 1994, ch. 3 for a review of research on this reaction, relevant for our discussion here.
90. See Goffman, (1974) about the origin of the concept of frame. However, it is Fogel who has given the concept its historical and temporal dimension.
91. Fogel (1993) points out that when the frame does not develop, that is, does not change in terms of exploring and enriching the theme, the frame will disintegrate as a dynamic system.
92. Thus there is no *Augenblick* for the child in this situation.
93. Lower-level, that is, relative to the higher-level organization of actions to return home.
94. As Searle (1983) would have it.
95. Myself included.

96. In the sense as proposed here above, cf. note 29, not in the sense as used by Fodor (1981)!
97. Cf. the German word griffbereit as equivalent to ready-at-hand on p. 115 here above.
98. I take Dreyfus to understand experiences as *Erfahrungen*. But what is the difference between brain structure and *Erfahrungen*, if not conceived as brain representations? I would prefer to lay down my history of *Erfahrungen* into my present situation as my current state of informedness, thus avoiding the idea of storage of *Erfahrungen*. *Erlebnisse*, on the other hand, as emerging mental representations, may have salience and order that impact on brain structure.
99. That means understanding as well as misunderstanding, dependent on what is to be considered as the "right" attractor basin!

REFERENCES

Barrett, L. (2011). *Beyond the brain: How body and environment shape animal and human mind*. Princeton, NJ: Princeton University Press.

Bergson, H. (1888). *Essai sur les données immédiates de la conscience*. Paris: Presses Universitaires de France. English translation: *Time and free will: An essay on the immediate data of consciousness*. Mineola, NY: Dover Publications, 2001.

Bergson, H. (1934). *La pensée et le mouvant. Essais et conferences*. Paris: Presses Universitaires de France. English translation: *The Creative Mind: An Introduction to Metaphysics. Essays and conferences* 1903–1923. Mineola, NY. Dover Publications, 2007.

Bergson, H. (1939). *Matière et mémoire: Essai sur la relation du corps à l'esprit*. Paris: Presses universitaires de France. English translation: *Matter and Memory*. New York: Cosimo, 2007.

Please write the full title in italics, both in French and in English.

Brooks, R. A. (1991). Intelligence without representation. *Artificial Intelligence*, *47*, 139–153.

Brugger, P., Kollias, S. S., Müri, R. M., Crelier, G., Hepp-Reymond, M.-C., & Regard, M. (2000). Beyond re-membering: Phantom sensations of congenitally absent limbs. *PNAS: Proceedings of the National Academy of Sciences of the United States of America, 97*(11), 6167–6172. doi:10.1073/pnas.100510697

Buzsáki, G. (2006). *Rhythms of the brain*. Oxford, UK: Oxford University Press.

Constantius, C. (pseudonym for Kierkegaard, S.) (1843). *Gjentagelsen: Et Forsøg i den experimenterende psykologi*. Copenhagen, Denamrk: C.A. Reitzel. Retrieved February 22, 2013, from http://sks.dk/g/txt.xml

Deleuze, G. (1994). *Difference and repetition*. New York, NY: Columbia University Press. (Original work published 1968)

de Saussure, F. (1977). *Course in general linguistics*. Glasgow, Scotland: Fontana/Collins. (Original work published 1916)

Dreyfus, H. (2002). Intelligence without representation—Merleau-Ponty's critique of mental representation: The relevance of phenomenology to scientific explanation. *Phenomenology and the Cognitive Sciences Special Issue: Hubert Dreyfus and the Problem of Representation,* 1(4), 367–383.

Fodor, J. A. (1981). *Representations: Philosophical essays on the foundation of cognitive science*. Sussex, UK: Harvester.

Fogel, A. (1993). *Developing through relationships*. London, UK: Harvester Wheatsheaf/University of Chicago Press.

Freeman, W. J. (1991). The physiology of perception. *Scientific American, 264,* 78–85.

Freeman, W. J. (2000). *How brains make up their minds*. London, UK: Phoenix.

Freeman, W. J., & Skarda, C. A. (1990). Representations: Who needs them? In J. L. McGaugh, N. Weinberger, & G. Lynch (Eds.), *Brain organization and memory: Cells, systems, and circuits* (pp. 375–380). New York, NY: Guilford.

Gallagher, S., & Meltzoff, A. N. (1996). The earliest sense of self and others: Merleau-Ponty and recent developmental studies. *Philosophical Psychology, 9*(2), 211–233. doi:10.1080/09515089608573181

Gibson, J. J. (1966a). The problem of temporal order in stimulation and perception. *Journal of Psychology, 62,* 141–149.

Gibson, J. J. (1966b). *The senses considered as perceptual systems*. Boston, MA: Houghton Mifflin.

Gibson, J. J. (1979). *The ecological approach to visual perception*. Boston, MA: Houghton Mifflin.

Goffman, E. (1974). *Frame analysis: An essay on the organization of experience*. New York, NY: Harper & Row.

Gibson, E. J. (1987). What does infant perception tell us about theories of perception? *Journal of Experimental Psychology: Human Perception and Performance, 13,* 515–523.

Gottlieb, G. (2003). Probabilistic epigenesis of development. In J. Valsiner & K. J. Connolly (Eds.), *Handbook of developmental psychology* (pp. 3–17). London, UK: Sage.

Haufniensis, V. (pseudonym for Kierkegaard, S.) (1844). *Begrebet Angest: En simpel psychologisk-paapegende Overveielse i Retning af det dogmatiske Problem om Arvesynden*. Copenhagen, Denmark: C.A. Reitzel. Retrieved from http://sks.dk/ba/txt.xml

Heidegger, M. (1927). *Sein und Zeit.* Tübingen: Max Neimeyer Verlag. English translation: *Being and Time.* New York: State University of New York Press; Revised edition, 2010.

Heidegger, M. (1989). *Beiträge zur Philosophie (Vom Ereignis). Gesamtausgabe Vol. 65, III Abteilung: Unveröffentlichte Abhandlungen. Vorträge - Gedachtes.* Frankfurt am Main: Vittorio Klostermann. English translation: *Contributions to Philosophy (From Enowning).* Bloomigton: Indiana University Press, 2012.

Heidegger, M. (1999). *Beiträge zur Philosophie (Vom Ereignis). Gesamtausgabe Vol. 65, III Abteilung: Unveröffentlichte Abhandlungen. Vorträge - Gedachtes) [Contributions to philosophy (from enowning)]*. Frankfurt am Main, Germany: Vittorio Klostermann. (Original work published 1989)

Husserl, E. (1998). *Ideas pertaining to a pure phenomenology and to a phenomenological Philosophy, Book 1.* Norwell, MA: Kluwer Academic. (Original work published 1913)

James, W. (1884). On some omissions of introspective psychology. *Mind, 9* (33), 1–26.

James, W. (1890). *The principles of psychology* (Vol. 1). New York, NY: Holt.

James, W. (1916). *Some problems of philosophy.* New York, NY: Longmans, Green.

Kauffman, S. A. (1993). *The origins of order: Self-organization and selection in evolution.* New York, NY: Oxford University Press.

Kauffman, S. A. (2000). *Investigations.* Oxford, UK: Oxford University Press.

Kelso, J. A. S. (1995). *Dynamic Patterns: The self-organization of brain and behavior.* Cambridge, MA: MIT Press.

Kierkegaard, S. (1980). *The concept of anxiety: A simple psychologically orienting deliberation on the dogmatic issue of hereditary sin. Kierkegaard's writings (Vol. VIII).* Princeton, NJ: Princeton University Press. (see Haufniensis, 1844)

Kierkegaard, S. (2009). *Repetition and philosophical crumbs.* Oxford: Oxford University Press. (see Constantius, 1843)

Klempe, S. H. (2014a). *Kierkegaard and the rise of modern psychology.* New Brunswick, NJ: Transaction.

Kugiumutzakis, J. (1985). The origin, development, and function of the early infant imitation *Acta Universitatis Upsaliensis. Comprehensive Summaries of Uppsala Dissertations from the Faculty of Social Sciences* (Vol. 35). Stockholm, Sweden: Almquist & Wiksell.

Maturana, H. R., & Varela, F. C. (1980). *Autopoiesis and cognition: The realization of the living.* Dortrecht, The Netherlands: Reidel.

Melamed, Y. (2012). "Omnis determinatio est negatio"—Determination, negation and self-negation in Spinoza, Kant, and Hegel. In E. Förster & Y. Y. Melamed (Eds.), *Spinoza and German idealism* (pp. 175–196). Cambridge, UK: Cambridge University Press.

Merleau-Ponty, M. (1942). *La structure du comportement.* Paris: Presses universitaires de France. English translation: *The structure of behavior.* Pittsburgh: Duquesne University Press, 1983.

Merleau-Ponty, M. (1945). *La phénoménologie de la perception.* Paris: Presses universitaires de France.. English translation: *Phenomenology of perception.* London: Routledge & Kegan, 1962.

Peirce, C. S. (1869). Grounds of validity of the law of logic: Further consequences of four incapacities. *The Journal of Speculative Philosophy, 2*(4), 193–208.

Peirce, C. S. (1892). The law of mind. *The Monist, 2*(4), 533–559.

Piaget, J. (1963). *The origins of intelligence in children.* New York, NY: Norton. (Original work published 1936)

Prigogine, I., & Stengers, I. (1984). *Order out of chaos: Man's new dialogue with nature.* New York, NY: Bantam.

Richardson, K. (2008). How dynamic systems have changed our minds. In A. Fogel, B. J. King, & S. G. Shanker (Eds.), *Human development in the twenty-first century* (pp. 25–40). Cambridge, UK: Cambridge University Press.

Rouse, J. (2005). *Heidegger on science and naturalism.* WesScholar, Division I Faculty Publications Arts and Humanities. Middletown, CT: Wesleyan University Press.

Runeson, S., & Bingham, G. P. (1990). *The state of informedness and the temporality of informational validity as the basis for the perception versus cognition mode distinction.* Unpublished manuscript.

Searle, J. R. (1981). The intentionality of intention and action. In D. A. Norman (Ed.), *Perspectives on cognitive Science* (pp. 207–230). Hillsdale, NJ: Erlbaum.

Searle, J. R. (1983). *Intentionality: An essay in the philosophy of mind.* Cambridge, UK: Cambridge University Press.

Smith, L. B., Thelen, E., Titzer, R., & McLin, D. (1999). Knowing in the context of acting: The task dynamics of the A-not-B error. *Psychological Review, 106,* 235–260.

Stoffregen, T. A., & Bardy, B. G. (2001). On specification and the senses. *Behavior and Brain Sciences, 24,* 195–261.

Thelen, E., & Smith, L. B. (1994). *A dynamic systems approach to the development of cognition and action.* Cambridge, MA: MIT Press.

Todes, S. (2001). *Body and world.* Cambridge MA: MIT Press.

Turvey, M. T. (1977). Preliminaries to a theory of action with reference to vision. In R. Shaw & J. Brandsford (Eds.), *Perceiving, acting and knowing: Towards an ecological psychology.* Hillsdale, NJ: Erlbaum.

van Geert, P. (2003). Dynamic systems approaches and modeling of developmental processes. In J. Valsiner & K. J. Connolly (Eds.), *Handbook of developmental psychology* (pp. 640–672). London, UK: Sage.

Vedeler, D. (1987). Infant intentionality and the attribution of intentions to infants. *Human Development, 30,* 1–17.

Vedeler, D. (1991). Infant intentionality as object directedness: An alternative to representationalism. *Journal for the Theory of Social Behaviour, 21* (4), 431–448.

Vedeler, D. (1993). *Intentionality as a basis for the emergence of intersubjectivity in infancy.* Acta Universitatis Upsaliensis. Comprehensive Summaries of Uppsala Dissertations from the Faculty of Social Sciences (Vol. 37). Stockholm, Sweden: Almquist & Wiksell.

Vedeler, D. (2012). *Affordance and intention: Complementary concepts.* Unpublished manuscript.

Vedeler, D. (in press). The development of intentionality during infancy and early childhood. In J. Wright (Ed.), *International encyclopedia of the social & behavioral sciences* (2nd ed.). London, UK: Elsevier

Vedeler, D., & Garvey, A. (2009). Dynamic methodology in infancy research. In J. Valsiner, C. M. Molenaar, C. D. P. Lyra, & N. Chaudhary (Eds.), *Dynamic process methodology in the social and devlopmental sciences* (pp. 431–453). New York, NY: Springer.

Vernon, J. (2008). Erfahren and Erleben: Metaphysical experience and its overcoming in Heidegger's *Beiträge*. *Canadian Journal of Continental Philosophy, 12,* 108–125. Retrieved February 28, 2013, from http://www.artsrn.ualberta.ca/symposium/files/original/632b380d4dca7b04a80081191763a616.pdf

von Uexküll, J. J., & Kriszat, G. (1934). *Streifzüge durch die Umwelten von Tieren und Menschen* (C. H. Schiller, Trans.). Berlin, Germany: Julius Springer.

Waddington, C. H. (1975). *The evolution of an evolutionist.* Ithaca, NY: Cornell University Press.

Wheeler, M. (2005). *Reconstructing the cognitive world: The next step*. Cambridge, MA: MIT Press.

Wyller, T. (2001). Wahrnehmung, Substanz und Kausalität bei Kant. *Kant-Studien, 92*, 283–295.

Wyller, T. (in press). *Was is Zeit?* Stuttgart, Germany: Reclam.

Zittoun, T., Valsiner, J., Vedeler, D., Salgado, J., Gonçalves, M. M., & Ferring, D. (2013). *Human development in the life course: Melodies of living*. Cambridge, UK: Cambridge University Press.

CHAPTER 8

TEMPORALITY, LIFETIME, AND THE AFTERDEATH

Case Studies From Hospice Patients

Meike Watzlawik

Four weeks ago, a close family member of mine had to be taken to the hospital. I was about 100 miles away when I got the telephone call. The caller said that the situation was serious. Immediately, I got in the car. While driving to the hospital, the only thing on my mind—repeating itself over and over again—was, "It is not his time yet!" Very fortunately, it was not. But when is it *time*?

In Europe and other Western countries, death is not something we usually talk about. In Germany, 80% of those who pass away today do not die at home, but in institutions (Gronemeyer, 2008). Death thus happens outside everyday life situations and almost becomes invisible. If it is talked about, then it is mainly in an abstract manner referring to life insurance, last wills, or goals we should pursue *now* since it might not be possible later. The paradox is that by the time adolescents turn 20, they usually have seen thousands of dead people on the news, in movies, on TV, and so on. Nevertheless, these deaths do not touch the personal environments of these

Temporality: Culture in the Flow of Human Experience, pp. 215–229

adolescents so that even with these experiences, they have not encountered death in a "tangible" manner. The same is true for many adults.

Death is kept outside society—a conclusion the German theologian Greshake had already arrived at in 1988. But that was not always the case. Especially in the Middle Ages and baroque times (up to the 1770s), death was very visible due to wars and epidemics, as well as short life expectancy. Death could not be denied and was even actively raised in the consciousness with, for example, *memento mori* objects (e.g., little coffins with skeletons worn as jewelry) that reminded their owners of life's finiteness, or religious paintings and writings known as *ars moriendi* that showed and described the appropriate behavior when facing death. Toward the end of the 18th century, all this changed. Even in art, the focus did not lie on the dying anymore, but on the bereaved (Knüttel, 2008). Gradually, the Church lost power and, with the 19th century, life expectancy doubled. In the 21st century, not only death but aging seems problematic. Watching advertisements, it seems obvious that we should use lotions, diets, injections, or even plastic surgery to fight the aging process. Staying young is of high value; growing old is not. With that, the integration of death in everyday life becomes even more unlikely.

I was not surprised when most of the students that wanted to become part of the Hospice Project stated that they had no personal experiences with death. After we were done interviewing hospice guests,[1] Lisa (age 25), one of the students, summarized her experiences as follows:

> At first, I was feeling a little uneasy knowing that we would interview hospice guests about their own dying and death. Usually, those are subjects you try to avoid—especially when you are young and you have little to do with it all. But quickly, I noticed that it is about way more than that. Everyone has his own story, his own life, his own and very personal topics; death and dying are not in the center of it all.

The Hospice Project

Why did we interview guests in the hospice? The primary goal was more a practical than a scientific one. And it was personal. A friend of mine who was a photographer moved into the hospice in Braunschweig, Germany, in 2010. Visiting her was my first encounter with a hospice. Even though Cicely Saunders (2000) had tried to change the image of the hospice as a "death house" years back, my personal expectations could probably be summarized as such. Nevertheless, I was positively surprised at the warm, accepting, and unexpectedly lively atmosphere in the hospice. In the new environment, my friend, who had been depressed for years, was suddenly full of energy again. She asked me to support her with one last project.

She wanted to take pictures of the hands of the other guests to document the different faces death can have; I was to interview the guests and write stories accompanying the pictures, from a psychological perspective if necessary. I agreed to do this. The doctors had told my friend that she would probably have a few more months to live; enough time to finish our project. But the estimation was flawed; she died within days after we had spoken. This came as a surprise to me, even though I knew that average life expectancy and risk information are not transferable to the individual level (Spiegelhalter, 2008).[2] Physicians are nevertheless faced with the dilemma of not knowing and still being asked by patients, How much *time* do I have left? The ways in which this dilemma is solved are quite diverse, as are the ways in which physicians' answers are interpreted. An example from our own data is the following. M. (57 years old) was diagnosed with cancer. She had a 4th gradient brain tumor, fast growing and inoperable. On being asked what happened when she and her husband were confronted with this diagnosis, they said,

> Husband: Our world collapsed. We were both just gone. I am still gone in a way; I cannot grasp or understand what is going on. First, we still had hope. Maybe we did not understand correctly what the doctors had been trying to tell us. Maybe they were still able to help her, we kept hoping, even after her first surgery and exposure.... Then, they told us that they would perform another surgery, with the danger of her being hemiplegic afterwards.
>
> M.: After discussing everything with my husband, I agreed since, even if I was hemiplegic, I would still live.
>
> Husband: The day of the surgery had almost arrived, when the doctor came—radiant with joy—and said that she would not have to undergo surgery, since it seemed as if the tumor had shrunk. They wanted to wait 2 months to observe the development of the tumor. We could not believe it, and left overjoyed. Two months later, the doctors told us that they had been wrong and that nothing could be done anymore.... What hurts me most is that, if they had told us the plain truth, we could have gone on vacation [instead of having to drive to the hospital regularly]; we could have visited places she always wanted to see. Now this is not possible anymore [due to her condition]. We could have used the remaining time differently.

After my friend had died, four students of mine and I decided to follow through with her idea, since, based on our own experiences and research in the literature, death and especially dying within a hospice still is a taboo subject connected to a vast variety of prejudices. With the help of an exhibition, we wanted to inform different audiences about the institution of hospice as well as about death and dying in general. What was on the guests' minds? Were they afraid? Did they expect something to happen after they died? How did they cope with the knowledge that the time they had left was limited? With the help of an interview guideline and after having trained the students in interviewing techniques, we interviewed 16 guests between 34 and 81 years of age (5 men and 11 women), at the hospice in Braunschweig and took pictures of their hands. The head of the hospice always informed potential participants about the project and, if the guests were interested, called us so that we could introduce ourselves. Informed consent was established before all interviews. One woman withdrew her picture and interview data from the project after talking to her husband, so the total number of interviews was 15. The interviews were recorded and partially transcribed afterwards. Based on this data, we focused on three major subjects that were addressed in all 15 interviews: (re)evaluating one's biography, accepting the finiteness of one's own life, and living in the hospice (expectations and experiences). The last topic will not be part of this chapter. We will instead focus on the first two.

Re-Evaluation: Integrity or Despair?

According to Erikson (1970) and many other researchers, growing up and growing old involve certain developmental tasks that are characteristic for different life phases. Erikson labeled the last developmental task or crisis in an individual's life *Integrity vs. Despair.* Retrospectively, older individuals evaluate their own life courses and, if the task is solved positively, are able to accept their individual life cycles and the significant people that are connected with them. Even the fear of death declines (Berk, 2009). The central question is thus: "Is it OK to have been me?" If the answer to this question is negative, despair is the outcome. One focuses on failures and the chances one either did not have or did not make use of, realizing that there is not enough time left to change things.

K. (62 years old) adds another aspect to this perspective in her interview. Overall, she says that she does not regret anything and that there is also nothing that she still wanted but now cannot do anymore. She says that she is ready to go and is not afraid of dying at all. Nevertheless, she starts crying when talking about her family and the way they always treated her, especially her sister, who seems to have hurt her in many different ways. An argument with her, K. remembers quite vividly, ended with the following literally translated dialogue (German original in italics):

„Ach, halt doch die Klappe!“	Sister: “Oh, hold your shutter!” [German expression for: “Oh just shut up!”]
„*Ich habe auch Augenklappen.* Und die mache ich für dich nicht mehr auf!“	K. answers: “My eyes also have shutters, and I will not open those for you ever again!” [with “eye shutters” she refers to her eyelids, but even in German this is an unusual expression]

When asked whether she tried to resolve the conflict with her family, K. adds that she did, but that all attempts have been quite frustrating for her. She does not wish to talk to her sister anymore, even now, knowing that time is limited. K. says that her sister will not understand her no matter how hard she tries to explain herself. That is a fact; a fact that she has learned to accept, she says. Nevertheless, she does not accept what has happened in the past. What her family did to her was not OK, but there is, from her perspective, nothing she can do about it. Accepting one's life course, thus, does not mean that one has to be content with all parts of it. Unresolved issues may have to stay unresolved issues. Possible scenarios may include the one described above. Another might be that others have died when one is finally ready to face them again, for example, after turning a blind eye to them as a consequence of seemingly unforgiveable incidents, thinking one would have enough time to resolve those issues later.

K.'s description of her experiences with her family and the conclusions drawn above became part of the exhibition we put together after we had interviewed all 16 guests. So as to provide a glimpse of the exhibition to the reader, K.'s exhibit is shown in Figure 8.1 as representative of all 20 exhibits we first showed to the public in March 2012. All exhibits follow the same logic: They show the photograph, the citation that triggered the motive, and an explanatory psychological text. The color was not chosen randomly, but is the favorite color of the person in the picture. The bottom part of the exhibit and a key word are printed in blue, K.'s key word being *Eye shutter*.

Accepting What Has Happened Versus What Is Not Going to Happen

In developmental psychology textbooks focusing on Western cultures, the stage *Integrity versus Despair* usually is said to start at about age 65. It is often the age of retirement; one's own children have moved out; physical changes can less easily be denied, all of which are markers for the aging process. Individuals become aware that more time has passed than lies

Figure 8.1. **Example exhibit: K.'s case.**

ahead of them. An example of a very positive evaluation of one's own biography is the way W. (78 years old, widow since 2009 and mother of four adult sons) summarizes her biographical narration:

> What I have managed, I'd like to see anyone else do that! I was always rock-solid, never smoked, never drank, never overindulged myself; was never overweight. My duties were the kids and later my own mother. I was never bored. My kids all do well, all of them have jobs; my two grandsons are still being trained. I had a very nice husband; we were married for 54 years. I can be content.

W. also says that it is time for her to go. She even says that it will be a relief, when she finally dies, since the cancer treatment has been very draining and she cannot take any more. She also states, with a smile, that her husband "has a spot saved for her next to him," leaving unsaid just *where* this spot is being saved for her.

Whereas W. mainly talks about her loved ones and people that accompanied her throughout life, C. (~ 70 years old, teacher, never married, no children) focuses on his own behavior and the consequences it had in the past. He says,

> I have been a very egoistic person in my life, I have ignored—for a long time—that there were people who wanted to be close to me.... I regret this deeply. I have hurt many people with my behavior. I don't mean physically, but often mentally by being solitary, by never wanting to be close to someone. I always turned them down. Today, I look at things very differently and am thankful and happy that there are still many people left that just like me. If I had been really mean, then this would maybe not be the case. [The reasons for my behavior in the past] were negative experiences; I have always been a late bloomer, I dragged that along; I did not want to get hurt.

Also, C. states later in the interview that someone is waiting for him and that someone will be there once he is dead. Being a very religious man, he is referring to God, however, and not a person he has met in his "earthly" life.

Both examples come from individuals who are older than 65, and who, even if they were not guests in the hospice , would, according to Erikson (1970), be expected to spend time reminiscing. But it may be assumed that the developmental task described as *Integrity vs. Despair* can occur out of sequence, for example, when younger individuals are faced with terminal disease diagnoses (e.g., Taghian, Smith, Erban, & Thomas, 2009). This is certainly true, but the case of B. (34 years old, married, father of two young children) shows that the retrospective evaluation of one's own biography can be of minor importance when the life one has lived has not been that long and when quite different developmental tasks or critical life events were thought to be next (e.g., being a good parent, a good husband, possibly facing the death of one's own parents). Having to face the developmental task of accepting the finiteness of one's own life had been anticipated neither by B. nor by his family at this point. Nevertheless, he has found a way to deal with it. B. says,

> It was a great shock at first [when he was told that he had to die]. But now, I got, well, what does that mean, "used to it," but I do not feel bad about it. One knows that that's how it is, one cannot change it. I tried for a very long time, but had to realize that I had lost, that I have to accept it.... Of course, it is difficult to leave the children behind; that one knows one has to leave soon and will leave wife and kids behind. One would have liked to see how they grow up and things like that. But no matter what, I live my life now the same way I used to; nothing will change until the last day [even in the hospice].... I am determined to write both of them [referring to his children] a letter, just a few lines, so that later, when they want to, they can have a look at it, that they have more than just pictures as reminders. [When I have died], time is up for me, the others have to cope with that. That will be very difficult. My wife is a very strong person, but by now she is wiped out. But when the chips are down, they will manage. Even if many things will change in their lives.

B. says he "does not feel bad about" the fact that he has to die; he does not mention regrets. He certainly would like to have spent more time with his children and wife, as he explicitly states later in the interview, but has accepted that this is not going to happen. Overall, he is less focused on his own biography but much more concerned with how his family will cope without him. He is trying to anticipate events that lie in an unknown future for his own family, and, at his age, is less concerned with looking back. He has to accept what will not be instead of what has been and, through the letters he is going to write, has found a way to have a potential future impact on the lives of his children and wife. Acceptance did not arrive suddenly, however, but took place through an active process of settlement, as hinted at above.

Terminal Disease Diagnosis: From Denial to Acceptance

The case of M. and her husband as well as the case of B. talking about his first reactions when confronted with a terminal disease diagnosis both show that shock and denial can be part of the grieving process. Kübler-Ross (1986) has described five typical stages of grief, of which denial is the first. The stages are not meant to cover all possible aspects of individual experiences; the model also does not claim to be chronological. But what are the possible reactions of terminally ill patients when they are told that they will die soon? According to Kübler-Ross, the patient might not be willing to accept the diagnosis at first (*denial*); this stage may possibly be followed by *anger* often connected to the question "Why me?" or maybe even "Why not somebody else?" Among our interviewees, R. (67 years old, widowed, two adult children) was the one asking exactly those questions. She says,

> I am being punished right now. Nobody knows how much time remains.... I am afraid of dying. I have this picture in my head that they just close the lid. Every morning, I am happy when I am still able to open my eyes. I prepared meals; I cleaned, as I was meant to. And suddenly it shall all not be possible anymore? I don't understand it. I am asking myself: Why you? WHY? Sometimes I have to shut that off, otherwise I cannot go to sleep.

The excerpt vividly shows that a terminal disease diagnosis can be understood as *disquieting experience* as described by Laskovski and Simão (this volume) or as *rupture*, as described by Freda, Picione, and Martino (this volume). The terminal disease diagnosis disturbs the individual's sense of continuity and is thus a "dramatic experience of discontinuity" (Freda et al., p. 233, this volume). R., as well as other guests in the hospice, were shocked by the unexpected diagnosis. They were not immediately able to or struggled with actively regaining a sense of continuity. How important it is to restore a feeling of continuity is shown by studies focusing on patients with posttraumatic stress disorder (PTSD). Patients with PTSD often relive traumatic events in fractures again and again, which is very stressful and highly disturbing. Nevertheless, details or certain aspects of the event can often not be recalled. Knaevelsrud, Kuwert, and Böttche (2013) believe that the lack of memory anchors prevents a coherent narrative of the traumatic event, which then inhibits its embedding into one's own biography. The result is the persistence of the PTSD. According to Knaevelsrud et al. (2013), therapy should aim at the defragmentation of memory to support a meaningful narration by the patient.

After the stages of *denial* and *anger*, Kübler-Ross (1986) describes a stage of *bargaining*, where the patient or (in our case, the guest) tries to negotiate, usually with a higher power, the time of death. One example would be "If only I can celebrate Christmas with my family one more time, I will leave without questioning your [referring to God] choices." When *bargaining* is given up, *depression* might follow, and finally, *acceptance* of the situation will be observed, as in the case of B., the young father introduced above. A circumstance that helped B. accept his situation was the hope that something would happen after he died; that he would not just disappear. He says,

> I am not sure what exactly, but I know: There will be something that comes next. Something lies ahead. I hope to see my family again. Many different things are said by many different religions, but in one aspect they all agree: Something comes next. What that is and what it looks like, we will see. Whether reincarnation exists, whether we turn into a star or angel or something else that buzzes around somewhere, or whether we go to the happy hunting ground, or somewhere else ... I believe that the soul is immortal [pauses to collect himself]. And it will become better, it can only become better. It will be a better world afterwards, a different, better being ... I do think that everything is good for something. That is why I am not afraid [of dying], I am just worried about the family, how they will handle it; but for myself, I am not afraid. It will all make sense, even if one is not able to grasp it yet.

Remarkable is that the projection into the future does not hold a clear picture for B. His narration thus has blank spots that are nevertheless not frightening, but meaningful, so that B. accepts them. He trusts in the fact that "it will all make sense" later.

On the contrary, S. (80 years old, widowed, one son) thinks that all the discussions about a possible afterlife are "humbug." He states,

> Dying is not difficult. Frankly, I tell you, I would like to go to bed and not wake up. Quickly. No funeral. Just dig a hole, put me in, and let grass grow over it. I was not born with a soul; I came with nothing and that is how I leave. Then I am dead; something like the soul does not exist. All this hocus-pocus. There is nothing to be afraid of. The medical college can even use my body for trainings. I will not be there to watch them anymore.

In both cases, the interviewees stated that they are not afraid of dying, even though their approaches are very different from one another. Surprisingly, almost all of our interviewees, except R., stated that they are *not* afraid of dying. It is likely though that those guests in the hospice who were afraid of dying, and possibly were even in denial, did not agree to talk to us. Self-selection might thus have taken place. However, quite a few of our interviewees admitted that they were afraid of pain. They did not want to suffer when time was up. W. (50 years old, in a relationship, no children) even admits to having considered suicide when he was first confronted with his diagnosis:

> First, you think when they tell you that you are seriously ill, that you will jump off a highway bridge or in front of a train. All nonsense. That does not work. You will only cause other people trouble. Think of the person driving the car or the train.... What do you want to do instead? You could blow up your house with gas; that would work. Gas is available. But how many people do you want to aggravate? On TV it all looks, well, I don't know.... What do you do now? Taking pills. You just visit three physicians, have them prescribe

> sleeping pills, they would not notice. And then you take 60 to 80 at once. But then you have to tell yourself: All of that does not make sense; wait for what the doctors say. You start hoping again.

According to the World Health Organization, about one million people commit suicide every year. That is equal to a global mortality rate of 16 per 100,000. In two thirds of the countries worldwide, suicide rates are highest among the male elderly. Mental disorders and physical diseases are additional risk factors for suicide in Europe and North America (Eisenwort, Heinrich, Schuster, Willinger, & Berzlanovich, 2007; WHO, 2012). One reason for *not* committing suicide after considering it (like W. did above) are negative effects the suicide would have had on others, for example, those who would find the body. In fact, every suicide has an impact on approximately six other people. In the case presented above, W. already includes this in his considerations. He describes how his reaction changed from wanting to end his life immediately to hoping again. Later, knowing that there is no hope anymore, he still does not consider suicide again but says that he will fight for every day and will try to make the best of it all. In fact, after the interview, his girlfriend was able to stay at the hospice with him to make the best of the remaining time—together.

Temporality, Finiteness, and Eternity—What Is the Bottom Line?

First, there are questions that cannot be answered by scientific examinations or experiments. We lack appropriate research participants when we try to answer questions like, "What comes after death?" Scientifically, we have to accept limits. There is no one we can ask; nobody has really been "on the other side," assuming there is one. Even those patients who were revived after being clinically dead can only provide limited data. And the data they provide is interpreted in many different ways. Just taking the description that some patients see a "light at the end of a tunnel" has given rise to a vast and diverse number of interpretations. With the physicist Niemz (2007), we might discuss the possibility that the tunnel impression originates in the soul's acceleration to light velocity; a physician might disagree and say that the tunnel impression is simply triggered by a lack of oxygen in the brain, and a priest might interpret the light as the door to a better world or even heaven. Who is right?

Second, I noticed that I did not use the term *temporality* throughout the chapter. It only appears in the title. The observation is surprising since, in German, one of the synonyms of dying is *das Zeitliche segnen*, which, using dictionaries, is figuratively translated as "meeting one's maker," but literally translates into *blessing the temporal*. The expression has its origin in the Christian church, when all that was left for the dying when facing the eternal was to bless the world and loved ones they would leave behind, which they did through or with the help of priests. Today, *blessing the temporal* has lost its religious notion; people use the term when talking about death, and this is what we wanted to do with our project more openly. The aim was to show that life is a temporal phenomenon and that death is an integrated and inevitable part of it, which makes it lose some of its frightfulness when seen as such. Another effect of the encounters we had in the hospice is described by Elli (27 years old), one of the students who supported the project:

> All humans have to die. At points in time that seem random. Lifestyle or character does not matter; there is no fairness in it that I might hope for ... Among the guests in the hospice are people my age—and I think my future is endless. It is not. Not naturally. That's what I became aware of. And it makes me more attentive with the life that is happening right now.

We were thus dealing with many different ideas of time and temporality: the finiteness of life; death, as an undeniably part of human life; the ideas of what will happen after death (see preceding paragraph); the continuity or discontinuity of life courses reflected in the stories (*narratives*) by the hospice guests; and, with this, the subjective and (un)conscious experience of time flow. Discussing the latter has a long history. If we only look at Strauss' and Hofmannsthal's opera *The Knight of the Rose*, which premiered in 1911 in Dresden, Germany, we find one of the main characters, the Marshallin, struggle with the following:

> She "concludes that Time possesses the sublime mystery of life: Time is that eternal measure of change that relentlessly alters one's body and soul. She realizes that each person copes differently with the inevitable progress of Time.... The Marschallin muses that others seem to know nothing of Time, nor care about it.... She acknowledges that Time relentlessly flows onward, and each of us bears its mysteries differently, nevertheless, Time, like the sand in an hour-glass, cannot be grasped or held because each irreversible moment of the present vanishes in an instant. Time becomes the recurrence of yesterday into today; the futile persistence of the younger self inside the inexorably aging body; Time viciously surrounds us, internally gnawing at us, and externally shifting our faces and torturing our bodies. (Fisher, n.d., p. 7)

Chronobiologists have tried to grasp internal "time markers" that are characteristic for all humans, showing, for example, that a basic precondition to being able to perceive time is the recognition of sequence[3] and the ability to remember events. Even though individuals usually do not question the fact that time "passes," the experiences of how fast or slow this happens differs tremendously between individuals and situations. The Marshallin, as described in the example above, struggles with the "relentless flow of time." Indeed, time does seem to pass quicker when we are waiting for something that we are trying to avoid but cannot. This is true for some of the guests in the hospice as well as for the Marshallin. Time also "flies" when we are not aware of it: when we are distracted or when we are caught in a certain activity (*flow* experiences[4]; Henk, 2013). Instead, time seems to pass very slowly when we are bored or have to do something monotonous. When these situations are remembered later, the *time paradox* can be observed: The individuals for whom the time flew feel like a long time has passed, whereas the others for whom it was slow-moving feel the opposite (Wittmann, 2012). These examples show that individuals experience situations and time very differently, as shown in many of the citations above. Not only understanding the physiological influences but grasping the individual experiences sheds light on how and what kind of support is needed in times of *rupture*—in our case for both, the guest and loved ones. The hospice is quite challenging for this endeavor since "time is limited." Guests who are still in denial or angry cannot just switch to acceptance and telling coherent stories. Some of them will die while still being disturbed by the experienced discontinuity, even if therapists are available at all times. Knowing this, the bereaved have to tell their own coherent stories, accepting the fact that they cannot positively influence the state of mind of the deceased.

Was being interviewed by our team helpful or stressful to the guests and their families in the hospice? I would say both. Some of them enjoyed telling their life stories; others became more aware of what they will miss since they could not distract themselves any longer. Being sad, nevertheless, was part of their story, so none of them took the offer to skip questions or end the interview early. It is important to note that narratives are always co-constructions. Our present and the questions we asked guided the course of the narratives. The stories would have been different ones if told at a different time, to a different audience, or for a different purpose. This observation does not diminish the validity of the data though. The interviews reflect the thoughts and meaning-making processes of the interviewees at the time, which are relevant to helping others (e.g., visitors of the exhibition, family members, future hospice guests) grasp the atmosphere in the hospice as well as the individual pathways of its patients, including all ups and downs.

Third and last, I must admit that it was odd at first to write this chapter as I have done it. Even though we collected "data," it did not make sense to follow the usual structure of a scientific article, starting with a theoretical introduction before introducing the research questions, methods, results, and discussion. The editors had explicitly asked the contributors to this book not to write a literature review, but to introduce our work. This is what I have done. Of course, I still write articles that are published in peer-reviewed journals and that follow a strict layout, but I am very thankful to have had more leeway in this chapter. This way, you get an idea of how we tried to get access to a whole different audience. And indeed, we have received very positive feedback from visitors to our exhibition, including an adult daughter of one of the women we interviewed. She said that by reading her mother's citation, she finally understood what was on her mother's mind shortly before she died. The mother had not dared to talk about it because she did not want to burden her daughter with it. We are hopeful that the exhibit provided a means of closure for both of them.

ACKNOWLEDGMENTS

This chapter could not have been written without the support of the guests in the Braunschweiger Hospice. We thank all of them for sharing their thoughts and stories so openly. The great teamwork of Birgit Sobiech, Lisa Rutter, Andrea Mannebeck, Elli Raabe, and Marie Müller, and, of course, the hospice staff is to be specially acknowledged.

NOTES

1. People staying at the hospice are intentionally not called patients, because they are neither in a hospital nor are they there for a particular medical treatment. They are in the hospice to die in a humane and person-centered manner, in an environment that allows them to do and experience things at their own pace. During this time, of course, everything is done to minimize their pain. For an overview of the modern hospice movement and philosophy, see Saunders (2000).
2. My thanks to Roger Bibace, professor at Clark University in Worcester, Massachusetts, who called my attention to this publication and with whom I discussed its different implications.
3. We can detect sequences when there is a minimum of 30 milliseconds between two events.
4. Researchers believe that flow is experienced when the brain is in a state of "transient hypofrontality", a transient downregulation of functional networks in the prefrontal cortex–the area of the brain in which the "most sophisticated levels of consciousness" are believed to be located (Dietrich & Stoll, 2010, p. 165).

REFERENCES

Berk, L. E. (2009). *Development through the lifespan*. New York, NY: Prentice Hall/Penguin Putnam.

Dietrich, A., & Stoll, O. (2010). Effortless Attention, Hypofrontality, and Perfectionism. In B. Bruya (Ed.), *Effortless attention: A new perspective in the cognitive science of attention and action* (pp. 159–178). Cambridge, MA: MIT Press.

Eisenwort, B., Heinrich, M., Schuster, A., Willinger, U., & Berzlanovich, A. (2007). Suizide im Alter. Demographische Faktoren, Motive, Methoden und Abschiedsbriefe [Suicide in old age. Demographic factors, motives, methods, and suicide notes]. *Rechtsmedizin, 17*(6), 359–362.

Erikson, E. H. (1970). *Jugend und Krise. Die Psychodynamik im sozialen Wandel* [Youth and crisis. Psychodynamics during social change]. Stuttgart, Germany: Klett-Cotta.

Fisher, B. D. (n.d.). *Der Rosenkavalier: A romantic comedy in three acts. Music by Richard Strauss. Adapted from the Opera Journey Lecture Series.* Online Bookstore: Opera Journeys Publishing. Available online at https://books.google.de/

Greshake, G. (1988). *Tod–und dann? End–Reinkarnatio –Auferstehung. Der Streit der Hoffnungen* [Death–and then? End–reincarnation–resurrection. Controversy of hopes]. Freiburg, Germany: Herder.

Gronemeyer, R. (2008). *Sterben in Deutschland. Wie wir dem Tod wieder einen Platz in unserem Leben einräumen können* [Dying in Germany. How death can be brought back into our lives]. Frankfurt, Germany: Fischer.

Henk, F. (2013). Spielend einfach die Zeit vergessen: Flow Erleben. In M. Watzlawik (Ed.), *Kreative Entwicklung: Beschreiben, Verstehen, Fördern* [Describing, understanding, supporting creative development] (pp. 199–213). Marburg, Germany, Tectum.

Knaevelsrud, C., Kuwert, P., & Böttche, M. (2013). Lebensrückblickstherapie bei Traumafolgestörungen [Life restrospection therapy treating patients with posttraumatic disorders]. In A. Maercker & S. Forstmeier (Eds.), *Der Lebensrückblick in Therapie und Beratung* [Life retrospection in therapy and counseling] (pp. 121–137). Berlin, Germany: Springer.

Knüttel, S. (2008). *Der Tod–Vom Alltäglichen zum Tabu* [Death–from the mundane to taboo]. Retrieved September 2012, from http://www.ard.de/-/id=899044/1xxvbyk/index.htm

Kübler-Ross, E. (1986). *Leben bis wir Abschied nehmen* [Living until we have to say good-bye]. Gütersloh, Germany: Verlagshaus Mohn.

Niemz, M. H. (2007). *Lucy im Licht. Dem Jenseits auf der Spur* [Lucy in the light. Tracing the afterlife]. Munich, Germany: Droemer Verlag.

Saunders, C. (2000). The evolution of palliative care. *Patient Education and Counseling, 41*(1), 7–13.

Spiegelhalter, D. J. (2008). Understanding uncertainty. *Annals of Family Medicine, 6*, 196–197.

Taghian, A. G., Smith, B. L., Erban, J. K., & Thomas, C. (Eds.). (2009). *Current multidisciplinary oncology: Breast cancer. A multidisciplinary approach to diagnosis and management.* New York, NY: Demos Medical.

Wittmann, M. (2012). *Gefühlte Zeit: Kleine Psychologie des Zeitempfindens* [Felt time: Little psychology of the sense of time]. Munich, Germany: Beck.

World Health Organization (WHO). (Ed.). (2012). *Mental health: Suicide prevention.* Retrieved September 2012, from http://www.who.int/mental_health/prevention/suicide/suicideprevent/en/

CHAPTER 9

TIMES OF ILLNESS AND ILLNESS OF TIME

Maria Francesca Freda,
Raffaele De Luca Picione, and Maria Luisa Martino

BEYOND THE ILLNESS NARRATIVE PERSPECTIVE

Every illness experience generates a discontinuity that impacts on some of the everyday life processes with different levels of intensity. Generally, the illness path is considered as something breaking the relations with others and the temporal dimension implied.

The difficulty, during illness, such as in carrying out simple actions, the frustrating inability to take part tin social life, the perceived and underlined condition of difference compared with the others, are all situations and experiences requiring the activation of new processes of sense making in which the sick role, unlike that the doctor's role, has to be learned and discovered, aimed at defining new ways of establishing relationships and expectations (Boesch, 1977). The outcome of such processes, in our opinion, is not something new and unrelated to any historical or social framework; it is rather the expression of the ways of facing the crisis generated by the bifurcation point represented by illness, within the construction of new processes of mediation with one's own story and the social context.

Temporality: Culture in the Flow of Human Experience, pp. 231–256

Thanks to the biopsychosocial model (Balint, 1957; Engel, 1977) and the integration into "evidence-based medicine" of the new-born "narrative-based medicine," the illness was considered as an expression of the person in his/her whole complexity rather than the condition of a defective organ. It would be pointless to make a prognosis without paying due attention to the patient, since it would be impossible to deeply understand the pathological process (Greenhalgh & Hurwitz, 1999; Hillmann, 1984; Kleinman, 1988). Our discussion has been inspired by this extended perspective and, specifically, by the *illness narrative*. Thanks to this approach, the medical practice and the care process have been connected with the patient's subjective experience that takes place within narration, intended as a device able to reconstruct the representations of the Self and of the world in order to encourage processes of "care" of the Self. We would like to underline that the illness of the organ and the illness of the person take place throughout different temporal dimensions: the temporality experienced by the person seems to be broken and interrupted by the actualization of the illness, whilst the temporal dimension related to the organ finds a new and systemic organization into the medicine's temporal system. The time related to the organ is measured through the complex temporal framework defined by the diagnostic processes and by the stages of the therapy, whilst the personal temporality seems to be suspended.

Using the approach of the illness narratives, a large amount of studies focused on the experience of time in illness, highlighting how the onset of a serious disease is narrated by the subjects as an autobiographical interruption, a temporal rupture (Simão, 2003; Zitton, 2006; Zitton, Duveen, Gillespie, Invison, & Psaltis, 2003). The terms used by the patients to narrate their illness experiences highlight how serious diseases can turn the temporal frameworks upside down (Bulow, 2003). Such conditions generate ruptures implying a representation of time characterized by a sense of uncertainty about the future (Hjelmblink & Holmström, 2006) and a sense of interruption of one's lifetime strong enough to endure over the years. The phenomenological perspective, predominant in the illness narrative literature, presents us with the observation of how, from a subjective standpoint, the time of illness is interpreted and represented as a broken life trajectory, a loss of the sense of continuity, replaced by the discontinuity due to the time of medicine, that is perceived as something external to the Self and to one's own field of life.

From this standpoint, narration itself expresses a "biographical disruption" (Bury, 2001), a sudden deviation of the hypothetical continuous line representing the temporal process in lifetime, and thus narration is assigned the task of healing this wound by reconstructing the sense of continuity that is interrupted and lost.

From a phenomenological point of view, then, illness breaks the sense of continuity and arises as a dramatic experience of discontinuity. On one hand, such a perspective underlines the crisis of the subject's sense-making systems and the necessity to pursue a meaning in order to interpret the ongoing experience; on the other hand, it seems to promote and to reify, in scientific terms, a continuous and linear conception of time, relying upon common sense, in which what needs to be fixed and reestablished is the sense of continuity interrupted by illness.

Narration is assigned several functions, such as the reconstruction of continuity story life, the relocation into one's story frame, the reinforcement of pieces of time, helping the patient to reestablish the state of the self, to fix the breaks and to realign present, past, and future (Murray, 2007). Such a perspective, within the dynamic between continuity and discontinuity, focuses on the processes of continuity and the pivotal role they play in securing the processes of construction of a sense of the Self.

OUR AIMS

In this chapter, we will bring forward some of the statements mentioned above, considering that the construction of the Self is always secured by the arising of some discontinuity within the flow of the events (Valsiner, 2001, 2011). According to this perspective, the crisis of linearity is assumed as a precondition to every process of sense making. Illness represents an element of discontinuity able to put a strain on the systems of continuity as well as the processes of mediation between continuity and discontinuity, impacting not only on the first but also, and maybe mostly, on the latter of these dimensions. Illness is an event of discontinuity that saturates on itself the processes of sense making (Freda & Martino, 2014). From our point of view, besides reconstruction, reinsertion, re-creation, realignment, relocation, we will assign narration also the function of a semiotic device mediating the connection between continuity and discontinuity, aimed not only at carrying out the previously mentioned functions, but at creating new dynamic relations among them as well.

In sum, we recognize the illness-narrative approach, the merit of which lies in having defined the importance of using a narrative device within illness contexts, underlining how the illness event generates a request of meaning by the ill person, along with the urgency of a sense making about the event in question. We are less willing to welcome, instead, in terms of temporal perspective, the sense of break, of reversal, of a broken and/or interrupted arrow, as it is proposed by the subject's experience.

Furthermore, we would like to explain the main topic of this chapter that is the idea of narration conceived as a process of mediation between continuity and discontinuity, leading to the pursuit of a new temporal synchrony.

Viewed in this way, we assume that the process of life itself relies upon the possibility to argue the intertwining between continuity and discontinuity of one's experience. Our aim is, then, to promote a model enriched by further theoretical cues about temporality (Sato & Valsiner, 2010) able to process a new and more complex vision of the phenomenon, leading to the opportunity to understand the role of semiotic reorganization (Valsiner, 2001, 2011) and of cognitive processing and sense-making operated by the ill subject.

From a Linear Perspective to a Complex Perspective of the Sense-Making Processes

With regard to our discussion, the linguistic indicators proposed by Brockmeier (2000) are interesting, since they are aimed at exploring the temporal perspective emerging from the narrations. He describes three different narrative orders (Brockmeier, 2002), corresponding to three different ways of meaning construction: linguistic, semiotic, and discursive.

Brockmeier considers such orders together as a *mnemonic system, a mnemonic space of memory and oblivion, where the temporal orders of the past and present are constantly recombined* (2002). The author also distinguishes six narrative models based on his studies (the linear, the circular, the cyclic, the spiral, the static, and the fragmentary), different in the way the narrator uses verbal tenses, figures of speech, time tropes, and other discursive and narrative features (Brockmeier, 2000).

We would like to underline, once again, that the construction (and consequently the analytic approach of study) of a narrative time is not solely a positive dimension visible through the terms and specific tropes used, but also a construction of meaning taking place starting from the opportunity to rebuild the continuity and the integration of different times.

In these terms, "what is said" has an *emergency relation* within the subjective field. "What is said" is a form, a gestalt acquiring and producing sense with regard to the field from which it arises. Speaking about time and temporality means then speaking about the way a semiotic network is built in relation to a phenomenon (Brockmeier, 1995). Anything about temporality is expressed by means of narration, and then it is conceived as a model of time, a linguistic and psychological framework attempting to organize in a synchronic way the diachronic dimension of our activities.

Narration appears as the most suitable device for the construction of time, representing the unique way it can be expressed (Brockmeier, 1995). In a synchronic perspective, narration expresses a process of semiotic connection among the different levels of generalization of time (diachronicity becomes synchronized).

Considering the autobiographical narratives, we see, furthermore, that they are not solely constructions on time but rather on times (Brockeimer, 2000), including and evoking different forms and orders of time, creating a multifaceted plot of the human temporality (Brockeimer, 1995).

In this representation, time stops being conceived as an *arrow* (linear model on which all the epistemology and the methodology of the positive sciences have been based), requiring then a new modelization. The TEM—Trajectory and Equifinality Model (Sato & Valsiner, 2010)—that defines as central the conceptions of *equifinality*, of a *bifurcation point* and of *obligatory passage point* (Sato & Valsiner, 2010) is interesting and full of stimuli. We consider this model suitable for a complex and articulated definition of the nonlinear development trajectories of the dialectic exchange between the subject and the contexts he is embedded in, consequent to the crisis of the connection between continuity and discontinuity.

Illness narratives report, in each singular situation, specific issues expressed in specific ways, and this is not the result of an incidental selection among infinite possible topics; it is a specific way for the meaning to emerge from a narrative in which the significance is built over the historical time. It is also the expression of a semiotic crisis concerning the processes that regulate the relation between continuity and discontinuity of the experience, generating a local and contingent significance able to interpret past and future through a new contextual framework. Narration assumes the function of a symbolic mediation device, aimed at managing the crisis generated by illness, serving a new transcription of the past and the possible future scenarios (Simão, 2003; Zitton, 2006; Zitton et al., 2003)

Accordingly, illness is only one among the several discontinuity situations people face during their lives. If we think of marriage, a child's birth, changes concerning the job position covered, we realize that we often deal with the crisis of the relation between continuity and discontinuity related to the processes of development.

Illness, as well as every other semiotic rupture, can activate the construction of brand new semiotic interfaces, more adjustable and able to mediate both intrasubjective and intersubjective relations with the purpose to create a new subjective synthesis of one's experiences within the specific intersubjective context, instead of reestablishing the social norm and standard.

TEMPORALITY, NARRATION, AND ILLNESS

Narration, in its quality of a semiotic process, can be considered as a process of organization and articulation of signs, all different from each other (generically, but not exclusively, words) aimed at defining a meaning with the capacity of being both an expression of the subject and an element of connection and relation with other people.

Benveniste (1966) reminds us that in any communicative situation, the subject makes use of the word and of the discourse to give representations of himself, as he wants to consider himself, calling for the others to ascertain. Then, the activity of a linguistic production is a semiotic device within which subjectivity and intersubjectivity always coexist. Consequently, the subjects, to be qualified as *speaking* ones (organizing signs and producing sense) need to share the same *symbolic space*. It is exactly the sharing of such a symbolic space that represents a *shared time*, characterized by a lower rate of transformation of the sense-making processes, allowing the synchronization and interaction among the speaking subjects. We need to recognize the need of a *shared cultural structure of time* in order to get the chance of narrating a personal experience. More specifically, the subjects are placed within a field made of signs that can be shared only when the times of the developing processes are synchronized.[1] It is specifically relevant when analyzing the change of sense, as well as the relation between continuity and discontinuity and the attempt to leave behind a linear representation of time, toward a more complex nonlinear, nonpredictable perspective.

Sharing meanings within a community means sharing a symbolic structure. The speaking activity of people is a way to make discontinuous, within the present time, the intersubjective field where its continuity is secured by a multipotentiality dimension (Salvatore, 2004).

We can interact with others thanks to the existence of a superordinate sense dimension, providing interaction channels previously made up. For instance, at the post office, at the refectory, even driving our car, with our partner, with our job supervisor, or obviously when we narrate a story, there are some symbolic superordinate dimensions that regulate the interactions and their times.

What is unsaid is not a lack of regulation but rather the sharing of something *taken for granted*, allowing the interaction within a specific relation. What is unsaid refers to a lower[2] temporality compared to single actions.

The Said and Unsaid in the Narrative Temporality

We find it interesting to observe the synchronic relation between what is *unsaid* and what is *said* in *narration*, from a semiotic and temporal point of view. In fact, we state that *what is said is the expression of the broken sense*. This statement helps us explain that not any sentence has the same value as any other one. When a specific sentence is expressed, we are encouraged to believe that there is a tension dynamic between continuity and discontinuity of the semiotic process.

Continuity exists in what is unsaid (it does not need to be said because it is a synchronic processuality among different subjects; it is what remains unaltered over time), while discontinuity arises from what is said (from *when* and *where*, from *how*, *by whom*, and *to whom* it is said).

Within the illness narration of a subject, we observe the passage from an organization of meanings taken *for granted* (culturally shared) that drives the trajectories of one's own development and of the contingent actions, to a moment when the sense seems broken and subsequently recomposed with regard to the organization element represented by the "illness."

Times and rates of the social life are no longer a dimension of guidance for the subject who gets deprived of the time of his own experience with the others. The experience of loneliness and social isolation generated by illness often reported in the narratives is observable in the emerging temporal disorganization. Times are organized through extremely subjective modalities appearing at first glance confusing, chaotic, superimposed, and inconsistent, but a deeper analysis such as the dyscrasia of the temporal order seems to be the result of the ongoing processes of sense making about what is happening at that moment. It is the attempt for the pursuit of a new intra-/intersubjective semiotic organization in social life (Laskovski & Simão, this volume). The time of succession, that is, of a linear nature, is not a time for the self, but rather a dimension emerging from a wider basic temporality, shared and omnipervasive. There is the need of a subject observing and separating in single events the continuous flow of transformation, providing them, at a subsequent stage, a new order.[3]

As a matter fact, in order to "catch" the time, we need a point of view, a perspective defined within space/time capable of giving time a start. For instance, looking from a train' s window, we can see closer objects moving away more fast than farther ones; such objects assume different temporal ways to be permanent within the perceptive-phenomenological field, starting from the observer's "perspective of the present" (Bateson, 1979). Which is to say that the narrator, by reorganizing and cognitively processing his/her own experience, builds a specific space-temporal system of reference, starting from which it is possible to provide the events with a narrative order.

CONCENTRIC CIRCLES OF TIME

General Hypothesis

Every single *relation* can be considered as a *temporal process* taking place among different systems by means of contact, going through several stages: a first phase of perturbation, followed by the selection of specific elements of fine-tuning; the production of effects, the establishing of a more pervasive relation and, finally, the conclusion.

The establishing of a relation contingent upon and local in the present time requires a time characterized by a more extended processuality, going on slowly enough to hold such an organization.

We believe that the transformation of a system as well as its relations within its specific context are intertwined by a precise temporal connection. In fact, the superordinate cultural dimension secures the continuity of the symbolic mediation devices and the possibility to utilize them. We are not stating that there is an absolute and abstract time of culture in which symbols and categories keep remaining pure and ideal; what we are stating instead is that the cultural artefacts can be utilized thanks to their temporality presenting an extension cycle wider than their contingent use within the lives of the subjects, so that they can utilize them in the present time, mediating between history and future. From this standpoint, memory, perception, enunciation, and sense making are intended as extremely cultural, ecological, and relational constructions.

We hypothesize then that the context of belongings (intended as macrosystem) must have a more extended temporality to allow each of its components to evolve as specific units. Such temporal organization makes sure that the single members are not simple components but rather units characterized by a specific developing identity. The superordinate system allows the lower levels to face discontinuity and change, securing a superordinate dimension of continuity.

The everyday experience provides a large amount of examples like the ones mentioned above, once we have hypothesized such a temporal relation. For instance, the cells in our body are born, live, and then perish generally within a few days, though they are in a temporal framework more extended than our lives. By the same token, we are born, live, and then die within a social and cultural community which exists before us and persists in respect of our individuality.

Modelization

Every single process, as we introduced in this paragraph, has its own life *cycle*, life and extinction made possible by the continuity of time secured by a superordinate level of temporality.

As represented in Figure 1, the temporal processes within system Z, creating changes and transformations, are allowed by the continuity that the belonging to a superordinate system, such as Y or X, guarantees. It means that Z's processes of transformation are constantly interfaced with superordinate temporalities, giving the whole system (Z) the chance to be a unit in change.

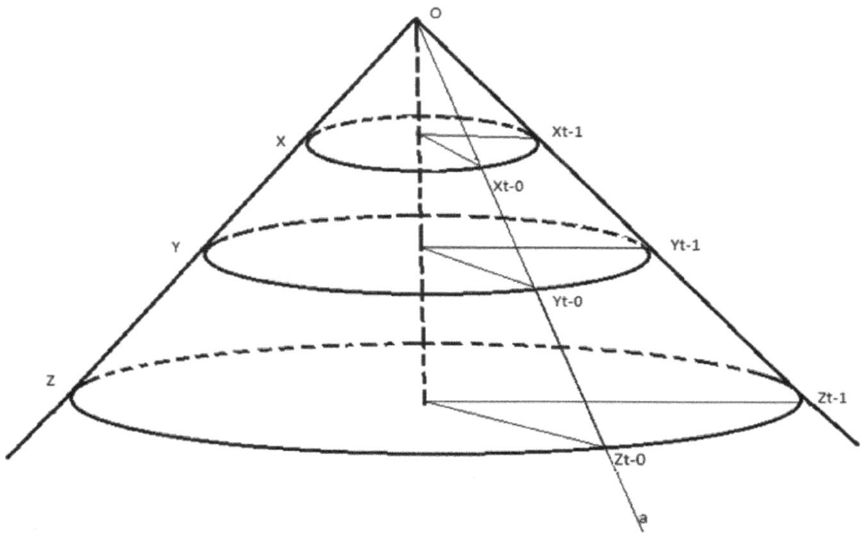

Figure 9.1. Concentric circles of sense-making processes.

We shall try to define clearly such a temporal relation. How is it possible for different temporalities to coexist within a unique system? In which ways do speed and length of some processes allow the interrelation with some others?

We consider as a reference the time interval t-0/t-1. We can immediately observe that, despite that the time interval is always the same, moving toward external circles, the space to cover gets enlarged and thus the speed at the level of system Z will be higher than Y level, and then higher compared to X or O level (in which it tends to be annulled). This way, even though the time interval remains the same (t-0/t-1), the concentric circles are characterized by different speed rates, increasing moving downward and decreasing moving upward (in point O the speed is annulled and there is no movement either).

Explanation of the Model

We will use this scheme to discuss the process of sense making in which human beings are engaged. In every moment of our life, we are involved in sense-making activities, made possible by our constitutive being in relation (in biological, subjective, intersubjective, and cultural terms). Such activities are connected by means of a relation of discontinuity or rather, of transformation, reestablishing, and change, since each experience is properly an interruption of continuity, an *ex-perior*, namely, an experience of crisis.

How is it possible for a crisis to occur without the risk of causing trouble for the person's survival?

According to the hypothesis that the sense-making systems are organized as concentric circles, the temporal continuity of the systems with increasing levels of abstraction and generalization secures the possibility of transformation for the less generalized and abstract ones, including the synchronization among different processes that would otherwise obstruct each other.

Considering the system Z, for example, as the one representing the everyday life, we can observe that a break of continuity presents a higher transformation rate compared with other systems.

In effect, while we are engaged in any action, the time interval t-0/t-1 produces a larger amount of meaning reifications (and with a shorter half-life) on system Z, with respect to other levels.

It is fundamental to remark that among the meanings in Z and in the other orders, there are different levels of abstraction and generalization. The meanings belonging to levels X and Y have a higher capacity of generalization and abstraction, able to support several transformations and changes taking place within level Z. In this sense, we state that the more abstract levels secure the continuity dimension so that the more contingent levels can frame a transformation (and/or some relations among them).

This way, according to TEM (Valsiner & Sato, 2010), the *broom of time*—giving a higher complexity to the linearity of the "arrow of time"—can be considered not solely as the series of future potentialities taking place, time by time, through bifurcation points (Valsiner & Sato, 2010), but as the contemporaneous existence of several time cycles, with different extensions and interactions between each other, able to secure both continuity and discontinuity of the experiences lived by the involved subjects. The chance of survival of an organism requires the capacity to make a synthesis of the different temporalities in action (appearing as episodes of discontinuity). Such synthesis is guaranteed in part by the superordinate wider systems giving cohesion, continuity, and relational opportunities and in part by a continuously broken sense.

When there is a possibility of relation (and then of episodes of discontinuity), it is due to the possibility of regulating two or more different temporal dimensions through a superordinate and supersystemic one so that their synchronization provides continuity. Also, the *hypergeneralized signs* (Salvatore & Valsiner, 2009), from our standpoint, have a wider and slower temporal extension (in their processes of self-transformation) able then to regulate the action development within the more external circles.

At the same time, though, we recognize that contingent and contextual signs are able to produce more generalized transformations. We can refer to the interesting semiotic study of the Russian author Lotman, entitled *Culture and Explosion* (1993/2009), discussing the relation between explosive processes and gradual processes. According to Lotman, all the explosive processes take place within a dynamic dialogue with the stabilization mechanisms. It is necessary to underline that gradual processes have a significant propulsive power. The relation between moments of explosion and episodes of gradual development is not to be represented only as happening through distinct and separate phases, since there is also a synchronic dimension of interrelation. Both explosive and gradual processes, within a structure functioning in a synchronical way, carry out important functions: they guarantee, respectively, innovation and continuity. In this way, the intersection of different structural organizations becomes the source of a dynamic; at the same time, each moment of explosion exhaustion can be considered as a turning point of the process (Lotman, 1993/2009).

We would like to highlight how our model is supported by a processuality, synchronically relaying upon both a *top-down regulation* (the top-down canalization of a specific abstraction and generalization relations, activating a field of attraction and continuity) and a *bottom-up regulation* (the single units interact with one another as they actualize and personalize the forces within the field, generating discontinuity), framing organizations of contingent and contextual processes of sense making. There is the creation, then, of areas of perturbation attempting to resettle on new balances in part already previewed/directioned, partially to be personalized.

Concentric Circles and Illness

Illness, in its quality of semiotic rupture, causes trouble to the relational flow between continuity and discontinuity of its processes of sense making. A break of the synchronization among different circles of sense takes place, generating a perturbation (with variable intensity) that claims to be held and absorbed by means of cognitive processing of new symbolic organizations and new trajectories of sense in order to guarantee the development of one's own identity and relational system.

Narration is the symbolic process fitting such a twofold need, constantly dealing with the issue arising from a temporal perspective. The experience of plotting a story is the attempt to reorganize the relation between continuity and discontinuity, whenever there is a crisis of the sense supporting, until that moment, the processes of development and maintenance of both subjectivity and intersubjectivity over time.

According to the graphic model here presented, we state that narration attempts to reorganize the relation among the concentric circles when, as a result of a semiotic rupture, some perturbations that can cause trouble to their synchronization are produced.

THE SEMIOTIC CRISIS BETWEEN CONTINUITY AND DISCONTINUITY: MODELS OF FACING THE CRISIS

The explosive process (of semiotic rupture) started by illness implies the need for a dialectical renegotiation among the different levels of abstraction and generalization, since the semiotic rupture needs a recomposition capable of guaranteeing a new dynamic balance between the continuity and discontinuity of the sense-making processes.

Illness represents a semiotic rupture able to create perturbations of sense at different levels (see Figure 9.2). The more disabling and serious the illness is, the more it will probably have consequences in the future, since it generates semiotic perturbations in the relation among the circles. For instance, a small wound (such as a scratch or hematoma), even creating a discontinuity to the experience, will not generate a crisis to the relation between the experience and the subject's identity and relational system (at most, the subject will reorganize his/her life throughout the period of prognosis and maybe he/she will create systems of generalization and abstraction according to which he/she will be more cautious and prudent). When a more serious disease occurs, instead, causing a strong sense of self loss, as well as a distortion of one's own corporal image and serious loss of abilities, the integration among the different circles strain, due to such pathemic-semiotic perturbations (Freda, 2008).

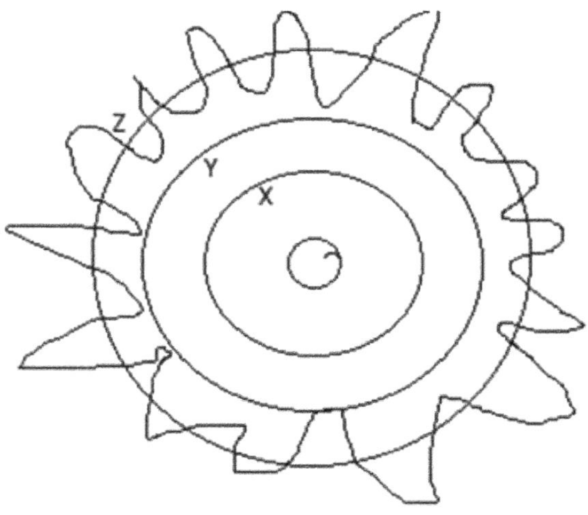

Figure 9.2. Semiotic perturbations.

We consider that long term, it is possible to create a new synchronization among the different concentric levels, once the sense-making systems have evolved as a consequence of the absorption of the semiotic perturbation generated by illness; whilst short term, it has generated a desynchronization and the semiotic survival is guaranteed by the capacity of the more generalized and abstract systems to hold the tumultuous variability of the ongoing experience.

We will report as an example the experience of a sportsman who cannot keep performing his competitive activity due to an injury and tries to face the crisis by utilizing the same beliefs and sense-making systems that worked until then in supporting his sport activity: to overcome his own limits, to accept challenges, to be engaged always at maximum, to never be discouraged. The continuity of some sense-making processes, then, is reread considering the new relation between continuity and discontinuity.

The phenomenology of the semiotic crisis produced by illness seems intertwined with multiple issues that, within our research, resulted as particularly significant. The wideness of the crisis seems to depend on the intensity of the shock wave impacting the semiotic processes that regulate the relation between abstract and generalizing systems and the variety of one's own life experience.

Serious diseases, causing significant inabilities (permanent or semipermanent) and risks related to survival, are the ones that properly bringinto question the symbolic systems of relation and definition of one's own identity. It does not represent a simple cognitive effort of redefinition of the subject's own adaptation domains; rather, we are dealing with an intense pathemic-emotional activity, engaging the diseased person in each field of his life, through an intense emotional oscillation among feelings of disillusion, depression, hope, need for support, regression, and so on.

Furthermore, it is interesting to observe the *dialectic between the subject and his/her social/cultural system of reference*. Such an area of interest seems relevant since, when someone experiences illness, being unable to take part in his/her social community life, they experience a deep sense of loneliness and exclusion. In this perspective, medicine (intended as the diagnostic and therapeutic system in a cultural and historical sense; Canguilhem, 1966; Gmerk, 1995–1999) can be conceived as the semiotic mediation device between the diseased subject and society, presenting him/her (in the best-case scenario) with categories of sense, bringing him/her back into the symbolic dimension of culture. In this way, the person affected by a disease, in a paradoxical way, is held in and put outside the society through the medical system. If on one hand, it defines a condition of deviation from a normal status, on the other hand, it is precisely this operation that links the person to the social and cultural system. Then, the *narration of an illness experience gives shape to processes of regulation and synchronization within the relation between continuity and discontinuity, transforming itself along a diachronic dimension*. The narrative activity gets transformed over time; in fact, in addition to being a synchronic process of reorganization of the experiences, it is also a diachronic process evolving and finding its organization over time.

For instance, narrations collected during the apical phases of illness, or within moments related to the diagnosis communication, are often characterized by confusion and semiotic fragmentation, whilst from the ones collected by the end of the illness experience often emerges the construction of criteria for the reorganization of the story, along with the cognitive processing of new semiotic organization devices aimed at reestablishing a coherence between the self and his relations. Such new semiotic forms, even not being the most suitable in an absolute way, constitute a process of resynchronization consequent to the illness experience.

It is important to us to underline the impossibility of defining linear causal variables determining the illness narration; we will rather try defining some areas of interest that we consider crucial to understanding, in a more refined and precise way, the process of narrative development experienced by the diseased person, from the moment of the semiotic rupture generated by illness to the pursuit of a new temporal perspective, able to synchronize once again the different circles of sense (see Figure 9.2).

The narrative path is aimed at setting a new form emerging from the crisis of the relation between continuity and discontinuity; what are the trajectories of sense contributing to the production of new forms? The amount of narratives collected is quite extensive, and we decided to choose among these samples five semiotic-narrative modalities, recurrent in the stories, in order to describe the organization of the relation between continuity and discontinuity.

We do not claim to create a taxonomy of narrative models, and thus we have not focused our analysis on narrative typologies, but rather we have observed the semiotic function carried out by the narrative sequences within the research of some solutions to face the crisis.

Semiotic Collapse

We hereby report some narrative sequences to better comprehend the process of sense making, considering different phases of illness (experienced by the subject or by his/her son), from the very first stages, going through the painful realization of the subject, to the beginning of the therapeutic path. Such situations can produce a serious difficulty in processing, in semiotic terms, the experience of broken sense (Watzlawik, this volume).

> As soon as I was communicated the diagnosis, the world fell apart around me, because I realized my daughter affected by physical disability would

> have missed her mum.... I felt like my mind was somewhere else, I felt a muffled atmosphere, my brain was able to perceive what was happening but, at the same time, wondering "is that possible?" This whirl of feelings and emotions lasted about one week. (Woman with breast cancer in chemotherapy)

Reading this narrative sequence, it appears that the categories of meaning, that until then had regulated the temporal integration between the subject and the external world, can neither be maintained any longer, nor are they capable of providing an appropriate reaction to the shock caused by such a traumatic event. After being communicated the diagnosis, the subject makes requests destined to be left without any suitable answer, concerning the necessity to "find the reason why," when facing a bifurcation point. The inability to find answers and reasons related to the illness experience, causing disorganization, introduces firstly a blank space of meaning, a semiotic collapse, showing as an absence of form and annulment of temporal perspective (being in a nontime), within which the different circles of sense become homogenized and then fall apart (see Figure 9.3).

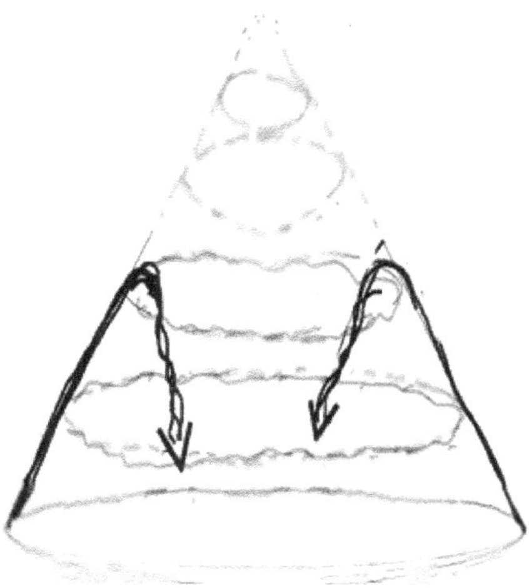

Figure 9.3. Semiotic collapse.

Salience of Moments

What appears as a lack of sense is actually the salience of the moment completing a process of sense making. Time appears as a chronological and linear dimension in its sequence of reified moments. The temporal perspective is meant to be focused on the more external circles within which events are presented, in their quality of lively concreteness, as salient moments. Narration is then settled in its function of chronology, a sort of *picture* of a static and hypostatized image of an event such as, for example, the day and place of the diagnosis communication, saturating the whole field of sense. In many of the narrations, especially in those produced during the heavier phases of the illness, the process of sense making is settled on the salience of the moment, or rather of a sequence of moments, presented as something concrete and as objects not transformed by the subjectivity of thought; the discontinuity generated by the detection of illness becomes something reified (see Figure 9.4).

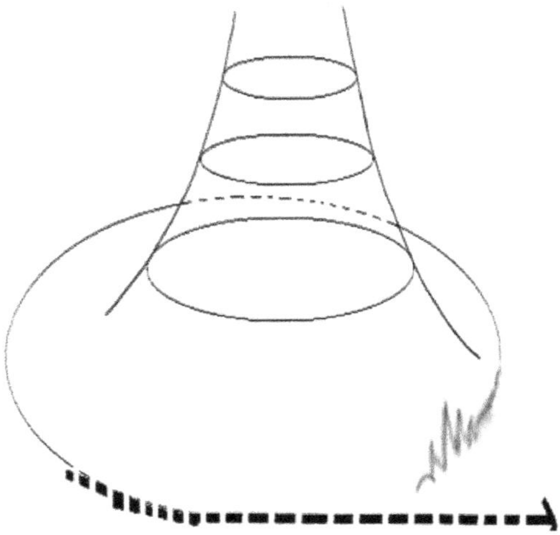

Figure 9.4. Reified discontinuity.

> Day by day I saw my daughter getting worse and worse, then one morning I decided on my own free will to let her take a blood count, despite her doctor kept saying her health condition was good. It was February 14th. When we went for the first aid request, we were suggested not to wait a minute longer. We were at the XXX Hospital and it was February 16th, 2007. On February 17th the doctor showed up at 12:00 P.M., introducing herself as a psychologist. Then I realized it was something serious. I was told the diagnosis would have been communicated by that evening, around 6:00 P.M. My heart was thudding desperately, because my pain was lasting since one month already. Finally, Dr. XXX communicated us it was a lymphoblastic leukaemia. (Mother of a girl with ALL in off-therapy phase)

In the wake of what is reported above, the subject gets sucked into a moment, into a concrete time becoming the space of memory characterized by the punctuation of single actions, single dates, single medical practices that still linger in the memories of these women and mothers.

When the time of the crisis experience gets stuck in the moment and in the fragmented activities of the medical practice, the future does not succeed in being a semiotic process, as it emerges from the narrative sequence we are going to report, since the reified temporal dimension cannot carry out any longer its function of regulating the relations (see Figure 9.5).

> I am not able to imagine any future, maybe I am afraid of what will happen to me. I can't see anything in prospect, neither positive, nor negative ... or I don't want to see anything. This is a moment of deadlock. I don't feel like making plans because there is a risk they will not be accomplished. I am trying to live hand-to-mouth. (Woman with breast cancer in chemotherapy)

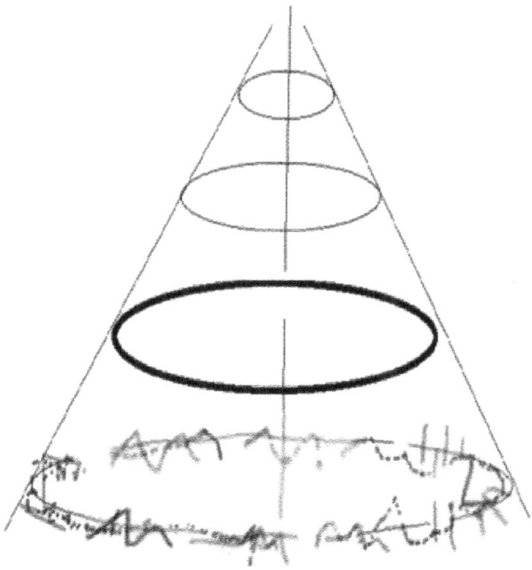

Figure 9.5. Temporal dimension cannot regulate relations.

A Point of Anchorage: The Salience of the Continuity Processes

Reading other narratives, we have observed how the subject reacts to the discontinuity of the rupture caused by illness through anchorage, hooking on trajectories of continuity able to secure, throughout the illness phase, the permanence of some points of reference and organization devices.

Within such a trajectory of sense, we would like to propose some narrative sequences where it is highlighted how the breaking and the disorganization, introduced by the illness detection, are faced through the opportunity to anchor to continuity processes, to the more generalized and abstract sense tendencies, in one's own life. We can think of more abstract and generalized circles able to hold the crisis of sense generated by illness, allowing, on one hand, the maintenance of a relation with a sense of continuity of the Self and permitting, and on the other hand, giving shape and meaning to a variety of events (housekeeping, working) introducing other levels of discontinuity and vitalizing the processes of sense making.

In other terms, illness, as a disabling and painful experience, becomes something to be faced thanks to the maintenance of a vital function, on a continuity level, by the processes of sense making, in order to deal with the vicissitudes of illness.

> I am living by pursuing a lot of projects that keep me alive: a new kitchen, the doors to be fixed, my dreams concerning my children's future, since I have also another son XXX. (Mother of a girl with ALL in off-therapy phase)

> The path has been a little hard, but I kept working.... They helped me, at work, making me feel needed. (Father of a girl with ALL in off-therapy phase)

The parents of ill children need to find a sense of continuity of the Self, in spite of the ongoing crisis, in order to get the chance to be in contact with the different circles, making the relation among them something dynamic. Referring to the examples mentioned above, we can observe how the more extended sense-making systems (to be a mother, to be a wife, as well as to be a worker, etc.) can carry out a significant function in facing the semiotic crisis.

With regard to our cone-shaped models, we can state that, even though some circles present a broken sense caused by illness, it is still possible to spot a dimension of support and of anchorage within more generic and abstract organization devices, characterized by a wider extended temporal perspective (see Figure 9.6).

The Homogenizing Coherence of the Negative Perspective

We will highlight in the following some narrative sequences showing what we have defined as a particular form of continuity of the experience, the continuity of the negative perspective.

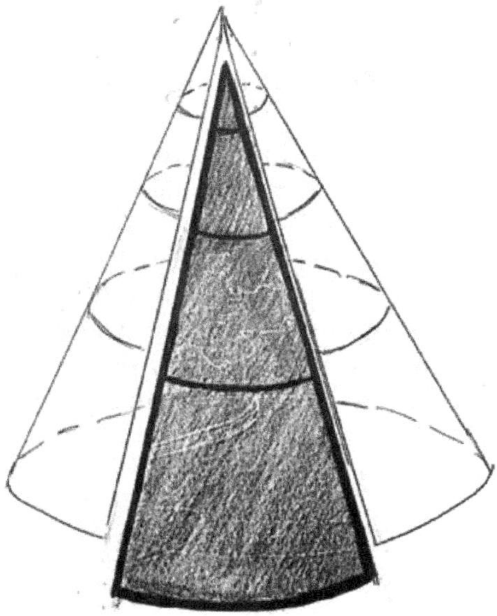

Figure 9.6. Dimension of support.

Narrative processes organize the relation among the circles relying upon the past experiences never sorted out nor processed. Illness and the diagnosis communication represent the sounding-board in which "the negative events of the past are summarized" into the moment of the diagnosis. It seems like a time spiral including past and present at the same time, overwhelming the two dimensions, making them adhere to a homogenization of the experience (Matte Blanco, 1975; Salvatore & Freda, 2011) in which it is reified; a continuous flow of the experience without difference, the time of pain.

> Previously, as a sign of destiny, my uncle died of cancer. (Woman with breast cancer in chemotherapy)

> When I look into a mirror, I see the same face my father used to have before passing away. (Woman with breast cancer in chemotherapy)

The continuity of the negative seems to become the semiotic organization device for these women; in the succession of life events, the negative experiences are constantly summarized at each turning point. From this point of view, it appears, as in a sort of continuous repetition, that the relation between continuity and discontinuity is based on the pain and on the return of something negative, as it was something already written in their stories over time. In other words, the dialectical relation between continuity and discontinuity, within the broken sense generated by illness, gets undone due to the emphasis put on meanings already known concerning past events with negative results.

Past events become relevant through the point of bifurcation of the illness. Then a fragmentation of the cone takes place (Figure 9.7), whilst a homogenization process only concerns some poor and scanty bases functioning as attractors of sense. The present is experienced as a recapitulation of the past, thus, there is a surplus of continuity in comparison with the small amount of images and memories concerning negative past events. We observe then, that intense and emotively significant symbolizations of past illness experiences cannot be processed in the light of the new present contexts.

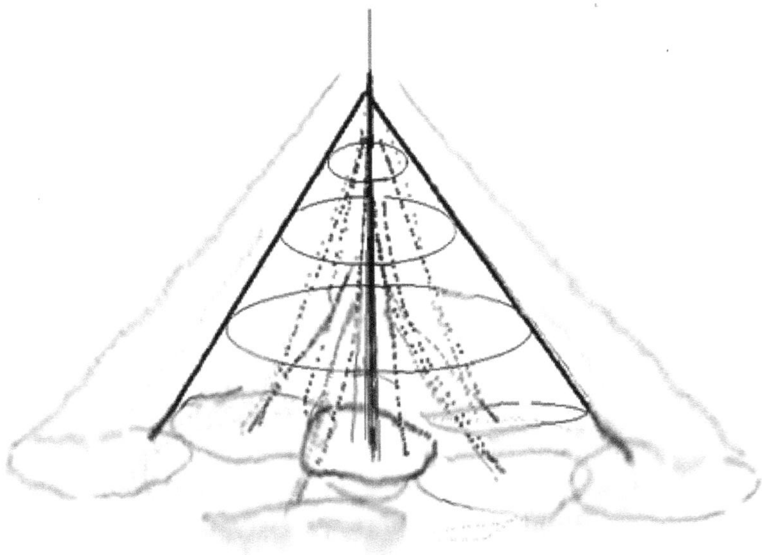

Figure 9.7. Fragmentation.

A Dialogue in Progress: Transformation of Relations

Lastly, we would like to describe another vector of sense through which it is possible to observe how illness (especially in case of chronic diseases) could be integrated, in semiotic terms, into more extended organization devices of one's own relational life and identity system.

For instance, within a study we are conducting on narrations produced by adolescents affected by multiple sclerosis, we have noticed how in some of them the narrated illness experience finds a fully functional collocation within the identity facets concerning the young subject.

> I did not properly realize; to me it was just a problem missing school and friends, since I didn't feel any pain, and I still don't, but now I am more aware of the risks I could take, and then I accept I have it and that's it ... it wasn't that difficult, maybe you realize later what you have gone through and then you are glad you were only 15 and ... well, the idea I have of myself is also based on this, now it is part of me. Nothing changed in my way of being, I just have to be more careful ... maybe if I didn't have this problem, I would have acted in a less cautious way.... I have to consider many things, for example last summer I couldn't go on vacation with my friends because I had the therapy ... but still I had some alternatives.... I just have to plan things better. (Young boy with multiple sclerosis)

In these cases, it is possible to observe how an adolescent organizes his/her everyday life and all of the related activities in an integrated way with respect to the therapeutic needs and the limits created by illness. Considering the everyday activities (such as doing sports, going to school, going out with friends, etc.) as supported by meanings belonging to more extended concentric circles, we observe how the "illness" meaning, absorbed within more abstract and generalized levels of the identity sense-making system, secures the autonomy and the care of the self to the young adolescent.

The integration of illness into one's own semiotic system of interface between the self and the world of the others becomes a resilience function (Garmezy & Masten, 1986) able to guarantee more complex and autonomous forms of development. Utilizing our metaphoric model (Figure 9.1), we could state that a higher level of resilience is made possible by the transformation of the whole cone, working on the true of its axis.

CONCLUSIONS

From a dynamic and morphogenetic standpoint (De Luca Picione & Freda, 2014, in submission), we consider narration as a *form* arising from a background. A form intended as a space-temporal expression of the forces in the field (Di Napoli, 2011). Its production has an effect on the field in terms of memory, since it organizes the forces in interaction. At the same time it is, on its whole, a product of the field. In this sense, words and narratives are, rather than the representation of what is visible, signs of the semiotic rupture affecting everything invisible until that moment requiring now a new organization, a restructuring of the field. Illness, in fact, brings to the fore, through its crisis of sense, the need to renegotiate the dynamic relation between continuity and discontinuity trajectories of one's own life.

That is the difficulty we encounter in reading narratives. There is an intertwining of diachronic and synchronic processes. On one hand, we deal with a story, articulating some salience throughout a symbolized time; on the other hand, we see the encounter among subjects in the present, defining the narrated story. The narrator, through his position (Hermans & Di Maggio, 2004), assumed with respect to who is listening, starts to plot a story. In this way, the diseased person brings to salience his/her own difference with the others, but he/she is able to narrate thanks to a community of significance shared dimension.

We state then that in order to *narrate*, it is necessary to communicate and to inform at the same time (Boesch, 1997).

1. *To communicate*, in the sense of *sharing* (sharing a symbolic space structured over a long time; settling on a certain processuality; having a time in common allowing the subjects to be part of the relational system).
2. *To inform*, in the sense of *creating a shape*, producing a content starting from the break/crisis of a relational system.

From this standpoint, the intersubjective semiotic *field* expresses in the shape of signs not solely the ongoing processes (revealing continuity), but also the ones dealing with a semiotic rupture (expressing discontinuity). Narration, seen as a process of dynamic exchange of salience and significance among different meanings, is both something attempting to provide a solution to the rupture of the canonical state (Bruner, 1990) and a way to keep staying in relation with the world.

When the more abstract sense-making systems are no longer able to interpret the discontinuity of some processes (when a particular event, such as an illness, cannot be signified) there is a time, a phase of perturbation related to some meanings that used to act, until that moment, in a silent way.

Time gets stuck, turned upside down; there is a crisis on a temporal perspective, intended as a process of sense making necessary to mediate the relation between continuity and discontinuity. Such a perturbation phase (seen as a bifurcation point) generates the need to reestablish a balance (or rather a new organization of the relations involved in the field), and that is the reason why only some meanings emerge instead of any other. It is the ongoing critical situation to define what has to be brought to salience.

In the light of the path we have outlined, illness narration tells us about a crisis of the temporal perspective; that is to say there is an illness of time, carrying out a significant semiotic function aimed at finding connections to construct a time of illness.

In this sense, narration is not only a meaning construction reacting to the rupture of canonical state, but also a proper process of appropriation of the cultural canonical state, through a movement of subjectification. Through the narrative experience, the subjects make an attempt at synchronization with the others and their culture. We need to comprehend that the narrative experience is a process of reinforcing a shared temporal perspective, within which the subject can find a space of belonging to live his present and future.

Illness has a sense, a meaning, and narration is the language to express it. The narrative discourse presents the subjects with a large space of connection with the others and reflection on themselves.

We could say that it is the unique useful space of experience, the only one necessary when living extreme conditions. Within the space of narrative, then, we have access to manifold ways to deal with existential challenges (Medved & Brockmeier, 2008).

When the crisis generated by illness can't undergo suitable cognitive processing, the processes of representation of one's own temporal perspective come to an *impasse,* disabling the support processes of development intended as an implementation of the systems of relation and interface with the related contexts. Illness is configured as a proper *illness of time*. It does not only consist of the loss of continuity within one's own life, as it was before the crisis, but rather it represents the loss of capacity of integrating the experiences.

When illness assumes instead the value of an experience integrated within the more extended framework of one's own life path, then we observe (despite painful memories of loss keep on existing) how the illness of time turns into "the time of illness," that is, a transition phase which has required a cognitive processing of the subject's way of life; a rethinking on his past, a reconsideration of his actions, and a revisioning of the images picturing his future.

Illness does not become a hypostatized device for the organization of the whole life experience, but rather it turns into a process of cognitive processing presenting, through the generated semiotic rupture, the opportunity to construct and to shape a new and more usable *form* of integration between trajectories of continuity and discontinuity.

Although the narrative experience is a spontaneous expression of the people and always contextualized within the social networks, we conclude with the importance of establishing a psychological support offered as an adequate space and time, addressed to the narrative of his own experience of illness that reconfigures an intersubjective space of sharing and development of new meanings and their integration. Such narrative space and time, placed at a proper distance by a traumatic acme of rupture, such as for example, in the remission phase of illness or in the off-therapy phase, allows acceptance of the uncertainties and feelings of confusion generated by the disquieting rupture (Simão, 2003) and a tolerance of the timing of the new integration.

The narrative of one's own experience becomes then not a mere description of the past experience or the illusion of a desire to restore the past before the illness but a social process of scaffolding (Freda & De Luca Picione, in submission; Wood, Bruner, & Ross, 1976) able to support the therapeutic process and the relationship of care that is being developed.

NOTES

1. By *development*, we intend the process of creation and implementation of new relations between the subject and his environment by means of semiotic devices characterized by a constantly increasing level of complexity.
2. As we will explain further on, "slowness" stands for wider abstraction and generalization.
3. A semiotic/temporal order is not solely a succession but also a basis of sense making and importance as it is demonstrated by the way our memory works. In fact, the memory of our past does not consist of a linear succession of events, but it is a semiotic modality of sense making about the events, moving from a present field of activation that builds pertinence.

REFERENCES

Balint, M. (1957). *The doctor his patient and the illness*. London, UK: Pitman Medical.

Bateson, G. (1979). *Mind and nature: A necessary unity (Advances in systems theory, complexity, and the human sciences)*. New York, NY: Hampton.

Benveniste, E. (1966). *Problèmes de linguistique générale [Problems of general linguistic]*. Pari, France: Gallimard.

Boesch, E. E. (1977). The medical interaction—A study in Thailand. *The German Journal of Psychology, 1*(1), 13–28.

Brockmeier, J. (1995). The language of human temporality: Narrative schemes and cultural meanings of time. *Mind, Culture, and Activity, 2*, 102–118. doi:10.1080/10749039509524692

Brockmeier, J. (2000). Autobiographical time. *Narrative Inquiry, 10*, 51–73. doi:http://dx.doi.org/10.1075/ni.10.1.03bro

Brockmeier, J. (2002). Remembering and forgetting: Narrative as cultural memory. *Culture Psychology, 8*, 15–43. doi:10.1177/1354067X09353212

Bruner, J. (1990). *Acts of meaning*. Cambridge, MA: Harvard University Press.

Bulow, P. H. (2003). In dialogue with time: Identity and illness in narratives about chronic fatigue. *Narrative Inquiry, 13*, 71–97. doi:http://dx.doi.org/10.1075/ni.13.1.03bul

Bury, M. (2001). Illness narratives: Fact or fiction? *Sociology of Health & Illness, 23*, 263–285. doi:10.1111/1467-9566.00252

Canguilhem, G. (1966). *Le normal et le pathologique, augmenté de Nouvelles réflexions concernant le normal et le pathologique [The normal and pathological, increased new reflections on normal and pathological]*. Paris, France: Quadrige.

De Luca Picione, R., & Freda, M. F. (2014). Catalysis and morphogenesis: The contextual semiotic configuration of form, function, and fields of experience. In K. R. Cabell, & J. Valsiner (Eds.), *The catalyzing mind*. New York, NY: Springer

De Luca Picione, R., & Freda, M. F. (in press). A discussion about the processes of meaning making starting from the morphogenetic theories of Rène Thom. *Culture & Psychology Journal.*

Di Napoli, G. (2011). *I principi della forma. Natura, arte e percezione [The principles of the form. Nature, art and perception]*. Turin, Italy: Einaudi.

Engel, G. L. (1977). The need for a new medical model: A challenge for biomedicine. *Science, 196*, 129–136.

Freda, M. F. (2008). *Narrazione e intervento in psicologia clinica. Costruire, pensare e trasformare narrazioni [Narration and intervention in clinical psychology. Build, think and transform narration]*. Naples, Italy: Liguori.

Freda, M. F., & De Luca Picione, R. (in press). *Scaffolding: A possible psychological intervention addressed to healthcare relationship.*

Freda, M. F., & Martino, M. L. (2014). Health and writing: Meaning-making processes in the narratives of parents of children with leukemia. *Qualitative Health Research.* doi:10.1177/1049732314551059

Garmezy, N., & Masten, A. S. (1986). Stress, competence, and resilience: Common frontiers for therapist and psychopathologist. *Behavior Therapy, 17*, 500–521. doi:http://dx.doi.org/10.1016/S0005-7894

Greenhalgh, T., & Hurwitz B. (1999). Narrative based medicine. Why study narrative? *British Medical Journal, 318*, 48–50. doi:10.1136/bmj.318.7175.48

Grmek, M. D. (1995-1999). *Histoire de la pensée médicale en Occident [History of Medical Thought in the West]*. Paris, France: Seuil.

Hermans, H. J. M., & Dimaggio, G. (2004). *The dialogical self in psychotherapy*. New York, NY: Brunner/Routledge.

Hillmann, J. (1984). *Le storie che curano [Stories that heal]*. Milan, Italy: Raffaello Cortina.

Hjelmblink, F., & Holmström, I. (2006). To cope with uncertainty: Stroke patients' use of temporal model in narratives. *Scandinavian Journal of Caring Sciences, 20*, 367–374. doi:10.1111/j.1471-6712.2006.00415.x

Kleinman, A. (1988). *The illness narratives: Suffering, healing and the human condition*. New York, NY: Basic.

Lotman, J. (2009). *Culture and explosion* (W. Clark, Trans.). Berlin, Germany: Mouton de Gruyter. (Original work published 1993)

Matte Blanco, I. (1975). *The unconscious as infinite sets: An essay in bi-logic.* London, UK: Gerald Duckworth.

Medved, M. I., & Brockmeier, J. (2008). Continuity amid chaos: Neurotrauma, loss of memory, and sense of self. *Qualitative Health Research, 18*, 469–479. doi:10.1177/1049732308315731

Murray, M. (2007). La psicologia narrativa per la comprensione del mondo e dei suoi cambiamenti [The narrative psychology for understanding the world and its changes]. *Rassegna di Psicologia, 3*, 103–117.

Salvatore, S. (2004). Inconscio e discorso. Inconscio come discorso [Unconscious and discourse. Unconscious as discourse]. In B. Ligorio (Ed.), *Psicologie e culture. Contesti, identità e interventi* (pp. 125–155). Rome, Italy: Edizioni Carlo Amore.

Salvatore, S., & Freda, M. F. (2011). Affect, unconscious and sensemaking. A psychodynamic, semiotic and dialogic model. *New Ideas in Psychology, 29,* 119–135. doi:10.1016/j.newideapsych.2010.06.001

Salvatore, S., & Valsiner, J. (2009). Idiographic science on its way: Towards making sense of psychology. In S. Salvatore, J. Valsiner, S. Strout, & J. Clegg (Eds.), *Yearbook of idiographic science* (pp. 9–19). Rome, Italy: Firera.

Sato, T., & Valsiner, J. (2010). Time in life and life in time: Between experiencing and accounting. *Ritsumeikan Iournal of Human Science, 20,* 79–92.

Simão, L. M. (2003). Beside rupture-disquiet; Beyond the other-alterity. *Culture & Psychology, 9*(4), 449–459. doi:10.1177/1354067X0394007

Valsiner, J. (2001). Processes structure of semiotic mediation in human development. *Human Devolpment, 44,* 84–97.

Valsiner, J. (2011). Constructing the vanishing present between the future and the past. *Infancia y Aprendizaje, 34,* 141–150.

Wood, D., Bruner, J. S., & Ross, G. (1976). The role of tutoring in problem solving. *Journal of Child Psychology and Psychiatric, 17*(2), 89–100. doi:10.1111/j.1469-7610.1976.tb00381.x

Zittoun, T. (2006). Transitions: Development through symbolic resources. Greenwich, CT: Information Age.

Zittoun, T., Duveen, G., Gillespie, A., Invison, G., & Psaltis, C. (2003). The use of symbolic resources in developmental transitions. *Culture & Psychology, 9*(4), 415–448. doi:10.1177/1354067X0394006

PART IV

TEMPORALITIES OF THE SELF

CHAPTER 10

HEIDEGGER, TEMPORALITY, AND DIALOGICAL SELF THEORY

Basia D. Ellis and Henderikus J. Stam

The question of time is important for cultural psychologists who study the unfolding of subjective life. For these scholars, human experience is observed as a dynamic process constituted by sociocultural structures and traditions as well as uniquely navigated by human agents (Kirschner & Martin, 2010). Selfhood is here regarded in cultural and temporal terms, observed as an ongoing practice of interpreting and reinterpreting varied cultural meanings. In recent decades, the dynamics of this process have been widely discussed via dialogical self theory, first proposed by Hermans and his colleagues (Hermans, Kempen & van Loon, 1992) and since then extensively elaborated (Hermans & Gieser, 2011; Hermans & Hermans-Konopka, 2010). The prominence of this theory makes it an important place for examining the relation of time and the self in cultural psychology, and this is our focus here.

In this chapter, we investigate how time is taken up in dialogical self theory and attempt to elaborate the current view with an onto-existential understanding. In our view, such an elaboration is necessary because time in dialogical self theory is for the most part regarded abstractly, as a background *clock time* according to which the dynamics of a primarily spatial dialogical self are traced. We in turn conceptualize the dialogical self as

Temporality: Culture in the Flow of Human Experience, pp. 259–282

temporal structure in its own right, unfolding *in lived time* with others in the world.[1] For this we take Heidegger as a contentious starting point, for with him we maintain that the question of time is not one of developmental accounts or narrowly conceived clock time but of our finitude in relation to Being.

The question of time and the self, more generally, is one that is philosophically and theoretically extensive. Both "self" and "time" are contested and the focus of limitless elaboration. Artificially separated through our notion of what we are here calling clock time and all that this implies for an instrumental world and a socially ordered life, psychologists limit the question of self and time by our ability to remember the past and predict the future. But this notion of cognitive "mental time travel" (Tulving, 1985) is premised on a functional account of the self, whose subjective life is beyond comprehension outside of a cognitive neuroscience. What if time were just part of what makes a self what it is?

Our chapter begins with a discussion of Heidegger's earliest and most recognized opus, *Being and Time* (1927/1962), from which we take *temporality* to constitute the ontological structure of existence. Next, to elaborate the sociocultural constitution of time, we examine Heidegger's later works, which theorize the relation between time and language in relation to Being. As a whole, Heidegger's philosophy allows us to develop an onto-existential, that is, an existentially founded ontology, and sociocultural conception of time, with which we then begin our critique of dialogical self theory. In our view, dialogical self theory assumes an abstract notion of time that fails to recognize the way existence is constituted by lived time. Accordingly, we draw upon Heidegger's philosophy to elaborate dialogical self theory with an onto-existential understanding. However, as Heidegger does not recognize the key insight of dialogical self theory, namely, the dialogical nature of conscious life, we evaluate the critical insights from both accounts and conclude our discussion with a defense of an onto-existential theory of dialogical selfhood.

Heidegger's Being and Time

The bulk of Heidegger's existential analyses are found in his major opus, *Being and Time* (1927/1962). It is important to note that these analyses are motivated by Heidegger's broader interest in the meaning of Being.[2] However, because Being can only be understood through our own relation to it, Heidegger begins with a study of the being of the questioner of Being, that is, with a study of our own existence. *Dasein* refers to the questioner of Being who already has some understanding of Being in general (Krell, 1993). Specifically, Dasein is "this entity which each of us is himself and which includes inquiring as one of the possibilities of its Being" (Heidegger, 1927/1962, p. 27).

Heidegger formulates the designation "Dasein" by nominalizing the infinitive *sein* "to be" and *da* (adverb) "here, there, since, then" (Griffiths, 2006, p. 54). Literally, then, Dasein means "Being-there" (Dastur, 1998/1990, p. 17). The term deliberately opposes traditional conceptions of the subject that treat it either as single entity or duality made up of substantive body and abstract mind. Dasein refers to existence as a *process* that unfolds prior to being thematized into subject/object dichotomies; it is neither a substance nor a subject but an elementary, existential unfolding that conditions all possibilities of experience in the world. Put differently, Dasein is an "I can" that precedes all possibilities of existence (Griffiths, 2006).

Given its primacy, Dasein cannot be understood by means of a presupposed essence or delineated idea; Dasein preconfigures all of these. Accordingly, Heidegger resists the analytical methods of philosophical anthropology and classic phenomenology for their ahistorical acceptance of preconceptions about existence. Because we ourselves are the beings to be analyzed and we are necessarily historical, Dasein's nature must be worked out through a phenomenology that is ultimately hermeneutical—what Heidegger calls an existential analytic (*existenziale Analytik*). Specifically, the phenomenology of Dasein *is* hermeneutics in the sense that it recognizes Dasein's historical formation and is the work of interpretation (Krell, 1993).

Being and Time (1927/2006) comprises Heidegger's existential analytic of Dasein, which begins as a study of the general Being of Dasein and moves progressively to more concrete interpretations of Dasein's structure: first, studying Dasein in its "average everydayness," Heidegger conceptualizes Dasein as Being-in-the-world; then, with further analyses, Heidegger translates Being-in-the-world into the more concrete structure of care; finally, the structure of care is exposed at the ontological level as temporality. In this way, moving from particular, existential determinants to the more general and concrete, Heidegger reveals an increasingly originary account of existence whose ontological structure is *temporality*.

Our discussion of Heidegger's philosophy aims primarily to defend the idea that existence at its fundamental level *is* ecstatic temporality and accordingly, it is always as *temporalities* that we engage with the world. Following his analyses, we conclude that existence achieves its unique form through the simultaneous work of all three temporal modes—past, present, and future—which together co-constitute the unfolding present. We live our past in the present and in the direction of the future. Temporality in turn refers to the unity of this tripartite structure that constitutes our basic mode of being, and as such, it forms the ontological structure of existence. From this, it also becomes apparent that abstract time, conceived as a succession of "now" moments, derives from a more basic, existential structure. We do not discuss the fundamental metaphysics implied by Heidegger's Being in temporality since that has been the subject of considerable debate.

Being-in-the-World

At the first level of analysis—in its general, everyday existence—Dasein refers to *Being-in-the-world.* As Being-in-the-world, Dasein is not a predetermined entity situated "inside" a predefined, external world but the unfolding process of "Being there." In Heidegger's words, "the Being which is an issue for this entity in its very Being *is* to be its 'there' " (Heidegger, 1927/1962, p. 171, original emphasis). As the "there" of being-there-in-the-world (Schürmann, 2008), "Dasein is its disclosedness" (*Erschlossenheit*) which also means that "the Being which is an issue for this entity in its very Being is to its 'there'" (Heidegger, 1927/1962, p. 171). The term *Erschlossenheit,* implies an "openness" and unconcealment of what otherwise remains concealed or closed off from coming into presence (Griffiths, 2006). Further, disclosedness relates to the question of truth as unconcealment of the world. But the world in Being-in-the-world is not a predetermined space external to Dasein; rather, it is a constitutive component of Dasein's being. As an onto-existential concept, the world unfolds as an undetermined *worldhood* that reveals itself (or unconceals) uniquely for every Being (Dastur 1998/1990, p. 21).

Disclosedness in general is seen to be constituted by four equiprimordial dimensions. *Attunement* or *disposition* (*Befindlichkeit*) implies that existence always unfolds according to moods that are not under our control. *Falling* (*Verfallen*) refers to the fact that in everyday existence, Dasein is absorbed in a world of other beings, existing alongside them. *Understanding* (*Verstehen*) describes our most basic ability to *grasp* the world in our own particular way—to actively "cease it" according to our own way of being. The latter is in turn made possible by *discourse* (*Rede*), which, as a fourth existential dimension, *articulates* understanding and explains how our existence is permeated by speech (Schürmann, 2008). Importantly, speech does not refer to spoken language but to the "*existential-ontological foundation of language*" (Heidegger, 1927/1962, p. 203, original emphasis). As such, speech provides the conditions for spoken language; more specifically, it articulates disclosedness, manifesting it *as* such and such regardless of whether this is orally expressed (Dastur, 1998/1990).

Constituted by these fundamental existential dimensions, Dasein is characterized as *Geworfenheit* (Being-thrown). Heidegger derives the term from the verb *werfen,* which means "to project, to cast, throw,"[3] and which is generally translated as "being-thrown" (Grifitths, 2006). Importantly, by "being-thrown," Dasein is both passive to the world and directed toward the future. Moods reveal how Dasein is Being-thrown in a passive sense, for through moods, Dasein recognizes that its plight is influenced by events taking place beyond its control; Dasein always finds itself "already in" a world that it does not itself create. At the same time however, moods also reveal our possibilities. Heidegger explains, "Dasein is always disclosed moodwise as that entity to which it has been delivered over in its Being; and in this way it has been delivered over to the Being, which, in existing, it has to be" (Heidegger, 1927/1962, p. 173). Accordingly, despite "being thrown" in a passive sense, Dasein always understands itself in terms of possibilities. Heidegger writes, "as thrown, Dasein is thrown into the kind of being which we call 'projecting'" (p. 185). Projecting has no relation to a thought-out plan but describes the very character of *understanding*. "As projecting," Heidegger explains, "understanding is the kind of Being of Dasein in which it *is* its possibilities as possibilities" (p. 185, original emphasis).[4]

Because Dasein relates to its Being in the mode of projecting, Dasein "is always more than itself in any given moment, since it holds its own Being before itself as possibility" (Schürmann, 2008, p. 88). That is, Dasein exists not as *totality* but as *potentiality*, as Being-toward-the-future. This experiential incompleteness does not signify that Dasein is not structurally whole however; rather, the incompleteness is its very structure. In Heidegger's (1927/2006) words, "das Dasein ist ihm selbst überantwortetes Möglichsein, durch und durch *geworfene Möglichkeit*" (p. 144), translated as "Dasein is Being-possible which has been delivered over to itself—*thrown possibility* through and through" (p. 183).

Care

Having outlined the existential features of average everydayness, Heidegger reexamines his account in relation to Dasein's concrete possibilities. Here, the experience of *anxiety* is paramount for revealing Dasein's broader structure as *care* (*Sorge*). Unlike in fear, where one is drawn to a particular threatening object, Heidegger argues that in anxiety, "*that in face of which one has anxiety* [*das Wovor der Angst*] *is Being-in-the-world as such*" (Heidegger, 1927/1962, p. 230, original emphasis). Accordingly, in anxiety, Dasein experiences itself as detached from itself, isolated, and individuated. The uncanniness of this experience constitutes anxiety's heuristic function, for it reveals that in its basic mode of Being, Dasein is *concerned* about its Being; that is, in its Being, its very Being is an issue for it. Treating anxiety as mode of Being-in-the-world, Heidegger writes,

> That in the face of which we have anxiety is thrown-Being-in-the-world; that which we have anxiety about is our potentiality-for-Being-in-the-world. Thus the entire phenomenon of anxiety shows Dasein as factically existing Being-in-the-world. The fundamental ontological characteristics of this entity are existentiality, facticity, and Being-fallen. (p. 235)

The experience of anxiety concretizes Dasein's more basic concern for its ownmost potentiality-for-being. "This potentiality," Heidegger (1927/1962) writes, "is that for the sake of which any Dasein is as it is" (p. 236). In anxiety, then, Dasein is revealed as "Being free for its ownmost potentiality" at a fundamental level. With this, Heidegger reinterprets Dasein's basic structure: "ontologically, Being towards one's ownmost potentiality-for-Being means that in each case Dasein is already *ahead* of itself [ihm selbst ... vorweg] in its being" (p. 236, original emphasis); in caring for its Being, Dasein is *being-ahead-of-itself* (*sich-vorweg-sein*).

But whereas Being-ahead-of-itself explains Dasein's existentiality, it does not yet involve the whole of Dasein's constitution. As the uncanniness of anxiety establishes, Dasein is always already thrown into a world already absorbed in it; and thus, in the same movement, the world constitutes Dasein's possibilities for Being. This fundamental entanglement constitutes Dasein's facticity and elaborates its care-structure as Being-ahead-of-itself-in-already-Being-in-a-world. Further, "Dasein's factical existing is not only generally and without further differentiation a thrown potentiality-for-Being-the-world; it is always also absorbed in the world of its concern" (Heidegger, 1927/1962, pp. 236–237). It is in this entangled *Being-together-with* that anxiety announces itself, revealing Dasein in its basic mode as "fallen," which means that Dasein is absorbed in the world with others. Taken together, these three existential determinants constitute Dasein as care-structure: "the Being of Dasein means ahead-of-itself-Being-already-in-(the world) as Being-alongside (entities encountered within-the-world)" (Heidegger, 1927/1962, p. 237). (Sich-vorweg-schon-sein-in-[der Welt] als Sein-bei [innerweltlich begegnendem Seienden]) (Heidegger, 2006/1927, p. 192)]. Heidegger (1927/1962) concludes, "This Being fills in the signification of the term '*care*' [*Sorge*], which is used in a purely ontologico-existential manner" (p. 237).

Care itself is understood in three senses: care (*Sorge*) refers to Dasein itself; concern (*Besorgen*) relates to Dasein's activities in the world; and solicitude (*Fürsorge*) refers to Dasein's Being alongside others. Despite these distinctions however, the three concepts are posited simultaneously in the care-structure so that the term "care" always refers to "concerned-solicitous care" (besorgend-*fürsorge Sorge*) for the Being that is at issue (Inwood, 1999). In sum, by designating "care" as the basic structure of Dasein, Heidegger broadens the scope of Being-in-the-world and translates its thrown/projectional character into Being-ahead-of-itself, where ahead-of-itself entails "Being-already-in" and "Being-alongside."

Notably, Being-ahead-of-itself is given a kind of phenomenological pre-eminence in the care-structure, for in characterizing Dasein's care for its own Being, it outlines the structure of the self (Schürmann, 2008). Heidegger (1927/1962) asks, "Who it is that Dasein is in its everydayness"

(p. 149) to which he responds, "Dasein is an entity which is in each case I myself" (p. 150). But *I* or *myself* does not mean to imply a substantive self that can be made objectively present in some way. Instead, the "who" that is experienced is *a way of Being*—a project and task to be achieved, albeit always already in the world and Being-alongside other entities.

Facticity (Being-already-in) means that Dasein cannot freely choose its project; as always already thrown into the world, Dasein's possibilities are constituted by that world. Moreover, as *Being-fallen* (or *Being-with*), Dasein's possibilities are also constituted by other beings. Referring to Dasein's collective constitution, Heidegger explains that the "who" in the question who it is that Dasein is "is not this one, not that one, not oneself [*man selbst*], not some people [*einige*], and not the sum of them all. The 'who' is the neuter, *the 'they'* [*das Man*]" (Heidegger, 1927/1962, p. 164, original emphasis). The concept of "the they" accounts for others whose meanings and practices permeate Dasein's existence and in this contribute to its making. So basic is their constitutive role that Dasein in its everyday existence is regarded "neuter" rather than any *one* in particular.

However, anxiety reveals that even as factical and fallen, Dasein is free for its ownmost potentiality-for-Being and this means that Dasein is free for the possibility of authenticity (*Eigentlichkeit*) or inauthenticity (*Uneigentlichkeit*((Heidegger, 1927/1962, p. 186). In anxiety, Dasein is stripped of its usual concerns and involvements and is exposed as bare freedom to choose itself (Inwood, 1999). This experience carries implications for Dasein's everyday existence, as it implies that while Dasein cannot escape facticity or fallenness, it can, in its existentiality, choose to live its *own* everydayness. That is, even as Being-already-in and Being-with other beings, Dasein can cease its Being in its *own* way, even if this too can only be done within a factical world and "in the way that others do" (Schürmann, 2008).

Authenticity thus requires that Dasein resists its tendency in everydayness to define itself in terms of "the They," and instead cease ownership of its Being. The motive for such a project cannot be found in anxiety alone, however. To will an authentic life, Heidegger explains, Dasein must come to recognize its own *finitude*. As we proceed to show, *impending death* constitutes another existential determinant, which reveals Dasein's basic structure as finite possibility "stretched along between birth and death" (Heidegger, 1927/1962, p. 426). Authentic being is predicated on Dasein's realization of its finitude, based on which Dasein can resist Being-toward-concrete-possibilities and transform itself into *Being-toward-death*.

Temporality

In his final analysis, Heidegger shows how Dasein's care-structure translates into a more basic, finite and ecstatic *temporality*. This formulation is motivated by another existential determinant, the experience of death.

Death, from the existential perspective, is something that *impends* without ever becoming objectively present, for its actual occurrence would require an end to existence altogether. Thus, death is something that is now, experienced as an ongoing "not-yet" (Heidegger, 1927/1962, p. 289). But whereas Dasein can never actually experience its death, "death reveals itself as that *possibility which is one's ownmost, which is non-relational, and which is not to be outstripped* [*unüberholbare*]" (p. 294, original emphasis). This is to say, death is something that is always mine, that cannot be performed by others in my name, and beyond which there is nothing more encompassing (Schürmann, 2008).

As Dasein's ownmost potentiality-for-Being, death exposes Dasein's *finitude*, revealing Dasein itself as "thrown Being *towards* its end" (Heidegger, 1927/1962, p. 295, original emphasis). With death, Dasein's existentiality finds its meaning in the future, that is, in "what is to come" (pp. 375–376). Since "what is to come" is Dasein itself (i.e., the completion of Dasein), Dasein "*stretches along* between birth and death" (p. 426, original emphasis). Heidegger writes, "Dasein does not exist as the sum of momentary actualities of Experiences which come along successively and disappear. Nor is there a sort of framework which this succession gradually fills up" (p. 426). Heidegger concludes, "Factical Dasein exists as born; and as born, it is already dying, in the sense of Being-towards-death" (p. 426). Death in this sense configures Dasein's existentiality, revealing its temporal structure by pointing to its inescapable finitude.

The recognition of finitude is significant, for through it, Dasein realizes that there must also be something like Being. Finitude is in this regard not an accidental or abstract cognition but a fundamental need to understand Being as well as submit to it (Dastur, 1990/1998). Only because Dasein realizes its finitude does it care for its Being and choose to engage in creative possibilities. Put differently, only by apprehending that there is both Being and freedom toward death, can thrown Dasein cease its Being authentically, albeit never in detachment from "the they" and its natural world.

With the significance of finitude revealed, Dasein's ontological structure can be defined as *finite temporality*, constituted by three central temporal aspects. First, *futurity* denotes Dasein's ability for projection. Futurity is not "the future" in the sense of a "now" that is yet to come but a basic constituent of Dasein's ontological makeup that conditions its ability to "let itself come toward itself" in projecting (Heidegger, 1962/1927, p. 372). Dasein must be *futural* in order to self-project for its own sake (p. 375).

But "to come toward itself futurally" is possible insofar as Dasein is as an "I-*am*-as-having-been" (Heidegger, 1927/1962, p. 376). Thus, another aspect of temporality, namely, that of *having been*, provides the basis of Dasein's facticity. Factical Dasein takes up its previous ways of being and projects them as present possibilities only because *its past remains alive in the present* rather than dissolve as something long gone.

Finally, the unique process that constitutes the *present* explains how Dasein exists alongside other beings: the present unfolds on a ground of potentiality, as a projecting and factical Dasein lets things come toward itself. Disclosedness ultimately takes shape through the three temporal modes: past, present, and future. Dasein, as this very disclosedness, is for this reason deemed temporality: "This phenomenon [that] has the unity of a future which makes present in the process of having been; we designate it as '*temporality*'" (Heidegger, 1927/1962, p. 374, original emphasis).

Thus, temporality constitutes Dasein's ontological structure. That is, at the ontological level, Dasein *is* ecstatic temporality, standing out ahead of itself as a radical openness, unified by a future that reformulates a "having been" into a unique present. Temporality, then, refers to the basic, tripartite process that configures past, present, and future into an existential whole.

What Heidegger argues is that the traditional conception of time, the tradition from Aristotle to Kant, is derivative from experience; it does not dictate experience. In his commentary on Kant and the problem of time he notes that

> although space and time as pure intuitions both belong "to the subject," time dwells in the subject in a more original way than space. Time immediately reduced to the givens of inner sense, however, is at the same time only ontologically more universal if the subjectivity of the subject exists in the openness for the being. (Heidegger, 1973/1997, p. 35)

Put differently, time is not a spatial, coordinating structure, thing, or entity but rather comes to us by way of living toward the future through the past.

Further, Heidegger argues that "world time" (taken as "pure sequence of now moments") is achieved via consensus with others: Dasein as ecstatic temporality first goes into the world, wherein it soon encounters natural, cyclical processes, which it then validates with others who experience them also. As it is only together with others that a world time is formed, lived time is attributed ontological priority (Heidegger, 1927/1962, p. 377).

In his focus on the primacy of onto-temporal structure of existence however, the early Heidegger does not yet elaborate what is sociocultural. This is because it is only after his study of Dasein that Heidegger reexamines the question of Being and addresses the question of our inherence in language. We find his later work relevant for this reason: it provides an understanding of how Being as *presencing* arises both in relation to humans and in the mode of language. After summarizing Heidegger's position, we draw upon it to reexamine the dialogical self, theorizing it as finite existence co-creating meaning with others.

The Later Heidegger

As early as 1930, Heidegger starts to think differently about the relation between Being and time, no longer conceptualizing the *meaning* of Being on the basis of Dasein's projection but, conversely, viewing humans from a broader perspective, as standing within the *truth* of Being to which they must respond (Dastur, 1998/1990, referring to Heidegger's "On the Essence of Truth"). This turning in Heidegger's thought refers to his recognition of Being as "historical *in itself*, and not only through the intermediary of Dasein" (Dastur, 1990/1998, p. 55, original emphasis).

Based on this realization, Heidegger elaborates his previous position from a new approach to the question of Being (Heidegger, 1947/1993). Having shown in *Being and Time* that Dasein has an implicit understanding of Being (as distinguishable from concrete "beings"), Heidegger maintains that Being itself can be expressly and conceptually apprehended. From a broader perspective however, Being is no longer "the ground of beings" understood via Dasein's projection; it is the unfolding of world through the co-belonging of "time and Being" in their reciprocal relation (Heidegger, 1969/1972).

Heidegger explains, "Being, by which all beings are marked, means presencing," where "to let presence means to unconceal, to bring to openness" (Heidegger, 1969/1972, p. 5). Being lets beings *come forth* in nonconcealment, allowing them to show themselves in the clearing. Notably, the clearing here no longer refers to Dasein's disclosedness but to the openness revealed by the co-belonging of Being and time. *Presencing* implies this precisely, referring to Being as temporal unfolding rather than as that which is present.

Heidegger (1969/1972) argues that presence is always given in a particular manner: it always involves an approaching "not yet present" (relating to the future) and a declining "what is no longer present" (relating to the past) that reciprocally bring about one another; as these processes always take place *at the same time*, their reciprocal bringing about gives rise to the present (p. 13). Presence is in this way "given" in the form of a three-dimensional temporal unity, always unfolding as a "mutual giving to one another of future, past, and present" (p. 14).

With Being now recognized as co-belonging with time, the later Heidegger reconceptualizes Being as *Ereignis*, which he translates from the German as "the event of Appropriation" that determines Being and time in their co-belonging (Heidegger, 1969/1972, p. 19).[5] Heidegger emphasizes that *Ereignis* is not an indeterminate power, so long as we consider Being as presence and time as the realm of the open, "where, by virtue of offering, a manifold presencing takes place and opens up" (1969/1972, p. 17). As the "the relation of all relations" that determines Being and time together,

Ereignis is not an "event" in sense of occurrence, but it is the appropriation that makes all occurrences possible (p. 19). Kockelmans (1970) goes so far as to argue that *Ereignis* is the unveilment that makes unnecessary the absolutes of history and relativism at once, leaving one only with the experience of Being.

But how does Ereignis occur? That is, how does Being let beings unfold in presence? The answer is found in language or specifically, in the mode of *saying* (*sagen, die Sage*) (Heidegger, 1959/1993b, p. 424). Importantly, for Heidegger, language is never merely speech in the sense of an expressive phenomenon, nor does it refer to mere concepts as representations of images (Heidegger, 1971, p. 191). Heidegger understands language in the broadest possible sense, as "everything by which *mankind* brings meaning to light *in an articulated way*" (Kockelmans, 1972, p. xiii, original emphasis); this equally includes audible speech as it does visual art and religious institutions.

Studying how language perjures *as language*, Heidegger maintains that it is not humans who speak and engage language but language itself that speaks and engages humans (Heidegger, 1971, p. 195). This is not to say that humans are not beings who speak; for humans are in fact distinguished from both plants and animals on the basis that speech enables them to be humans (Heidegger, 1971, p. 187). What Heidegger means to establish is the prevalence of language that comes from its correspondence with Being. Earlier in his writings, Heidegger famously states, "Language is the house of Being" (Heidegger, 1947/1993a, p. 237), claiming that humans exist by dwelling in language, and in this way inhering in the truth of Being and guarding it. Later, Heidegger returns to this statement, emphasizing, "we are within language, at home in language, prior to everything else" (Heidegger, 1959/1993b, p. 398, original emphasis).

In the context of Ereignis, the essence of language as a whole is *the saying* (*Sagen*) that shows (*Zeigen*)[6]: "What *unfolds essentially in language is saying as pointing*. Its showing does not culminate in a system of signs. Rather, all signs arise from a showing in whose realm and for whose purposes they can be signs" (Heidegger, 1959/1993b, p. 410). Importantly, showing cannot be attributed exclusively or even definitively as human doing.

> Self-showing as appearing characterizes the coming to presence or withdrawal to absence of every manner and degree of thing present. Even when showing is accomplished by means of our saying, such showing or referring is preceded by a thing's letting itself be shown. (Heidegger, 1959/1993b, p. 410)

It is because of this priority of the thing's letting itself be shown (in language) that in speaking we are also listening. The two are not opposed to one another, but neither do they happen simultaneously; rather, speaking is a hearing *in advance*: "Such listening to language precedes all other instances of hearing, albeit in an altogether inconspicuous way. We not only *speak* language, we speak *from out of it*" (Heidegger, 1959/1993b, p. 411, original emphasis). ["DIeses Hören auf die Sprache geht auch allem sonst vorkommenden Hören in der unscheinbarsten Weise vorauf. Wir sprechen nicht nur die Sprache, wir sprechen aus ihr" (Heidegger, 1985/1959, p. 243)] Language speaks by *saying*, which is showing; humans in turn hear this saying in advance, in this way speaking *by listening to language*.

Finally, speech can only let itself be told the saying if we ourselves are "granted entry" into the saying; that is, we must belong to the saying in order to hear it (Heidegger, 1959/1993b, p. 411). Heidegger thus redefines the intimate relation between humankind and Being: "Man is only man because Being itself addresses itself to him in the unfolding of presence, and not because he is, in virtue of his ecstatic essence, the 'place' of the self-address of Being" (Dastur, 1990/1998, p. 67). Being thus calls on humans, requiring them to "bring the soundless saying into the resonance of language" (Heidegger, 1993b/1959, p. 418); and humans must respond to Being by speaking, which is also to say, by listening to language. Heidegger (concludes that as mortals, we are "those who are needed and used for the speaking of language" (1959/1993b, p. 423).

Language, on Heidegger's account, is never just an expression of thinking, feeling, and willing. It "is the primal dimension within which each man's essence is first able to correspond at all to Being and its claim, and, in corresponding, to belong to Being" (Heidegger, 1952/1977, p. 41). For Heidegger, this too is *thinking*, the express carrying out of corresponding.

To summarize Heidegger's account, whereas his initial writings in *Being and Time* (1927/2006) reveal temporality as our ontological structure, his later works help us understand how, within the greater context of *Ereignis*, presence unfolds through the co-belonging of humans and Being. While the continuity between Heidegger's early and later philosophy is subject to debate, it is his ongoing concern with an onto-existential understanding of time as well as its relation to language that we take to be especially relevant to the current discussion. Having shown how Heidegger develops these ideas, we think through them to address time in dialogical self theory.

Dialogical Self Theory and Its Temporal Elaborations

Thus far only a handful of scholars have expressly addressed time in relation to dialogical self theory (Baressi, 2011; Bertau, 2011; Bhatia & Ram, 2001; Hermans & Hermans-Konopka, 2010; Linell, 2010; O'Sullivan-Lago & de Abreu, 2010; Raggatt, 2010; Valsiner, 2007), and their approaches to the question have been varied. Some situate the dialogical self as a phenomenon germane to the contemporary, postmodern era (Bhatia & Ram, 2001; Hermans & Hermans-Konopka, 2010); others focus on the dynamics of I-positions as a means to make sense of the self's development (Hermans, 2003; O'Sullivan-Lago & de Abreu, 2010; Raggatt, 2010); still others address the development of the dialogical self as it transforms more generally over the lifespan (Baressi, 2011; Bertau, 2004, 2011; Linell, 2010). However, few researchers expressly theorize the dialogical self as it unfolds in *lived* or *existential* time (Baressi, 2011; Valsiner, 2007), and of their accounts, none theorize selfhood as temporal structure at the ontological level.

We argue that without an ontological grounding, the dialogical self's meaning-making processes tend to be treated abstractly and functionally, studied in an abstract time rather than one that is lived.[7] As we have seen with Heidegger, time is fundamentally neither abstract nor functional; it is what makes consciousness possible, or rather, temporality precedes our consciousness of time. Accordingly, we argue that dialogicality must be understood as an onto-existential process that unfolds in lived time for a finite temporality. What is needed, then, is a theory of dialogical self*hood*, one that studies how *finite, temporal existences co-create meaning with others in the world*. Here dialogicality can be seen as a feature of a more basic, onto-temporal existence and conversely, onto-temporal existence too can be elaborated with a dialogical configuration.

Our attempt at such a theory is discussed below. Specifically, after disclosing the limited notion of time in dialogical self theory, we draw upon Heidegger's ideas to show how the self is not just temporal but coexistent with temporality, and dialogicality is itself a feature of time, or in Heidegger's sense, it is the self's coming to awareness of time in language. We conclude that temporality and dialogicality co-constitute the (temporal) unfolding of (dialogical) subjective life.

Time in Dialogical Self Theory

Hermans (2002) defines the dialogical self as a "dynamic multiplicity of I-positions in the landscape of the mind, intertwined as this mind is with the minds of other people" (p. 147). According to this view, the self emerges from a dialogue between multiple, spatially positioned I-positions, and it is their ongoing dialogue that constitutes the very structure of the self. Such a conception of "I-positions in conversation" arises from the theoretical pairing of late 19th century Western accounts of self with notions of dialogue put forward by European thinkers, especially Bakhtin.

From James and Mead, dialogical self theory borrows the distinction between the subject-self and the object-self, wherein self experience is conceived as a felt *I* always observing an *object-Me*. It is this I-Me distinction that constitutes the structure of the dialogical self's I-position and provides the first extension of the self in space. Adding the reflected sense of "Me" distinguishes the dialogical self from other Cartesian theories that treat the self as an autonomous, volitional center (Hermans, 2003). The dialogical self understood in its I-Me structure includes within it all things that the self appropriates and calls "mine"; in the Jamesian version, these include one's body, others, and things in the world.

Drawing on Mead, Hermans (2003) stresses the social constitution of the self. Specifically, whereas the *I* in the I-Me composition stands as "agentic capacity" that continuously appropriates and rejects a multiplicity of Me's, the Me's are necessarily acquired from the given individual's society. Hermans and Gieser (2011) elaborate that within a globalized world with increasing intercultural contacts, a diversity of perspectives exist that can be appropriated by any given *I*, making the case for a complex and even contradicting multiplicity of Me's within a single self.

Whereas the basic structure of the I-position provides the first extension of the self into physical and social space, Bakhtin's notions of dialogue and polyphony multiply the I-Me structure and create a self that is made up not of a single I-position but of many I-positions who voice their concerns in ongoing conversation. Each I-position is said to speak its world from a unique ideological stance and it is the I-positions' ongoing dialogical exchange—through tension and détente, agreement and disagreement, questioning and answering—that defines the self's unfolding *in time*.

Bakhtin thus gives further impetus to a dynamic understanding of selfhood, since in dialogue, the Jamesian and Meadian self grows from a single voice to a vibrant cacophony of voices who engage one another in animated and continuously developing relations. Dialogicality here implies that the self is dynamic and communicative, unfolding not as a monological train of thought, but as a vibrant cacophony of utterances directed at others, whether those others are real or imaginary. Furthermore, the self only comes into being through the other, as it is always with an other that the self exists *as a self*.

While such a conception of selfhood is readily thought *dynamic*, it is significant that this dynamism is presumed according to an abstract understanding of time and is not grounded in the lived time of a given ontological being. In our view, I-positions are abstracted from lived experience as soon as they are relocated within the spatial metaphor of a mental landscape. In their most recent statement on dialogical self theory, Hermans and Gieser (2011) argue that Bakhtin's dialogical construction "makes it possible to *contract temporally dispersed events into spatial oppositions* that are simultaneously present (in this way creating a 'landscape of the mind')" (p. 6, emphases added). This is possible because the dialogical construction translates the inner thought of a given character into an utterance directed at an other and in this brings together temporally divergent voices into a spatially envisioned mental landscape.

This landscape stands as a defining image for dialogical self theory (Hermans, 2002, 2003) and with this recapitulates the spatial metaphors of time dominant in Western thought from Aristotle to Kant (Bambach, 2011). This metaphor regards time to be movement in space and lays the foundation for the creation of clock time, that is, a time that is measurable. What we claim is that through this metaphor, temporally divergent voices are reduced into spatial oppositions conceptualized within an abstract, linearly unfolding time.

In emphasizing space, Hermans and Gieser (2011) treat time as an infinite series of "now" within which a configuration of spatial oppositions could be mapped. Time, from their perspective, unfolds objectively and independently from the lives of particular individuals; it is not a lived structure through which I-positions are organized and experienced, but an independently unfolding, Newtonian succession. The self is then studied as a dynamic system whose transformations can be mapped *in time*; whereas the self as a particular temporal structure is left unaddressed.

Similar concerns can be expressed about Raggatt's (2010, 2011) temporal elaboration of dialogical self theory as a triadic process of mediation. Drawing upon the semiotics of Charles Peirce, Raggatt maps the dynamics of "dialogical triads" (composed of an I-position, a counterposition, and an *ambiguous third*) into unique positioning histories, illustrated as a chain of triads unfolding along a temporal axis. The final image achieved is that of a "personal chronotope" that depicts the self as a series of transformations meaningfully connected by triadic relations. However, despite Raggatt's special emphasis on the self's dynamism, time in the personal chronotope remains abstractly understood: time is the temporal axis along which chains of dialogical triads can be mapped and different configurations of I-positions located.

Recognizing time as undertheorized in dialogical self theory, Baressi (2011) proposes a revised, three-dimensional model of the dialogical self, which elaborates the self's I-positions by attributing to them unique temporal dimensions. Baressi argues that the self grows increasingly aware of itself as temporally extended, developing in this increasingly complex I-positions that are capable of observing their dialogical activity across time (p. 49). In his developmental theory, it is when children first gain the capability for self-reflection in their fourth and fifth years that they begin to move imaginatively in time; then, as further development motivates conflict among dispositional and situational selves, the self resolves these inconsistencies by structuring them into coherent narratives. This begins in adolescence, when various temporally distributed positions grow integrated into metapositions; whereas more stable and minimally conflicting narrative positions are achieved in adulthood.

Baressi's (2011) elaboration of time in dialogical self theory is significant, in our view, since it observes time from the first-person perspective. However, his elaborations refer to the temporal dimensions of I-positions and not to the self as a single, ontological structure. Baressi's position relies on the development of an *awareness* of oneself as temporal being, but it does not explain the temporal structure of the ontological being for whom this awareness develops. Moreover, despite questioning time from the first person perspective, Baressi ultimately observes the self along an objective, developmental trajectory and not within the context of a *lived* time.

We take Valsiner's (2007) elaboration of the dialogical self's temporal dimensions as the most comprehensive to date, as Valsiner describes selfhood expressly *as temporal structure*. This is already the case in his well-known theory of semiotic mediation, but since Valsiner also connects this theory with that of the dialogical self, we consider both of them here.

Semiotic mediation describes how persons construct signs in their ongoing experience. Valsiner (2007) explains, "The whole of our psychological system is set up to make distinctions in the field within which we are constantly moving" (p. 127). Thus, Valsiner studies personal experience as an ongoing, selective, and semiotic process. The oppositional nature of signs is a central feature in this process, for new meanings are formed when sign oppositions are galvanized into dynamic, temporal relations. This "galvanization" does not unfold merely within an abstract flow of time however; sign oppositions come to life always within a perpetually unfolding, *personal* present.

More succinctly, Valsiner (2007) states that our sensory apparatus recognizes forms-in-movement whose perceptual inputs are further "sieved" (p. 127) via unique attention mechanisms; the attention mechanisms in turn make the sieved inputs available for semiotic reconstruction and presentation, and in this, prepare us for future meaning formulations. The personal present is here regarded as an ongoing, selective, and semiotic process, one involving a continuous reconstruction of acquired meanings within a particular context and according to a unique orientation toward the future.

The entire process of semiotic mediation thus unfolds within a present that is essentially liminal. By definition, the present is liminal in that it "takes place" on the border between the past and future. But taken as a lived, personal experience, the present, explains Valsiner (2007), is an infinitesimal moment that *unites* the past and future through a basic temporal tension. Whereas the past imposes itself onto the present and organizes experience into a particular configuration of an "as-is" (p. 88), the unknown future pulls the present toward further possibilities, elaborating it with a goal-oriented "as-if" (p. 88). The present is thus a temporal opposition between an "as-is" and "as-if"—a perpetual moment of experience uniting past and future within a single structure.

Valsiner (2007) integrates this understanding with dialogical self theory in order to account for the social character of the semiotic process. This is possible because dialogical self theory too conceptualizes experience as an *ongoing tension*, albeit between differing I-positions, each of which is ascribed a unique *past* and distinctive *goal orientations*. Valsiner identifies within this tension between differing I-positions a temporal opposition between an "as-is" and "as-if" that characterizes the personal present and instigates the production of new meanings. The opposition between the "as-is" and the "as-if" can thus be translated into the opposition between *an I-position in dialogue with an other I-position*, and in this way the social dimensions of the semiotic process are elaborated. Ultimately, the tension between the I-position-that-is and the I-position-that-is-to-be reveals "the birthplace of becoming—the movement into a new state" (p. 149).

By conceptualizing the semiotic process within the perpetually unfolding present of a dialogical self, Valsiner (2007) explains the microdynamics of meaning generation as they relate to the lived time of a single existence. In this way, he avoids the individualist problem and instead offers a way to think about how the dialogical self unfolds in lived time. However, because Valsiner gives priority to the sociocultural constitution of experience, treating selfhood primarily as semiotic process unfolding "at the level of meanings" (p. 154), it is the nature of signs that ultimately comes to define the possibilities of existence, not temporality.

Notably, Valsiner's (2007) position is comparable to Heidegger's in that Valsiner too regards the present as an ongoing "taking up" of the past in the direction of the future. However, with semiotics as the explanatory framework, Valsiner's account of time serves as a coordinating formula for sign construction. Having related the semiotic process to a particular existence, in focusing on the microdynamics of a perpetually unfolding present, Valsiner's account does not integrate these processes within a broader conception of a finite, onto-temporal existence at the level of personal lived experience for which it is meaningful. Time serves as an organizing principle for semiotic processes rather than as an ontological structure of selfhood, and accordingly, selfhood achieves its unique form not because of the nature of everydayness but because of the nature of signs—and particularly, their oppositional structures. In sum, in Valsiner's account, it is semiotics and not temporality that determines the unfolding of experience.

Discussion of Dialogical Self Theory

Generally speaking, dialogical self theorists tend to prioritize the constitutive role of semiotic and linguistic dimensions, since they study subjectivity primarily as an interplay of variably opposed voices in dialogue, continuously agreeing and disagreeing, questioning and answering, and creating new meanings in turn (Hermans, 2003). This kind of approach is noteworthy for having taken seriously the sociocultural constitution of human experience, which, despite being widely recognized during the linguistic turn in the social sciences, has not been addressed in mainstream psychology. However, in theorizing subjective life primarily in semiotic and linguistic terms, dialogical self theorists have paid less attention to how these processes are grounded in the lives of finite, onto-temporal existences. Indeed, in our view, the nearly exclusive focus on dialogue and sociocultural meanings suggests an implicit dualism between the self as a linguistic phenomenon, socioculturally understood, and the self as a category of experience, ontologically understood.

It remains undetermined in dialogical self theory how a cultural subjectivity comes to be experienced at all. By studying the dynamics of meaning as they unfold in abstract time, researchers have left unaddressed how unique sociocultural experience emerges as a result of its association with a particular ontological Being. Accordingly, we argue that what is needed is an understanding of "dialogical selfhood" that addresses how *finite existences in the world co-create meaning with others*.

As we have seen with Heidegger, existential time involves a threefold process wherein a "having been" is taken up in the "present" and transformed into a new possibility directed toward the future. Furthermore, it is only on the basis of this primordial temporality that abstract time can be derived together with others. Given such an analysis of temporality and subjectivity, what might a dialogical self look like when considered in this light? Recognizing that the dialogical self is not a full-fledged theory,[8] nor is it meant to carry the weight of a metaphysics, we can nevertheless make some suggestions for theoretical advances.

Already we have argued that time in dialogical self theory cannot be regarded as an abstract, unidimensional series of "nows" along which transformations in I-positions could be mapped. Instead, we argue for a "thick" and existential consideration of time that ought to be seen as constitutive of the dialogical self's organization. Specifically, the configuration of I-positions ought to be related to a *single* and *finite* temporality that chooses its future in the present on the basis of its having been. I-positions then are unfolding conversations whose horizons are shifting and labile because they envelop the past. Notably, our suggestion to relate I-positions to a single existence does not imply a return to a Cartesian understanding of a single self that autonomously chooses its course, nor to a neo-Kantian transcendental subject. Rather, we want to *ground* the multiplicity of I-positions by configuring them within a single temporality ontologically understood.

Such a grounding is necessary if we are to understand why a given *I* chooses to appropriate particular *Me*'s. As per Heidegger's (1927/1962) account, it is only because we recognize ourselves as finite existences directed toward death that we are motivated to choose creative possibilities directed at the future. Similarly, we argue that theorizing the dialogical self as a single, onto-temporal existence is necessary for justifying the particular choices of a given self's I-positions, for otherwise there is no *one* for whom their developmental trajectory can be meaningful—without a single, temporal existence, I-positions belong to no one who *cares* for them or their particular configuration.

To ground the dialogical self in a single ontological existence means reconceptualizing the dynamics of I-positions as a *lived process* rather than an abstract unfolding taking place "in the landscape of the mind," or in "the realm of meaning." In our view, the latter accounts are inadequate because they imply a dualism between an ontological being and a mental realm of psychological phenomena and in so doing they separate the self from the real, ontological consequences of its thinking. Not only the later Heidegger but the entire linguistic turn has focused on our fundamental inherence in language. However, in Heidegger's terms, it is not language alone that is constitutive of Being, for we too participate in unveiling presence by speaking. Heidegger explains that speaking *names*, and in naming, it *calls* or *invites* things into presence, bringing the things it calls closer, "so that they may bear upon men as things" (Heidegger, 1971, p. 197). Therefore, what we speak—or think or *dialogue*—has a bearing on what is cleared in the opening of presence. We are in this sense co-*responsible* for what appears in presence by virtue of our relation with Being, for we are *in speaking*, allowing the *saying* to unconceal itself.

Heidegger indirectly addresses the theoretical shortcomings of dialogical self theory, as researchers thus far have not theorized how the dialogical self unfolds as an onto-temporal process (in language). However, what Heidegger does not address in his elaboration is how existence unfolds *with others*. Whereas in his early works he explains the social constitution of experience through the concept of "the they," in his later works he maintains that presence is unveiled in our speaking with one another: "to speak to one another ... implies a mutual showing of something, each person devoting himself or herself to what is shown" (Heidegger, 1959/1993b, p. 409). But neither "the they" nor the mutual engagement of "each person" expresses the chief insight of dialogical self theory: that experience itself involves a multiplicity of others in dialogue. Conversely then, dialogical self theory can also elaborate Heidegger's account by revealing in the "presencing" of presence a social dimension that goes unaccounted for in the philosopher's view. From a dialogical perspective, the unconcealing, ontological process that is "speaking" emerges in each case through a lived dialogue with an other, whether that other is real or imaginary. Presence in this sense is populated by others who maintain dialogical relations. Using Heideggerian terms, the meaning of Being unconceals historically through *dialogues with others*.

In our view, that dialogicality characterizes selfhood carries implications for its onto-temporal structure. Accordingly, if we see dialogicality as existentiality, we may think of existence itself as thrown-Being-*in-dialogue*-toward its end. Our conception does not mean to suggest a simplistic bridging of Heidegger and dialogical self theory but invites an understanding of dialogical selfhood as a genuinely (onto-)temporal process constituted through dialogical relations with others. For whereas death discloses the dialogical self as a temporal whole (existing as potentiality), dialogicality (as existentiality) exposes its temporal structure as Being-multiple.

With this perspective, to care for one's Being means to care for one's dialogical relations with others, and conversely, it means that these relations constitute the structure of care. With impending death, it is the entire self-system (ontologically whole and dialogically unfolding) that is at issue and not one or even any number of I-positions. Accordingly, to respond to one's finitude entails reconfiguring the multiplicity of relations that constitutes the self and not simply transforming any one position taken independently. Furthermore, the self only comes into being through and with others, which means that its creative responses always entail a multiplicity of divergent voices.[9] We realize that this complicates the question of authentic existence, but it is beyond the scope of this chapter to address it here. What we have proposed is a way to think of dialogical selfhood as an onto-temporal process, maintaining in this how it is always *single and finite existences* that co-create meaning with others in the world.

FINAL REMARKS

In this chapter, we have examined the dialogical self through a rudimentary study of Heidegger's work. We realize the difficulty in both interpreting and applying a Heideggerian position to a concept that has been broadly influenced by numerous contemporary perspectives in the social sciences. Furthermore, we have been mute on the critical work both on Heidegger and his philosophy that has flourished since the 1920s. In addition, we do not expect to come to any rigorous conclusion for Heidegger's philosophy demands the acceptance of a philosophical language that would exclude most work in contemporary psychology. Hence, this is not an attempt to create a "fusion" of Heideggerian philosophy and dialogical self theories. Instead, we hope we have mined Heidegger's insights for a conception of lived time that is at odds with the tradition of mechanical, spatial, clock time. In so doing, we hope to have shown that in reconsidering time, we also reconsider the dialogical self as a project of and in time.

Notably, if cultural psychologists are to take seriously the temporal structure of subjectivity as we have argued it here, then the basic view of subjectivity as deeply cultural must be complemented with a corresponding view of culture as deeply temporal. That is, culture cannot be treated as mere system of signs and meanings whose transformations can be traced along the line of an abstract temporal axis. Rather, culture ought to be studied as a lived process, one whose developments emerge through the onto-temporal organization of selves involved in dialogical relations with others. One way to do this, our study may suggest, is to consider culture in the broad and dynamic sense of Heidegger's language; from here, culture may be thought to engage ecstatic temporalities that, through this engagement, structure the nature and development of cultural meanings.

To summarize, our discussion of Heidegger, temporality, and dialogical self theory aimed to address the limited notion of time in dialogical self theory and elaborate it with insights from Heidegger's approach. In so doing, we have proposed to ground the self ontologically, claiming that selfhood unfolds as finite temporalities speak the world with others in dialogue. By presenting the dialogical self in this way, we have argued that the self is coexistent with temporality, and dialogicality is itself a feature of time. In dialogue with others, the self comes to awareness of time in language.

ACKNOWLEDGMENT

We thank Lívia Mathias Simão, Jaan Valsiner, and Danilo Silva Guimarães for their astute comments and helpful suggestions on an earlier version of this chapter.

NOTES

1. The distinction between "objective" and "subjective" time is also discussed by Cornejo and Olivares (this volume), who explain its importance for advancing a cultural-psychological approach to time. Specifically, these authors distinguish between (objective) *time* as "linear sequence of moments" and (subjective) *temporality* as "the sense of time," tracing the roots of this distinction to 19th century phenomenology. Whereas our discussion will arrive at a similar distinction, because we come to it by way of Heidegger, "temporality" and

"time" are introduced in their unique, Heideggerian senses. Accordingly, "temporality" is explained as the ontological structure of existence, whereas "time" (discussed in Heidegger's later work) is understood in terms of its co-belonging with Being. Once these terms are clarified, we will consider their implications for the nature of time assumed in dialogical self theory.

2. Being does not refer to any being in particular but means "that which determines entities as entities" (Heidegger, 1927/1962, p. 25) and refers to "the Being of entities" (p. 26)
3. *Geworfen* is the past participle of werfen, whereas *Geworfenheit* reflects Heidegger's propensity to nominalize verbal forms (Griffiths, 2006).
4. Understanding does not grasp thematically that upon which it projects (i.e., possibilities) for "in projecting, project throws possibility before itself as possibility, and as such lets it be" (Heidegger, 1927/1962, p. 183).
5. Heidegger (1969/1972) writes, "What determines both, time and Being, in their own, that is, in their belonging together, we shall call: *Ereignis*, the event of Appropriation. *Ereignis* will be translated [from the German into English] as Appropriation or event of Appropriation" (p. 19). Inwood (1999) translates *Ereignis* as "event" and relates it to sich ereignen, "to happen, occur." Until the 18th century, the latter were spelt *Eräugnis, eräugnen,* arising from Auge (meaning "eye") and meaning "placing/to place before the eye, becoming/to become visible" (p. 54).
6. For Heidegger, showing is implied in saying, "Yet what is it we call saying? To experience this, we shall hold to what our language itself calls on us to think in this word. Sagan means to show, to let something appear, let it be seen and heard" (Heidegger, 1959/1993b, pp. 408–409). In his translation of this essay, Krell recognizes the difficulty in connecting in English "saying" (*Sagen*) with "showing" (*Zeigen*); Krell suggests that "the Latin *dico* brings both senses together: 'I say' originally means 'I show through words'" (footnote, p. 409 in Heidegger, 1959/1993b).
7. Here our discussion corresponds with that of Cornejo and Olivares (this volume) who, after tracing the historical development of the distinction between "objective" and "subjective" time, discuss its relevance to cultural psychology.
8. Hermans and Gieser (2011) refer to dialogical self theory as a "bridging theory."
9. Readers may be interested in Ricoeur's (1990/1992) *Oneself as Another* for an alternative phenomenological theory of selfhood that comes into being with others.

REFERENCES

Bambach, C. (2011). The time of the self and the time of the other. *History and Theory 50*, 254–269.

Baressi, J. (2011). Time and the dialogical self. In H. J. M. Hermans & T. Gieser (Eds.), *Handbook of dialogical self theory* (pp. 46–63). New York, NY: Cambridge University Press.

Bertau, M.-C. (2004). Developmental origins of the dialogical self: Some significant moments. In H. J. M. Hermans & G. Dimaggio (Eds.), *The dialogical self in psychotherapy*. New York, NY: Brunner-Routledge.

Bertau, M.-C. (2011). Developmental origins of the dialogical self: Early childhood years. In H. J. M. Hermans & T. Gieser (Eds.), *Handbook of dialogical self theory* (64-81). New York, NY: Cambridge University Press.

Bhatia, S., & Ram, A. (2001). Locating the dialogical self in the age of transnational migrations, border crossings and diasporas. *Culture & Psychology*, *7*, 297–309.

Dastur, F. (1998) *Heidegger and the question of time* (F. Raffould & D. Pettigrew, Trans.). Atlantic Highlands: Humanities Press International. (Original work published 1990)

Griffiths, D. B. (2006). *The keywords of Martin Heidegger.* Lewiston, NY: Edwin Mellen.

Heidegger, M. (1962). *Being and Time* (J. Macquarrie & E. Robinson, Trans). Oxford: Blackwell. (Original work published 1927)

Heidegger, M. (1971). Language. In M. Heidegger (Ed.; A. Hofstadter, Trans.), *Poetry, language, thought* (pp. 185–208). New York, NY: Harper & Row.

Heidegger, M. (1972). Time and being. In M. Heidegger (Ed.; J. Stambaugh, Trans.), *On Time and being* (pp. 1–24). New York, NY: Harper & Row. (Original work published 1969)

Heidegger, M. (1977). The turning. In M. Heidegger (Ed.; W. Lovitt, Trans.), *The question of technology and other essays* (pp. 36–49) . New York: Harper Torchbooks. (Original work published 1952)

Heidegger, M. (1985). Der weg zur sprache. In F.-W. von Herrmann (Ed.), *Unterwegs zur sprache* (pp. 227–257). Frankfurt, Germany: Vittoriao Klostermann. (Original work published 1959)

Heidegger, M. (1993a). Letter on humanism. In D. F. Krell (Ed., Trans.), *Martin Heidegger: Basic writings* (Rev. & exp. ed., pp. 217–265). New York, NY: HarperCollins. (Original work published 1947)

Heidegger, M. (1993b). The way to language. In D. F. Krell (Ed., Trans.), *Martin Heidegger: Basic writings* (Rev. & exp. ed., pp. 397–426). New York, NY: HarperCollins. (Original work published 1959)

Heidegger, M. (1997). *Kant and the problem of metaphysics* (5th ed., R. Taft, Trans.). Bloomington: Indiana University Press. (Original published 1973)

Heidegger, M. (2006). *Sein und zeit*. Tübingen, Germany: Max Niemeyer Verlag. (Original work published 1927)

Hermans, H. J. M. (2002). The dialogical self as a society of mind. *Theory & Psychology*, *12*, 147–160.

Hermans, H. J. M. (2003). The construction and reconstruction of a dialogical self. *Journal of Constructivist Psychology*, *16*, 89–130.

Hermans, H. J. M., & Gieser, T. (2011). Introductory chapter: History, main tenets and core concepts of dialogical self theory. In H. J. M. Hermans & T. Gieser (Eds.), *Handbook of dialogical self theory* (pp. 1–22). New York, NY: Cambridge University Press.

Hermans, H. J. M., & Hermans-Konopka, A. (2010). *Dialogical self theory in psychotherapy*. New York, NY: Cambridge University Press.

Hermans, H. J. M., Kempen, H., & van Loon, R. (1992). The dialogical self: Beyond individualism and rationalism. *American Psychologist*, *47*, 23–33.

Inwood, M. (1999). *A Heidegger dictionary*. Malden, MA: Blackwell.

Kirschner, S. R., & Martin, J. (2010). *The sociocultural turn in psychology: The contextual emergence of mind and self*. New York, NY: Columbia University Press.

Kocklemans, J. J. (1970). Heidegger on time and being. *Southern Journal of Philosophy, 8*, 319–340.

Kocklemans, J. J. (1972). *On Heidegger and language*. Evantson, IL: Northwestern University Press.

Krell, D. F. (1993). *General introduction: The question of being*. In D. F. Krell (Ed.), *Martin Heidegger: Basic writings* (Rev. & exp. ed., pp. 1–35). New York, NY: HarperCollins.

Linell, P. (2009). Rethinking language, mind, and world dialogically. In J. Valsiner (Ed.), *Advances in cultural psychology: Constructing human development*. Charlotte, NC: Information Age.

O'Sullivan-Lago, R., & de Abreu, G. (2010). Maintaining continuity in a cultural contact zone: Identification strategies in the dialogical self. *Culture & Psychology, 16*, 73–92.

Raggatt, P. (2010). The dialogical self and thirdness: A semiotic approach to positioning using dialogical triads. *Theory & Psychology, 20*, 400–419.

Raggatt, P. (2011). Positioning in the dialogical self: Recent advances in theory construction. In H. J. M. Hermans & T. Gieser (Eds.), *Handbook of dialogical self theory* (pp. 29–45). New York, NY: Cambridge University Press.

Ricoeur, P. (1992). *Oneself as another* (K. Blamey, Trans.). Chicago, IL: Universityof Chicago Press. (Original work published 1990)

Schürmann, R. (2008). Heidegger's being and time. In S. Levine (Ed.), *On Heidegger's being and time* (pp. 56–131). New York, NY: Routledge.

Tulving, E. (1985). Memory and consciousness. *Canadian Psychology, 26*, 1–12.

Valsiner, J. (2007). *Culture in minds and societies: Foundations for cultural psychology*. New Delhi, India: Sage.

CHAPTER 11

ON TIME AND TEMPORALITY FROM A CLINIC AND PSYCHOANALYTIC POINT OF VIEW

Nelson Ernesto Coelho Jr.

We know from a phenomenological point of view that being and time are not external to each other. There should be no distinction between being and time. Time is neither external nor contingent on subjectivity. If we direct our attention to a more psychological point of view, we shall find time as a constitutional part of experience. So we could say that time is a concrete element of subjectivity as well as of every psychical feature.

If we now turn to psychoanalysis, we shall find time as one of the most central figures in clinical and theoretical contemporary debates. We can say that psychoanalysis has its own time. What is the idea of time we are focusing on here? Much has been written on what time means to psychoanalysis, much less on what it means to experience time in a psychoanalytical setting, as a patient and as an analyst. Psychoanalysis is first of all an emotional experience, as suggested by Thomas Ogden (2001). In this chapter, I would like to explore the experience of time in the way I experienced it in a recent session with a male patient in his third year of analysis.

Temporality: Culture in the Flow of Human Experience, pp. 283–291

THE SESSION

Patient: Last night I had a dream. Now I'm not sure if it was last night. Well, it doesn't matter. I was in a house, my grandparents' house, in which I lived when I was five or six years old. It was a small house and I saw myself in the garden with two of my younger sisters. But the strange thing is that I was like I am now, 29 years old and my sisters were also the age they are now. The house does not belong to my family anymore and I don't recall having ever been there since then.

Me: Do you recognize any feelings attached to this scene? [I've formulated this question and at the same time I was trying to picture the house. For some reason, I used a familiar image of a house I used to live in when I was 7 or 8 years old. I was transported to a very familiar and reassuring place of my childhood.]

P: This was something like nostalgia for a time when I experienced the feeling of being in a familiar house. Feeling safe and reassured. It's strange, but after all these years, I don't remember ever being in a house where I have experienced this kind of feeling, of belonging and protection.

Me: You were thinking of buying a house, weren't you? [Then I began to picture my own house, where I live nowadays. I realized that with this question I was also taking the risk of interrupting his line of associations and his plunge into the past. Something conservative captured me and I noticed that I was using my own parameters to figure out what his image and his memory could mean. Or even more, I was infusing his life with my own desires instead of listening to him. I must be more careful about what I say. I must keep the information generated in my mind only to myself. Perhaps I should listen to him a little more. Yes, that's what I'll do.]

P: Yes. But I'm not sure if this is a good idea. You know, my mother is always telling me I should find a better use for my money, perhaps buying a house and settling down. I'm still not sure if I've achieved the things I want and I'm afraid of losing my youth before I find the things that are really meaningful to me. I'm good at what I do, but I'm always checking to see if this is the best I can do. My father seemed to know very well what he wanted from life.

For him, knowledge and his academic career came first, he never cared about money or status. I remember asking my father when I was about 10 why we had such an old car when all my friends' parents drove brand new and above all, expensive cars and lived in big houses with swimming pools. He had never given me a straight answer. But I felt that he was trying to tell me something about his life and choices with his silence.

Me: You have never told me that you had had this kind of conversation with your father. From my perspective, there is something important about this memory: by always being so quiet and sure about his choices, your father gave you a life example, helping you to develop your own values. Since our first meeting, I've noticed that you've been struggling with role models that have always fallen short. But now I hear something different.

P: Yes, I've never thought about my father this way.

Me: Our time is over. See you next session.

The patient left and I was alone with my thoughts. I was glad that he could feel safe with me after all the years we've been working together. He felt safe enough to remember things that he has been avoiding for so many years. But in terms of the experience of time, what really happened during this session?

CHRONOLOGY AND "DREAM TIME"

The duration of the psychoanalytical session is defined in minutes (45 or 50 minutes). It must be said that consistency in the number of sessions per week and their duration is one of the most critical elements for the psychoanalytical setting, or more precisely in this case, for the psychoanalytical frame (it is not only an external condition, but also a symbolic condition). It enables a special type of psychical liberty. Liberty in terms of time and conscious control that occurs inside the structure of the psychoanalytical setting, enabling the patient to experience a different time, something that could be called *dream time*. The patient is aware of the chronological time, he is conscious of coming to the consulting room at a precise hour of the day. For instance, he comes two times a week always at 8 o'clock in the morning. He knows that the session will take from 45 to 50 minutes. But when he lies down on the couch, a different time experience begins. Past, present, and future come together in one and the same construction and expression of a complex process of emotional and ideational experiences. The tension between conscious and unconscious levels of representation appears (or can appear) in each sentence, in each new expression. And especially when a patient tells a dream, we will probably be in touch with something that can be called an unconscious process, where temporality is experienced in a very distinctive way. The reference here is the classical Freudian idea (1915/1985a), that

> the processes of the *system Unconscious* are *timeless*, i.e., they are not ordered temporally, they are not altered by the passage of time; they have no reference to time at all. Reference to time is bound up, once again, with the work of the system *Consciousness*. (p. 191, original emphasis)

Things, persons, and experiences that belong to the past appear as belonging to the same time of present events. In a dream, for instance, my grandfather who is dead is suddenly alive and talking to someone I met for the first time yesterday.

In Freudian terms (1915/1985a, original emphasis), this is a different type of logic, the logic of the unconscious, characterized by its "*exemption from mutual contradiction*" (p. 191). But we can go a step further. In the session above, the patient starts telling me about a dream he had the previous night. In the moment of the session, which happens in the present time, he tells me something about his recent past, but uses the present tense. The image that is produced in the dream (and reproduced in words during the psychoanalytical session) concerns an experience even farther back in time, a memory of his childhood. In the same context, something from the future appears, represented by his mother's idea about how he should spend his money by buying a house. Emotions and ideas are mingled together; experiences from the past, the present, and the future are in some way united in the construction of a special kind of temporality, a dream time.

Something similar occurs with the analyst. We could call it the reverie time. Thomas Ogden (2001) defined the analyst's reverie in terms of what "presents itself obliquely to awareness in the most unobtrusive, quotidian of forms: as ruminations, daydreams, sexual fantasies, snippets of films, 'audible' musical phrases or lines of poetry, bodily sensations, and so on."(p. 5). In the session, my own reverie began with the image of my childhood home. This is something that appears to me, related to my personal history, but it has also a deep connection to what the patient is experiencing. It informs me about some of the patient's unconscious ideas, images, and feelings.

TIME AND THE THIRD SUBJECT OF ANALYSIS

In a psychoanalytical setting, we have two persons and, in most situations, two different experiences of time. There is the time of the analyst and the time of the patient. Ordinarily there is nothing in common between these two times. There is the time when the patient expresses his free associations and the time when the analyst evenly suspends his attention. We have the time when one speaks and the time when the other listens. They are not the same time, but they need to merge to some extent, otherwise there will not be any conversation. The psychoanalytical setting enables an ordinary conversation, which is also asymmetrical. The patient does most of the talking and the analyst listens. But there is a way for him to take part in the conversation. Thomas Ogden (2001) describes it as conversations at the frontier of dreaming. It is simultaneously a conscious and an unconscious conversation: "The unconscious conversation that in sleep we experience as dreaming, in the analytic setting we experience as reverie. The analyst's reveries are his waking dreams"(p. 5).

The psychoanalytical setting as defined nowadays is a transference-countertransference field. The patient's transference (the unconscious form with which the patient displaces emotion and images from parental figures and from the past to the analyst, in the present) is received by the analyst's countertransference (emotions and images generated by the patient's transference). Emotions from both participants form a characteristic field of experiences. It's a field that involves the conscious and the unconscious experiences of the analyst and of the patient. We could add to that, as Thomas Ogden (2001) stated, that this field also involves a co-created experience that is

> jointly, but asymmetrically, constructed by the analytic pair. This unconscious intersubjective construction is the "subject of analysis": a third subject with a life of its own, generated by the analytic pair and standing in dialectical tension with patient and analyst as separate individuals. (pp. 11–12)

So we have the temporality experienced by the patient, the temporality experienced by the analyst, and the temporality of this third subject, co-created by both. In the session described above, I try to show how this temporality works in the mind of the analyst. The analyst must be simultaneously in touch with the time experience of the patient and with his own experience of time. If he is able to do so, something of this third subject will appear as a transforming experience for both participants of the psychoanalytical session. More precisely, what I've tried to show above is the difference between a more empathically constructed *reverie* (my reassuring image of my childhood home) and a typical "countertransference invasion" (when I started thinking of my own house nowadays and wished that the patient also had a house in which to settle down). The reverie enabled some kind of new experience by the patient, and fortunately the countertransference invasion that is almost always a disturbance within the field didn't disturb the emergence of a new memory regarding his childhood. He didn't leave the time of the past in which he was so interested in investigating.

At the end of the session, when we talked about his father and the role models (and/or the absence of them) in his life, we experienced something relevant together, something that could be regarded as the appearance of the "subject of analysis," the third subject.

TIME OF REGRESSION, TIME OF REPETITION

Regression, repetition, and rhythm—these are three of the most important psychoanalytical concepts related to the experience of temporality. They are concepts that indicate a psychoanalytical conception of human development and of the time experience in a psychoanalytical session. For the psychoanalytical theory, the solution to the temporal evolution problem in human development does not come from a notion of time that relies on an idea of an ordinary temporal succession.

Psychical processes are always bidirectional, which means that they can be seen as progressive and regressive processes. The experience of present time can be influenced by elements of the past and by elements of the future, or even by both simultaneously. Thus, we should refuse any notion of linear determinism, as in statements like "a past event always determines a subsequent one." But we also know through psychoanalytical practice and experience that many achievements in a psychoanalytical process depend on whether the patient can or cannot live an emotional and an ideational regression to past experiences. The psychoanalytic setting is especially oriented to permit regression, conceived not as a defensive mechanism but basically as a way to enable more innovative psychical organization.

A temporal regression brings back emotions linked to experiences of earlier life times, as well as images, desires, and wishes that will take part in the transformation of reluctant behaviors that keep insisting and repeating.

There is a psychical defensive way that is based on repetition. That's why, as analysts, we keep a question in mind: In this patient's life, what is repeating itself and why? In Freudian terms (1920/1985b), we can say that a patient "is obliged to *repeat* the repressed material as a contemporary experience instead of ... *remembering* it as something belonging to the past" (p. 288, original emphasis). But repetition in analysis, in some specific situations, can also be a way of elaborating what has not yet been transformed in representation or in meaning if analyst and patient are engaged in the construction of an intersubjetive field of analysis.

In the clinical situation shown above, the dream was the first step on the path of regression where the patient was enabled to remember something. I followed his regression with my own regression, which, I believe, enabled something new to come in his working through, stopping a repetition that had prevented him from remembering. But, as I've stressed earlier, the psychoanalytical setting is also structured by another kind of repetition. For instance, in the case described above, we have two sessions a week at the same time and at the same place with the same analyst. We could call this repetition rhythm. It's a rhythm absolutely necessary to enable the patient to be in two different times, living a strange experience of past and present simultaneously, transforming by regression a defensive repetition into a new insight about his life. Rhythm and continuity are essential forms of temporality in the psychoanalytical process, as it is to the constitution of a baby's psychic structure. But in many clinical situations, especially with very disturbed patients, we are confronted with experiences of abolition of any perspective of future, and it is difficult to establish rhythm. It's a face-to-face with emptiness and despair. A traumatic past, a present that is beyond bearing, and no perspective of future.

Instead of rhythm, we have paralysis. It is what we can call a coagulated time. Here, even more, we must rely on an intersubjective field of construction, where patient and analyst converge to the creation of new psychical and temporal beginnings.

NACHTRÄGLICHKEIT

We now can direct our attention to the most important and innovative of the psychoanalytical concepts regarding time. The word Freud used in German was *Nachträglichkeit.* In English, it was translated by James Strachey as *deferred action:* Lived moments that could not immediately be integrated in a intertwining of meanings (establishing psychical representations) but have a potential for significance that can't be completely consummated. The meaning of these lived moments will be supplemented in a *deferred action.* New experiences will enrich the first marks and will offer unknown aspects not yet discovered. It's not just a simple reconsideration of the first moment. It's a constant transformation by means of the supplement brought by new moments that will bind to them. The deferred action implies that there is a temporal distance between the lived situation and its psychical effect (or the constitution of a proper psychical representation of the lived situation). That's the reason why, in psychoanalytical terms, meaning and psychical consistency come afterwards and not at the same time as the lived situation. So we should refuse common developmental, maturational, and genetic ideas regarding the psychical unconscious experience. Here we find a great disagreement between Freudian theory and most of the psychological theories regarding time. As psychoanalysts, we don't believe that we can empirically observe development and be sure of a direct relationship between facts of life and its psychical meaning. As André Green (1999) once suggested, "*process related to time are those that escape observation and most of them have to be deduced retrospectively.* Why? Because they took place intrapsychically, reorganizing the results of perception, affects, phantasies, wishes, etc. This is the basis for transference" (p. 28, original emphasis). I think this statement is controversial, but I do agree with Green that we are never sure to observe objectively a fact that depends on psychical processes; and I also agree with him that transference is the way we keep reorganizing psychically the facts that have been previously lived. So it is really uncertain and even wrong to state that a trauma, for instance, took place when the child was 2 or 3 years old, relying only on the parent's observations or memory of observations or even self-observation. This doesn't mean that there is no relationship between facts of life and psychical meaning or between external reality and psychical reality or even between historical truth and narrative truth. The whole question here is whether, as analysts, we should rely on or believe in the reality obtained by observational methods—by empirical methods of research—concerning psychical experiences. Psychoanalysis has its own research method, a method based in the clinical experience, in what emerges from the analytical setting. This is evidently an incomplete construction of the fascinating theme of research methods in psychoanalysis, but I hope to be forgiven, since the development of a more complete argument would require a whole new chapter.

Finally, I would like to stress that in consonance with the idea of a deferred action, there is what Green (2000) called "the time path of a psychoanalytical process." It is a time path that produces different effects of retroactive reverberation and of anticipatory expression, which will enable new forms of irradiation of meanings in a psychoanalytical setting. In the clinical vignette that opens this chapter, I've tried to show a moment of this time path that is inherent to every psychoanalytical process.

REFERENCES

Freud, S. (1985a). The unconscious (J. Strachey, Trans.) *The Pelican Freud Library* (Vol. 11, pp. 159–222). Harmondsworth, UK: Pelican. (Original work published 1915)

Freud, S. (1985b). Beyond the pleasure principle (J. Strachey, Trans.) *The Pelican Freud Library* (Vol. 11, pp. 269–338). Harmondsworth, UK: Pelican. (Original work published 1920)

Green, A. (1999). The greening of psychoanalysis: André Green in dialogues with Gregorio Kohon. In G. Kohon (Ed.), *The dead mother. The work of André Green* (pp. 10–58) London, UK: Routledge.

Green, A. (2000). *Les temps éclaté.* Paris, France: Editions Minuit.

Ogden, T. (2001). *Conversations at the frontier of dreaming.* Northvale, NJ: Jason Aronson.

CHAPTER 12

TEMPORALITY

Expectation and Futurity in Physiotherapy Patients

Larissa Laskovski and Lívia Mathias Simão

In this chapter, we understand *temporality* as a personal experience of the time lived. The notion of *temporality* is rooted in the phenomenological analysis of the human experience with and in time, beginning with Husserl, followed by a number of developments, particularly the work of Heidegger and Merleau-Ponty (see also Baerveldt, Coelho, Ellis & Stam, Simão, this volume).

For the purpose of the analysis proposed in this chapter, it is important to note the fact that temporality is inherent in the subjective processes for building personal I-world relations. It is within temporality, as one of the dimensions of human subjectivity, that people build and manifest themselves intra- and intersubjectively, creating their momentary organizations of I-world relations. Contribute to this process the manifold meanings that we (re)elaborated with respect to that which has occurred before now, directly and indirectly affecting us in the present, leading us to a certain future. In this sense, the past never ceases to be, thereby leading to a conception in which time is not a cause-based sequence of "before" and "after,"

Temporality: Culture in the Flow of Human Experience, pp. 293–310

given that this would require the objectification of the passing of time as externally independent of the subject that experiences it (Simão, 2011, this volume).

From this perspective on lived time, this chapter will touch on the role of expectations and futurity in physiotherapy sessions with patients suffering from neurological damage and different medical prognoses and therefore is limited by the clinical circumstances of their neurological system's functions. Our main concern regarding the research, which shall serve as a basis for this chapter, was to understand how patients who have suffered injuries to their neurological system, and thus suffer from motor damage, experience time, in an effort to integrate the unforeseen traumatic experience with future expectations, fears, hopes, and uncertainties. For these patients, the fulfillment of their expectations depends in large part on rehabilitation professional, including their physiotherapist, given that it is within this relationship that possibilities and limits for the patient's future are laid out.

This chapter is based on the context of a disquieting experience (Simão, 2003):

> By *disquieting experience*, I mean experience that hurts our *expectances*, prodding the subject cognitively as well as affectively to feel, think and act. The affected person may be the actor who lives the experience itself, or another peson who co-experiences the actor's disquiet through verbal dialogue or joint silence. (Simão, 2003, pp. 450–451, emphasis added)

Giving coherence within life situations imposed by disquieting experiences requires that one distinguish and, at the same time, seek to integrate, *experiences as they happened* from the subject's perspective, their desire for how *things should have happened*, and the *affective-cognitive gap* that is formed as a *non-sense field*, thereby generating disquiet (Simão, 2003).

Seeking to examine and understand temporality in the context of the relationships between physiotherapists and patients with neurological damage, we took as a central the fact that sudden changes have altered in meaningful ways the previous situations that were once seen as consistent and permanent by people who are now physiotherapy patients. In this new situation, the patient seeks to recover the continuity lost in the course of their life (see Freda, Picione, & Martino, this volume).

In cultural psychology, the complexity of relations between patients and healthcare professional was already examined by Boesch (1977) in his research involving the doctor-patient dyad. In this sense, Boesch emphasized a paradoxical aspect of the doctor-patient relationship and that, in a sense, we consider analogous to the physiotherapist-patient relationship focused on herein. The fact is that in this type of relationship (doctor-patient; physiotherapist-patient), the professional

> deals with basic threads to a person's ego (i.e., incapacitation and death) in a detached, matter-of-fact, institutionalized way. Yet it is precisely this tension between the medical professionalism, on the one hand, and the patient's deep-felt expectations and need for concern and security, on the other, which accounts for many of the complications involved in this particular kind of relationship.
>
> Doctor and patient interact in the framework of a joint action. Yet the individual actions each of them perform differ fundamentally from each other: as to the actions themselves (i.e., what is done), as to their aims and motives, and as to the total situation (perceived and real) in which they are performed. (p. 13)

Physiotherapy is usually sought out by those who seek treatment as one of the ways to overcome the problem caused by their neurological injury. Therefore, physiotherapy activities are symbolically tied to the achievement of the patient's desire to overcome an adverse reality. In this situational context, the physiotherapist is one of the "main others" with whom the patient is in constant interaction in their quest to reach these expectations.

For physiotherapy professionals, the rehabilitation process optimizes the individual's physical, mental, and social conditions in the face of a physiological and anatomical deficiency, becoming an educational means for the improvement of abilities, participation, autonomy, and quality of life (Delisa & Gans, 2002; Thompson, 2007).

It is well recognized in the area that the success of rehabilitation, including rehabilitation of patients suffering from neurological illnesses or damage, depends on the participation of families and caretakers, the involvement of the multidisciplinary team that cares for the patient, and the patient's own desire to get better. With regard to the latter, for a successful treatment, the patient's desire to get better should be actualized in their actions, which in turn are modulated by their desires and expectations concerning the treatment.

In general, excessive attachment to the idea of things "going back to how they were before the injury," together with a belief in this possibility, often going beyond what is clinically possible, as well as attachment to the present situation in a "conformist" manner to avoid the risk of deception if more difficult goals are not met, are all considered by physiotherapists as aspects that make success in treatment more difficult.

Therefore, it is widely understood that patients' desires and expectations can have a significant impact on the course and success of treatment. It is from this understanding that certain professionals working in the area of physiotherapy often choose to work in its boundaries with psychology.

Goals, desires, motives, and expectations are not merely situated in time, but also occur within the ambit of temporality:

> Time experience is based on the experience of the course of action; the distancing separating a goal from our present situation must be overcome by progressive steps of action, and it is this progression which constitutes a necessary condition for the formation of time as an *existential* dimension. However, to experience time requires more than the simple awareness of sequences in our activity: I have to perceive a difference between *now* and *then* and, for an anticipated *duration*, to "suspend" my consumption tendencies. The particular, time-specific activities filling a duration are, therefore, *waiting*, and *striving*. Thus "duratio," "waiting" and "striving" appear to mark the specific quality of the time experience.... Yet, in the "objective" sense, i.e., as a rational construct, time consists in a juxtaposition of two movements, one of them being constant (just as the movement of the sun or of a watch). Much research has been carried out on this *conceptual* time (e.g., Piaget, 1946; Fraisse, 1957; Ornstein, 1969); however, our main concern here is not conceptual but experiential time. (Boesch, 1991, pp. 159–160, emphasis added)

The perspective of temporality faces us, among others, with the fact that acting in the present is making choices that are projected into an unforeseeable future, based on memories and reflections of a past that has already existed—at least to us—intrasubjectivelly.

Continuity as a Desire and Rupture as a Disquieting Event in the Life of Patients With Neurological Damage and Their Physiotherapists

The following will present certain possible ways of elaborating disquieting experiences, in the articulation of present condition and expectations, which can be learned through an analysis of interviews with patients suffering from neurological damage and undergoing rehabilitation treatment in physiotherapy, participating in a research conducted by the writers of this study in the interface between the areas of physiotherapy and psychology.

As stated at the beginning of this chapter, this research was focused on knowing more about how patients deal with the disquieting experiences brought on by changes in the course of their life, produced by neurological damage, and possibilities opened by the treatment, elaborating on and reorganizing this tension-generating situation.

Four people volunteered and consented to participate in this study; namely, three males and one female, with ages varying between 18 and 39 during the time of the study, all of who were diagnosed with a neurological disease or damage (acquired; progressive or degenerative) and attended conventional physiotherapy on a regular basis. All of the participants possessed the physical conditions necessary to understand the questions being asked of them as well as to answer them using spoken language.

The methodological procedures began by the researcher (first author) following the routine of the physiotherapy sector of a multidisciplinary rehabilitation clinic in the city of Londrina, in the State of Paraná, in Brazil, in an effort to become familiar with the routine and professionals of the clinic, as well as to seek out volunteers to participate in the study.

In the next step, the researcher began to interview participants. The interviews (all recorded) were made up of three phases conducted during the patient's physiotherapy sessions: at the beginning, during patients' rest periods between exercises, and at the end of sessions.

In the first phase of the interview, the researcher showed each participant three cards carrying the following words, *rehabilitation*, *physiotherapy*, and *deficiency*, asking the patient to say what came to mind upon reading the words in question, and then talked to the patient about the theme when given the opportunity to do so by the patient. In the second phase, the researcher asked about the patient's disposition to go to sessions, their expectations for treatment, and the meaning given to the physiotherapist. In the third phase, conducted during intervals for rest between exercises or immediately at the end of the session (depending on the length of the session and the possibilities offered by the physiotherapist who led the session), the researcher asked the patient to talk about how they felt and/or what they thought upon doing their exercises.

The information obtained was analyzed in an effort to capture changes in how the participants saw their clinical condition and the conditions of their life, as well as their relationship with the physiotherapist, emerged from that process, which could be taken as indicators of some reorganization in the meanings and senses of their *disquieting experiences*.

The research theoretical-methodological presupposition was, then, that patients' symbolic actions, directed to themselves and at others, are *constrained* (Valsiner, 1998), particularly by their *disquieting feelings* (Simão, 2003) with regard to their new current situation and their expectations and desires to overcome it, often expressed in narratives directed at "their significant others" at the time.

Each participant in the study had a different manner for elaborating and reorganizing their symbolic relations among feelings brought up by memories of their previous condition, thoughts on their clinical condition and current life and desires, fears, expectations, and hopes for the future, channeled by and emerging from their I-other-world relations, in which the physiotherapist could be included as one of the "significant others" for the patient.

Below is a transcription of one of the researcher's interviews with the participant CD, male, 25 years old during the time of the interview, who had worked as a carpenter and waiter before his traumatic event. CD's medical diagnosis was neoplastic astrocytoma (medulla), with a physiotherapy diagnosis of paraparesis (damaged segment: L3-L4). The interview occurred 2 months after the participant's injury. CD utilized a wheelchair and required assistance to bathe.

The interview with CD will mainly serve as an illustration to analyze ways of elaboration of the disquieting experience generated by the traumatic event, in the articulation among current expectations and desires. In the following analysis, certain portions of the transcript will be used to illustrate each aspect being focused on. At other moments, interviews with other patients who participated in the same study shall be used and will be duly indicated as such (to read the whole interview with participant CD, see the Appendix).

The Immediate and Abrupt Character of the Past, the Latency and Uncertainty of the Future, and the Urgency of Transformative Actions in the Present

In general, the interview with CD has provided us an insight into some of his manners for attempting to articulate past events that, when they were present, imposed on him, in an abrupt fashion, a unique, unmatchable situation with strong meaning for future consequences, never before imagined by the patient. Part of the elaborations and expectations regarding treatment appear, at least at the beginning of the interview, in the desire to return to the clinical situation and life before the lesion, in the sense of "rehabilitating the damage":

> 3 Re: And this word here? [rehabilitation]
>
> 4 CD: Wow, that one is really hard, isn't it! To go back to the way things were.... To get your job back ...
>
> 14 CD: Well, I end up thinking, how can this be so hard, you know? It's because before I used to do this ... I end up thinking, how do you lose this so fas ... because I stopped being able to do this from one day to the next, you know?! From one day to the next.... And how come it is so easy to stop, to lose it, but to go back ... you know?! I keep thinking, will it take a long time? When will I be able to walk again?

Here, we have the manifestation of the tragic as one of the modes of temporality, as discussed by Romilly (2009) in her study on the meaning of time in Greek tragedy. The tragic installs itself in the relation between time and action (not necessarily intentional or controllable by the acting subject), giving birth to a new event, which disturbs and disorganizes the I-world relation in effect up to that point. Thereby, a sense of urgency is created for new affective-cognitive organization, requiring unequivocal actions by the person involved in the tragic event. Time is therefore felt in a relationship of emergency and tension between a previous, short-lasting present— now past—that brought the unexpected event and, with it, the demand for urgent actions going into the future—now the present. These actions should be "unequivocal," given that they will have a decisive, pervasive, and long-term effect on the person's life:

> 4 CD: To be honest, I saw them, those people in wheelchairs on the bus, and I thought, "Wow, how do they do it?" I thought to myself, "Whoa, you have to be dedicated, right?! Because.... You have to be dedicated to overcoming a problem like that...." But I never saw myself, you know, in a wheelchair like that, you know? And when I saw those people, I'd say, "Shoot, is their life that easy, is it ... what is it like, how does it work?" I'd sit there imagining it! "How do they ...," I can feel it in my skin, it's not easy at all, it's very complicated. You have to really feel it yourself to value it.
>
> 14 CD: Because I stopped being able to do this from one day to the next, you know?! From one day to the next.... And how come it is so easy to stop, to lose it, but to go back ... you know?! I keep thinking, will it take a long time? When will I be able to walk again? When I can walk again, will I start walking before my daughter? I keep thinking, you know? It's hard ... I keep thinking, who will walk first? Me or her?
>
> 16 CD: I would always get nervous when it was time for physiotherapy, the bottom half would hurt, at the beginning I thought ... I would say, darn, I'm not going to make it ... I suffered a lot, man ... I wanted to quit physiotherapy, but I got it into my head.

Therefore, actions in the symbolic field (Boesch, 1991) of physiotherapy can feed and weave into new symbolic relations in a number of different contexts of the patient's life, as a path for the reconstruction of knowledge of themself, now in this new and previously unknown situation, which is inevitably part and parcel of the comparative reflection on the situation before the event leading to the damage.

This reconstructive process, in which symbolic relations are elaborated and reorganized among memories of the prior condition and desires, fears, expectations, and hopes for the future, is the ambit of temporality. It is a process that requires to (re)signify the past event in the context of the I-other-world relation, given that the inherent unpredictability of the future, which becomes present via the damaging event and is now belonging to the past, unexpectedly "broken and entered" in the person's life, requiring her elaboration of new meanings based on the "as ... if" (Valsiner, 1998). The "near future, if not predictable, must be at the very least trustworthy. In this sense, the physiotherapist appears as a "significant other," possessing a "knowhow" in favor of the trustworthiness of the future:

> 5 Re: And this word here? [physiotherapy]
>
> 6 CD: Important, right?! [laughter] It's these time that.... It's like the doctor told me, "You can leave your wife, but you cannot leave physiotherapy if you want to walk, because...." Because it is important. I think.... For me it has been ... verrrry important. All I need now is to be able to move my leg, right!?... By the end of the year M. [Ph.] will get me there!
>
> 18 CD: Every day that I do my exercises, I see some evolution, you know?! I get excited, right!? Every day that I leave here, I leave excited ... because I got better, more exercises, every day that passes, I feel happy the entire day after physiotherapy.... You can't lose that, right? Because if you do, you're lost.
>
> 11 Re: Do you always feel like going to physiotherapy? Do you feel this desire?
>
> 12 CD: I do. I depend on her, you know! I mean, one of our biggest wins so far is me just being here, walking again. I love coming, I don't come just because I have to.

Temporality is a personal experience of time lived in a dual relational way; on one hand, it emerges through I-other-world relationships, embracing intersubjective time and the subjective duration, here included the interlocution with the oneself as other; on the other hand, it implies in the construction of a plurality of contextual meanings, relating past, present, and future circumstances (Simão 2010a, 2010b, 2012, this volume). It is in the interplay between intersubjective time and subjective duration that the possibly and momentary "attunement to the attunement of the other" happens, as referred by Rommetveit (1992):

> Reciprocal adjustment of perspectives is achieved by an "attunement to the attunement of the other" by which states of affairs are brought into joint focus of attention, made sense of, and talked about from a position temporarily adopted by both participants in the communication.
>
> A dyadic state of intersubjectivity with some state of affairs S is attained by verbal communication when some aspect or set of aspects A_i of S is brought into focus by one participant and, as a consequence, jointly attended by both of them.
>
> A dyadic state of shared social reality with respect to some state of affairs S is attained when both participants take it for granted that S is A_i and each of them assumes the other holds that belief.
>
> What is made known and claimed to be true about our "external" world by an assertive utterance at a given stage of a dialogue (its "propositional content") is contingent upon the dialogically established shared social reality at that stage (p. 23)

Within the physiotherapist-patient relationship, *different events of doing in the session and being done in the session have a place in accordance with this attunement,* which touches from the times and duration of the session to the rhythm of the exercises of each session and the level of difficulty of activities, estimated by the physiotherapist, upon proposing them to *that* patient.

Most importantly, however, it is the fact that the patient and the professional let themselves be touched by one another; that is the key to the physiotherapist becoming a "significant other" in the reconfiguration of a shared reality in the sessions, which in turn transcends the immediate reality of the session. It is from this point that may emerge different movements of articulation among the impossibility of reversing the neurological injury, its consequences that "make themselves heard" in the damage and the steady construction of a new sense for themself, in what it could come to be:

> 5 Re: And this word here? [physiotherapy]]
>
> 6 CD: Important, right?! [laughter] It's these time that.... It's like the doctor told me: "You can leave your wife, but you cannot leave physiotherapy if you want to walk, because.... " Because it is important. I think.... For me it has been ... verrrry important. All I need now is to be able to move my leg, right!?... By the end of the year M. [Ph.] will get me there!
>
> 7 Re: If you had to explain to someone what physiotherapy is, what would you say?
>
> 8 CD: [pause, 10 seconds] umm.... A person that ... I guess [pause 5 seconds] Person....
>
> 9 Re: You can say what is on your mind, no need to worry.

10 CD: Something that is a lot of work ... it's only for those that have a lot of love and dedication. I don't think that just any physiotherapist would do this, if it was just an obligation, right?! You have to love it, right? I think that, I've even seen it in the hospital ... the person would be like "Ah, time to get back to work." A lack of desire, you know?! They were doing it out of obligation, not out of love! Because there are physiotherapists out there that work out of love, right?! Because they love it. And when I was in the hospital I was.... Not just physiotherapists, but others too.... But there are some people that work, you know, with no desire, they don't even want to be there. So physiotherapy serves as an example for other people that only work like that.

17 Re: CD, what goes through our head when you are done? When it is time to go home?

18 CD: I keep thinking, will it take a long time . . . for my legs to work again, because ... the only thing that I think is about them now. Every day that I do my exercises I see some evolution, you know?! I get excited, right!? Every day that I leave here, I leave excited ... because I got better, more exercises, every day that passes, I feel happy the entire day after physiotherapy.... You can't lose that, right? Because if you do, you're lost.

Nonetheless, time in its linearity does not cease to impose its presence on patients and physiotherapists, when they compare current states with previous states before the beginning of the process of physical rehabilitation, indicating the patient's improvements or the lack thereof; or then, when patients observe others and exchange thoughts on their respective progress in rehabilitation, or lack thereof:

16 CD: I wanted to walk, but I'm good, at physiotherapy, because you see, you feel ... you see ... you think that your life is hard, but then you see other people in wheelchairs on the bus, you see ... you realize you have it bad, but there are people that have it worse off than you out there, I was even watching, when she went to pick me up ... [referring to the moment in which the physiotherapist went to get the patient from the waiting room]

These are also ways of articulating the past, present, and future, which do not lose their value in reducing the anguish and uncertainties regarding the effective result of treatment, and also inherent in all human endeavors:

> 14 CD: Well, I end up thinking, how can this be so hard, you know? It's because before I used to do this?! I keep thinking, will it take a long time? When will I be able to walk again? I keep thinking, you know? It's hard.

The very same is true for WS[1]: "I keep thinking....Will I get better results, right? Will it take a long time ... If it will work or not."

Nevertheless, the course of physiotherapy sessions also brings into focus, in a very unequivocal manner, the relational pair of "quickness—slowness," a central dimension of temporality, given that this pair addresses the relationship between performance that *may* be achieved at a given moment, given the patient's conditions, and hope as *latency* and betting on the *potential of what is to come*, which is actively sought out (or not), compared with the past situation and must find a plausible place in the present reinterpretation:

> 13 Re: What goes through your head when you're doing your exercises? When M. [Ph.] is helping you stretch, or when you are doing the exercises that are the hardest for you?
>
> 14 CD: Well, I end up thinking, how can this be so hard, you know? It's because before I used to do this ... I keep thinking, will it take a long time? When will I be able to walk again?
>
> 17 Re: CD, what goes through our head when you are done? When it is time to go home?
>
> 18 CD: I keep thinking, will it take a long time ... for my legs to work again, ... You can't lose [hope], right? Because if you do, you're lost.

The same is true for WS: "In the beginning, I wasn't able to come here often, no; because I thought I would get better quickly, you know? I keep thinking.... Will I get better results, right? Will it take a long time.... If it will work or not."

And also for MM[2]: "because one thing is to want it, you know, as soon as possible, understand? The thing is, that's not how it works."

Therefore, beyond the immediate meaning brought up by the linear relationship between past-present and future, often indicated by the narrative forms of the patients' and physiotherapists' speeches in their conversations during the sessions, certain aspects are revealed concerning the process of searching—and often finding—self-understanding and reorganization of the self in the face of a new, adverse condition, of which temporality is an inherent part:

As MP[3] said, in his interview: "Ahh, that there [when shown a card carrying the word physiotherapy] helped me so much. So much so that I used to grab on to my shorts to pull myself up, onto my pants so I could get up . . . ; today, I don't. Now, through that I have gained balance, strength, I've gained ... you know, stability. I have gained a lot and things are changing, you know?! Brushing my teeth, shaving, changing my shirt, putting on sandals, transferring ... moving around [moving from one place to another to meet someone] ... yeah, that's it."

I-Other-World Relations in the Temporality of Physiotherapist-Patient Relations: Final Considerations

Physiotherapy, more than exclusively and simply a "treatment for possible rehabilitation of physical performance," constitutes itself a constructed symbolic field of action (Boesch, 1991) from whence new articulations of meaning of the self may emerge in the context of temporality.

This therefore implies in possibilities and limits of self-understanding of the patient and the professional, at each moment, in the constant and mutual evaluation of the treatment that takes places through the ongoing interpersonal relationship.

Given the centrality of desire and affectivity, together with rationality, in the reconstructive cognition of our autobiography, affections, emotions, and desires not only take part in a sort of "assessment of what happened and what we achieved in the past," but also constrain (Valsiner, 1998)[4] present actions that project the patient and physiotherapist into the future (cf. Boesch, 1991; Valsiner, 2006).

Desired future states mean the present quest for transformation in the I-world relationship, which also include I-other relations (cf. Simão, 2010). Desiring and hoping are, in most part, to imagine a conceived future as possible, even if not given as certain. This is the symbolic meaning of the goals of actions, from which the patient and the physiotherapist cannot escape.

As explained by Boesch (1991), the goal of an action is imagined by the person as something that can happen, given certain circumstances that the subject will seek to ensure occur, with the aid of weighting and rationally planning (not necessarily consciously) in the present situation. But this imagining anticipation of the future, when in the here and now of the present, will always occur in different ways from what is imagined. Therefore, we are always and constantly seeking to reduce the perceived difference between what was desired and imagined and what is happening to us in the now, understanding that difference, making plans or even giving up on them, which inaugurates another imaginative anticipation of the future, and so on and so forth. At the same time, desirable anticipations of the future allow the person to confront the unforeseeability of the future in a less threatening fashion. These situations are *susceptible* to interaction with others, when they tend to become more explicitly and rationally elaborated.

It is quite typical, at the beginning of physiotherapy, that the physiotherapist has indications of the patient's desires as imaginative anticipations of his future physiotherapeutic conditions, and his expectations, as anticipations of the success of the treatment, which may often fail to consider neurological possibilities, even if they had become clear through an explicit medical prognosis. During treatment, in interactions with the physiotherapist, these expectations are often ruptured by the very nature of activities proposed by the physiotherapist. Previous goals for future treatment gain new meaning in the here and now of rehabilitation sessions.

A fairly common situation that illustrates this aspect is when, during physiotherapy sessions, it becomes clear that the patient effectively realizes that walking is no longer possible, given the severity of the neurological lesion, even when the patient had already been informed of the fact.

In this situation, the consequences of the past event are made present, breaking with the imaginative anticipation of walking again, which means the possibility of coming and going—autonomy—concerning free will.

Considering that it is often in the here and now of physiotherapy that this rupture effectively occurs, it is fair to suppose that it is also during physiotherapy that the patient will be able to confront this fact. In other words, it does not appear that merely hearing an explicit prognosis given by a doctor/physiotherapist allows the patient to confront and ressignify the treatment and their own *dévir* as a person, that is, their own personal future.

Creating new goals, in the sense mentioned herein (Boesch, 1991), does not mean giving up the desire to walk. When the physiotherapist proposes motor activities during treatment to add autonomy to the patient's coming and goings, and not necessarily to walk again, the professional is acting in the range of *change* and not in the ambit of "giving up on a desire." As the value of locomotion, of which the patient was previously capable and of which others are still capable, is not questioned, but rather reiterated according to the new conditions of the present, in an effort to improve the patient's locomotion, what gains new meaning, even painfully, is the manner in which to improve locomotion, rather than what could not be achieved ("independent walking").

For Costa, Melo, Garanhani, and Fujisawa (2010), an individual suffering from the loss of the ability to walk and who is apparently stable in the new condition, for example, dependent on the use of a wheelchair, transforms the manner in which to reach autonomy and thereby becomes capable of perceiving new possibilities for themself, making choices and rebuilding their own future goals.

At the same time, the same implications of this perspective apply to the healthcare professional. Given that during the initial evaluation, objectives for treatment are determined based on expected improvements in the clinical condition, the physiotherapist is also forced to change their treatment plans to reach new goals, particularly in accordance with evolution in the patient's clinical state.

Moreover, the physiotherapy implicitly outlines a range of "nondefinition" for that which is impossible to foresee with regard to the prognosis and the patient's performance in treatment. This zone of indefinition reduces anguish as to the professional's uncertainties, allowing for greater ease in changing goals, while also reducing the explicit emotional impact on the patient.

The patient's availability to adhere to exercises proposed by the physiotherapist is clearly guided by the foreseen possibilities the patient foresees concerning the efficiency of the exercises for their imagined future state, and not exclusively or necessarily by their current condition. In other words, patients opened to treatment seem to act on the base of "as ... if" (Valsiner, 1998): even before the physiotherapeutic condition is clinically modified, the patients act as if the change will actually happen or is actually, to a certain extent, happening. Thus, their openness to feeling different is part of the movement of creation of the present, in which the patient imagines future possibilities beforehand, what fosters their organization at the moment in the direction of their goals. The patient, in a sense, must feel like he is better than he actually is to truly adhere to physiotherapy.

It is in this delicate path of walking that the physiotherapist must go.

APPENDIX

Interview With Participant and Patient CD:

Key: Re = researcher; Ph = physiotherapist; CD = patient/participant

1 Re: So, first things first, what goes through your head when you see this word [deficiency].

2 CD: Umm, I'm not sure.... You have to feel it yourself to see,

to understand how difficult it is, right?

3 Re: And this word here? [rehabilitation]

4 CD: Wow, that one is really hard, isn't it! To go back to the way things were.... To get your job back.... To be honest, I saw them, those people in wheelchairs on the bus, and I thought, "Wow, how do they do it?" I thought to myself, "Whoa, you have to be dedicated, right?! Because.... You have to be dedicated to overcoming a problem like that.... " But I never saw myself, you know, in a wheelchair like that, you know? And when I saw those people, I'd say, "Shoot, is their life that easy, is it ... what is it like, how does it work?... " I'd sit there imagining it! "How do they ... ," I can feel it in my skin, it's not easy at all, it's very complicated. You have to really feel it yourself to value it.

5 Re: And this word here? [physiotherapy]

6 CD: Important, right?! [laughter] It's these time that.... It's like the doctor told me: "You can leave your wife, but you cannot leave physiotherapy if you want to walk, because...." Because it is important. I think.... For me it has been ... verrrry important. All I need now is to be able to move my leg, right!?... By the end of the year M. [Ph.] will accomplish it!

7 Re: If you had to explain to someone what physiotherapy is, what would you say?

8 CD: [pause, 10 seconds] umm.... A person that ... I guess [pause 5 seconds] Person.

9 Re: You can say what is on your mind, no need to worry

10 CD: Something that is a lot of work ... it's only for those that have a lot of love and dedication. I don't think that just any physiotherapist would do this, if it was just an obligation, right?! You have to love it, right? I think that, I've even seen it in the hospital ... the person would be like "Ah, time to get back to work." A lack of willingness, you know?! They were doing it out of obligation, not out of love! Because there are physiotherapists out there that work out of love, right?! Because they love it. And when I was in the hospital I was.... Not just physiotherapists, but others too.... But there are some people that work, you know, with no willingness, they don't even want to be there. So physiotherapy serves as an example for other people that only work like that.

11 Re: Do you always feel like going to physiotherapy? Do you feel willing?

12 CD: I do. I depend on it, you know! I mean, one of our biggest wins so far is me just being here, walking again. I love coming, I don't come just because I have to.

13 Re: What goes through your head when you're doing your exercises? When M. [Ph.] is helping you stretch, or when you are doing the exercises that are the hardest for you?

14 CD: Well, I end up thinking, how can this be so hard, you know? It's because before I used to do this… I end up thinking, how do you lose this so fast … because I stopped being able to do this from one day to the next, you know?! From one day to the next.… And how come it is so easy to stop, to lose it, but to go back … you know?! I keep thinking, will it take a long time? When will I be able to walk again? When I can walk again, will I start walking before my daughter? I keep thinking, you know? It's hard … I keep thinking, who will walk first? Me or her?

15 Re: I see.

16 CD: It's just like my physiotherapist said, right. A.P. [Ph.], who helps me at home, some men take this harder than others. This situation, you know? Women are more accepting.… But men are, you know!? She told me it is difficult … I would always get nervous when it was time for physiotherapy, the bottom half would hurt, at the beginning I thought … I would say, darn, I'm not going to make it … I suffered a lot, man … I wanted to quit physiotherapy, but I got it into my head. Now for me, it's easy to come here, no problem. I wanted to walk, but I'm good, at physiotherapy, because you see, you feel … you see …you think that your life is hard, but then you see other people in wheelchairs on the bus, you see … you realize you have it bad, but there are people that have it worse off than you out there, I was even watching, when she went to pick me up.

17 Re: CD, what goes through our head when you are done? When it is time to go home?

18 CD: I keep thinking, will it take a long time … for my legs to work again, because … the only thing that I think is about them now. Every day that I do my exercises I see some evolution, you know?! I get excited, right!? Every day that I leave here, I leave excited … because I got better, more exercises, every day that passes, I feel happy

> the entire day after physiotherapy.... You can't lose that, right? Because if you do, you're lost.

ACKNOWLEDGMENTS

This chapter is part of a research conducted by the first author, presented as a master's dissertation to the Institute of Psychology of the University of São Paulo, Brazil, under the supervision of the second author.

NOTES

1. Male;18 years of age; professional occupation: carrier; student; medical diagnosis: Traumatic brain injury and fracture of the D humerus; physiotherapeutic diagnosis: left hemiparesis; 2 months after the lesion at the time of the interview; independent locomotion and independence in all activities.
2. Female; 39 years old; occupation: salesperson; student; medical diagnosis: Aneurysmatic bone cyst on the T5 vertebra; physiotherapeutic diagnosis: segment damaged T4 paraplesia; 7 months after the lesion at the time of interview; utilizes wheelchair and is dependent for all activities.
3. Male; 39 years old; occupation: bus driver; medical diagnosis: spinal injury; physiotherapeutic diagnosis: tetraparesis; damaged segment: C8-T1; 2 years after the lesion at the time of interview; utilizes wheel chair and requires aid in bathing and eating.
4. "The notion of constraint touches upon the philosophical issue of the inescapable concomitancy of determinism and free will in human action in general, and in human symbolic action in particular. As such, the concept allows—and even demands—a focus on the permanent tension between the *possible (now)* and the *not possible (now or not yet)*, between *what can be* and *what exceeds* meaning in current human experience" (Simão, 2003, p. 450, emphasis added). We are here in the context of relations between should and is value in cultural psychology in symbolic action by Boesch. For more details, see Boesch (1991) and Simão (2010a).

REFERENCES

Boesch, E. E. (1977). The medical interaction—A study in Thailand. *The German Journal of Psychology, 1*(1), 13–28.

Costa, V. S. P., Melo, M. R. A. C., Garanhani, M. L., & Fujisawa, D. S. (2010). Representações sociais da cadeira de rodas para a pessoa com lesão da medula espinhal. *Rev. Latino-Am.Enfermagem, 18*(4), 1–8.

Boesch, E. E. (1991). *Symbolic action theory and cultural psychology*. Berlin, Germany: Springer-Verlag.

Delisa, J. A., & Gans, B. M. (2002). *Tratado de medicina de reabilitação: Princípios e prática* (3rd ed.). São Paulo, Brazil: Manole.

Romilly, J. de (2009). *Le temps dans la tragédie grecque—Eschyle, Sophocle, Euripide*. Paris, France: Vrin.

Rommetveit, R. (1992). A dialogically based socio-cognitive approach to human cognition and communication. In A. H. Wold (Ed.), *The dialogical alternative—Towards a theory of language and mind* (pp. 19–44). Oslo, Norway: Scandinavian University Press.

Simão, L. M. (2003). Beside rupture—disquiet: Beyond the other—alterity. *Culture & Psychology, 9*, 449–459.

Simão, L. M. (2010a). *Ensaios dialógicos: Compartilhamento e diferença nas relações eu-outro*. São Paulo, Brazil: Hucitec.

Simão, L. M. (2010b). *Temporality: Some questions for psychological research and professional practice*. 6th DS Conference. Athens, Greece.

Thompson, A. J. (2007). Principles of neurorehabilitation and its application to chronic neurological disorders. *Neurology Supplements, 3*(7), 1–8.

Valsiner, J. (1998). *The guided mind: A sociogenetic approach to personality*. Cambridge, MA: Harvard University Press.

Valsiner, J. (2006). Human development as migration: Striving toward the unknown. In L. M. Simão, & J. Valsiner (Eds.), *Otherness in question: Labyrinths of the self* (pp. 349–378). Charlotte, NC: Information Age.

CHAPTER 13

TIME OR NOT TIME IN MIND

What Is Temporality?

Ruggero Andrisano Ruggieri and Anna Gorrese

> I constantly get the impression that from this established fact beyond any doubt of the inalterability removed by time, we have drawn too little profit for our theory. However, there seems to be a gap which enables us to access the maximum depth. Unfortunately, I went even further on this point. What we would give to be able to understand these things.
>
> —Sigmund Freud (1932, p. 186)[1]

INTRODUCTION

It is well known that the question of what time actually is has been widely discussed in philosophical science, resulting in various conceptions of "time"—time as order, measurable of movement, time as intuitive movement, time as the structure of possibility (Abbagnano, 2006; Galimberti, 2006)—that due to brevity will not be dealt with in this chapter. However, it cannot not be acknowledged how these concepts have without doubt affected science in general, including psychology. The concept of time is

Temporality: Culture in the Flow of Human Experience, pp. 311–330

not usually studied in psychological science per se, although it has been widely used to explain several psychological phenomena. In fact, it becomes significant for psychological science due to it being considered a property of the nature of psychic phenomenon, or in other words, whether or not it is a constituent element of psychic phenomenon.

Thus, for example, the *time perspective theory* of Zimbardo and Boyd (2008), segments the flow of human experience into six time perspectives: a negative attitude toward the past, a positive attitude toward the past, fatalistic to the present, hedonistic to the present, looking at the future, and finally to the transcendent future. These time attitudes determine various psychic phenomena with obvious repercussions on a behavioral level. In fact, the authors show how "timing is an ideal equilibrium condition in which the past-positive attitude is high, the orientation to the future and that of the hedonistic present are moderately high, while the negative past and fatalistic present orientations are low" (p. 12). This profile appears to meet increased human adaptability to the environment by offering more coping skills. This theory is based on the recognition of the importance of time in the constitution of psychic phenomenon with a clear impact on the behavior of the subject under consideration. In contrast, psychoanalytic theories have always postulated the timelessness of the primary process to the secondary, as the only true psychic reality. Therefore, the spontaneous functioning of the psyche (primary process) has been considered the true psychic reality without the function of time, highlighting the effects of this timelessness on the behavior of the subject under consideration.

Based on these assumptions and adopting a cultural perspective, this chapter aims to investigate the relationship between mind and time in a clinical setting, highlighting how the psychic process is constituted by elements of temporality as well as elements of timelessness. Several examples will be discussed.

Mind and Time in the Classic Psychoanalytic Tradition

Bateman and Holmes (1995), in an attempt to provide a framework to psychoanalytic theory, define psychoanalytic science as (a) a theory of the functional model of the mind, (b) a theory of evolution, (c) a theory of intervention, and (d) a theory of continuity between normality and pathology. This scheme provides a clear and useful conceptual input to guide the reading of psychoanalytic texts, highlighting the common elements among the various authors. At this point, it is worth asking, In the models of mental functioning proposed by psychoanalysis, is time recognized as a property of the psychic content? What is the relationship between time, the mind, and evolution? In what way does the concept of time enter the technique of psychoanalytic treatment?[2]

Obviously, the answers to these questions would require a detailed discussion of each author, an objective that goes beyond this work. However, the conceptual framework of Bateman and Holmes (1995) is adopted, limited to some of the contents of Freudian theory aimed at understanding the relationship between mind and time.[3]

Working Model of the Mind

Freud (1915) was able to theorize a model of how the mind is apparent in the articulation of "primary process" and "secondary process." The first process is characteristic of dreams and fantasies, and is predominant in childhood. It is characterized by the absence of the laws of time and space, due to the absence of contradiction, thus the distinction between opposites. Past, present, and future cannot be distinguished, which is why multiple events can occur simultaneously and in the same place. The second process, however, follows the logic of time and space and is based on an assumption of rationality.

This distinction remains valid in the first and second issues, in both a topographical perspective, therefore in a measurable space, of the mind (divided into conscious, preconscious, and unconscious), as well as a structuralist perspective (articulated in Id, Ego, and the Super-ego). In fact, Freud specifies that there is nothing in the dynamic drive of the unconscious, and then the Id, which looks like time: "The processes of the *Unconscious* system *are timeless*, they are not temporally ordered, not altered by the passage of time, they have, in fact, no relationship to time" (Freud, 1915, p. 71, emphasis in original). Ultimately, according to Freud, "the unconscious is *primarily* out of time" (1901, p. 293, emphasis in original).

In the *Interpretation of Dreams*, Freud (1899) distinguishes between the "manifest content" of the dream (what people remember of the dream) and its "latent content" (its true meaning) to indicate the operating mode of dreaming." Freud seeks to identify the laws of mental functioning during the dream state, laws that can be related to the operation mode of the unconscious mind. In this case, albeit implicitly, Freud comes across the timelessness of the psychic content, apparent in the latent content, where any syntactic connection, normally due to the phrases "why," "when," "where," and so on—seem to be fused and confused, failing any relationship system between the content of the dream work, including that of time. Condensation, representation, free association, displacement, and symbolization are ways in which syntactic connections are lost. Thus, the Freudian idea is that the unconscious mental activity has its own language and that this language is apparent more than ever in dreams.

In fact, Freud says much more. The interpretation of dreams is, for the father of psychoanalysis, a clinical way to access latent dream content through the analysis of the manifest and, as pointed out by Eagle (2011), is a general process through which the unconscious becomes conscious, or rather, the unconscious is translated into the conscious.

This observation has a considerable heuristic fertility, as it clearly highlights a Freudian intuition that has remained in the background of the psychoanalytic perspective. Maintaining the distinction between the primary and secondary processes, Freud seems to attribute to the interpretative function of the dream the purpose of translating the unconscious into the conscious, in other words, "timelessness" into time. It therefore seems that the *interpretation is a function of the mind through which temporality is constructed in the mind.*[4]

In summary, the model of the functioning of the mind proposed by Freud suggests the coexistence of both a temporal and timeless dimension of psychic content.

Development Model

Consistent with his biological background, Freud (1905) developed the idea of a psychological development of the mind divided into several stages (oral, anal, phallic, and genital). These stages go along the road to maturity. Each of these stages has its own peculiarities due to an erogenous zone that is the source of the libidinal drive.

Beyond the specificity of the Freudian model, it is interesting to note the notion of an evolution in stages covering to some extent the idea of temporality. This correspondence implies how the theoretical model proposed the idea of time consisting of successive sequences. This is the proof that the disorder, according to Freud, is the result of a "stopping" of development at one of these stages. In this sense, he considers the "compulsion to repeat," in order to show this dimension of arrest for which the person reifies an inner world that does not correspond to the outside world due to it belonging to another time. "Fixation" and "regression," for example, are defense mechanisms which provide for the use of this idea of time. In this case, there is the dichotomy of the timelessness/temporality of psychic content.

Psychoanalysis has subsequently modified this idea, replacing the concept of stages of development with those of phases of development. For example, Klein (1932) discusses the paranoid-schizoid position and the depressive position as the pre-oedipal phase of development, while Mahler, Pine, and Bergman (1973) consider child development as phases in a process of separation/individuation. Winnicott (1965) shows the construction process of the true self ranging from total dependence on the mother to the construction of articulated autonomy through the emergence of specific needs at certain times. The examples go on. However, when considering stages or phases, it is impossible not to note how the polarity of temporality and timelessness in the construction of psychic content remains constant even in relation to the idea of development.

Intervention Model

Psychoanalysis, by definition, is constituted as a science of the analysis of the psychic process. Some of the contents of the technique, just as some of the tools, however, make it possible to understand in this case the duplicity of the use of the concept of time, now recognized as an element of psychic content, now denied.

To clarify this issue, the concept of transference is referred to. At first, Freud considered such content as a form of resistance to the analytic work as well as to change. Then he saw it as a factor of production, due to it being able to capture the dynamic unconscious of the patient through the relationship with the analyst. *Transfert* is a reflection, a recurrence of infantile conflict but in a therapeutically analyzable form, or better still, usable.

Thus, transference seems to be a dimension of the past still active and/or belonging to another time. In clinical terms, something "*there and then*" active in the "*here and now*" of the therapeutic relationship. In other words, transference reflects a timeless psychic content that is expressed in a temporal psychotherapeutic encounter. In fact, psychoanalytic technique uses time as an organizer of the intervention beyond which it is no longer useful to go. The duration of a psychoanalysis session, generally around 45–55 minutes, is considered essential as a framework for the analysis of psychic content. Beyond that, any analytical work would lose its effectiveness as it would be similar to any other social practice, such as a chat between friends. If intervention on psychic content requires specific time, for methodological issues, as well as ontological, it is obvious that temporality is to some degree a property of the psyche. Even in this case, there is a condition in which psychic content appears temporal/nontemporal, and the analytical technique is presented as a method of construction of temporality on the basis of access to psychic contents removed from the unconscious.

In psychoanalytic tradition, temporality and timelessness seem to be two sides of the same coin. They both belong to psychic content. Post-Freudian psychoanalytic literature has certainly enriched the theoretical corpus, providing rich and varied explanations of what the unconscious and conscious are, how disorders are formed, and so on, in other words, how time and the "absence of time" are articulated in the mental process. However, these conceptual differences, present in authors such as Melanie Klein, Donald Winnicott, Jacques Lacan Wilfred Bion, and Matte Blanco, have only indirectly addressed this issue, and in particular, the theme of the timelessness of the unconscious content (Pulli, 2008), leaving the Freudian legacy mentioned in the introduction of this work unfinished.

TIME AND TIMELESSNESS AS AN EXPRESSION OF BI-LOGICAL FUNCTIONING OF THE MIND

Temporality in Matte Blanco's Theory

A useful contribution to understanding the dynamics of the temporality/timelessness of psychic content is set out in the theory of the bi-logical mind by the Chilean psychoanalyst Matte Blanco (1975). He maintains the fundamental distinction between the primary and secondary processes (Freud, 1932).

According to Matte Blanco (1975), the mind operates on the basis of the *co-presence* and *coexistence* of *bi-logical* structures, both conscious and unconscious. The unconscious is essentially governed by symmetrical logic, based on the principle of symmetry (with a *generalizing* and *homogenizing* function). This logic assumes that (A > B) = (B> A); for which (A > B) = (A = B) as (B > A) = (A = B). Therefore, this logic does not distinguish between the elements. On the contrary, each element appears to be equivalent to another. Thus, if we say that Mario is Marco's father, it is also true that Marco is Mario's father. A graphical representation of such a way of operating is through models of mathematical sets (Figure 13.1), where a single element (A) is treated as a member of a set that contains other elements (B), and this in turn, as a subset other classes of elements (C) and so on ad infinitum, determining classes of symmetry.

Conscious thought, on the contrary, is governed by the principle of asymmetry (with a *dividing and heterogenic* function). It follows the criteria established by Aristotelian logic, if A > B then B < A, if A > B then B < A; resulting, however, that A ≠ B. In this case, it operates by means of a division of the elements (*dividing function*, consisting in the extraction of classes of symmetry), which are considered by their diversity (*heterogenic function*). A graphical representation of thinking is apparent in the following, where each element once extracted from the classes of symmetry (Figure 13.1), is captured by its specificity (Figure 13.2). Therefore, A is not B, B not C, A not C, and so on:

The idea of a bi-logical mind gives the human mind the simultaneous coexistence of two logics that basically appear to be irreconcilable and incompatible due to being opposites.

Conscious logic is a bivalent logic that allows for rational thinking to take shape through the separation and distinction between the different categories of objects based on which reality is built. *It is the principle of non-*

contradiction that governs its expression, which is why it is possible to hold a thought in time and space. This logic *treats the relationship between objects as irreversible*. It implies, therefore, a mental function based on *cognitive codes*. In contrast, the unconscious logic, based on the absence of any distinction between the elements to which it applies, is *what is both irreversible as well as reversible* (Oneroso, 2003). It is a mental function based on *affective codes*, which Matte Blanco (1975) coincides with emotions.

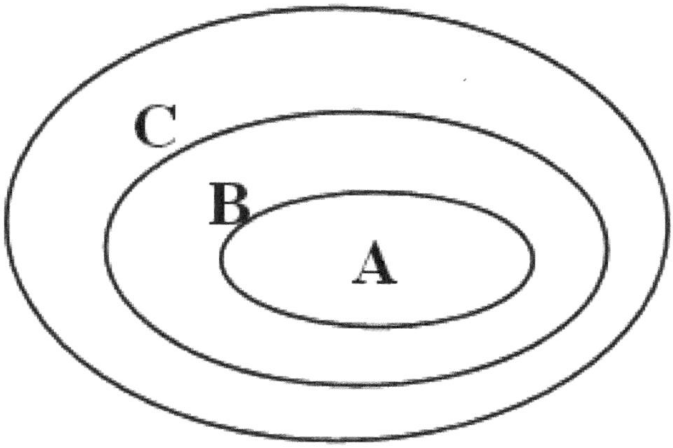

Figure 13.1. For symmetrical logic, therefore, the contiguous elements, although different become identical (Ruggieri & Pecoraro, in press). In other words, the properties of an element become the properties of another (homogenization) as well as others, until infinity (generalization). Then, reasoning in terms of time, time t_1 is equal to time t_2 which is equal to time t_3.

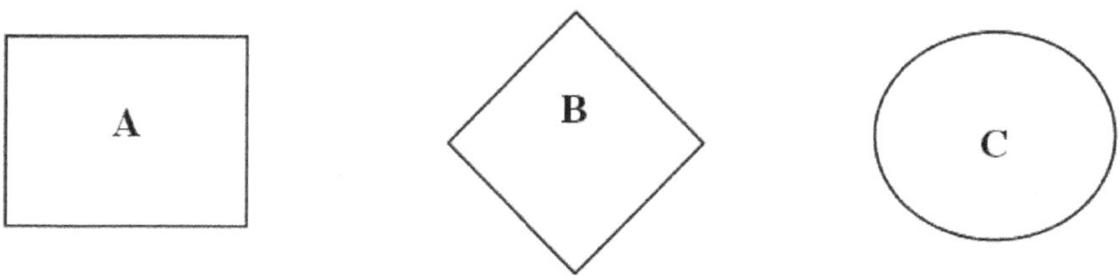

Figure 13.2. Reasoning in terms of time, t_1 is different from t_2 which is different from t_3. Then, t_1 is different from t_3.

The innovation of the thought of Matte Blanco (1975) is right here, in having reversed to some extent the terms of the philosophical discussion on the concept of time,

> which implies that the past is irreversible and cannot go back to being the present it was, if not when it is recalled and relived in consciousness. In this case, it returns transformed into something other than it is due to having acquired a new being, the being of time linked to consciousness. Time regained, therefore, is not the past. (Oneroso, 2003, p. 31)

This conceptualization finds correspondence in Kant's philosophic thought. Studying the *simul,* he conceptualized "the time" not in terms of empirical evidence but in terms of phenomenological and transcendental concepts. For Kant, simultaneity is the objective evidence of unit-multiplicity of the world (Valent, 2003). Simultaneity should be defined through the reversibility of the casual relationship. To be together, to be at the same time means coexistence and exchanges of precedence and succession, namely, A precedes B as B precedes A. B follows A as A follows B; thus, simultaneity implies a flow, a dual direction of the only serial path of to and fro where "the before" and "the after" are coincident. This coincidence builds simultaneity through a relation of identity between "the before" and "the after." Time is not only a succession or serial series—one-dimensionality of time—but is also a complexity, as a unity of manifold as bidimensionality of time, namely, reversibility, where simultaneity refers to a relation of succession in terms of concomitance of two sequences in the opposite direction: reversibility of the time process (Kant, 1787; Valent, 2003).

Thus, unlike with Carnot's (1824/1897) second law of thermodynamics, which states the irreversibility of the process of transmission of heat from a warmer element to a cooler one, on the basis of the degradation properties of the heat, with which to some extent the irreversibility of natural phenomena is determined, Matte Blanco (1975) highlights how *the psychic process is as much temporal as it is timeless* on the basis of the coexistence and copresence of bi-logic. There is no human act that is free of this coexistence of the two logics that does not always contain symmetric and asymmetric components in thought, in language, in art, as in dreams and disorders.

Thus, *temporality can be understood as a time delay of human experience* that becomes so, and only when an asymmetricalization of the classes of symmetry is adopted, that is, when building the conditions in which relationships can be established between the elements that make it possible to think experience and to separate it to some extent from the experience itself. This operation is to a certain extent due to the intervention of psychotherapy, or more generally, to clinical intervention, when it is based on the ability to think about emotions (Carli & Paniccia, 2004) by, for example, reformulations, interpretations, and so on, just as Freud had guessed.[5]

It is evident that in Matteblanchian theory, this operation maintains and still contains a dimension of both timelessness/reversibility as well as temporality/irreversibility. It would be impossible, in a psychoanalytic perspective, to consider a completely and totally asymmetrical mental function, as this would require an idea of man as a being totally rational that operates on linear models based on precise sequences, devoid of symbolic content, and therefore emotionality.

COLLUSION AND CULTURE AS INTERSUBJECTIVE AND INTERINDIVIDUAL WORKING OF THE MIND

Building Temporality in a Cultural Prospective

The idea of a mind equipped with its own language, its own laws in articulating the relationship between the inner and outer worlds, between the temporal and timeless dimensions of the psychic process is the central point of the basis of psychoanalytic theories, as illustrated by the studies of Bion (1962) and Lacan (1977), to name a few. However, unlike other contemporary psychoanalytic theories (Salvatore & Zittoun, 2011; Storolow, Orange, & Atwood, 2001), Matte Blanco (1975) offers a very systematized model of the functioning of the mind, adopting a structuralist perspective (conscious/unconscious) which equates functional dimensions of the mind (homogenization/generalization, dividing/heterogenic). In this case, the model makes it possible to focus on the idea of *a mind generating symbols as an expression of affection underlying the cognitive process, that is, as an expression of affective and cognitive codes.* In other words, this model considers the mind in terms of meaning and significance operations (Salvatore & Zittoun, 2011), highlighting time and timelessness aspects. However, his model remains anchored to an individualistic dimension of psychological behavior and mental process.

In order to develop a psychoanalytic theory that had the social[6] relationship and not the individual as its unit of analysis, Carli and Paniccia (1981, 2004) proposed the concept of collusion, anchoring in this way, the theoretical model of Matte Blanco (1975) to a cultural perspective. In fact, the concept of collusion is based on the *shared affective symbolization* of a context that builds the relationship between those who share the same context. This operation is made possible by the bi-logical mind that, in this case, loses the individualistic dimension proposed by the Chilean psychoanalyst, and assumes a social one. In other words, it is *systems of shared symmetrizations that are always active and to use of the actors of the case.*

Collusion, therefore, indicates the symbolic and emotional relationship (therefore timeless) that characterizes the social dynamic,

> it is a "cultural" and not structural context. The co-occurrence of affective symbolization *does not require, therefore, the presence of those who contribute to the phenomenology of collusion within a defined space or time*. Collusion is a phenomenon that goes across groups, organizations, cultural dimensions. It is a phenomenon that establishes social relationships. (Carli, 2006, p. 1, emphasis added)[7]

Therefore, it can collude with a given situation without ever having taken part in any way. In other words, the process of collusion is what allows each individual to relate to any other individual in the widest possible sense, making every conceivable practice social. It is evident that the concept of collusion causes a paradigm shift in the psychology of psychoanalysis, as it defines the idea of a psychic process free from a solipsistic aspect and anchored within a social practice.

While not addressing the subject directly, studies by Carli and Paniccia (1981, 2004) as well as Salvatore and Scotto di Carlo (2005) lead to a reassessment of Matteblanchian thought, placing it within *the idea of an intersubjective and interindividual mind,* namely, within a cultural psychology perspective (Bruner, 1990; Cole, 2003; Vigostsky, 1962), in particular, in a part of cultural psychology that recognizes the sense-making dynamic as the main psyche function (Bruner 2009), seen as a semiotic process where social and individual dimensions give a shape to the culture dimension (Rosa & Néstor, in press; Salvatore, 2012; Salvatore & Venuleo, in press; Valsinier, 2012, in press).

Focusing again on the objectives of this work, it is worth highlighting another aspect: *the collusive process in timeless terms*; that is, people belonging to different spatial and temporal situations involved in the same collusive process.[8]

This is clear when Carli and Paniccia's model of collusion (1981, 2004) finds an operative form through the concept of culture, which has become a scientific construct through which it becomes possible to observe and measure the shared psychic processuality underlying the manifestation of the behavior of individuals, groups, and organizations (Carli & Paniccia, 1981, 2004; Salvatore & Scotto di Carlo, 2005).[9]

In the authors' opinion, culture is the result of a system of collusion, that is, of the same attribution process of meaning and significance; a shared matrix of cognitive and affective codes for the consumption of the actors involved in a given situation. Technically, it is a system of *shared affective symbolizations, hierarchically organized and object dependent* as well as an expression of the bi-logical model (cognitive/affective, conscious/unconscious) of how the mind of Matte Blanco works, and the shared nature of the knowledge processes (Fornari, 1981). Therefore, culture is a specific outcome of the processes of coexistence and copresence of affective and cognitive codes; a feeling and understanding, the temporal and timeless content, where the affective code (feeling/timelessness) restricts, limits, or defines the cognitive one (understanding/temporality).

Culture, therefore, is distinguished by the collusion in that it requires an object. Collusion, on the contrary, is the means by which the cultural process is enabled; the context in which this process comes to life. This suggests that culture cannot exist without context and is a form of historicization of the context evoked by the actors. In this sense, *it is bound to a local form through which the temporal and timeless psychic contents* are shaped in the social positioning of each individual (Oles & Hermans, 2005; Varisco, 2002), that is, in the manner in which each is in relation to others to something in a given context. In other words, the context becomes a form through which *to experience a delay of the social dynamic due to it binding the cultural process to operative objectives. Like the setting of a psychoanalytic intervention, the context makes it possible to immediately understand the appropriateness of the process carried out by the collusive cultural dynamics of it.* Our thesis is that every time you come across a cultural process that is not relevant to the context in which this dynamic is active, there is then a perception of the lack of temporality of psychic experience. This gives the impression of belonging to a different and another time. It is, therefore, of paramount importance to identify the absence of temporality of the cultural process in the *asynchronicity* of cultural models with respect to the positioning required by the context. Temporality is configured as the way through which to express the congruity of its symbolic model with respect to the context, that is that is, when the context and culture are setting in complementary or *synchronic* way in terms of symbolic positioning.

To better understand this relationship, it is appropriate to try to operationalize the concept of culture with appropriate examples in order to make its functioning clear. Culture works like a mental software with which information and behaviors built on the basis of a precise symbolic hierarchically organized order are decoded and then interpreted (Cesaro & Ruggieri, 2011). Since culture is a dependent phenomenon, when the objects change, so do the symbolic orders, that is, the cultures associated with them or, if you will, the emotions, the affective code associated with them, resulting in a change of behavior apart from those that are the individual characteristics of the actors.

On an operational level, this translates into the fact that there are many people who cognitively understand several aspects of reality but are only able to use them in some cases. In a recent study of the professional culture of the students attending the degree course in educational science at the University of Salerno, it has been highlighted how students seem to be unable to use professor's office hours and exams as teaching tools. In this case, the student culture expresses a specific role identity, reason for which the professor's office hours, for example, are considered only to be used when working on the thesis or to request information on the exam texts. It does not represent an opportunity to ask for further explanations. If this aspect is elicited by the students, they are able to understand it, but they are unable to operate in this direction, that is, they are not able to define actions consistent with this objective. In other words, they are bearers of a cultural model that is not congruent with the learning context for which the exam and professor's office hours are teaching tools through which to test their learning.

However, when the objects and/or contexts are modified, there is a different behavior. In this case, for example, the same student would be able to suggest to a fellow student to go to the professor to ask for further explanations of the course content; he would also be able to accompany his colleague as well as to speak in his place, but would be unable to do it for himself, within his own faculty, university, or professors.[10]

It is interesting to note that this collusive dimension is effectively identical for all students of different years. This aspect, in addition to confirming the effectiveness of the model of collusion, highlights the active presence of the *timeless* content of this cultural process. In fact, the different students, from the first, second, and third years, while participating in a quantitatively different university experience retain the same symbolic position with respect to the "office hours" tool, regardless of the stress received by teachers in several years, showing an inability to build temporality of the university experience.

In other words, students in the first year, despite not having experience and contact with the university system, behave the same way as those of the second and third years. They therefore have not taken part in the collusive phenomenology but recognize it as their own; although, in this sense, belonging to a different time and space, that of the school upon entry into university.[11]

We are therefore arguing that the object as a discriminating element of culture in the cultural process is the element that makes it possible to capture the active sense of hierarchy, allowing to capture the temporality expressed or not expressed in the psychic process. Taking the example of the student's culture, it is evident that they are evoking contexts different from the one proposed by the university system. It is as if they were elsewhere, in another time and another space as a function of the learning tools required by the university curricula. They share the same symbolic order: lesson, clarification, question, thesis, office hours, and such; a symbolic order different from that of the professors: lesson, office hours, clarification, question, thesis.

In this case, it is possible to speak of *asynchronicity* of the collusive dynamic because of the fact that each actor (student and teacher) is the bearer of a different social position in which different university contexts are evoked, both expressions of a bi-logic: the first where it is not possible to think of using the "office hours" tool to study in greater detail the lesson content; the second where it is. Therefore, they play on different time dimensions, while sharing the some content. However, it is precisely this sharing of content that allows for the construction of a temporality that is reflected in the complementarity of symbolic orders due to the frame of sense given to the context.

Therefore, the context becomes the instrument capable of activating interpretation functions (asymmetrizations) that create stable syntactic connections in collusion processes, that is, capable of creating a temporality of the psychic process, alias culture.

OPERATIVE ASPECTS OF THE CONGRUENCY/INCONGRUITY DYNAMIC BETWEEN CULTURE AND CONTEXT

Collective Memory as an Example of Lost Temporality

It may not have gone unnoticed by the reader some clarification of the dynamics of symbolic orders and that it may be worth considering them due to their social and practical importance. We refer to the *mobility and changeability of cultural processes at play.*

This peculiarity, or if you will, the operating mode of cultures based on the variability of the same by reason of objects and contingencies, or rather the diversity of contexts evoked by active cultural patterns, if recognized, explains many reasons the why of history. It explains why even after many years, there are, albeit in different forms, identical episodes as well as similar social and cultural phenomena. In summary, in our opinion, this characteristic can explain the gap between history and memory, that is, it makes it possible to understand *why history hardly ever coincides with the memory of man*.

It is worth considering the contents of the film *The Wave* (*Die Welle*) by the director Dennis Gansel (2008). During a theme week, a German high school teacher has to deal with the subject of the autocracy, although he would have preferred anarchy, which is closer to its ideals. The didactic purpose is to show the students that Nazism can come back at any time. Faced with this goal, openly stated, the students clearly express their opposition to the Nazi regime, asserting with vehemence and superficiality that they could never become Nazis, due to the memory and the pain of that period still being vivid in Germany. They also ensure that people have learned from the mistakes of the past.

The teacher decides to organize an experiment, in order to prove the contrary. The experiment involves the entire class itself and begins with the selection of a leader and the imposition of some basic rules. The layout of the desks is changed so that the small groups of friends are broken up and less able students can be nearer to the better ones in order to improve the overall performance of the class in a logic of mutual aid and unity of the group. The next step in the identification of the group is to give it a name. "The Wave" is chosen and a special logo designed. Each student has to wear a sort of uniform, consisting of white shirt and jeans, so as to remove any distinctions and class individual. A greeting is also invented that simulates the logo. The students began to spread the Wave logo throughout the city and have parties in which only members of the movement are allowed to participate, fighting and discriminating against all others.

In summary, the experiment carried out under the guidance of the teacher consists of the establishment of a *role-play* that lasts a week in which it is proposed to live as if you were under the Nazi regime. Students adopt Nazi attitudes and behaviors in the space of a week without realizing it, finally becoming caught up in a tragic episode that will act as a collective alarm for everyone.

How is it possible that people who say they are not Nazis find themselves to be so? How is it possible to forget?

The answer to this question is precisely in the characteristics of cultures and how they relate to contexts. In fact, *role-playing* is a psychological intervention tool used in the educational process in order to promote a change in the symbolic order. It is an accelerator of the inherent dynamism in cultures. Therefore, the starting point is the very dynamism of cultures.[12]

Activating certain cultural devices rather than others means addressing one side rather than another of the cultural process *insofar as it migrates to other contexts and thus loses is consistency with the initial context*. It is precisely in this case that there is a temporal asynchronicity that highlights active meanings that are no longer congruent with the original context. Thus, a symbolic past, Nazism, all of whom claim to remember the atrocities becomes current, present.

We are therefore discussing the loss of temporality to indicate the release the cultural process has from the initial original context. In fact, the *role-playing* constructed by the teacher in the film clearly shows this migration of affective symbolizations, which lose their original instructional/educational context to migrate to another context, that of a Nazi culture which makes the *in-group/out-group* dynamic, *friend/enemy* hierarchy element that gives meaning to action. Ultimately, this timeless content indicates a psychic processuality far from its dialogical dynamic context in which it occurs.

It is worth recalling several properties of the culture in order to clarify what has just been mentioned.

It is worth remembering that cultures are meaning systems that reify and realize reality through actions, behavior, communication and common narratives.[13]

In this sense, they are *situated*, *pragmatic*, and *performative* (Cesaro, Ruggieri, & Pecoraro, 2010; Salvatore, Forges Davanzati, Potì, & Ruggieri, 2009).

In fact, culture, being a useful way to represent reality in order to act in it, is a situated system, in the sense that it changes on the basis of the objects and contexts that vary depending on the circumstances, history, and narrative to which it refers, at the same time maintaining a strong pragmatic value. In other words, it has a strong practical implication. In fact, it does not qualify as impartial or neutral. On the contrary, it expresses some form of intentionality that allows the actor to position himself in the context in which he is operating, thus producing a statement about his own interests or own idea of the world and identity.

At the same time, due to the recursion between action and thought, culture, in addition to influencing behavior, makes it possible to reproduce thoughts through actions. Since they are agitated, they are made real. In this sense, culture seems to be performative, because of its ability to make its very existence socially acceptable.

The film clearly highlights all these features. The choice of the logo, the leader, the rules, the uniform, and the roles determine a symbolic movement, with different behaviors and actions being assumed. These actions recursively preform the symbolic content of a reality that evokes a context that is different from that of the school, making the reappearance of a Nazi community possible.

CONCLUSION

In this chapter, we have tried to investigate the relationship between mind and time within a psychoanalytic frame of a cultural matrix, highlighting how the psychic process is constituted of elements of both temporality and timelessness.

Moreover, we tried to highlight how time plays a constitutive role in the psychic process. "Think of the psychoanalytic principle of free association: according to this principle, the associative chains created by the temporal contiguity between the signs are a carrier of unconscious meaning" (Fornari, 1983). The principle through which the association between signs, related to temporal contiguity, is in turn the sign of a more global pattern of meaning is not confined to the rest of the field of psychoanalytic theory; it is in fact the basis of the semantic differential (Osgood, Tuci, & Tannenbaum, 1957) that interprets patterns of association between the different connotations of the objects as an indication of the salience of a latent dimension of affective meaning. "Time, in the end, rather than being a container in which to explain meanings, is a vector for the creation of meaning" (Salvatore, Laura-Grotto, Gennaro, & Gelo, 2008, p. 391).

In this regard, we have defined temporality as a function of the consistency of the symbolic order, or in other words, the sequence of signs with the context within which this sequence is the event. Thus, the lack of contextual consistency determines the idea of a content belonging to a different time. Finally, it is worth highlighting one aspect that has remained in the background in discord brought forward and left between the lines of this last sentence. To claim that temporality is a model of consistency of the symbolic order with the context of belonging highlights two aspects: The first is related to the fact that the timing function primarily belongs to those who observe the psychic process, while the second is that the bearer of psychic process seems to be prone to a loss of temporality.

This aspect, though written in the bi-logic of the mind, in our view needs further study, as it is the bearer of a heuristic fertility that deserves to be explored to its full potential.

NOTES

1. This is our translation.
2. The reader will have noticed that among the questions posed, the one relating to the relationship between continuity and pathology in the operating model of the mind is missing. This is consistent with the conceptual framework adopted, since the category under examination is comparable to some extent to the question relative to psychoanalytical treatment. Furthermore, the development of this point would require a clinical treatment that is beyond the scope of this work.
3. The reference to Freud is due to the fact that he is not only the founder of psychoanalysis, but he is also more well known.
4. This point is also considered differently in the following paragraphs. In fact, for the psychoanalytic theories, the ability to reflect on the experience enabled through narrative processes in the analytic *setting* lead to a restoration of lost temporality, or rather a translation in terms of time of events.
5. This content is clearly highlighted in the theory of psychological intervention known as Ananlysis of Demand (Carli & Paniccia 2004; Salvatore & Scotto di Carlo, 2005; Carli & Govannioli, 2011). This model relates to a suspending system of the action—agitated by the active emotional process—and the construction process of thought on action, in order to verify its correspondence in terms of definition of aims and strategies adopted.

This model implies the recognition of the foundational role of emotions in mental processes, due to the primary process.

6. The social relationship here should be understood as a unit and not as the result of individual characteristics in relation to each other. It is precisely in order to overcome the latter perspective that the studies by Carli and Pianiccia were carried out.
7. Carli (2006) highlights how the etymology of collusion (from the Latin *cum ludere,* play together) maintains its various facets of "allude," "deceive," "disappointing," and "collude," a social dimension strictly anchored to a symbolic emotional dimension.
8. The proof of what has just been stated can be seen through a series of empirical studies, apologizing if referring to the examples a few lines later, but first it is necessary to introduce the concept of culture and its relationship with collusion.
9. Culture thus becomes the empirical system through which the collusive dynamic is measured.
10. This observation opens up to further considerations: whenever coming into contact with objects foreign to their symbolic order, cultures tend to perceive these objects as threatening, then behave like computer software: "they cannot open the file because they do not know the application that created it." Culture, therefore, does exactly what software that can open the file does and works on them only when recognizing the application that created them. Thus, the "office hours" tool is read and integrated according to a logic of its own that makes it unusable for the purpose of studying in greater detail the lesson content.
11. In this sense, Salvatore and Scotto di Carlo (2005) discuss the self-referential quality of culture in reference to its tendency to assimilate anything in itself according to a principle of adaptability to the environment in which it operates; these elements can be recognized in the *transference of culture.*
12. Besides, if cultures were immutable, they would never change their position, customs, or ideas, as well as political and economic processes.
13. Cultural devices are symbolic sources capable of producing texts that feed contexts through the symbolic richness of what they carry. For example, the exercising of the power of the role takes on a different symbolic meaning if accompanied or not by the power of competence. Think of the news on television of a bomb made by the President of the Republic or by a citizen. It is evident that the first moves a much more powerful symbolic device than the second. With the example of the film, the choice of the logo, the leader, rules, and so on moves symbolic contents of a certain type rather than another.

REFERENCES

Abbagnano, N. (2006). *Dizionario di filosofia* [Dictionary of philosophy]. Turin, Italy: UTET.

Bateman, A., & Holmes J. (1995). *Introduction to psychoanalysis. Contemporary theory and practice*. Milan, Italy: Raffaello Cortina.

Bion, W. R. (1962). A theory of thinking. *International Journal of Psychoanalysis, 43*, 306–310.

Bruner, J. (1990). *The acts meaning*. Cambridge, MA: Harvard University Press.

Bruner, J. (2009). Culture and mind: Their fruitful incommensurability. *Psicologia Culturale, 1*(1). Rome, Italy: Firera.

Carli, R. (2006). Collusion and its experimental basis. *Rivista di Psicologia Clinica*, 2–3. Retrieved from http://www.rivistadipsicologiaclinica.it/english

Carli, R., & Giovagnoli, F. (2011). Cultural approach to clinical psychology. Psychoanalysis and analysis of the demand. In S. Salvatore & T. Zittoun (Eds.), *Cultural psychology and psychoanalysis in dialogue. Issues for constructive theoretical and methodological synergies* (pp. 117–150). Charlotte, NC: Information Age.

Carli, R., & Paniccia, R. M. (1981). *Psicosociologia delle organizzazioni e delle istituzioni* [Psychosociology of organizations and institutions]. Bologna, Italy: Il Mulino.

Carli, R., & Paniccia R. M. (2004). *L'analisi della domanda. Teoria ed intervento in psicologia clinica* [The analysis of demand. Theory and intervention in clinical psychology]. Bologna, Italy: Il Mulino.

Carnot, S. (1897). Reflection on the motive power of Heat and on Machines fitted to develop that power. In R. H. Thurston (Ed. & Trans.), *Reflection on the motive power of Heat*. Retrieved from https://archive.org/stream/reflectionsonmot00carnrich#page/n7/mode/2up (Original work published 1824)

Cesaro, M., & Ruggieri, R. (2011). Costruire convivenza civile. Il suo tempo, la sua dimensione attraverso la governance politica. In F. Freidiani & F. Gallo (Eds.), *Ethos repubblicano e pensiero meridiano* [Republican ethos and meridian thinking] (pp. 96–125). Bologna, Italy: Diabasis.

Cesaro, M., Ruggieri, R., & Pecoraro, N. (2010). Meaning systems. Their practical implications due to a recursive function between action and thought. In S. Salvatore, J. Valsiner, S. J. Travers, & A. Gennaro A. (Eds.), *Yearbook of Idiographic Science* (pp. 129–135). Roma: Firera & Liuzzo Group

Cole, M. (2003). *Cultural psychology: A once and future discipline*. Cambridge, MA: Harvard University Press.

Eagle, M. N. (2011). *From classic to contemporaneity psychoanalysis*. New York, NY: Taylor & Francis.

Fornari, F. (1981). *Il codice vivente* [The living code]. Turin, Italy: Bollati Boringhieri.

Freud, S. (1899). *Die Traumdeutung* [The interpretation of dreams] OFS, vol. 3 (C. L. Musatti, Trans.) Ed., 1980, Turin, Italy: Bollati Boringhieri.

Freud, S. (1901). *Zur Psychopathologie des Alltagslebens* [The psychopathology of everyday life] OFS, vol. 3 (C. L. Musatti, Trans.) Ed., 1984, Turin, Italy: Bollati Boringhieri.

Freud, S. (1905). *Drei Abhandlungen zur Sexualtheorie* [Three essays on sexuality] OFS, vol. 4 (C. L. Musatti, Trans.) Ed.,1984, Turin, Italy: Bollati Boringhieri

Freud, S. (1915). *Das Unbewusste* [Introductory lectures on psycho-analysis and other works] OFS, vol. 8 (C. L. Musatti, Trans.). Ed., 1980, Turin, Italy: Bollati Boringhieri

Galimberti, U. (2006). *Dizionario di psicologia* [*Dictionary of psychology*]. Turin, Italy: UTET.

Gansel, D. (Director). (2008). *Die welle* [Motion Picture]. Germany: Costantin Film Produktion.

Kant, I. (1787). *Kritik der reinen Vernunft*. Bari, Italy: Laterza.

Klein, M. (1932). *The psycho-analysis of children*. London, UK: Hogarth.

Lacan, J. (1977). The four fundamental concepts of psychoanalysis. London, UK: Hogarth.

Mahler, M., Pine, F., & Bergman, A. (1973). *The psychological birth of the human infant*. New York, NY: Basic.

Matte Blanco, I. (1975). *The unconscious as infinite sets. An essays in bi-logic*. London, UK: Gerald.

Oles, P. K., & Hermans, H. J. M. (Eds.). (2005). *The dialogical self: Theory and research*. Lublin, Poland: Wydawnictwo.

Oneroso, F. (2003). Reversibilità e bi-logica. *Parenklisis, 1*, 31–45.

Pulli, G. (2008). *Freud e Minkosowski. L'inconscio e il tempo* [Freud and Minkosowski. The unconscious and time]. Naples, Italy: Liguori.

Rosa, A., & Néstor, P. (in press). *Semiotic methodology for the analysis of the cultural and individual dynamics of social representations: A view from cultural psychology*. Paper on Social Representations.

Ruggieri, R., & Pecoraro, N. (2014). Family business dynamic: Generation change as identity transition. In A. Gennaro, S. Salvatore, & J. Y. Valsiner (Eds.), *Yearbook of idiographic science, Volume 5*. Charlotte, NC: Information Age.

Salvatore, S. (2012). Social life of sign: Sense making in society. In J. Valsiner (Ed.), *The oxford handbook of culture and psychology* (pp. 241–254). New York, NY: Oxford University Press.

Salvatore, S., Lauro-Grotto, R., Gennaro, A., & Gelo, O. (2008). Per una psicologia dinamica. In S. Di Nuovo & G. Falgares (Eds.), *Per una psicologia psicologica. Scritti in onore di Franco Di Maria* [For a psychological psychology. Essays in honor of Franco Di Maria] (pp. 377–397). Milan, Italy: Franco Angeli.

Salvatore, S., Forges Davanzati, G., Potì, S., & Ruggieri, R. (2009). Mainstream economics and sense-making. *Integrative Psychological and Behavioral Science, 43*(2), 158–177. doi:10.1007/s12124-008-9084-5

Salvatore, S., & Scotto di Carlo, M. (2005). *L'intervento Psicologico per la Scuola* [The psychological intervention for the school]. Rome, Italy: Firera.

Salvatore, S., & Venuleo C. (2013). Field and dynamic nature of sensemaking. Theoretical and methodological implications. *Papers on Social Representations, 22*, 21.1-21.41.

Salvatore, S., & Zittoun, T. (Eds.). (2011). *Cultural psychology and psychoanalysis in dialogue. Issues for constructive theoretical and methodological synergies*. Charlotte, NC: Information Age.

Stolorow, R. D., Orange, D. M., & Atwood, G. E. (2001). World horizons. A post-Cartesian alternative to the Freudian unconscious. *Contemporary Psychoanalysis, 37*(1), 43–61.

Valent, I. (2003). Simultaneità e reversibilità in Immanuel Kant. *Parenklisis, 1*, 7–22.

Valsinier, J. (2012). Introduction culture in psychology. A renewed encounter of inquisitive of mind. In J. Valsiner (Ed.), *The Oxford handbook of culture and psychology* (pp. 3–24). New York, NY: Oxford University Press.

Valsinier, J. (2013). Creating sign hierarchies: Social representation in its dynamic-context. *Papers on Social Representations, 22,* 16.1-16.32.
Varisco, B. M. (2002). *Costruttivismo socio-culturale* [Sociocultural constructivism]. Rome, Italy: Carocci.
Vigostsky, L. (1962). *Thought and language*. Cambridge, MA: MIT Press.
Winnicott, D. W. (1965). *Maturational processes and the facilitating environment: Studies in the theory of emotional development.* London, UK: Hogarth.
Zimbardo, P., & Boyd, J. (2008). *The time paradox. New psychology of time that will change your life*. New York, NY: Free Press/Simon & Schuster.

PART V

COLLECTIVE-PERSONAL TEMPORALITIES

CHAPTER 14

LIVING WITH THE BELIEF IN CYCLICAL TIME

Collective and Personal Constructions of Hindus

Nandita Chaudhary

> Within infinite myths lies the eternal truth,
> Who sees it all?
> *Varuna* has but a thousand eyes,
> *Indra* a hundred,
> And I, only two[1]
>
> —Devdutt Pattanaik (2006, p. ix)

INTRODUCTION: TIME, TEMPORALITY, AND CULTURE

Time and temporality is a critical area of the human experience and understanding. The way in which time is conceptualized is culturally constructed and socially shared. Temporality is also addressed in philosophical study within which the nature of time is explored since it has impact on our lives

Temporality: Culture in the Flow of Human Experience, pp. 333–351

physically and biologically. Culturally, the Hindu understanding of time has several distinct beliefs. These beliefs influence the ways in which issues of temporality are approached and understood. This chapter deals with the ways in which time is constructed in Hindu ideology along with a discussion of examples of the ways in which time is understood. Further, an investigation into the meaning-making of a collectively shared reality is also used as an instantiation of the fact that the discipline of psychology has to engage with cultural processes in order to expand the outreach of the discipline and include alternative ways of understanding the world.

Human beings have a relationship with time that is simultaneously physical, psychological, social, biological, historic, geographical, and technical. In fact, lived time is irreversible since it only moves forward. Imagined time on the other hand moves from the present to remembered time, envisioned time as well as the here and now. The capacity for the human mind to contemplate what is not, in fact, permits the characteristic play with the notion of time. In spite of its central role in our lives, temporality is a complex construct to grasp and explain. Thus, language, logic, and myths are nodal players in people's understanding of time.

The notion of the Ratchet principle (Tomasello, 1999) to exemplify the restrained ability of developmental and cultural processes also applies to the passage of time. Time is unidirectional; the only way it moves is forward, therein encompassing culture, history, development, and change. Not only does time move forward, unlike the mechanical ratchet, the movement is spontaneous and uncontrollable. There is also a cumulative effect where the past constantly adds on toward the future. As a person grows older, age is accumulated, whereas the remaining life is reduced. Memories of the past accrue as a person grows older, and developmental changes have an irreversible, cumulative impact on human beings. The human mind makes sense of the world around it by reconfiguring objects and events mentally. Temporality plays a critical role in this reconstruction, and narratives assist in the meaning-making of this activity (Bronckart, 2005). Regarding human experience, temporality also frames social activity in the sense of the development calendar of every individual, basically placing constraints and affordances that people must consider (Wrosch & Heckhausen, 2005). The deadlines that we face in everyday life cause restrictions on our activity to which it is essential to adapt. Experienced time is time that is marked within a social frame that is contained within a culturally shared ideology (Valsiner, 1993).

Human beings have always had a desire to measure the passage of time with increasing accuracy and detail and speculate about its nature. One of the important areas of study has been the relationship between temporality and developmental changes in the growing individual. The Hindu understanding of time and temporality is cyclical. The pervasive cultural reality that profoundly determines individual belief and conduct, even in contemporary Indian society, is bound to a meaning-making process within which time is believed to be cyclical, and context as central to conduct, providing a serious challenge to the academic notion of "identity" as a consistent and enduring sense of selfhood. The notions of *dharma* (loosely translated as righteous or appropriate action) and *karma* (loosely translated as action or conduct) are basic to the belief system and provide an important example of what Sato and Valsiner (2010) label as mesogenetic organizing principles of human action.

MYTHOS, LOGOS, AND TEMPORALITY

The classical divide between mythos and logos characterizes Greek mythology, wherein mythos consists of intuitive narratives and logos is its complement; deliberations based on reason. The notion of time is also framed within this very divide. Lived time is prone to the illusory, imperfect experience that persons live through; whereas divine time is framed by ultimate truth. In Hinduism, this play between reason and intuition permits the dance between gods and mankind, between belief and practice, between truth and Truth. The history of Indian culture over centuries has allowed the entry of and acceptance toward multiple ways of living, leading to a confounding amount of plurality of expression and belief. In this chapter, a specific focus will be placed on the notion of time as it is manifested in literary and mythical sources as well as lived reality among Hindus. The passage of time, the duration of a life, the belief in reincarnation, and the notion of karma are core issues in this phenomenon. But before that, it is important to give some background of Hinduism in India.

HINDUISM AND OTHER RELIGIONS OF INDIA

India is a political democracy with a secular constitution; it is important to mention here that secularism in India does not imply the distancing of the State from religion, but acceptance of all religions, wherein citizens of the country are free to pursue their own religion. In a country that has been intensely religious throughout its history, characterized by heterodoxy (Sen, 2005), this remains a characteristic feature of its people. Table 14.1 provides us with the numbers of people following different religions in India.

From Table 14.1, it is evident that although the proportions may indicate a dominance of the Hindu community, the absolute numbers of people belonging to other faiths are significant, making their presence very visible. Further, religious activity is mostly public and shared, adding to the prominence of religion in the lives of people, although to view the nation purely through a spiritual lens is an error in judgment (Luce, 2006), adding that this perspective may even be a burden for India since this reflected view of the civilization became very popular among Indians as well (Sen, 2005). In consonance with this view of India from the outside, Sen found that every book on India was under the category "Religion" when browsing at a bookstore at Harvard University.

Table 14.1. Religions of India

Religion	*Number*	*%*
Hindu	827,578,868	80.5
Muslim	138,188,240	13.4
Christians	24,080,016	2.3
Sikhs	19,215,730	1.9
Buddhists	7,955,207	0.8
Jain	4,225,053	0.4
Religion not stated	6,639,626	0.6
Other	727,588	0.1

Source: Government of India (2010–2011).

India's reputation as a land of spiritual bliss is popular, but often misrepresentative of the intensely complicated lives of Indians. The presence of large-scale economic despair among many of its people who live on the edge of subsistence seems like one of the strongest violations of this image. However, at the community level, religion is actively practiced by all religious groups. Religion and poverty are not the only enduring symbols of India any longer. More recently, India has also become known as a rapidly developing nation, a member of the BRICS countries, the association of emerging economies consisting of Brazil, Russia, India, China, and South Africa.

CENTRAL BELIEFS OF HINDUISM

In order to contextualize the Hindu view of time, it is important to present a brief overview of its beliefs. This essay is best regarded as introductory and not a summary since ideas in Hinduism are as rich and varied as its history. Regarding the peculiarities of Hindus, what are some of the ways in which they think and behave differently? How do they approach, confront, interpret, and act? What are the implications of these differences? Kumar (2008) attempts to identify certain typical features of the "Indian" psyche, features he claims are common the people living in India, most of whom are Hindus. The historical confluence of people from different places over the history of the subcontinent has been formative in the sense that Indian is best understood as a conglomeration of several people rather than a singular race. The variety of historical influence is staggering, and Indians are best understood as a people with "multi-cultural minds" (Hong, Morris, Chiu, & Bennet-Martinez, 2000).

There are many great and little traditions in India. The important divides are based on religion, region, language, ecology, modernity, and economic status. Taking the example of language itself, the Indian constitution recognizes 22 official languages where more than a 1,000 more are sustained. Many scholars have attempted to identify an Indian way of thinking (Kakar & Kakar, 2007; Kumar, 2008; Ramanujan, 1989); this endeavor toward a grand narrative is as elusive as it is seductive. Perhaps we *want* to believe in the unity of Indians and then search for commonalities, just as psychology is perhaps an exploration of answers to an equivalent question, "Is there a human way of being?"

India as an idea is ancient, and yet, as an independent nation and as a sovereign state, a young country having gained independence from British colonial rule in 1947. A key person in attaining freedom was Mahatma Gandhi, whose title reflects the adulation of a nation and the world—great soul (Mahatma). Gandhi's role in the shaping of modern India is crucial as his influence still lingers in spite of economic reform and globalization. Three generations have passed since India's independence, but malnutrition among children is still a serious problem, and as many as half the women are illiterate, and accessibility to essential services remains a challenge for a large majority of India's rural and tribal areas. Economic disparity is another stark issue. Alongside this profile, religion remains an intensely expressive feature; private and public lives are framed by religious plurality, marred only by politically motivated crimes toward communities, despite the large number of people who express no affiliation to any religion (see Table 14.1). Regarding its geography, Eck (2012) argues that

> what is clear from the study of Hindu India is that its geographical features … no matter how precisely rendered, mapped or measured, are also charged with stories of gods and heroes. It is a resonant, sacred geography. But it is also a landscape, in that these features are connected, linked to a wider whole. While I use the term "imagined landscape," it is far from imaginary. (p. 11)

Most stories from mythology are intensely woven into cosmology in the sense that they often begin with the creation of the universe and then proceed to explain the structure of the universe before weaving in geographical features identified more by their place in the order of things rather than their uniqueness (Eck, 2012). Space and time thus frame the minds of the Hindu Indian in a way that is intensified by the myths by which they live. The "organic ontology" (Eck, 2012, p. 25) of Hindu theory employs the symbolism of the human body to create a view of the entire world. The Vedic image is of a cosmic person (*purusha*) sacrificed at the time of creation into time, space, and order in this world. From the cosmic creation's eye was born the sun, the moon from its mind, and from the ears directions were made, it is believed (Eck, 1981).

There are several important distinctions between the Biblical tradition and Hinduism; that the creation of man is not seen as a fall, the power of the human being is in fact believed constituted from the same divinity, albeit unbeknownst to the ordinary person on account of the failings of the human mind. These failings can in fact be overcome by contemplation and concentration, it is believed. Further, the variety of Hindu gods are not seen as the ultimate truth, only figures who indicate toward that truth (Campbell, 2001). Campbell further discusses the popular belief in Hinduism, that this reality we experience is simply a manifestation of Vishnu's dream (one of the holy trinity: Shiva, Vishnu, and Brahma). It is the fluency of time that permits all the changes in manifestation—in form. Time, is thus considerred divine!

Regarding some of the common features of the Hindu way of life and thought, here is a tentative list (Chaudhary, 2010):

1. The true self is believed to be beyond the mind and its senses; the material world is argued as having a "mirage-like" quality.
2. Ultimate reality is beyond description; it can only be experienced.
3. Notwithstanding the intense relationships during the life course, detachment or nonattachment from people and material things is the true objective of human life.
4. Consciousness rather than matter is the ultimate reality.
5. The concept of the "flux of all things."
6. Nonduality of substance (nature of a thing) and conduct or code of law.
7. Nonduality of mind and body.
8. Hylozoistic thinking—that all matter has life, and there are no clear distinctions among matter, life, and mind. As a consequence, subject and object, thesis and antithesis, all emerge from the unitary absolute and its differentiation.
9. Ideology is "unitive" as different from "dualistic" thought in other cultures; qualitatively different from Western thought.

10. A tendency to idealize and focus on context-sensitive rather than context-free rules in almost all situations.
11. Reason is subordinated to intuition; matter to spirit.
12. *Karma* is an important guiding principle.
13. Belief in the transmigration of the soul and reincarnation.
14. Evidence of a pan-Indian predisposition for context-sensitive rather than context-free thinking, as can be seen in the notion of *dharma* (Menon, 2003) and *jati* (Ramanujan, 1990).

These features have been assembled to contrast the proclivities of an "Indian" way of thinking to that of other cultures and form the basis for several themes in Indian society, sometimes even transcending social and religious differences.

Among Hindus, myths are regarded as *mithya*, truth that is picturized through a frame of reference, to be distinguished from *sat*, truth that transcends time and space. This distinction between lived reality that is spun in stories and ultimate truth that is beyond the grasp of an ordinary person lies at the crux of Hindu ideology (Pattanaik, 2006). Contextualized truth, or *mithya*, is accepted as being illusory and open to correction since eternal truth is impossible to communicate; it is believed that it can only be realized through concentrated spiritual activity. Thus, mythology becomes a strong vehicle for the expression of ideas, even today. There is a sense of prevailing "other-worldliness" which is sometimes seen as responsible for the tolerance of extremely difficult circumstances among the poor; that this world is only a passing reality. Some even justify poverty by saying that the poor suffer the consequences of a past life, making the notion of reincarnation central to living in this lifetime. The prevailing belief in reincarnation is critical to understanding the idea of a passing of the cycle of life.

THE PAST AND THE PRESENT

According to Hindu philosophy, the primary predicament of the human being is change and impermanence with the passage of time, while in our minds, we long for continuity and permanence. The sense of self, its ego, then becomes the stabilizing factor that our minds cling on to, it is believed (Rao, 2008), leading to identity, selfhood, and attachments with others. This is the basic cause for suffering of the human mind, and spiritual awakening is aimed at precisely this feature of our minds, the ego or the self. Dissolution of the self and self-obsession is the primary aim of all Hindu and Buddhist spiritual activity, thereby leading to the freedom from materialism and worldliness, freedom from the cycle of birth and death (Rao, 2008). Although the modern Indian may seem extremely "worldly" and even "Westernized," there is no denying that ancient ideas still reverberate in the context of everyday lives (Kumar, 2008; Thapar, 2000).

The term "Hindu" was first used by Muslim invaders in India's ancient past to categorize the "idol-worshipers" whom they encountered and ruled over. Hinduism is historically not a coherent consolidated religion like Islam and Christianity. In the Indian subcontinent, the years of coexistence of different religions has resulted in a sort of diffusion of cultural practices leading to much similarity in family life as a grounded reality in the face of external religious differences (Mullati, 1995). Traditional forms of family among the Hindus carry a deep resonance with the presentations of society depicted in ancient Hindu texts. The gender specificity of divine relationships provides a persistent template for everyday relations. Despite contemporary changes, these divine stories find expression in today's world through public culture (Kumar, 2008); the ways in which the past, even the ancient past, finds its way into the world of Indians today is fascinating, adding yet another dimension to the confounding complexity. This does not imply a homogenous, monolithic structure of a consolidated, coherent way of living. If anything, culture and family life in India provides us with the exact opposite of that which is implied above. The number of people, the diversity in languages, ethnic practices, stories, and spirituality makes such statements rather impossible. Even the past is textured with plurality, variety, and multiplicity. India "does not have one past, but many pasts" (Ramanujan 1989, p. 187). To add to the texture of plurality is the fact the past always has a constant presence in our present lives (Anandalakshmy, 2010). As Thapar (2000) observes, "At certain levels there are aspects of cultural traditions in India that can be traced to roots as far back as a few thousand years, but such a continuity must not be confused with stagnation" (p. xxv). Thus, we find that there is an intricate interweaving and interlocking of the various pasts, such that "every description of tradition can be as a negation or contradiction of another" (Anandalakshmy, 2010, p. 19). As Rich (2010) writes of her days as an American student of Hindi living in the city of Udaipur,

> This town is in a time warp, though it is hard to say which time. When neighbours wave and ask "Who was that who walked you home?" I think the '50s. When the *rikshaws* whiz by, looking like Model Ts from the back, it's the muted Indian '20s. But when I go to the doctor and find myself next to a tribal woman with an enormous gold nose ring, or when I talk to a Rajput, a member of the autocratic warrior caste—the men have knife-straight backs and villainous handlebar moustaches—I can see the links to the Middle Ages so clearly, its startling.... In the centre of time's web ... is the sly tugging present. (pp. 49, 52)

This sense of parallel historicity underlying technological advancement in every sphere of life is enormously difficult to comprehend, especially for people who live in homogenous-looking environments of technologically advanced nations. There is thus a collective "multivoicedness" that characterizes Indian cultural reality. It is like living in a time machine which simultaneously creates periods of history along its journey. Perhaps the metaphor of the "time machine" would be a favourable to describe this diachronicity of the present.

Despite the distinctions between different religious groups in the country, family practices remain strikingly similar, deeply bound by tradition in the face of change. Perhaps it could be said that that ideology of family life remains largely unchanged, it is the practical adjustments that allow for the transformations that we see. The beliefs of people with regard to family relationships, role allocations, and gender relations continue to rest actively on tradition, despite practical adjustments that are attributed to everyday challenges, not to the change in the form of the family (Chaudhary, 2009). This divergence between ideal constructions and real practices does not seem to create too much dissonance in a culture which has, through the course of history, been able to sustain complex and even contradictory practices on account of several sociohistoric factors like plurality, conflict, invasions, and ecological and regional variations. This is, in my judgement, the reason for the "experiential gap" between research studies that focus on what people believe as opposed to what they actually do. In a study of beliefs and practices, we have found that mothers tend to offer traditional solutions as a first response to a question about aspirations or care of their children. However, on further probing, practical constraints and actual solutions emerge (Tuli & Chaudhary, 2010). This divergence is probably responsible also for the investigations that are based in different traditions, like sociology, anthropology, family studies, or gender. Primed toward looking within different layers of life, different approaches render divergent fragments of a reality that continues to evade simple explanation. We are labeled as a collective culture while expressing extreme self-centeredness on occasion; if one looks within the home, there are deep and dedicated commitments to emotional closeness that children are taught from a very early age (Raval & Martini, 2011; Raval, Raval, & Deo, 2011), whereas we may be quite inconsiderate toward unknown people on the street, as can be seen on urban roads every day. There are many linked contradictions in Indian society that can be mentioned here; we worship women as goddesses and may perpetuate the abuse of women leading to one of the worst records related to women's lives. There is technology for the future that reaches the streets in the company of extreme poverty; children are believed to be a gift from the divine, and also bought and sold and abused without flinching as newspapers discuss every day; there is religious plurality accompanying serious ethnic difference and intolerance. In a recent article, Kanjilal (2012) highlights the ambiguity in Indian society and its governance. Despite several rather outdated laws, there are some advances in family and inheritance law that are noteworthy. Kanjilal quotes a Cambridge scholar having said something that is strikingly similar to A. K. Ramanujan's statement about India: "It has been said more than once, that whatever you can truly say about India, you can also say the exact opposite with equal truthfulness" (1990, p. xv). The plurality of Indian culture is visible also in the history of its calendars. The Official Calendar Reform Committee gathered shortly after Indian Independence in which more than 30 distinct calendars that were in systematic use in the country were identified and further reduced to a list of 7 (Sen, 2005).

Another popular reputation that Indians have is of being systematically tardy, not giving much importance to being on time in the modern world. This trait appears despite the fact that otherwise, timeliness in ritual observances and consciousness of time is an important detail in social practice. There is much discussion about this phenomenon in public spaces and literature. As a *Washington Post* reporter who lived in and wrote about India for 4 years, Bumiller (1991) describes her experience of one Indian wedding where she arrived half an hour after the designated time for the reception, only to find herself meeting up with the caterers. Waiting for others to arrive is a common pastime in Indian cities, whether it is the plumber or an electrician or an official or for guests to arrive at your place for dinner (Nellore, 2004). There does not seem to be much respect for punctuality. This trait is often sarcastically referred to in common discourse as the IST not "Indian Standard Time," but "Indian Stretchable Time!"

KARMA, REINCARNATION, AND TIME

The belief in "ontological individualism," the idea that the individual exists prior to society (Menon, 2003) has through the history of the subject, tended to underestimate the power of the social. In this chapter, I have focused especially on socially shared phenomena to emphasize the cultural element in the understanding of time and its course and how this influences the ways in which people think about their worlds. Another related comment was advanced by Moghaddam (2010). According to him, conventional research adopts the "embryonic fallacy" that as soon as a child is born, he or she becomes a source of psychological experience (p. 465). The notion of "interobjectivity," which he defines as the "understandings that are shared between and within cultures" about social reality (Moghaddam, 2003, p. 221), becomes a pertinent domain for psychological investigation in addition to the intersubjective. The socialization process is a strong influencing factor in the way in which people approach the world around them and therefore becomes an important consideration in research on human understanding.

Since the distinction between the social and natural is rarely ever made among Hindus, it is difficult, if not impossible, to separate convention from morality (Menon, 2003). Everything is considered to be material, including time and space, since these have relational properties. Time can be considered auspicious or inauspicious, and great care is taken to determine the exact moment at which it is best to perform a particular ritual. Elaborate representation of the Hindu calendar is consulted for important social events like marriages, temple visits, departures, and arrivals. Time can also be evaluated as "good" or "bad," sometimes even making people do things differently. *Vinaash kale, vipreet buddhi* (Destructive time; reverse/ destructive mind) is a common saying, which is often used as an explanation for uncharacteristic behavior of an individual. When someone behaves in a fashion that completely departs from his or her regular conduct, this statement would apply. Interestingly, although auspicious and inauspicious timings are socially shared and calculated on the basis of the Hindu calendar and the movements of the stars, each person's individual "timing" will be different depending upon the person's own horoscope. Very detailed speculations about person, time, and space are made for alliances like marriage or other partnerships.

The Sanskrit word *karma* means doing, causing, and making, leading to three distinguishable meanings of karma in Hindu theory. Firstly, it refers to the concrete actions of a person during the life course. Secondly, it refers to the accumulation of "rewards and retributions" that carry over from previous lifetimes (Menon, 2003, p. 435) leading to preexisting proclivities or *samskaras* on account of past lives (Kakar & Kakar, 2007). Human beings at birth are believed to be "carrying" the karma of their past lives that are likely to have consequences that are most likely to be unexplained within any single lifetime. The rapid eye movement of young babies is explained by remembrances of a past life by grandmothers. Unexplained affection or enmity between otherwise unrelated people can also be explained by a past connection. Mostly it is things that defy simple explanation that are attributable to karma. The third meaning of karma is its active, future orientation, where a person has the capacity to actually improve his or her fate by performing righteous action or vice versa. Karma is also believed to be shared between close relations, especially between parents and children since it sets limits to the situations and extent of benefit or otherwise that a person is able to accomplish. It is the broad course of a life and not the details that are believed to be influenced by karma. The temporality of karma thus distinguishes the Hindu view of life from others. That this life is not the sum and substance of our experiences on earth, and if there is suffering now, it is on account of a past that we are unable to estimate. This clause has often been taken also as an explanation for the extreme resilience and tolerance for adversity that can be experienced among Indians, whether it is situational or self-imposed. Since there is no separation between the different forms of life, plants, animals, and humans, transmigration of souls can also, it is believed, take place across species. An example of this can be seen in the advice that older Oriyas[2] give to the young: "Don't live this life as if it was your last. Reflect on the consequences of your actions/bad actions before you do anything" (Menon, 2003, p. 436).

THE LIFE COURSE

As per Hindu theory, a single life passes through several sequential stages. Each stage has specified codes of conduct and social obligations attached to it. Family life among Hindus is characterized by adult-child continuity (Saraswathi, 1999). There is a fluctuating sense of interpersonal distance described between a person and others at different stages; preceded by the dependence of infancy and childhood, the first of the four *ashramas* or stages of life is that of the student, *brahmacharya*, wherein the young person learns the rules of the community, ideally in the company of a guru. Following this is the householder's stage or *grihastha ashrama*, when a person has to fulfill obligatory duties toward society by establishing a home and continuing the next generation. This is prescribed as an intensely social phase of life, which is followed by *vanaprastha*, where gradual detachment from others and detachment from the household is supposed to be accomplished. Finally, during old age, *sanyasa* arrives, which is a stage during which a person is supposed to completely renounce the world and all its pleasures and attachments in preparation for *moksha* or freedom from the cycle of birth and death. Although transmigration of souls is believed, freedom from the cycle of birth and death with the merging of the soul into divinity is an ultimate aim.

The four basic aims of life according to Hindu theory are believed to be realized through the *ashramas*. These include *dharma* (righteousness), *artha* (wealth), *kama* (pleasure), and *moksha* (liberation) (Kakar & Kakar, 2007). Very clear guidelines are described for each of these stages, and prior to the age of 2 (*shishuawvastha*), the child is believed to be free from moral codes, whereas some learning tasks are prescribed for the child above 2 during *balavastha* and *kishoravastha* (early and middle childhood). This sequential template is clearly laid down, but followed only as an ideal principle (Kakar, 1982). Most ordinary people know about each of these stages, and socially guided expectations exist, but it is certainly rare to find anyone living strictly by them. Yet the *ashrama* theory is a strong element of the ambient belief in temporality and codes of conduct.

THE ELASTICITY OF TIME

Psychological experiments with the perception of time have indicated that our capacity to perceive time is impacted by the amount of attention that a task needs: The more we are occupied, the faster time appears to pass (Macar, 2005). Human beings have receptors in the brain that are involved with the perception of time that are sensitive internal factors like body temperature and intake of drugs as well (Hoagland, 1935; Meck, 1996). The knowledge of time can be also discussed at three different levels: personal, social, and natural. Although deeply interconnected, these different manifestations of temporality permit the play that leads to variations in perceived and constructed time.

A myth is told and retold to express the belief in the elasticity of time among Hindus. A king who sought a perfect match for his daughter even travelled to the world of the gods in this pursuit. Having returned to this world, he finds that hundred years have passed and everyone he knows has died. Time in the divine world is believed to pass faster than it does in human experience. The contraction and expansion of time in Hindu mythology is a common theme. Crossing over from the divine to the daily—a day is a lifetime, and vice versa—thereby predisposes the Hindu mind toward the essential elasticity of time (Pattanaik, 2012).

As opposed to the passage of time, subjective time is therefore recognized as elastic. Indian psychology (psychological principles from Indian philosophy), unlike its Western counterpart, has always recognized the existence of subjective consciousness as a critical quality of the human mind (Rao, 2008).

TEMPORALITY AND DEVELOPMENT

Age has been an important consideration in development, although not always without problem. Early developmental theories rested very heavily on the chronological age of the subject, until the time that the focus on the large individual differences in the pace of development forced researchers to abandon age as a simple independent variable. Although changes with age are still popular in research, their sequence has gained greater focus, and age is often used now as a descriptive variable (Flammer, 2005). Temporality remains a central element in development, although changes are manifested not simply by the movement of time, but this movement in culture. When age differences in measured intelligence showed a decrease with age after 25, the shift from cross-sectional to longitudinal designs proved that simply taking cohorts of different ages does not seem to provide a sufficient assessment of developmental changes since different cohorts experience different socialization practices and different life conditions, evidently impacting group scores (Flammer, 2005).

In a study on urban Indian children's understanding of truth, Pillai (2012) demonstrated how children's responses could be interpreted as similar or dissimilar at different ages and how older children's responses, although superficially in the same category as those of 5 year olds, the actual meaning of the answers were much more complex. During the course of her study, she asked children about their understanding of the concept of God. The answers from children within one year age gaps between 5 and 10 years of age demonstrated that the youngest and oldest age groups tended to accept the presence of God in ways that could be interpreted as very similar; God is omnipresent, all-powerful, and so on, tending to romanticize the existence of a God. It was only after exploring the responses in detail, that it was discovered that although the younger children saw God as an essential being who could do miraculous things, older children's responses tended very much toward accepting the notion as a social obligation, one that could also be denied by some. At the age of 10, it was accepted after having known that it could be denied, whereas at younger ages, God was taken as a given. In assessing children's beliefs in the omnipotence of God, they were also asked if God answers every prayer. Most children indicated the affirmative. However, when two sorts of prayers were listed (that it would rain and that a friend would fall and get hurt), older children clearly distinguished between the two. A negative prayer would not be fulfilled so easily because the desire is wrong, whereas God would certainly answer the prayer for rain. The researcher then questioned the children who responded in the affirmative, whether it would then rain on the same moment as the prayer. To this, older children were very clear that it may not rain immediately as you pray, but in a few days or in the monsoon season, it would definitely rain!

Here we find clear trends that are discernible with age even though the frame was not longitudinal and age was taken as a descriptive (rather than independent) variable. Temporality is thus a necessary (although maybe not sufficient) criterion for development to take place, and changes with age are an interesting area of study in developmental psychology. Yet it is critical to examine how age and time are incorporated in the research design and how the results are interpreted so as not to gloss over individual differences among people of the same ages.

POVERTY AND SPIRITUALITY IN INDIA

Many scholars have written about and attempted to explain the sheer deprivation within which many in India live. Some explanations relate to the "otherworldliness" of Hindu beliefs. The living conditions of the poor in India and economic disparity is such that it is sometimes difficult to figure out how people live in such circumstances. Some explanations have related to the belief in karma—destiny—that this life course is not the only one that impacts us. As Kakar and Kakar (2007) write,

> If spirituality has been at the centre of the Indian world, it would be reasonable to expect that it has continued to condition the Indian mind, colouring its intellectual, artistic and emotional responses in certain distinct ways. In other words, there are various consequences of this belief. One of these is the pervasive presence of hope, even in the most dismal of circumstances. For centuries, Indian civilisation has conveyed to the growing child the almost somatic conviction that there is an order, even if hidden and unknown, to our visible world. There is a design to life that can be trusted in spite of life's sorrows, cruelties and injustices. The Indian mind, then, tends to convert even the slightest ray of hope into a blaze of light. (p. 183)

Thus, the "other worldliness" of the Hindu mind actually permits the presence of constant optimism and is, according to Kakar and Kakar (2007), the reason for this seemingly unrealistic perspective. The belief in a "higher reality" and the presence of gods and goddesses is intense and pervasive.

In a study of children's understanding of truth statements and falsity, Pillai (2012) investigated young children (5 to 10 year olds) for their responses to several statements relating to natural phenomena. Although contrasting high-fee-paying schools and government-aided schools was not the primary intention of the study, one of the trends she found when she asked children if a child had one birthday every month how much older would she or he be at the end of the year, most of the children said that was not possible! An exploration of the reasons behind the assumed incredulity of such a situation seemed to indicate the enormity of economic constraints for children from aided schools. They were more likely to insist that such an eventuality could not be realized since it would be very "expensive" to have a birthday every month. Although older children even in these schools were aware that this was an impossible outcome; even otherwise, younger children from poorer families were certainly entangled in the financial impossibility of such an outcome over and above the more fundamental one. Children from high-fee-paying schools showed an age trend toward a more stable understanding of time as expected, but economic constraints did not find any mention. Piaget's extensive research (1969) on children's understanding of time has shown that the stability of time is discovered by children only after the age of around 5 years. The findings of this study also indicate toward that fact, but the additional consideration of income and expenditure seems particular to the poorer families.

CONCLUSIONS

In this chapter, we have visited some of the particular traits that characterize a Hindu way of life, with a focus on issues related to temporality and how these intersect with the larger belief system. The corresponding presence of different ways of living with regard to modernity and technological development is one of the many criteria due to which Indian society can be labeled as heterodoxy, seeming somewhat like a time machine with different periods of human existence found in a single subcontinent. Religious plurality in India has sustained and encouraged secularity to mean the freedom and pride in expressing one's own religion rather than reducing the importance of religious practice in everyday life. The significant tenets of Hinduism, namely, the belief in *karma* and *dharma*, have found an importance place in the issue of living in cyclical time within the Hindu worldview. Regarding the course of a person's life as a cyclical loop in the journey of a soul traversing through the path to the final merger with cosmic reality, the Hindu worldview provides us with a theoretically unique perspective on time and space that has characterized the Indian way of life for centuries and still persists as an acceptable premise of life as it is understood even today. This persistence of the past in contemporary reality is an enduring feature of the ambient ideology of Indian ways of understanding their world. Regarding the implication of this worldview on the discipline of psychology, perhaps the most important contribution of this discussion is to understand and appreciate that there are alternative ways of understanding time and space which depart fundamentally from the way these are understood in the Euro-American tradition, and in doing so, be able to also review and reflect on the meaning-making process of cultural processes and human conduct with a renewed understanding and appreciation. For instance, if there is an acceptance of and a belief in reincarnation, there is likely to be a profound difference in the way the notion of identity or "self" is constructed, to use a single example. Since culture is primarily about meaning-making, the entrance of the collectively shared ideology in the psychological reality of individuals is undisputed (Valsiner, 2007). Therefore, the discipline of psychology has to consider collectively shared ways of understanding the world in order to fully grasp the ways in which individuals think about their lives.

NOTES

1. Varuna, god of sky and rain; Indra, god of war and thunderstorms, king of the gods.
2. People living in the State of Orissa.

REFERENCES

Anandalakshmy, S. (2010). *Through the lens of culture: Centuries of childhood and education in India*. Monographs of Bala Mandir Research Foundation. Chennai, India: Bala Mandir Research Foundation.

Bronckart, J. (2005). The temporality of discourses: A contribution to the reshaping of human action. In A. Perret-Clermont (Ed.), *Thinking time: A multidisciplinary perspective on time*. Cambridge, MA: Hogrefe & Huber.

Bumiller, E. (1991), *May you be a mother of a hundred sons*. New Delhi: Penguin.

Campbell, J. (2001). Hindu wisdom. *Joseph Campbell Foundation*. Retrieved from http://www.hinduwisdom.info/articles_hinduism/12.htm

Chaudhary, N. (2009). Families and children in poverty: Objective definitions, subjective lives. In A. C. Bastos & E. P. Rabinovich (Eds.), *Living in poverty: Developmental poetics of cultural lives*. Charlotte, NC: Information Age.

Chaudhary, N. (2010). Book review. Handbook of Indian psychology by K. R. Rao, A. C. Paranjpe, & A. K. Dalal (Eds., 2008). *Journal of Cross-Cultural Psychology, 41*(2), 284–292.

Eck, D. (1981). The dynamics of Indian symbolism. In P. Berger (Ed.), *The other side of God*. New York, NY: Doubleday.

Eck, D. (2012). *India: A sacred geography*. New York, NY: Random House.

Flammer, A. (2005). Time in developmental psychology: Development in time. In A. Perret-Clermont (Ed.), *Thinking time: A multidisciplinary perspective on time* (pp. 95–97). Cambridge, MA: Hogrefe & Huber.

Government of India. (2010–2011). Census of India. *Office of the Registrar General and Census Commissioner, India*. Retrieved September 18, 2012 from http://censusindia.gov.in/Census_Data_2001/India_at_glance/religion.aspx

Hoagland, H. (1935). *Pacemakers in relation to aspects of behavior*. Experimental Biological Monographs. New York, NY: Macmillan.

Hong, Y., Morris, M. W., Chiu, C., & Bennet-Martinez, V. (2000). Multi-cultural minds: A dynamic constructive approach to culture and cognition. *American Psychologist, 55*(7), 706–720.

Kakar, S. (1982). *Inner world: A psycho-analytic study of childhood and society in India: Psychoanalytic study of childhood and society in India*. New Delhi: Oxford University Press, India.

Kakar, S., & Kakar, K. (2007). *The Indians: Portrait of a people*. New Delhi, India: Penguin.

Kanjilal, P. (2012, February 10). We're one big, happy family. *Hindustan Times*. Retrieved from http://www.hindustantimes.com/News-Feed/ColumnsOthers/We-re-one-big-happy-family/Article1-809689.aspx

Kumar, S. K. K. (2008). Indian thought and tradition: A psycho-historical perspective. In K. R. Rao, A. C. Paranjpe, & A. K. Dalal (Eds.), *Handbook of Indian psychology*. (pp. 19–54). New Delhi, India: Cambridge University Press.

Luce, E. (2006). *In spite of the gods: The strange rise of modern India*. London, UK: Little, Brown.

Macar, F. (2005). Time passing, attention and internal timers. In A. Perret-Clermont (Ed.), *Thinking time: A multidisciplinary perspective on time* (pp. 15–21). Cambridge, MA: Hogrefe & Huber.

Meck, W. M. (1996). Neuro-pharmacology of timing and time perception. *Cognitive Brain Research, 3*, 227–242.

Menon, U. (2003). Morality and context: A study of Hindu understandings. In J. Valsiner & K. Connolly (Eds.), Handbook of human development (pp. 431–449). London, UK: Sage.

Moghaddam, F. (2003). Interobjectivity and culture. *Culture & Psychology, 9*(2), 221–232.

Moghaddam, F. (2010). Intersubjectivity, interobjectivity and the embryonic fallacy in developmental psychology. *Culture & Psychology, 16*(4), 465–475.

Mullatti, L. (1995). Families in India: Beliefs and realities. *Journal of Comparative Family Studies, 26*(1), 11–25.

Nellore, R. (2004, July 31). Indian standard time. *Indian Currents, The complete Indian American Magazine*. Retrieved October 1, 2012, from http://www.indiacurrents.com/articles/2004/07/31/indian-standard-time

Pattanaik, D. (2006). *Myth=mithya: A handbook of Hindu mythology*. New Delhi, India: Penguin.

Pattanaik, D. (2012). *Elastic time*. Retrieved September 26, 2012, from http://devdutt.com/articles/leadership/elastic-time.html

Piaget, J. (1969). *Children's conception of time*. London, UK: Routledge and Kegan Paul.

Pillai, P. (2012). *Children's understanding of truth*. Unpublished doctoral dissertation of the Department of Human Development and Childhood Studies, Lady Irwin College, University of Delhi, India.

Ramanujan, A. K. (1989). Is there an Indian way of thinking? *Contributions to Indian Sociology, 23*(1), 41–58.

Ramanujan, A. K. (1990). Is there an Indian way of thinking? An informal essay. In M. Marriot (Ed.), *India through Hindu categories*. New Delhi, India: Sage.

Rao, K. R. (2008). Introducing Indian psychology. In K. R. Rao, A. C. Paranjpe, & A. K. Dalal (Eds.), *Handbook of Indian psychology*. New Delhi, India: Cambridge University Press.

Raval, V., & Martini, T. S. (2011) Making the child understand: Socialization of emotions in urban India. *Journal of Family Psychology, 25*(6), 847–856.

Raval, V., Raval, P. H., & Deo, N. (2011). Mothers' socialization of emotion, child emotion regulation and child's social emotional functioning in urban India. Cited in Rawat, V. & Martini, T. S. (2011). Making the child understand: Socialization of emotions in Urban India. *Journal of Family Psychology, 25*(6), 847–856.

Rich, K. R. (2010). *Dreaming in Hindi: Coming awake in another language*. New Delhi, India: Tranquebar.

Saraswathi, T. S. (1999). Adult-child continuity in India. In T. S. Saraswathi (Ed.), *Culture, socialization and human development: Theory, research and applications in India*. New Delhi, India: Sage.

Sato, T., & Valsiner, J. (2010). Time in life and life in time: Between experiencing and accounting. *Ritsumeikan Journal of Human Sciences, 20*(2), 79–92.

Sen, A. (2005). *The argumentative Indian: Writings on Indian history, culture and identity*. London, UK: Allen Lane.

Thapar, R. (2000). *The Penguin early history of India*. New Delhi, India: Penguin.

Tomasello, M. (1999). *The cultural origins of human cognition*. Cambridge, MA: Harvard University Press.

Tuli, M., & Chaudhary, N. (2010). Elective interdependence: Understanding individual agency and interpersonal relationships in Indian families. *Culture and Psychology, 16*(4), 477–496.

Valsiner, J. (1993). Making of the future: Temporality and the constructive nature of human development. In, G. Turkewitz & D. A. Devenny (Eds.), *Developmental time and timing* (pp. 13–40). Hillsdale, NJ: Erlbaum.

Valsiner, J. (2007). *Culture in minds and societies: Foundations of a cultural psychology*. New Delhi, India: Sage.

Wrosch, C., & Heckhausen, J. (2005). Being on-time and off-time: Developmental deadlines for regulating one's own development. In A. Perret-Clermont (Ed.), *Thinking time: A multidisciplinary perspective on time* (pp. 110–123). Cambridge, MA: Hogrefe & Huber.

CHAPTER 15

TEMPORALITY IN CULTURAL TRAJECTORIES

A Psychological Approach on Semiotic Constructions

Danilo Silva Guimarães

> The mind which the psychologist studies is the mind of distinct individuals inhabiting definite portions of a real space and of a real time. With any other sort of mind, absolute Intelligence, Mind unattached to a particular body, or Mind not subject to the course of time, the psychologist as such has nothing to do. "Mind," in his words, is only a class name for minds. Fortunate will it be if his more modest inquiry results in any generalizations which the philosopher devoted to absolute Intelligence as such can use.
>
> —William James (1890, p. 183)

The present discussion concerning the notion of temporality from a semiotic, cultural, and constructivist perspective emerged from a very real and situated psychological issue related to the project Amerindian Network Support, which I have been coordinating since 2012. It is supported by the Rectorship for Culture and Extension of the University of São Paulo,

Temporality: Culture in the Flow of Human Experience, pp. 353–380

Brazil (*Pró-Reitoria de Cultura e Extensão da USP*). Our activities with the autochthon peoples, who live in the city of São Paulo, mainly consist in organizing meetings where they talk about the psychosocial vulnerabilities they are facing. A group of undergraduate students from different fields of study (psychology, history, and social sciences) participates in these meetings under my supervision. The project is considered a service to the autochthon communities that live in São Paulo, and we do not collect data for scientific purposes. Nevertheless, it offers a catalytic background for a reflection on the meaning of temporality in interethnic situations.

Time and temporality are notions approached in myriad myths traditionally collected by ethnographers who worked with non-Western peoples, evincing how different the meanings of time and temporality are for these cultures when compared to Western cultures (cf. Scheurmann & Tuiávii, 1920/2003). Our weekly activities with the Mbya Guarani communities of São Paulo are also pervaded by distinct meaningful experiences of time and temporality that demand continuous adjustments and repairs so we can connect ourselves and establish the partnership we need for the development of a shared work. It is argued here that distinct temporalities take place in the interethnic dialogues, as the idiosyncratic way each culture deals with temporal issues in the process of sharing meaningful actions and discourses. Therefore, this chapter addresses how distinct temporal experiences, apparently irreconcilable, can be articulated. I presume that this reflection can help cultural psychologists to find a way of integrating the notion of temporality as a cyclical process with the conception of time as irreversible.

This chapter also brings a discussion on how semiotic-cultural constraints, actively constructed by psychological open-ended systems related to the environment, guide life trajectories. The focus is on the dialogical relation of perception and imagination in the developmental process of semiotic construction in which temporality takes place. The psychological notion of temporality that concerns this study refers to a cultural-personal way of living and giving meanings to the transformations in the course of life. These transformations can be stabilized in symbolic constructions a person may use to describe his/her historicity in the flow of experiences, thoughts, and feelings. Personal and collective life trajectories entail cyclical and rhythmic processes that interact in different and contrasting levels, that is, within the self, in the boundaries of prepersonal dimensions and in the boundaries of the self, the environment, and its multiple others.

At the end of the chapter, we present an analysis of a concrete situation—the construction of the "*Tenonderã* letter" by the Mbya Guarani people from São Paulo, Brazil. The analysis aims to show the pertinence of the descriptive diagrams presented and discussed along the chapter and to offer a systematization of theoretical and methodological reflections in the framework of semiotic-cultural constructivism in psychology. This approach to the empirical phenomena allows an accurate observation of interethnic dialogues by integrating linear and cyclical dimensions of temporality, showing how different trajectories of life affect each other and guide the heterogenetic process of semiotic constructions.

Myths as Narratives of Change

To introduce our topic of discussion, I will present a Kaxinawa[1] myth on the origins of time translated by Capistrano de Abreu, edited and discussed by Els Lagrou (2007). This myth prefigures a rhythmic/cyclical approach on temporality:

> [In primordial times] all beings lived in the root of the sky, each group or house protected its own cave (Abreu, 1941, p. 436). The features that constitute the basic rhythms of the earth, sky and waters were kept in these separated caves, protected by primordial beings, and there was not a predetermined rhythm in their use by their owners. One of the caves protected the warmth; another hosted the cold; another one contained the dawn and a fourth one housed the darkness.
>
> The darkness did not fall there, the weather did not get cold, and there was no sun. My people lived well and had many houses. One of the houses was close to the hole of the morning, another house was close to the hole of darkness, another one was next to the hole of the sun and the last one was near the hole of the cold. Nevertheless, when someone was angry, he/she would open the hole of cold, and the other would react by opening the hole of the sun, another yet would open the hole of the night and someone else, the hole of the dawn (Abreu, 1941, pp. 436–437).
>
> In sum, the vicissitudes of social agreements determined the time and the pace of day and night alternation. There is an account of a shaman (*mukaya*) who decided to open the hole of night, and the sudden darkness of the sky surprised all people during their daily activities. They had to spend the night where they were when the darkness aroused. Those who were swimming in the river to fish had to try finding the river border without seeing anything. Those who were carrying water home, had to stop on their way. Those who were in the forest for any reason, had to stay where they were.
>
> The chaos brought about by the shaman (*mukaya*) when he opened the hole of the night caused the owner of this hole to hold a meeting with the owners of the other features. They agreed to introduce a regular rhythm to the alternation of light and darkness, and then the order was established in the world. Once the rhythmic alternation took effect, the features were also mixed up. Before, when each feature was housed separately in the root of the sky, darkness meant impossibility to see anything, the sun light meant blinding light, the warmth burned and the cold was frosty. The myth ends with the

following utterance: "If they had not done so, we would have to sleep with the fall of the sun." (Lagrou, 2007, pp. 256–257)

Lévi-Stauss answered a question concerning the utility of the myth in an interview with Didier Eribon, published in 1988, saying that "the mythical thinking when confronting a particular problem, sets it in parallel with others, using diverse simultaneous codes" (1988/1990, pp. 179–180). He continues saying that "it serves to explain why things were transformed in what they are and why they cannot be different now" (p. 180) and concludes, "The myth works using images and events, which are rustic objects, presenting the phenomena in a contrastive and compact way, but reflecting basic conditions of the exercise of thinking" (p. 183). From this point of view, myths are symbolic constructions that articulate cognition of reality and imagination in a complex way, as one among other types of development of high-level thinking.

The mythological narrative does not present chronological correspondences of past events in history. It is a narrative that addresses a moment that is neither past nor future, but expresses the process of transformation between different relational configurations among beings in the cosmos. For Amerindians, the transformative process is mainly accessible with the development of some skills and only under specific conditions. Dreams and inebriation, for instance, allow the shaman to access the mythical flow considered to be the background of the usual, socially shared experience of the world. Myths are constantly confirmed in social meetings of each community, constructing "a pattern of intelligibility that allows the articulation of the world, society and history understandings, which are hidden in the thresholds of consciousness" (Lévi-Strauss & Eribon, 1988/1990, p. 182). They guide social symbolic practices, providing values for the personal actions of those who belong to a particular cultural field.

The recursion of mythological speeches, social values, and practices associated with them constrains the ways we psychologically feel and think the reality: "Human practices, while repeating thousand millions of times, become fixed in the consciousness as logical pictures. These pictures have the rigidity of a prejudice, its axiomatic aspect, precisely, is only due their thousand millions repetitions" (Lenin as cited in Vygotski, 1934/2001).

Rituals are human practices continuously repeated, and each culture has a set of characteristic rituals that promote the internalization of valuable conceptions for a community or a people. From this perspective, the concept can be understood as a symbolic elaboration maintained due to thousands of millions of repetitions, that is, as a result of culturally ritualized practices. Therefore, knowledge is constrained by the ontological route that creates the illusion of intersubjective sharing in a social situation. In interethnic dialogues, the sharing of rhythmic experiences given by ritualized situation is the key for understanding the conceptions of the other as dimensions of thinking articulated to the mythical narratives belonging to specific cultures.

Cyclical Temporality and the Irreversibility of Time

Culture cultivates psychological systems through investments in the construction of personal and collective bodies. These investments involve the establishment of a social order, which implies mutual regulation of activities in the environment through the construction of a shared temporality inhering in the pattern of activities in the lived environment. The anthropologist Tim Ingold argues that temporality is neither chronology nor history, that is, it is neither a regular system of dated time intervals nor a series of events that can be dated in time according to their occurrence in one or another chronological interval. It is related to the reciprocity of activities in which "people, in the performance of their tasks, *also attend to one another*" (Ingold, 2000, p. 196, emphasis in original). These activities are coordinated according to periodical recurrences in different spheres of life. For instance, in some families, we find patterns of personal meetings in breakfast, lunch, or dinner; and in some companies, the patterns of regulated activities between workers, targets expected to be reached recorded weekly, monthly, and annually.

Time is not just cyclical in terms of intersubjective and interobjective regulation of activities. The cyclical regulation is a property of organic systems as the human body, which needs to regulate its cycles of wakefulness and sleep, hunger, thirst, and its satisfaction, and so on. All these cyclical demands for the personal body and for social tasks need to find a rhythmic harmonization of mutual attention in order to keep working. Nevertheless, "there is no reason why the domain of interactivity should be confined to the movements of human beings" (Ingold, 2000, p. 199). Our perception of the passing of time is also connected to our perception of the landscape as a whole. We can attend to animals and plants as well as people. Additionally, "the rhythms of human activities resonate not only with those of other living things but also with a whole host of other rhythmic phenomena—the cycles of day and night and of seasons, the winds, the tides, and so on" (Ingold, p. 200).

Therefore, the cyclical approach also inheres the Western experience of time (not only the Kaxinawa experience, for instance), although each culture organizes it in a very specific way. A major evidence of cultural organization of temporal experience is found in our scientific conception according to which a second is the duration of 9,192,631,770 cycles of microwave light absorbed or emitted by the hyperfine transition of cesium-

133 atoms in their ground state undisturbed by external fields. From the solar cycle to the cesium atoms, the temporal dimension bestows the cyclical resonance of elements on the environment.

As far as our experience of time is connected to the cyclical resonance of elements belonging to the environment, time, and temporality are also perceived as irreversible and linear, that is, continuously moving from the past to the future. The notion of the irreversibility of time has been largely discussed by Valsiner (1994, 2002), who emphasized the singularity of events in developmental trajectories and the perceptual synthesis of the present through semiotic stabilization "out of basic flux of experiencing" (Valsiner 2002, p. 50). Through semiotic mediation, the present is constructed as an adaptive illusion of stability; the past and the future become reflexive in the consciousness. Valsiner's notion on the irreversibility of time is grounded in Bergson's notion of duration. "For Bergson, duration is not merely the succession of one lived-through instant replacing another, but '... continuous progress of the past which gnaws into the future and which swells as it advances'" (Bergson, 1911a, p. 7). (Valsiner, 1994, p. 31).

Valsiner also takes into account the Bergsonian emphasis on the semiotic mediation of the human consciousness:

> Consciousness separates the flow into units, by way of symbols—and perceives the reality only through the symbol (Bergson, 1910, p. 128).... Language performs a dual function: on the one hand, it generates self-reflexive stability in the face of duration (thus eliminating the real "flow" of irreversible personal experience by translating it into symbols reflecting stability. (Valsiner, 1994, p. 31)

Confronting the notion of the irreversibility of time with that of the cyclical experience of time discussed before, we face a challenge concerning the viability of articulation between these conceptions: how the cyclical can be systematized with the irreversible in the path of achieving a broader conception of temporality, able to conciliate different cultural conceptions of time. I argue that a way to find a possible integration between cyclical and irreversible time can be reached through the observation of the relation between the singularity of developmental events and the cyclical repeatability of developmental processes:

> If the irreversibility of duration is taken seriously, each unity of analysis may be constructed by way of three sequentially distinguishable processual sub-units: *emergence, steady state and dissipation*. Adjacent units would have temporally overlapping sub-units (e.g., the dissipation phase of one unit is the same as the emergence phase of the next). Furthermore, it may be in principle impossible to determine the exact beginning "point" of an emergence phase (or the end "point" of the dissipation phase). The three-part units can

> be strictly definable by their middle ("steady state") part, but the relevant material from the perspective of developmental psychology is not in that definable part of the unit of analysis. The "steady state" may merely serve as a basis for detection of a developmental process that has arrived as such "steady state." The focus of analysis needs to be on the fuzzy sub-parts of the unit, and on the transition from one unit (that dissipates) to the next (that emerges). (Valsiner, 1994, p. 37, emphasis in original)

From this assertion, it is possible to observe that the process of emergence, steady state, and dissipation is cyclically repeated over irreversible time. It carries some theoretical and methodological implications to the psychological understanding of the experience of temporality:

1. The process of "emergence, steady state, and dissipation" is cyclically repeated over the irreversible time.
2. Nevertheless, when a cycle is finished, it does not return to the same starting point (recursive principle).
3. The cycle is intercepted by the irreversible time and backs recursively, producing novelties over the sedimented past.

Considering Ingold's (2000) reflections concerning temporality's relation to the reciprocity of activities according to which people in the performance of their tasks attend to one another; and considering that one of the main sources of transformation in human development is the subject's quest for I-Other sharing (Guimarães & Simão, 2007), we observe that

4. The rhythmic harmonization of mutual attention in irreversible time is related to Self, others', and the world's influence on mutual living cycles.

When Self, others', and the world's mutual living cycles affects each other, they dislocate each other from their starting points, guiding the movement from the past to the future. From this point of view, irreversibility of time is considered an outcome of the encounters of the reciprocal cyclical emergences of different beings in the world. Figure 15.1 is a synthetic diagram of this discussion:

The attunement of cyclical emergences in the construction of a shared temporality demands efforts for perspective adjustments in which each process in interaction places a semiotic element (artifacts, utterances, melodies, etc.) to the other. From this perspective, one's action will always regard an object that already has its cultural history made by multiple others. This makes the object also active, as it would never serve as a *tabula rasa* for the symbolic action of a "just arrived subject" (Guimarães & Simão, in press; Simão, 2010). Cultural objects, that is, a song, a book, a painting,

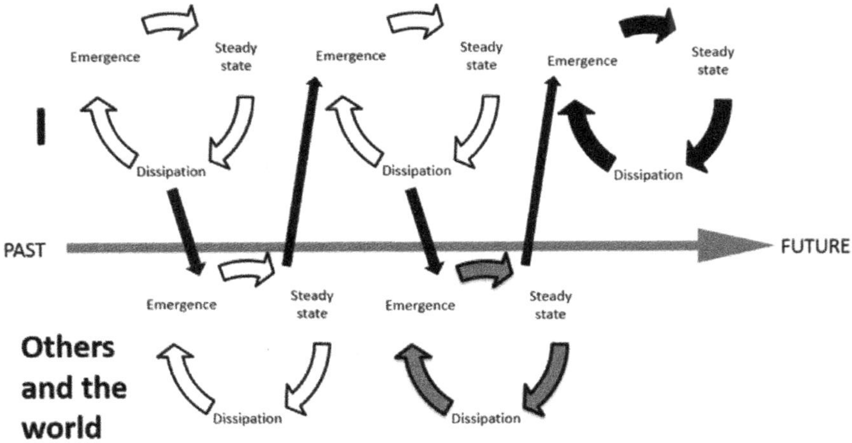

Figure 15.1. Semiotic-cultural constructions as an outcome of rhythmic harmonization of mutual attention in irreversible time.

industrialized goods, and such, are relatively independent/alienated from who produced them. They are constructed on the grounds of a seductive effort of an agency addressing the other. The duality creation-and-alienation is always relevant in the communicative process. Cultural expressions, meaning,s and artifacts are, in this sense, vestiges and products of subjective elaborations of experiences (Berger & Luckmann, 1974/2003)

The rhythmic dimension of the communicative process organized in a compositional setting is able to structure symbolic forms of human feeling (Langer, 1953). From ancient Greek philosophers to Lukàcs, music is strongly related to the "inwardness of humankind" (Johansen, 2010, p. 185).The rhythmic experience in Self-other-world relationships, previous to the verbal interaction, is on the basis of the heterogenetic process that leads to the emergence of multiple perspectives in the cultural relationships. Temporality, in this sense, is an outcome of the intersubjective stream of mutual affection.

The philosopher Emmanuel Lévinas (1993) gives an important contribution to the understanding of the process involved in alterity relationships. According to him, the alterity of the other cannot be immediately apprehended, but it is mediated by vestiges. Vestiges are any perceived, comprehensible aspect related to the communicative experience. Nevertheless, taking into account the alterity of the other implies regarding its foreign dimension as a fundamental aspect of the relation, because the "other" is always bonded to peoples and worlds that were not equally experienced by the "I". Therefore, Lévinas stresses the *ipseity* of the I as a particular aspect for the elaboration of the experience in mediated world:

> The world of perception expresses a face: the things affect us as *belonging* to otherness.... Things, as merely things, have their first independence because they are not mine, because I´m in relation with humans from whom they come. Hence, the relation of the I with the totality is a relation with human beings I recognize the face. (Lévinas, 2004, p. 39, emphasis in original)

The subjectivity of the other, nevertheless, is never fully accessible or comprehensible through the fragmentary expression or artifact presentation. A person has to deal with some devices for attracting one's attention and logically articulating feelings in a powerful shape. The act of communicating implies an intervention in the world. Therefore, the independence of the object embraces different degrees of relativity. Anyway, one's expression or object mostly remains opened to multiple meanings. Feelings have an important role to the active elaboration and understanding of the communicative expressions. As a dynamic mediation of the person-world relationship, feelings leave their traces in any human cultural production. Discussing some Dissanayake findings, Johansen (2010) claims that the persuasiveness and emotional impact of poetic language—melody and rhythm, patterns and repetitions, hyperboles, elaborations and simplifications—have their origins in the extremely affective talk between mothers and their babies.

Figure 15.2 is a systematization of the I-otherness relationships as implying the emergence of cultural expressions/objects. It presupposes that the symbolic object is a result of creative symbolic elaboration of feelings, addressing the production of affective-cognitive impacts on the social environment. The symbolic elaboration of the listener acts recursively on the affective field of the speaker, who can be disquieted (i.e., ruptured in his/her expectancies). It demands a reorganization of the affective field through the creation of a new objective structure, leading the other/listener/speaker to a novel affective-cognitive experience.

Temporality and Mutual Regulations Between Psychological Cycles: Articulating Perceptions and Imaginations

Vygotski (1934/2001) proposed that it is impossible to correctly know the reality without a certain content of imagination, therefore, proposing dialogism between perception and imagination in the process of cognition. When apprehending the reality, the individual is constantly in between the perceptive and imaginative process, as far as between affective and cognitive processes because cognition is understood as the possibility of feelings becoming distinguished from an originally affective relationship between the I and the world (Josephs, 2000). The affective-cognitive dialogism between perception and imagination allows an enriched and more elaborated apprehension of reality.

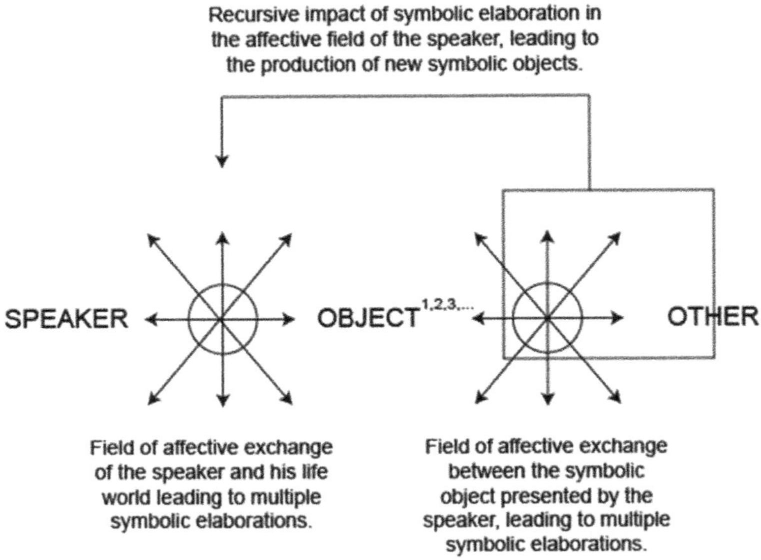

Figure 15.2. Intersubjective flow of feelings leading to symbolic elaborations and construction of cultural expressions/artifacts/utterances.

Figure 15.3 shows the articulation between perceptions and imaginations as a field of tensions in which semiotic elaborations take place. The triad imagination, perception, and cognition of reality resonates in other conceptual triads found in cultural psychology. For instance,

1. The dialogic opposition between the possibilities (as-IF) and realizations (as-IS) in life trajectories, leading to the emergence of symbolic resources that guide human futurity (Valsiner, 2007);
2. The situation as lived (IS value) and what is desired (SHOULD value) articulated in the potential for our symbolic actions in the world (Boesch, 1991);
3. The personal (intrasubjective) elaboration of the experience and the social sharing (intersubjective) experience of the lifeworld, resulting in interobjective elaborations (Latour, 1996); and
4. The articulation of what is lived as concrete (real) and what is projected as illusory (virtual) in a process that involves the configuration of the personal body (body fabrication) in a communicative process (Lima, 1996).

In sum, all these triads lead us to conceive that reality, in a semiotic-cultural sense, is a plan of consistency constructed in the interval of openness between the person and the environment. With the aim of filling this interval, the semiotic-cultural process articulates cosmological values that give consistent basis for the interpretations of the experience.

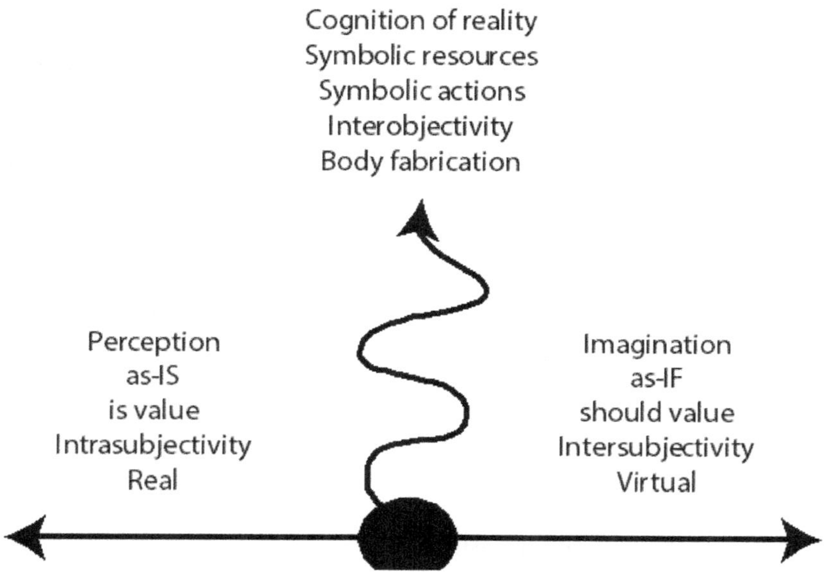

Figure 15.3. Field of tension between perception and imagination articulated in the cognition of reality and its semiotic resonances.

Semiotic constructions depend on the way each one internalizes perception and elaborates it through the sphere of imagination. Perception and imagination constitute the double face of the consciousness related to its objects: When perceiving, the consciousness apprehends the object as exterior to itself; when imagining, the consciousness apprehends the object as independent of its external presence. During the ontogenesis, perception and imagination relate recursively as such that much of our perceptive approach to the exteriority is developed by our imaginative internal elaborations and reciprocally. Besides, there is a fundamental relation between action and meditational means, the developing person articulates personal symbolic resources and actions, embraced by Self perceptive-imaginary apprehensions.

Semiotic constructions articulate the perceptive experiences of the exteriority as it is, including intersubjectivity, with imaginary experiences of the intrasubjective psychological dimension, converting life trajectories into self-narratives of personal transformations in time, hence elaborating nebulous, affective-cognitive experiences. Therefore, the process of emergence of symbolic elaborations in personal trajectories occurs through the adjustment of cyclical intrapersonal processes involving perception and imagination and concerns the cyclical process that occurs with others in the world in a similar way. Reciprocal adjustments develop in bidirectional ways and continuously, dislocating each other personal cycle of symbolic elaborations in the construction of an intersubjective space.

Reciprocal adjustment of perspectives is achieved by an "attunement to the attunement of the other" by which states of affairs are brought into joint focus of attention, made sense of, and talked about from a position temporarily adopted by both participants in the communication (Rommetveit, 1992).

The mutual affection between Self and others, Self and the world, does not allow the cycle of emergence, steady state, and dissipation enclosed in a merely repetitive cycle. On the contrary, mutual adjustment guides personal changes in the symbolic trajectories by dislocating the sedimented symbolic cycle that happened in the immediate past demanding the emergence of new symbolic resources and actions.

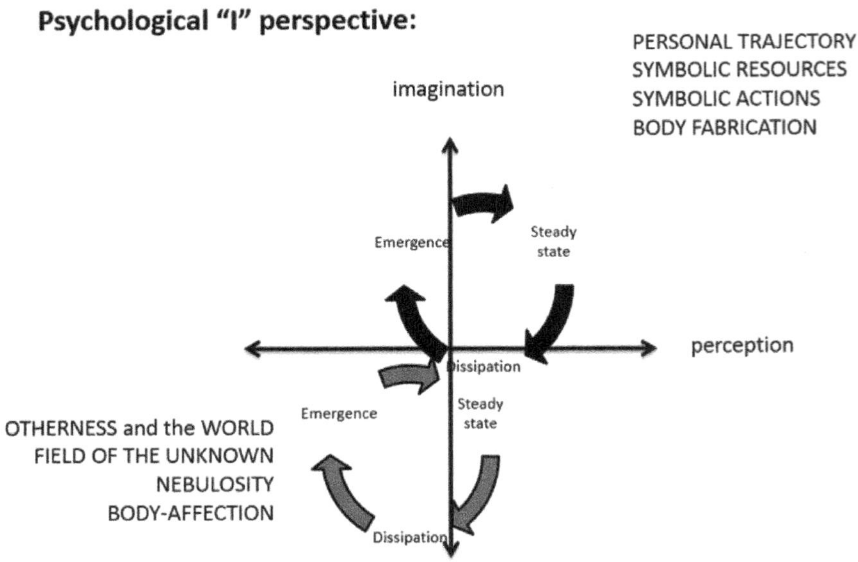

Figure 15.4. Personal trajectories as outcomes of the mutual attunement self-others-worlds.

Based on what is discussed here, our interpretation of the stream of semiotic-cultural elaborations in life trajectories implies that each agent in a relation is an array that coordinates a set of personal tunes, constituting a proper rhythm out of selective intrapersonal and interpersonal resonances. What is considered noisy or harmonic in the process of Self, other, and world mutual regulation is differently evaluated by each part in relation. The harmonization depends on the persistence and regularity of the relation and works as an active process of semiotic stabilization. Once emerged, semiotic stabilizations can be aggregated to the experience and worked as hierarchic regulators that enable the psychological experience to distinguish the past from the present.

In the following section, these four steps will be taken into account in order to interpret how a discourse of the Mbya Guarani people was constructed as a semiotic stabilization that guides an effort for the harmonization of interethnic relations in Brazil.

Regarding the Future: Dimensions of Time and Temporality Expressed in the Discourse of Mbya Guarani Youngsters

For the present discussion, I have selected a concrete situation concerning the process of construction of the "*Tenonderã* letter" by the Mbya Guarani People of São Paulo. This process was registered in the documentary "*Tenonderã: Um Olhar para o Futuro*" (Ideti & Duwe, 2010). Tenonderã was a meeting held on May 21, 2009, at the Tenondé Porã community, located in the region of Parelheiros, São Paulo. It involved 200 young Mbya Guarani from 14 communities of the State. They aimed to discuss their perspective about the environment, culture, education, sport, and territoriality. In addition to the Amerindian participants, some researchers and specialists were invited to the event, as were the geographer Aziz Nacib Ab'Sáber (1924–2012), anthropologists, and environmentalists.

At the end of the meeting, the participants organized a group of five Mbya Guarani delegates to exchange ideas with the communities that had sent their representatives. This group aimed to publicize the discussed proposals in order to collect the feedback of people from the communities who could not attend the meeting, especially the thoughts of Xeramo (elder people) and of Nãnderu-ixá (leaders) of communities within the state of São Paulo. As a result, the Tenonderã letter was formally published in a ceremony at the State Legislative Assembly, held on February 25, 2010, and the documentary was produced. Tenonderã, in the Mbya language, means a look to the future, and it conveys a particular concern regarding temporality in life trajectories.

The documentary begins with images of life in the city of São Paulo, of its skyscrapers, traffic jams, and crowds walking fast in the streets, and then contrasts it with the image of an Mbya community on the periphery of the city, a quiet place with simple houses among green bushes. After that, it shows the best regards of the participants in the meeting and the following speech is delivered:

> For us, Guarani Mbya people, the forest is very important, that is why we are always in contact with it. But we have been accused of destroying the forest because we use some of its resources for our subsistence. These accusations demonstrate how Juruá people [non-Amerindians] do not comprehend us. And to change it we, the young Guarani from São Paulo together with invited people from other Guarani villages, decided to express ourselves with an open letter to the society.

The sequence of images and the discourse emphasize a distinction between the ways of life of indigenous and nonindigenous peoples (Juruá), defining two different stable rhythms of life harmonization, one faster than the other. Also, it presents different arrays of cultural selections to the mutual attendance of each other. On one hand, the Juruá people are presented as attending to the urban context with its specific aspects such as the crowd, the work, the cars. The Juruá's concern with the destruction of the forests is also remarked, a concern that is somehow shared with the Amerindians in general. On the other hand, the Mbya Guarani people are presented as concerned with the subsistence inside the forest. Both cultures are presented as not harmonized; there is a misunderstanding between them because the Juruá people do not understand the Mbya people. The aim of the Amerindians is therefore to construct a common ground for a possible dialogue through the elaboration of a semiotic device—the Tenonderã letter. The movie continues with a sequence of utterances selected from the speeches of the participants of the meeting, within a "House of Prayer," a sort of Guarani temple:

Many things have been discussed about the natural environment, but where are we in this context?

> Guarani culture needs to be respected in its particularities.
>
> Biodiversity is the synergy of species from 'n'.
>
> Differing schools.
>
> Global warming.
>
> What are the main difficulties young people are facing nowadays?

This set concludes with the phrase: "Tenonderã says: a look to the future," so it started with a claiming for Mbya participation in the contemporary debate concerning the environment and presented some topics of discussion that are constitutive of the Juruá context. These topics are focused in the documentary from the perspective of the Mbya notion of Tenonderã.

The movie follows with the testimony of Jerá Poty Mirim from Tenondé Porã village:

> Amerindian communities in Brazil have been accused many times. I think as a Guarani that these accusations are good for some Governmental institutions, for some politicians, for instance because it helps to create a negative image of the Amerindian population when an Amerindian community is struggling to have its territory formally demarcated, for instance. These accusations create a situation in which the majority of the society also does not support the Amerindian causes.

In the middle of Jerá's testimony, the following statement by another Mbya woman is presented: "I think that it is important to value education, both the non-Amerindian education and our traditional education." This set of utterances shows two arrays that evince a coordination of intercultural resonances: some people are concerned to meet Governmental demands, resonating the negative image of the Amerindian population, and do not support the Amerindian causes; other people are "struggling to have their territory formally demarcated." This expression resonates with the utterance of the woman who stresses that Amerindian people need to pay attention to both the demands that come from the Amerindian tradition and from the non-Amerindian Education. Therefore, she addresses a way of integrating these different cultural demands that we can understand as cultural rhythms, since Mbya tradition expresses a particular way of understanding the cycles of life and social organization:

> (Marcos Tupã, from Krukutu village)
>
> Since the time the Portuguese invaded Brasil.... When Cabral came to Brazil he was full of intentions from Portugal. When non-Amerindian peoples came to our land, they soon thought of a way for earning money, and as a consequence, all the redwood [Pau-Brasil] in the coast was chopped. They took it and sold everything. It is because of these trees that were chopped that the non-Amerindian gave the name of Brazil to this land. That's how Brazilian history started.

> (Pedro Vicente, *Xeramõi* from Tenondé Porã village)
>
> I'll never forget when my grandparents talked about the slaughter of our ancestors carried out by the non-Amerindian people. You have butchered the

> Guarani and all of the other Amerindian peoples. You have killed children and elderly people. It hurts me. That this is what I have received from you.

In the Marcos Tupã discourse, we note a conflict between attending to money versus attending to the forest. Pedro Vicente emphasizes the opposition of interests bringing the internalized memory of the elders that resonates affectively within his Self. They cannot attend to their parents because they were killed. Nevertheless, the few persons who survived actively worked on the maintenance of the Amerindian rhythm of life, that is, their dynamic cultural stability is focused on keeping the culture alive through active, regular, and persistent rituals. The next sequence of images in the documentary shows a group of Mbya dancing and singing the following lyrics:

> Let us strengthen ourselves, altogetherYes, let us sing our songs.
> And we will be happy.
> Let us strengthen ourselves, altogether
> Yes, let us sing our songs
> And we will be happy.
> Yes, let us sing our songs
> And we will be happy.

The effort for constituting a harmonic tune between people of the ethnic group is then emphasized with the message expressed after the lyrics: "I thank you for the opportunity of dialoguing throughout these three days. Here we are all equals. Everybody has the same value." This message emphasizes the process of cultural harmonization between the partners, trying to construct a common ground for intercultural exchanges. Nevertheless, Mbya culture emphasizes not just the attunement with other people, for the cyclical regulation also depends on the relation with natural environment:

> (Pedro Vicente, *Xeramõi* from Tenondé Porã village)
> In São Paulo we don't have fruits to eat anymore. Here there is no Gabiroba, there is no Jaracatia, there is no Araucaria, there isn't any of these species here.

> (Aziz Ab'Saber, geographer, professor at the University of São Paulo)
> The forests were virtually replaced by condominiums, houses, and streets with commerce, etc.

> (Ivandro Tupã Mirim, Tekoa Pyau village)
> To Mbya Guarani people, nature is the place where culture lives. Here we no longer find the fishy rivers, the forest, here there is no way of hunting and

gathering. The young Guarani are concerned. From Guarani perspective, he himself is the guardian of the nature, so, the villages are always located at the boundaries of the Atlantic forest.

Considering themselves able to construct a harmonic relation with the natural environment, the Mbya people assert in the sequence of the movie that the source of noise to their cultural system is the intervention of the foreigner. The non-Amerindians way of organizing society and the relation with nature is based on dissonant cyclical and semiotic regulations when contrasted with the Amerindian cyclical and semiotic regulations:

> (Paulo Karay Mirim, Tenondé Porã village)
> We live in the small residual Atlantic Forest. Most of the researchers that come here say that wherever there is an Amerindian village the forests are more conserved, but the State does not acknowledge it.
>
> (Athaid Guarani, from Tenondé Porã village)
> The nature to the white men is something to be subjugated.
>
> (Pedro Vicente, from Tenondé Porã village)
> Ah, this is my property, nobody can pass through it, nobody can chop anything.
>
> (Nicolau Tupã, from Tenondé Porã village)
> To the Guarani people, the land is their life. Nhanderu [Our Father] created us to live in the land to produce food so the children would not get hungry. But Juruá people [non-Amerindians] have put their hands on everything, they didn't let us live as our ancestors.
>
> (Marcos Tupã, from Krukutu village)
> In the past we could freely move from one place to the other. Today our lands are small, sometimes demarcated, sometimes not, and it has deep, serious consequences to the Amerindian populations that have resisted and are surviving nowadays in this land that is named Brazil.

These testimonies also reveal a nostalgic feeling regarding the past, at the same time that the resistance of the Mbya Guarani survivors denotes that, for them, a situation closer to the past can be recovered, through the maintenance of their culture. The regarded future, Tenonderã, is an image inclusively addressed to the past:

> (Maria Inês Ladeira, anthropologist CTI)
> All the territory that Guarani peoples used before European people arrive, that is, what we nowadays call Paraguay, Argentina, Uruguay, Brazil . . . Guarani people were there previously.

(João de Quadros, Paraty Mirim village)
I've listen many times to Juruá people (non-Amerindians) saying, Amerindian people have rights over their territory, but it doesn't really happen. They don't let us walk in the forests. If we enter the forest, we walk scared. Sincerely I think that they never respected all of our rights.

(Jerá Poty Mirim from Tenondé Porã village)
Amerindians don't need territory, they are vagabonds, they do not produce anything, they want the territory only to mess around. These ignorant people think like this because they don't see Amerindian people trying to produce 2, 3, 4, 5, 10 thousands hectare of soybean to sell, to become rich individually. Moreover, today we don't think the following anymore: you [non-Amerindian people] have to pay everything you took. Because we know that a small piece of land costs 20 thousands, so imagine 8 billion hectares [roughly 20 billion acres].

These utterances denote the discrepancy between the evaluation of what is considered noisy or harmonic in the process of Self, other, and the world and mutual regulation when Amerindian culture is contrasted with non-Amerindian cultures. Amerindian peoples are accused of being lazy and unproductive; nevertheless, they argue that there are other more relevant aspects of the world to be attended to, instead of the productivity emphasized by the Juruá people. Based on what was presented, we can observe that mutual attendance regulates the relations between the beings in the environment and animals and plants, between peoples, families, and persons in a society, between intrapersonal embodied cycles, as hunger and thirst, and between the substantive phenomena of nature, as seasons, winds, and tides. Additionally, Pedro Vicente addresses another dimension related to their cosmology concerning what the Mbya Guarani people actively chose to attend:

(Pedro Vicente, *Xeramõi* from Tenondé Porã village)
They call us vagabonds—Amerindians vagabonds. What about them? If they weren't vagabonds they would not exploit the land without paying to Nhanderu [our Father]. Juruá people exploit the land, many tons of clay are taken to make bricks, after that they sell it. Are they paying Nhanderu? All things, all resources that are in the ground, in the forest, and in the rivers, everything is being sold today. You change the rivers. Who is doing this? Are the Amerindians doing it? Are the Amerindians changing the rivers? Building great factories? Who is doing this? The Guarani people? Are Juruá people paying Nhanderu to change the course of rivers and to exploit everything? No, they don't pay anything. When it rains, why do the streets look like rivers? Who did it? Did I do it? Was it us, the Amerindians? The unique owner of the land is Nhanderu, it isn't a property of the State Governement. The land does not belong to the Mayor, to the President. The unique owner is

> Nhanderu. I'm not the owner of the land, I'm here only as a passenger. It would be nice if you would admit it and say that you destroy the lands, that you changed the rivers, that you take the resources of the land to build houses to make money, but you don't assume what you do.
>
> (Marcos Tupã, from Krukutu village)
>
> Nature itself imposes limits, rules, norms that all of us need to respect. So, when we use any natural resource a spiritual compliance is necessary. Spiritual compliance is . . . when we are going to build a house, chop the wood for the house, it cannot be done any day. It needs to be chopped according to the moon—the sun and the moon determine the good period for chopping the wood, for building houses, and for preparing the arable land. We also have a period to fish and a period to hunt. There is always the season of the year, that for us is divided in two cycles and these cycles are what sustain the natural rules. So, we always need to respect and also consult the Great Shamans and they will decide the moment, the correct day to do our activities in the nature. What the non-Amerindians need to think is that even with small changes, we haven't destroyed the forest

The speech of Pedro Vicente is presented in the documentary intertwined with the speech of Marcos Tupã, who expresses clearly what guides Mbya Guarani practices in the natural environment. Considering that temporality inheres in the pattern of dwelling activities, the movie demonstrates how, to the Guarani people, temporality can be observed as a result of a selective array of Self-other-world relations that does not correspond to the array of Self-other-world relations belonging to the non-Amerindian cultures. This difference imposes challenges in order to find a mutual regulation between these cultures.

The first part of the movie closes with a synthesis of Mbya meanings about their dwelling activities, in contrast with the non-Amerindian dwelling activities:

> We are the ancient inhabitants of the Atlantic forest. Now, everything was occupied, divided, and sold. The small portions of land that remained conserved are within our domains, because for us, Nature is sacred, it is the house of Nhanderu, it cannot be divided into lots. Life is in everything, it is in the animals we hunt to our survival and also in the stones that eternally guide our paths. We respect the creation and love the nature in silence, using its resources with wisdom.

The documentary follows presenting the process of constructing the Tenonderã letter. It shows the path of the delegates to the communities from São Paulo to the coast and a sequence of images from the villages (Pindo-ty village in the municipality of Pariquera-Açu, Sapucaia village in the municipality of Angra dos Reis, Rio Silveira village in the municipality of Bertioga, and Itaoca village in the municipality Monguaguá):

> As a result of the meeting, we decided to write the Tenonderã letter. After the letter was written, we went to the villages and presented it. Then the chiefs, elderly people, and the people from the communities introduced their ideas and corrections to improve the letter.

The lyrics of a song remember the history of the meeting between the Amerindians and Europeans in the past:

> If someone wants to know me, with pleasure my name I'll say.
> In the village I'm Okuendju, descendant of Sepé Tiarajú.
> I'm the guardian of all the beauty of the nature created by Tupã.
> I'm the son of Pindorama, land of the palms or Brazil.
> I'm the grandson of who has once friendly hosted you,
> when you arrived our land, Pindorama.
> They offered you gentleness and affection,
> without thinking that the danger was close,
> letting the foreigner sleep in the same house.

After that, a young Mbya Guarani states that "This document is being collectively constructed" and another one follows claiming, "We constructed this letter according to the speeches pronounced in the meeting, according to the reported needs, what is the main need. We know how Juruá peoples are: to them, we cannot express something only verbally," then the first one continues by saying, "The aim of the letter is to bring improvements to the Guarani people, to the Mbya Guarani, and other peoples that have an objective, that is, those who don't want to destroy anything." The documentary follows with the testimony of Renato Tupã, from Pariquera Açu village: "Our main focus was the environment, but we don't even know the processes and environmental laws that Juruá people created. We need to be strong and courageous to be able to talk face to face with the Juruá and and tell them why we exist until nowadays."

Here we noticed again a contrast between the constructed meanings about the way the Juruá people deal with the environment through the creation of laws and the way the Amerindian peoples deal with the environment. To the Mbya peoples, the legislation could be a meaningless semiotic device in the process of environmental conservation. Instead, the legislation is perceived as a device used to create an asymmetry between who knows and who does not know the law, that is, creating obstacles to a fair dialogue. The Tenonderã letter, on the other hand, can be an instrument to circumvent the obstacles, allowing the Mbya people to be listened to.

The issue of attending to the legislation opposes the legal exploitation of the environment. Frequently, through this legal exploitation, that is, strictly attending to the law, the environment is not necessarily attended to:

> (Luís Eusébio, chief of the village Peguaó-Tỳ)
> What is more relevant is to protect the forest around us, more and more the landowners destroy it.

On the other hand, attending to the families instead of attending to the economic system makes sustainability possible:

> (Jerá Poty Mirim from Tenondé Porã village)
> When an Amerindian community doesn't have enough territory for its Nhanderekó [traditional way of life], many families in the community starve. Any person, any father that sees their children starving will try to find a way to feed them, and one of the solutions they find is the extraction of palm cabbage. But it isn't made as ... I've never seen a Guarani community, for instance, carrying five trucks of palms so, the risk of extinction of Jussara palm, for instance is mainly the responsibility of the Juruá, because of the deforestation, because of the Juruá's illegal extraction, with more and more sophisticated equipment and organization that allow them to extract many, many, many palm cabbages.

Finally, Athaid Guarani returns to the issues presented at the beginning of the documentary, as the common challenges for the Amerindians and the non-Amerindians peoples concerning, for instance, global warming. Nevertheless, despite the fact that we share the problems, it is stressed in the movie that the reflections on the solutions are not similar, and different directions are designed regarding the future:

> (Athaid Guarani from Tenondé Porã village)
> In the Tenonderã meeting, the elders and the Chief Timóteo explained why we have reached this point. Nowadays Juruá people are mobilizing themselves around the world because of the global warming, but they think that they can use only the science to reduce the damages caused by them. However, what is lacking is the awareness of the consequences of their choices, that is, their notion of progress to generate wealth.

In spite of the divergences, the effort for harmonization is remarked on again:

> (Athaid Guarani from Tenondé Porã village)
> We don't want to fight. We want to obtain, to fetch the collaboration of the Federal government, the State, and municipal governments.

(Jerá Poty Mirim from Tenondé Porã village)

We also want to establish partnerships with simple people that can come to the village and help us to construct a project, people that can help to mobilize others, organize meetings. Because the villages are open there are no Guarani with bows and arrows to kill whoever comes to the village, the villages are always open for visiting, for conversation, for collective reflections, for establishing partnerships and ideas that, as I've said, will be good for everybody.

The movie constantly shifts from a closer to a more distant perspective between the Mbya and Juruá peoples; this movement is deemed responsible for setting a rhythm to the relation between the Self and the other. It seems that shifting also needs to be harmonious in order to portray the differences without losing the partnership perspective.

The documentary closes with the young Mbya Guarani going to the State Legislative Assembly of São Paulo for the ceremony in which the final version of the letter was to be read and officially included in the staff of the State agenda as a semiotic device constructed by a native tradition:

Nowadays we notice the alert signals of Nature and we are observing the Juruá peoples mobilizing to solve the problems caused by their progress. But we, Guarani, do not believe that the environmental problems will be solved only with Juruá science. On the contrary, we believe in the need to comprehend Nhanderu creation. With the support of our faith, we can teach them to live in harmony with nature.

[Reading Tenonderã letter]

The Karaguy Avy is a sacred place, the home of Nhanderu, the creator of life. The destruction of the Karaguy Avy by Juruá people has been observed by us, Guarani, for a long time. The Yvy Rupá, as we call the traditional Guarani territory, is being divided into lots and deforested, leading to the exhaustion of the natural resources that are in the Karaguy Avy. Nowadays, we are observing Juruá peoples mobilizing themselves to solve the environmental problems created by their development models. To Guarani people, what is happening is not a novelty. The Xeramõi, that are our spiritual authorities, had alerted us since a long ago that one day Juruá peoples would realize the consequences of their activities in the environment. Juruá world treats the nature only as an economic asset, but our ancestors have taught us that the natural resources need to be used with wisdom.

(Pedro Vicente, *Xeramõi* from Tenondé Porã village)

The great stones, the forests. We need to strengthen them. Why do you have consciousness? Why do you have feelings? It seems that you have a poorer understanding than mine.

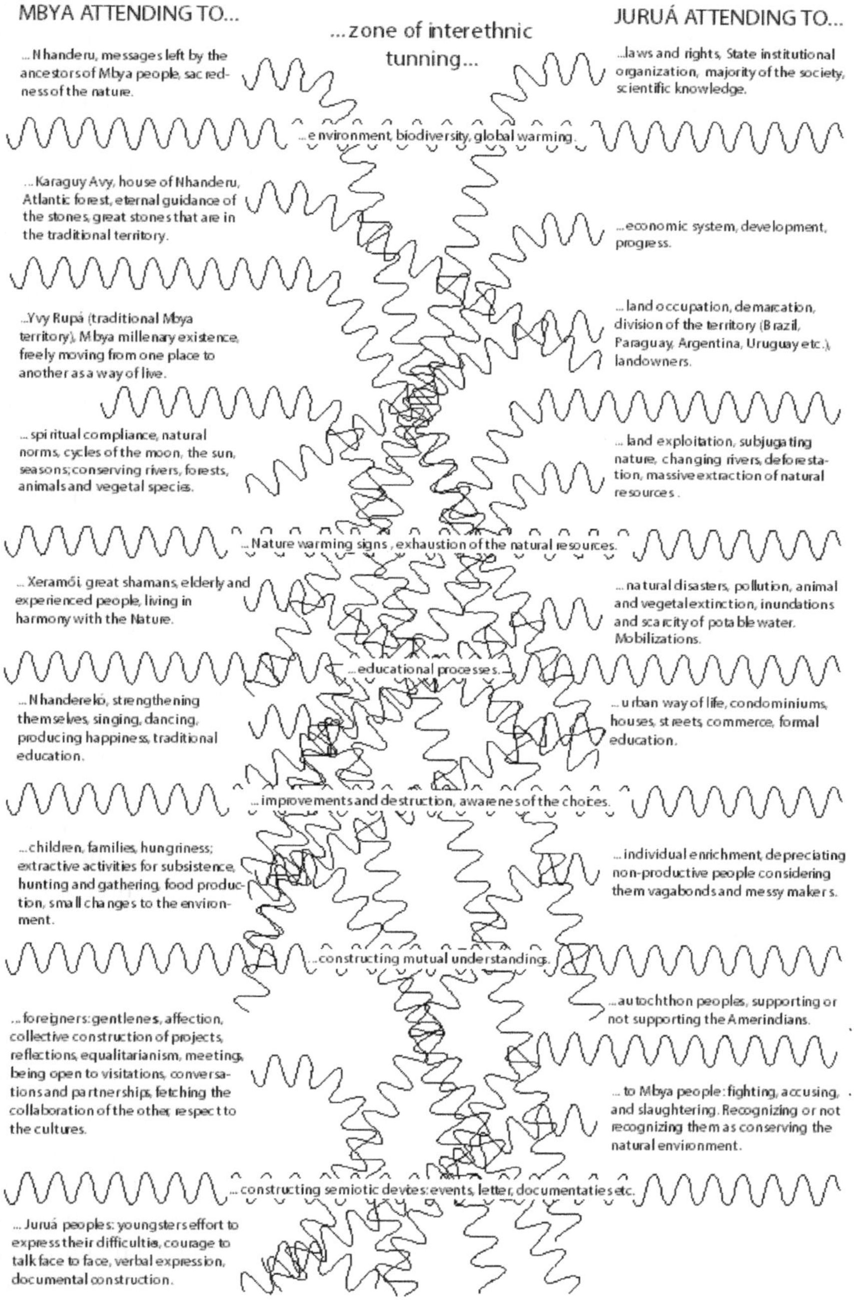

Figure 15.5. Distinctive temporalities: Bridges and noise between Mbya and Juruá cultural tunes.

world relevant elements that coordinate the characteristic rhythm of life in a specific culture, that is, its temporality is different from the correspondent array in other cultures, or other temporalities. Nevertheless, by observing the content of the movie, it is possible to find some shared references that demand the attention of both Mbya and Juruá peoples. These shared references of mutual attention were inserted in Figure 15.5 on top of the noisy and nebulous articulation that emerges in the interethnic boundary.

For a psychologist who works in the interethnic field, it is very important to take into account the different cultural temporalities as the ways in which each culture constructs harmony. It can be achieved through the identification of which social, environmental, and cosmic elements are activated in order to coordinate mutual actions in a cultural field. The cultural psychologist in an interethnic situation needs to be able to understand how the claim for attending to elements that do not fit the cultural temporality of the people he is working with produce noise or rupture. It demands the reorganization through semiotic elaboration of the experience with the other. Part of these noises or rupture is constructive and contributes to the construction of mutual understanding; another part is perceived as the reproduction of historical slaughtering, leading to the avoidance of relationships.

One of the main challenges therefore is to find a common tune between the cultures. The alternation between movements of approximation and distancing we see in the movie is similarly experienced by our support group in relation to the Mbya community. Our reflections in the construction of the project lead us to consider that when constructing the interethnic dialogue with Amerindian peoples, we need to be very attentive to these movements. Our experience has been showing especially that, from the perspective of the professional who directly works with the people in the interethnic situation, the attention needs to focus on three aspects that permeate the affective bond construction:

1. The regularity of the meetings in time;
2. the asymmetry due to the difference and noticed singularity of each part in the relation; and
3. the availability to listen to the other in the coexistence.

The regularity or the frequency of meetings in time allows each part to create expectancies about the future since they are aware that they will meet the other soon, thus opening the possibility of imagining or constructing meaningful plans for expected reciprocal actions. This sort of imagination is constructed from the perception of the regularity of the reciprocal actions and guides the intersubjective stream of feelings as a consequence of the activity of the participants who are previously informed of the rules of the setting. On the other hand, these meetings are only sustainable in the sense of a respectful coexistence, once the other is perceived as someone who can contribute to the construction of a shared projected future. To be able to contribute depends on the identification of the variety the Self can offer to the other and on whether it fits the expectancies of the other at any level.

Final Considerations

This chapter started with a discussion on the notion of temporality, concerning an Amerindian reflection on time articulated by the Kaxinawa myth. It emphasized the reciprocity of activities in the regulation of the temporal dimensions that pervade the beings in the cosmos and the community's environment. Such recursive notion of temporality is discussed in some anthropological and phenomenological frameworks (cf. Ingold, 2000), and it was compared to the notion that the irreversibility of time is present in the semiotic-cultural constructivist approach of Valsiner (cf. 1994, 2002). As an effort to integrate both cyclical and irreversible dimensions of time, I presented and discussed the process of the rhythmic harmonization of mutual attention in irreversible time, leading to the emergence of semiotic elaborations that become stable and sedimented in the Self-other-world relations. From an intrapersonal perspective, the chapter showed how the psychological cognition of reality, such as the stream of feelings and thoughts, is built as a continuous attendance to the spheres of perception and imagination in the flow of personal experience. The personal flow resonates in a singular way in each person and in each cultural field, articulating a proper array of perceptions and imaginations of what is considered relevant in the environment. From this psychological perspective, the process of harmonizing an array of perceptions and imaginations allows us to speak in terms of personal or cultural temporalities and to observe dissonance between the temporalities of different persons and ethnic groups.

These reflections on the issue of temporality in interethnic dialogues offered support to our approach concerning an empirical situation, which was the construction of the documentary *Tenonderã: Um Olhar para o Futuro* by Mbya youngsters from São Paulo. The documentary allowed us to observe the distinction between different cultural temporalities as noticed by the Mbya people and to focus on the noise or dissonance in the meeting between them. For each culture is constituted by people reciprocally attending specific dimensions of their meaningful environment perceived and imagined through cosmological constraints; although it is a hard challenge for the Mbya people to harmonize the relation with the non-Amerindian peoples who invaded their lands centuries ago. Some bridges are being constructed thanks to contemporary issues that concern both cultures:

the macro-consequences of the Western cultural changes to the natural environment.

It is fact that time is irreversible and that an experience can never be repeated in the stream of temporality. However, some rhythms of life both culturally and personally constructed and experienced can be rescued from a reference in the past and collectively harmonized in the present cultural field of action that is disturbed by disruptive temporalities. Possibly, similar challenges concerning a process of cultural harmonization take place in different interethnic conflicts around the world.

The presented discussion on the temporality in interethnic situations is connected to the project entitled Amerindian Support Network we developed with Amerindian communities that live in São Paulo, Brazil. The semiotic-cultural constructivist approach our communitarian activity is based on considers that it is important not to reproduce the same authoritative manner to relate to the peoples and that this was responsible for them shunning the contact with the non-Amerindians for centuries. To know the singularities of the otherness' temporality in a respectful manner through a responsible dialogue implies being able to express our contrastive meanings, taking into account the proper rhythm of the community. It is relevant to comprehend that Amerindian peoples are attending to other issues, some of them unknown for the psychologist or any other participant on the supportive setting and beyond the primary interest for sharing in the dialogue.

Therefore, the availability to listen to the other in conviviality appears as an essential dimension of the relation, allowing the reconstruction of the horizons for shared actions through the perception that other intentions are present rather than the usual oppressive role that the foreigner assumes. Together, these attitudes support the constitution of a shared temporality in interethnic dialogues, which does not mean a fusion between the Self and the other, neither a strict separation. The temporality that emerges psychologically out of the rhythm of reciprocal activities engaging the self, the other and the world is a dimension extremely relevant to our work. Additionally, these reciprocal activities leave their traces or sediments all over the world, which we call semiotic elements, differently taken into consideration, valuated, or disregarded by each culture.

ACKNOWLEDGMENTS

This work was realized with the support of the National Council for Scientific and Technological Development (CNPq), Brazil. I would like to acknowledge the Guarani people, who are teaching me how is the life in their cultural-collective temporality, especially Laurindo Veríssimo, Sonia Barbosa, Roberto Verísimo, Ivandro Martins da Silva, and Emerson Souza.

NOTE

1. Kaxinawa is the ethnonym of a people from the Amazon rainforest in the regions among the east Peru, Acre State, and the South of Amazon State (Brazil). Their native language belongs to the Pano linguistic family, that is, the native language of several peoples from Amazon (Sharanawa, Mastanawa, Yaminawa, among other nawa) (ISA, 2015).

REFERENCES

Berger, P. B., & Luckmann, T. (2003). *A construção social da realidade: Tratado sobre a sociologia do conhecimento* [The social construction of reality: A treatise in the sociology of knowledge]. Petrópolis, Brazil: Vozes. (Original work published 1974)

Boesch, E. E. (1991). *Symbolic action theory and cultural psychology.* Berlin-Heidelberg: Springer-Verlag.

Guimarães, D. S., & Simão, L. M. (2007). Intersubjectivity and otherness: A stage for self strivings. In Simão, L. M., &Valsiner, J. (Eds.), *Otherness in question: Labyrinths of the self* (pp. 317–349). Greenwich, CT: Information Age.

Guimarães, D. S., & Simão, L. M. (in press). Mythological constrains to the construction of subjectified bodies. In M. Han (Ed.), *The subjectified and subjectifying mind.* Charlotte, NC: Information Age.

Ideti (Producer), & Duwe, E. (Director) (2010). *Tenonderã - Um olhar para o Futuro* (Documentary). São Paulo, Brazil: Retrieved from http://vimeo.com/20263900

Ingold, T. (2000). *The perception of the environment*: Essays on livelihood, dwelling and skill. Oxon, UK/New York, NY: Routledge.

ISA. (2015) Identificação > Huni Kuin (Kaxinawa). Retrieved January 15, 2015, from http://pib.socioambiental.org/pt/povo/kaxinawa/393

James, W. (1890). *The principles of psychology.* New York, NY: Holt.

Johansen, J. D. (2010). Feelings in literature. *Integrative Physiological and Behavioral Science, 44.*

Josephs, I. E. (2000). Feeling as movement from a person-centered standpoint: Going beyond William Stern. *Theory & Psychology, 10*(6), 815–829

Lagrou, E. (2007). *A fluidez da forma*: Arte, alteridade e agência em uma sociedade amazônica (Kaxinawa, Acre). Rio de Janeiro, Brazil: Topbooks.

Lévinas, E. (1993). A significação e o sentido. In E. Lévinas (Ed.), *O Humanismo do outro homem* (pp 19–80). Petrópolis, Brazil: Vozes.

Lévinas, E. (2004). O eu e a totalidade. In E. Lévinas (Ed.), *Entre nós: Ensaios sobre a alteridade* (pp. 34–65). Petrópolis, Brazil: Vozes.

Lévi-Strauss, C., & Eribon, D. (1990). *De perto e de longe: Reflexões do mais importante antropólogo do nosso tempo.* Rio de Janeiro, Brazil: Nova Fronteira. (Original work published 1988)

Rommetveit, R. (1992). Outlines of a dialogically based socio-cognitive approach to human cognition and communication. In A. H. Wold (Ed.), *The dialogical alternative: Towards a theory of language and mind* (pp. 19–44). Oslo, Norway: Scandinavian University Press.

Scheurmann, E., & Tuiávii, J. (2003). *O Papalagui: Comentários de Tuiávii, chefe da tribo Tiávea, nos mares do Sul.* São Paulo, Brazil: Marco Zero (Original work published 1920)

Simão, L. M. (2010). *Ensaios dialógicos: Compartilhamento e diferença nas relações eu outro* [Dialogical essays: Sharing and difference in I- Other relationships]. São Paulo, Brazil: HUCITEC.

Valsiner, J. (1994). Irreversibility of time and the construction of historical developmental psychology. *Mind, culture and activity, 1*(1/2), 24–42.

Valsiner, J. (2002) Irreversibility of time and the ontopotentiality of signs. *Estudios de Psicologia, 23*(1), 49–59

Vygotski, L. S. (2001). La imaginación y su desarrollo en la edad infantil. In L. S. Vygotski, (Ed.), *Obras escogidas II* (pp. 423–438). Madrid, Spain: A. Machado. (Original work published 1934)

CHAPTER 16

REPETITION, DURATION AND PERSISTANCE

Temporality in the Performing Arts

Juliano Casimiro de Camargo Sampaio and Lívia Mathias Simão

Since Aristotle and, from another perspective, Saint Augustine, with a defining role in the work of Bergson, but also Kant, Husserl, Heidegger, and most recently, that of Benjamin and Gadamer, reflections on time and temporality have been intensely focused on the dimension of *duration* as an essential element in a subject's understanding of an experience.

In touching on and seeking to articulate a portion of these reflections on time and temporality, on the border between the performing arts and psychology, this chapter will delve into questions regarding the experience that man has in and around time using one of our primordial activities: the performing arts. It is from this angle that *repetition*—a primordial temporal category in the act of performance—becomes the central focus of the discussion herein. In this, *duration* is recognized in the context of the performing arts, in the subject's *persistence* in symbolic action, with regard to future desired studies.

Temporality: Culture in the Flow of Human Experience, pp. 381–411

This text is divided into four topics: the first, *Repetition and Aesthetic Experience—Aspects of the Now in Theatrical Phenomenon,* presents the general context for the affirmation of repetition as the temporal category of the performing arts. This topic also contains a reflection on the differences between repetition and replication (and of the facets thereof), copy and reproduction, based on the work of Gadamer, in articulation with the work of Benjamin; this then leads to a reflection on the author's intentions, in the case of the process of artistic creation. The act of repetition, when approached with emphasis on the acts of copying and reproducing, serves as a field for discussion of restrictions, resistance, and disengagement, which the artist-subject faces in their symbolic actions and the implications thereof in structuring the subject's field of action. The idea of *duration* is presented via a theoretical analysis of the thoughts of Bergson and Gadamer on the concept in question. This topic is closed by comparisons and analogies of the performing arts, games, and periodic festivals.

The Game of Art—From Expectations to Forgetting: The Presence of Absence, touches on discussions regarding the similarity between playing a game and the act of artistic creation, taking unpredictability and forgetting as drivers of the subject's perception of their actions and being acted on in and over time; these discussions on the unpredictability-forgetting duo are in large part analyzed via Gadamer's idea of the *claim* of duration and the duration of *claims* in artistic symbolic actions.

The third topic, *Persistence and Duration—Temporal Inclinations of Aesthetic Configuration,* delves into discussions regarding the concept of *duration,* articulating the thoughts of Husserl, Heidegger, and Kant on the relationship between present, past, and future within the first topic. The discussion then turns to the recognition of performing arts from the dual aspect of duration, as a lapse of time, in the case of a single theatrical performance, and *persistence,* the continuation of that which is the same in different presentations, through the subject's effort to ensure that something always remains present; this assumes the present of performing arts, since Saint Augustine, as a form of distension. The theatrical phenomenon, in this sense, is seen as guided by the act of *persistence,* as the duration of the phenomenon, and units of *duration,* in the case of single performance presentations.

Last but not least, *Eternal and Infinite—Perspectives of Temporality in Theatrical Phenomenon,* recognizes the possibility of the subject experiencing a metaphor for eternity when in the context of a theatrical performance. This topic however provides a reflection on the sensation that the subject feels within the metaphor for eternity, in which things, through their aesthetic configuration, only exist when acted upon by the subject and that the subject, as the creator, may in some cases, inhabit a plane without time, while always forced to live within time.

Thus, this chapter presents possible fields for investigation of the meaning of time and its development to become a symbolic agent, considering: being in time, but also the sensation of not being a part of time; the clarity that time is linear and does not repeat and being able to experience time cyclically; the perspective that events are finite and their *persistence* in the metaphor of eternity; the present, on one hand, as distention, and, on the other, as ephemeral; the past as a memory, as a remnant and forgetting, all operating in the *repetition* of the present; and futurity, expectations and desires, as one of the guides for symbolic actions in the present.

As such, this work seeks to recognize that theatrical phenomenon, in *repetition*, radicalizes the subject's experience of time and what is done during this time in the subject's *persistence* in the act of radically confronting himself during the *duration* thereof.

REPETITION AND AESTHETIC EXPERIENCE—ASPECTS OF THE NOW IN THEATRICAL PHENOMENON

Repetition in the professional work of the performing arts is associated with several constructive stages of artistic creation as well as the maintenance of the artistic product. Readings, rehearsals, bodily and vocal preparation exercises, and presentation seasons are illustrative enough to allow for an understanding of the intrinsic relationship between theater and repetition. Furthermore, in this artistic (theatrical) context, repetition goes beyond the mere act of *doing once again* and becomes a part of the existential nature of the performing practice. Even if the possibility of unforeseen circumstances is seen as a path to the creation of the public presentation, the piece will always have a minimum structure: actors, space for improvisation, a theme, soundtrack, technical team, which remains and lasts. The alteration/substitution of all elements and the entire team within a given artistic practice therefore implies in the end of that one and the beginning of something that is no longer what previously existed.

Even so, partial alterations and substitutions of the creative team are often necessary in maintaining the work. Long seasons, for example, are often faced with the impossibility of certain actors or technicians to remain with the original project. The reconfiguration of the team of creative subjects reorganizes the creative process and the work in and of itself, establishing new limits and possibilities for what had already existed as a process and/or work. That said, the existing process and work, in the act of reconfiguration, establishes a range of possible symbolic actions for the creative subjects. In this sense, creation requires that subjects adapt their intentions to the emerging intentions collectively shared in the original project, negotiated by the creative peers.

From this perspective, repetition inherent in the theatrical phenomenon allows for procedural adaptation of the desires and expectations of a subject in relation to the creation from their symbolic actions regarding the creative group, while allowing the work to become a constant process of construction as a unit of meaning in and of itself.

In symbolic action (Boesch, 1991) within a theatrical repetition (rehearsals and public presentations), the subject will most likely be confronted with resistance to the creative process regarding their symbolic actions; this resistance, on one hand, is the fruit of constant negotiations between creative peers regarding what may or may not be most effective for the creation of a unit of meaning that makes up the work as an entity of its own. One must still consider, however, that, in repetition and faced with the requirements of the creative process, the creating subject must face himself: their body—technical and expressive limitations and possibilities, desires, expectations, uncertainties, and doubts. This intrasubjective dimension, for its part, reorganizes the possibilities for negotiation with the creative peers. This resistance may become, for the creative subject, either barriers or frontiers. According to Boesch (1991),

> A *barrier* is an area that is difficult to overcome, and may need particular interposed actions for the passage; once the barrier is crossed, however, the action can proceed more or less as before. A *frontier*, in contrast, marks the separation between two areas of behavior which requires an area-specific adaption in the form and direction of action taken by the individual. (p. 113, emphasis in original)

Thus,

> barriers give rise to diversification of action, not meaning, however, that this diversity will be maintained. In short, the changes to the subject are temporary. Frontiers, however, give rise to diversification in action that tends to be maintained, with a certain permanency in the direction of change. (Simão, 2010, p. 167)

Thus, repetition, apart from being phenomenologically inherent to theater performance, is also a temporal dimension that guarantees that the performing artist must always revisit, evaluate, and reorganize their potential for symbolic actions in order to maintain the unit of the creative process and their own pertinence within this process. Rehearsals and presentations, therefore, and once again, as repetitions, explain to the creative subject the need to confront himself over and over again with the same (rehearsals and/or work) partially altered at each event, whether during rehearsals or during the presentation season.

This act of repeatedly confronting the self, even when faced with resistance from the (re)occurancing misalignment of desires and expectations of the creative group and subject, as a member of a larger group, leads the subject to create new possibilities for himself and the other, to try new actions and situations that hold meaning for their relationship with the creative process, but also with the world as a whole.

From this point, repetition allows the subject to prospect for the authorship of the work of art, allowing the creator to constantly revisit his own (self)process, while driving contexts for the subject's growth in situations of negotiation and sharing knowledge of the art of theater, but also, and most importantly, of I-other-world relations.

In a complementary sense, it is mainly due to the fact that the subject constantly reorganizes his field of symbolic actions (Boesch, 1991) that repetition may occur, in theatrical phenomenon, as the same item presented in different ways. The mere fact that the subject is never the same at chronologically diverse times (Simão, 2007) guarantees that a presentation will never be the same as the first presentation, even in theatrical practices that look upon improvising as absolutely undesirable, thereby calculating and training to no end the timing/rhythm of scenes, tones of voice, accuracy of actions, and micro-actions, illumination, sound, and so on. In this final case, it seems that the creators almost intend to direct the work to close in on itself, becoming something that can be replicated at chronologically different times (given that its elements of aesthetic constitution and composition are not, even in the least bit, altered).

In this sense, the desired replica is, in itself, if seen as a reproduction, impossible, given that, as explained above, subjects are never the same between the first performance and its (re)presentation in time. Thus, the replica may be seen, in the sense that its creators seek a whole with the minimum of alterations (the least amount possible) through the course of its presentation over time, as one of the possible manifestations of repetition as a temporal category of the theatrical art.

From this perspective, to take repetition as a part of the theatrical phenomenon requires that one understand what separates a copy from a reproduction. A copy seeks to become identical to the original; the subject in the act of copying seeks out the possibility the original being recognized in the copy. Thus, the subject that copies seeks to ensure that the copy is capable of reminding others and the subject of that which drove the subject to copy (i.e., the original). Therefore, the copier does not seem to show any interest in establishing any form of authorial process in relation to the work of art but rather shows interest in the object being copied. Quite to the contrary, it seems that the most important thing to the copier is to demonstrate his technical abilities in creating the copy. The act of copying, at times and in the sense explained herein, is much closer to the idea of work—production of something with a specific function and contextual use—than the idea of artistic creation itself.

The subject that copies therefore does not seek to build, as in repetition, a history with a past moving towards a future that is completely beyond reach, but rather, to build the means to be as close (fully rebuild) as possible to an experience in the past that, often, was never experienced by the subject in question. This means that the contact that the subject that copies has with the past, with the original work, does not seek any form of substantial transformation of the present in which they "live"—both the subject and the copy. The copy does not establish any form of conflict between the origin of something within arts or history and, thus, beyond any technical questions in the act of copying, does not force the subject to face any barrier or frontier (Boesch, 1991) that may cause any form of reorganization of their potential for symbolic action.

Boesch (1991) states, in a manner quite similar to that of Baldwin, that, in the case of imitation, the individual choses a model to be followed as a goal; the presence of a model to be observed and followed may lead to the erroneous idea that this is a simple process; in truth, the model to be followed must, first and foremost, be selected by the actor-subject from a wide range of possibilities; secondly, the possible models are, as a rule, multi-purpose: they include both positive and negative aspects, which must be weighted and coordinated by the actor; thirdly, not all of the aspects of the model selected may be accessible for observation, requiring the actor to fill the "gaps;" this entire process, mainly due to the different forms of models that may be imitated, is marked by the subjectivity and emotionality of the actor (cf. Boesch, 1991).

A unilateral relationship between original and copy would be, therefore, a paradox: the original was created as an unmeasured existence, which seeks to, through the dynamics of selectivity and the artist's technical and expressive abilities, say something about the *founding phenomenon*, which only the existence of the work is able to make visible; the copy, in turn, seeks to mediate – to create access to the original work. The copier, according to Benjamin (1997), in addition to not seeking authorship, cannot produce something that can be called authentic, given that authenticity requires material duration and historic witness of the process that led to create the work; in the case of performing arts, repeating as an experience of time. One must also consider that the copy, but also the reproduction —to be touched on at length at a later moment—does note imply in the tension of unpredictability in the process of artistic creation experienced by the creator. According to Benjamin (1994), not matter how faithful the copy of something is, within artistic creation, the *in loci* presence of the original is irretrievably lost.

Furthermore, the copy in its essence, quite the opposite of repetition, is not experienced by the subject as an experience of time; the copy is merely an organization of actions in time. In this sense, the relation between the *real world* and the *original work* cannot be considered as a process of copying; the work is established as a positive manifestation of aspects intentionally selected by the creating subject from the said "real world"[1] and intensely negotiated by the creator. The work is, therefore, within the idea of Gadamer, based on Hegel, an "*appearance*" of the idea that drove the creation that reveals the creator-subject and leads the subject to confront himself at every symbolic action, driving the artistic process as a whole.

The reproduction, as an opposite of the copy, seeks to not-be; to-be-without-autonomy, given that it depends on something that has been produced to occur, the mirror image of which it reflects. In not making up an entity in and of itself, the reproduction is incapable of sustaining any form of claim. Without persistence, there is no art (work). To Gadamer (1975, p. 112), "The play of the art does not simply exhaust itself in the ecstatic emotion of the moment, but has a claim to permanence and the permanence of a claim." This is not to claim that duration, in its chronological aspect, is fundamental to all creation, but rather, persistence as a lack of depletion in and of itself; there must be a tension of affective-cognitive nature that goes beyond the work and that provides the user, simultaneously, with the sense of being with the work (absolutely), without undoing the being-in-the-self. In Gadamer (p. 113), this "simultaneousness" implies in the measurement (in the relation of the work with the user) "dissolved in total presentness." According to Gadamer, with regard to reproduction,

> This means that it is its nature to lose its own independent existence and to serve entirely the communication of what is copied. Thus the ideal copy would be the mirror image, for its being can effectively disappear; it exists only for someone who looks into the mirror, and is nothing beyond its mere appearance. But in fact it is not a picture or a reflection at all, for it has no separate existence. The mirror reflects the picture, i.e., a mirror makes what it reflects visible to someone only for as long as he looks in it and sees his own reflection or whatever else is reflected in it. It is not, however, accidental that we still speak here of an image, and not of a copy or illustration. For in the mirror image what exists appears in the image so that we have the thing itself in the mirror image. But a copy always requires to be seen in relation to what is meant by it.... A copy cancels itself out in the sense that it functions as a means and, like all means, loses its function when it reaches its goal. It exists by itself in order to cancel itself out. This self-cancelling of the copy is a purposive element in the being of the copy itself. (p. 122)

The subject that reproduces, therefore, seeks to produce the means that make the reproduction possible, and nothing more. The value sought

(Boesch, 1991) by the subject that reproduces a work is focused on the exposition thereof; there is no intent to create. According to Benjamin (1994, p. 173), "as works of art are emancipated from their ritual use, the occasions for their exposure increase" and therefore, occasions for reproduction. One must pay close attention to the fact that according to Benjamin, all works of art are a form of reproduction (*wiedergibt*) and what varies is the content reproduced. We have no interest, however, in discussing this dimension of the nature of art at this point, but rather technical reproduction as a means for the transition of cult value (ritualistic) to the value of exposure of a work of art. It is from this perspective that this term becomes problematic within the performing arts from the temporal confrontation with Gadamer's notion of repetition. Nonetheless, theater, as we will see later, has throughout its history been phenomenologically implicated in a certain ritualistic dimension, which, in essence, prevents the act of reproduction. If, according to Benjamin (1994, 1997), this historic passage of the cult value of art to the value of exposure enables access to the "position" of artist; theater, however, given the continued ritualistic dimension (as previously mentioned) establishes resistance to this perspective.

The nature of the performing arts makes it nearly impossible for a creative subject to utilize reproduction techniques, excluding in certain situations, those used as instruments for creation of a scene; for example, shows that utilize multimedia projections. In short, the reproduction, effectively, in the space of the arts, is reserved for the use of technical, mechanical equipment specifically for that purpose, as in the case of cinema; a work of art created exactly for reproduction (Benjamin, 1994). Reproduction makes "the enormous technical device of our time the object of human innervations—it is the fulfillment of this historic task that gives cinema its true meaning" (p. 174).

Just as with a copy, therefore, the subject that reproduces is not subject to an experience of time. While repetition is a theatrical phenomenon and allows the creative subject to maintain the vivacity of the artistic process, the copy and reproduction are fated by the very opposite nature of the phenomenon, even when used as instruments. Therefore, it is only repetition that provides the creative subject with the possibility of authorship and to play, in the case of performing arts, the role of artist. Moreover, according to Benjamin (1994), quite to the contrary of repetition, "generalizing, we may say that the technique of reproduction highlights the reproduced object from the domination of tradition" (p. 168).

Repetition as the original that it is, particularly with regard to the performing arts, may be directly experienced, without mediation, removing it, beforehand, from an existence as a copy; furthermore, in being an entity-in-itself, repetition prohibits recognition as a reproduction. It is in the festive character of the theatrical phenomenon (and that of other performing arts such as dance, for example) and the approximation thereof with the play (of art) that repetition takes on its true existence and function. According to Gadamer (1975),

> We started from the position that the work of art is play, i.e., that its actual being cannot be detached from its representation and that in the representation the unity and identity of a structure emerge. To be dependent on self-representation is part of its nature. This means that however much it may be changed and distorted in the representation, it still remains itself.... The representation has, in an indissoluble, indelible way the character of the repetition. Repetition does not mean here that something is repeated in the literal sense, i.e., can be reduced to something original. (p. 109)

Repetition, taken here as an original, represents the possibility of the process/work to be presented in a number of different manners, including the sphere of improvisation as a change inherent in the festive character of the phenomenon. The subject, in repetition of performing arts, is ready to deal with the unpredictability of the future of the process for constitution of the presentation, for example. Two presentations of the same theatrical piece will always be two different presentations of the same work. Therefore, symbolic actions that the creative subject executes during the process of repeating the creation are, in some way, constantly focused on the dimension of something that cannot be understood—otherness in its essence. With this, it becomes clear that the temporal category of repetition in the performing arts, with the creative subject as the focus of the investigation, is imbued with the dimension of otherness. According to Simão (2010, p. 258), "the relationship with otherness requires the subject to attempt to overcome himself, in tension between what is, what is not and what may come to be in the relation with the other." In this sense, the work repeats itself in time, however nuanced by dissimilar aspects—the fruit of negotiations between the (co)creators and the (co)creators with the objects they work with—organized using a common, identical basis.

There is yet another factor that expands the dimension of otherness in the performing arts: A theatrical work only exists, effectively, in the moment it becomes concrete in front of a given public. At each encounter between the creator and the public, through the work, the former is able to evaluate the effectiveness of their interactions to create a unit of meaning in the work and to satisfy their desires and expectations. It seems, therefore, that in the case of the performing arts, the procedural unit of meaning of the work is made possible by the creator's ability to satisfy, even to the most minimal extent, himself. The occurrence in repetition of the (theatrical) phenomenon produces, in the relationship between the creator and the work, a cyclical movement of restructuring of subjectivity of the agent and form in which the work is manifest. This movement is fed by what goes beyond the creator in the work.

In this sense, the artist, faced with proof of otherness in the creation, acts symbolically in the process to create scenes, for example, as if the artist were able to share with the spectator the moments when they are along at rehearsal. The creator, therefore, tends to act as if it were possible to foresee negotiations of meaning that will only become clear in the presence of a spectator. According to Simão (2010, p. 62), based on Valsiner (1998), "The 'as ... if' is, therefore, a subjective construction based on the desire for future states, which gives meaning to the present world, organizing I-world relations in an effort to reach the desired state."

> To Valsiner (2007), the symbolic action in the base of "as ... if" is something inevitable, given the irreversibility of time, which imposes on us the duality of our psychological functions, as our actions and reflections are situated "between" the present that we are living and the imminent future, which "flows over us in all of its uncertainty" (p. 351). (Simão, 2010, p. 62)

This anticipation, in turn, produces new expectations not only that the work be included in a unit of meaning (thought up by the creator), but also that negotiations of the unit of meaning move in the direction desired by the creator. The relationship between creator, work, and spectator is, therefore, a relationship of tension and cooperation. Tension tends to momentarily dissipate at each event seen by the artist as satisfactory, in which artistic cooperation seems to function. The criteria required to reach the conclusion that a given presentation was satisfactory are many and vary from context to context. Even so, some of these criteria may be considered to appear more frequently, namely, spectator reactions recognized by the artist as positive with regard to scenes at the moment of their occurrence; execution of all actions and propositions brought up and rehearsed during the preparation of the show, with the expressive intensities desired for them; later conversations with spectators; a capacity for improvisation, by the creator, in moments when the action is necessary; the creator's involvement in a show beyond the mere mechanical execution of previously agreed-upon creations. In short, the universe of the performing arts is, in an analogy with the world of ordinary life and in accordance with Boesch (1991), symbolic meaning and meaning through action. A sense of satisfaction guides the artist's selection when reorganizing and maintaining the performance; or rather, it constrains, in Valsiner's (1998) terms, the artist's intentions. Even so, there is still persistence, duration of tension, in the possibility of other performances for other publics and in new contexts (even technical contexts) given that the artist is constantly faced with a gap between what they desire to create, artistically, and what they effectively perceive themselves capable of producing (and what is produced) (Boesch, 1993). By letting himself be challenged by the otherness inherent to the repetition of the performing arts, the artist makes himself available to a formative process of *coming-to-be* as a dynamic for overcoming the self (Gadamer, 1975).

In this sense, even if repetition is innocently seen as a constant future, this orientation toward the future corresponds to what Gadamer (1975) assumes, since Kierkegaard (1986), as the *duration of a claim*. The performative artist persists, through symbolic action, in creating a requirement: the public's encounter with the world and with the performer. A new requirement results from the *duration of claim*; the work's unit of meaning (co) created by artist and users, but first and foremost validated by the spectator, must last beyond the act in which it is presented to the public, creating a form of *claim of duration.* In a certain sense, it is the claim that makes the continuity between past, present, and future indivisible, in the case of the performing arts. Nonetheless, according to Bergson (2006), duration only exists when there is consciousness, and before anything else, one can only think about duration because the subject naturally feels that they last:

> There is no doubt that our consciousness feels that it lasts, or that our perception is part of our consciousness, or that something in our body or from the material that surrounds us enters into our perception: thus, both our duration and a certain participation felt, lived, of our material environment in this interior duration are facts of the experience. (p. 53)

Therefore, the work does not last in itself; it lasts as a coincidence that the subject is capable of recognizing, through the work, between continuity and heterogony of experiences. One must recognize here, however, an important limitation to Bergson's (2006) relationship between the past, present, and future, to allow for thoughts of the duration in terms of claim in the symbolic actions of a performing artist. If Bergson is able to recognize that

> the present *is not*; it would be foremost pure future, always outside of itself. The present is not, but *acts*. Its main element is not ebbing, but the active and the useful. [And that] From the past, to the contrary, one must say that it has stopped acting and is no longer useful. But it has not stopped being. Useless and inactive, impassive, it *IS*, in the total sense of the word; it is confused with being in itself. This is not to say that the past "was," because it is the in-self of being and the form on which the being can conserve itself (through opposition with the present, which is the form on the being consumes itself and surpasses itself). On the limit, common determinations are exchanged: it is of the present that we must say, at each moment, that it "was" and, of the past, that it "is," that it is eternally, the whole time. (Deleuze, 1999, emphasis in original)

In the specific case of performing arts, it seems correct to affirm that the past is configured as *being*, but it is, in the case of rehearsals (the process of creating the presentation), that the constant coming of the present operates, therefore it acts. Thus, the artist (actor) acts in the present, oriented by the past (rehearsal). The past becomes, in this specific case, a repeatedly reorganized draft directed to a future, the essence of which is revisited when it becomes past and is revised by new futures. Therefore, this is not, in the case of rehearsals, the past as a memory; rehearsing goes beyond the act of producing memories, even if it does so; it is something like (self) revealing the symbolic action for a desired future. In summary, the present-rehearsal is thought, acted, and experienced, so that when it becomes past, new present-rehearsals and present-shows may operate. It is in this direction that the performing artist has a *claim* for duration. Thus, in this context, the past does not lose its function and use in the continuity of the present in the artistic process.

From this perspective, if in ordinary life, Bergson's (2006) *durée* is the coexistence of the present and the past, the former as an agent, the latter as a being; within the performing arts, the *claim* is thus anchored in the perspective that the past is intentionally operated (to operate) by the creative subject, removing the future of this privilege. A present is operated to become the past and thereby operate new presents, aimed toward a desired future that becomes the intentioned present. The duration of the *claim*, therefore, is mostly based on the dual intention of operation and the operation of intention that the performance artist, particularly the actor, executed with a focus on the past and future.

Even so, in this duration, there is a perspective for futurity that may be recognized by both Bergson (2006) and Gadamer (1975) with regard to the *claim* of duration of a work; the subject must constantly deal with the denial of access to the future and recognizes the self in this universal trauma (Bergson, 2006, pp. 51–79).

In a sense, and based on what has been presented up to this point, a theater piece, in only truly existing when performed, is not aimed toward the future or past but rather toward maintenance "*in the presents.*" Gadamer (1975) considers this to be a common aspect between performative artistic manifestations and periodic festivals. Periodic festivals are, by nature, constantly modified, just like theater, each time it occurs, given that "there are always other things going on at the same time" (p. 110) which also include the subject.

In this sense, the maintenance of the self "*in the presents,*" both in theatrical phenomenon and periodic festivals, is only possible because these phenomena are cyclical constructions of the future and common returns to repetition, in the terms covered herein. That said, one might affirm that repetition guarantees that the theatrical phenomenon is "temporal in a more radical sense than everything that belongs to history" (Gadamer, 1975, p. 110). That then destroys the possibility that the subject, when involved in a theatrical phenomenon, or even a periodic festival, experiences a natural and uninterrupted course of time. It may be that the repetition, in this sense, guarantees the possibility for maintenance of the ritual character (original) of the theatrical phenomenon, by promoting a given temporal cohabitation of presentation, expectation, and memories while also suspending the temporal experience.[2]

The ritual dimension present in the festive character of theater, maintained by the possibility for repetition, removes the temporal experience from its linear dimension for a circular/cyclical perspective by breaking the flow of daily life. Thus, repetition becomes an intentional relationship between the interplay of that which is immediate and that which lasts, into a single being in its several occurrences. In doing so, the theatrical phenomenon as repetition is the constant possibility of an experience[3] in the complete sense, and from this perspective, finds the means to ensure the persistence lasts.

> If something is called or considered an experience its meaning rounds it into the unity of a significant whole. An experience is as much distinguished from other experiences—in which other things are experienced—as from the rest of life in which "nothing" is experienced. An experience is no longer just something that flows past quickly in the stream of the life of consciousness—it is meant as a unity and thus attains a new mode of being one.... We mean the lasting meaning that an experience has for someone who has had it. (Gadamer, 1975, p. 60)

In these terms, once again, nowness is the temporal category that sustains the possibility of existence of theatrical phenomenon. The experiential phenomenon speaks for itself and confronts itself, even as it occurs in different ways as the repetition it is; this is the game that floats in *the presents* of its maintenance. This confrontation reorganizes its limits and possibilities as a power that acts in the interior of the phenomenon, based on the phenomenon itself; this process occurs in the phenomenon as a whole and in its units of composition, the subject being one of these units.

To Gadamer (1975, p. 113), "Everything that is experienced is experienced by oneself, and it is part of its meaning that it belongs to the unity of this self and thus contains an inalienable and irreplaceable relation to the whole of this one life." As a part of the work, the user is also called to confront himself in the encounter with that which is different and remains in and due to the work; the aesthetic experience as an experience, in the immediate destruction of the flow of daily life leads to a use of existence as a whole as a reference.[4]

> Every experience is taken out of the continuity of life and at the same time related to the whole of one's life. It is not simply that it remains a living experience only until it is fully integrated into the context of one's life-consciousness, but the very way in which it is "preserved" through its being worked into the whole of live-consciousness, goes far beyond any "significance" it might be thought to have. Because it is itself within the whole of life, in it too the whole of life is present. (Gadamer, 1975, p. 62)

In this sense, in Gadamer (1975), the work of art takes on its true being upon becoming an experience of now that leads to a transformation in those who experience it. It is clear, then, that the encounter with artistic language is always an unfinished event, which requires that the user/spectator be a true part of the event, meaning he must be outside of himself without clearly nullifying the self. The user acts on the work as he is acted on by it. For Gadamer, the difference between playing together in art and the mere curiosity resides in the fact that the latter provides no opportunity for leaving the self and effectively returning to the self, and in such, all experiences of temporality that the work may create disappear.

Leaving the self, however, is always leaving some self to return to the self, which may no longer be the same self in relation to … ; this requires the understanding that being with … requires that the entities do not fuse, to allow for a (mutual) effect between parties. It is exactly the being outside of the self of the spectator that guarantees that he is completely in the world. It is necessary to reiterate that, in the act of aesthetic experience, faced with the repetition of phenomenon, the user intentionally gives himself away to this temporary forgetting of the self, given that the intention creates the experience. Similar to periodic festivals, theater, and playing, this intentional process of self-forgetting, in the case of the user, "is determined by his being there present. To be present does not mean simply to be in the presence of something else that is there at the same time. To be present means to share" (Gadamer, 1975, p. 110).

Thus, what is proposed herein with regard to the phenomenon of subjects living time and in time, as one of mankind's most perennial and constructive activities, is that the performing arts, or human symbolic actions, are always susceptible to internal and/or external restrictions. These restrictions may require changes in the direction of actions, making them more or less temporary or more or less permanent. In this sense, when exposed to situations of repetition, the subject must face himself, even as a noncoincidental phenomenon. This noncoincidental nature, in turn, implies in the subject overcoming contextual mismatches that force him to reorganize his mode of symbolic action. In short, in symbolic action, the subject experiences the world that goes beyond him, affectively and cognitively, forcing the subject into constant subjective restructuring. In this sense, the world is experienced by the subject through relationships of continuity and heterogency. Thus, the subject's potential for symbolic action is guided by the duration of claims or, even more so, by the claims of duration; and it is in the sensation of satisfaction or insatisfaction of the subject's desires and expectations that he finds an anchor for guiding the selectivity of his future symbolic actions. In short, a subject's experience with symbolic actions is, first and foremost, a confrontation of the self in time.

THE GAME OF ART—FROM EXPECTATIONS TO FORGETTING: THE PRESENCE OF ABSENCE

The following reflection, as a development of the first part of this text, is based on the approximation of the artistic process with the game to improve on the understanding of the creator-work-user relationship as a temporal relationship—of the experience in and of time. This will be based on the constructive dimension of unpredictability in creative processes, together with its wholeness in the dynamics of forgetting.

Even if seeking an encounter with the user to complete the work and if depending, to a certain extent, on the existence of a being beyond himself, the creator, to ensure that he exists as a work—as a unit of meaning, in the game—the being in the work of art is not interested in the identity of who is play, but rather what is played, "The players (or poets) no longer exist, but only what of theirs is played." (Gadamer, 1975, p. 167). Nonetheless, one must remember that without the interaction of the user, the player, the artist, or those who celebrate, there is not art, game, or festival. In that sense, however, one must seek to recognize that the aesthetic configuration is not passive with regard to the subjectivity of artists and users (co-authors); the configuration, just as in the game, requires, in the process of creating the work, mutual affection between subjects (players and/or artists-users), between themselves and with regard to the diverse materialities resulting from the unpredictability of the relationship. For Gadamer (1975),

> No matter how much the variety of the performances or realizations of such a structure goes back to the conception of the players—it also does not remain enclosed in the subjectivity of what they think, but it is embodied there. Thus it is not at all a question of mere subjective variety of conceptions, but of the possibilities of being that the work itself possesses, which lays itself out in the variety of its aspects. (p. 106)

In their interaction with unpredictability, the creator-subject is always faced with the possibility of inconsistency in the map of their symbolic actions with regard to their claims that make the process last. Having a user validate the work's unit of meaning requires the artist to feel his symbolic actions in relation to the creation as a minimally coherent whole to allow for understanding (artistic, in this case) by the other. According to Simão (2010), based on Boesch,

> Internal consistency is a much more complex and elusive phenomenon than external consistency, given that the former requires symbolic coordination, in addition to factual coordination, coordination between aspirations and reality; imagination and action; and the physical world and the social world. The subject's phenomenology will oscillate, then, between attempts to widen internal consciousness, seeking self-realization, and attempts to protect the internal consistency that appears to have been achieved, seeking self-protection. (p. 226)

The creative I-other relationship, in these terms, works due to two complementary poles of tension: tension from the subject with regard to their desires, expectations, goals, and claims, in relation to the context in which they survive or fail to do so; and the tension from relationships which the creative subject feels forced to establish with others who surround them in creative processes—users, objects, spaces, and such—overall, a cultural relationship. The subject's actions, therefore, begin to have dual meaning for the subject—consciousness by the agent of the relationship of minimal coherence between expectations—the current action is always prospective, while an executed action is comparative (Berger & Luckmann, 2005).

The creator-subject, therefore, is constantly attempting to reduce tensions and ensure coherence between dynamics to widen and protect his own field of symbolic action. In the words of Simão (2010),

> This opens up a field of relational tension between consistency and inconsistency, in which the tension may only be overcome by rearranging the system of relationships between expectations and experiences, which will require adjustments of expectations that the subject held upon first involving themselves in the experience. (p. 226)

The indefinition of the work up to the moment of its true execution allows co-authors an excess of expectations; it is because of this plurality of expectations, of desires and fears, that the duration of claims in the performing art process is organized. The range of possibilities for the combination of given materialities within the organization of a concept, such as the process of artistic creation, will reveal the work's potential as a project and a trajectory. The inevitable presence of selectivity in any process for artistic creation is a direct result of this fact.

In the opposite sense, a group of expectations not contemplated in the work's trajectory will certainly have a structuring power over the work in question; not in the sense of what the work could have been (or, maybe, not only in this sense), but rather from the perspective of what the work is not, of all of the absence made evident by the existence of the work. This is not to say that the work should be understood as a negation; much to the contrary, the work of (performing) art is taken as the presence of one (or more) absence(s). Thus, based on Gadamer via Heidegger, given that this text has already affirmed that the subject acts symbolically in the present (in performing arts) to allow for the past to operate the present in certain moments and contexts, it becomes clear that the past of a work is not only a memory—a reminiscence—but foremost the possibility of *forgetting*[5] (Gadamer, 2011, pp. 159–173). In this sense, the work of art, just like the game, is tied to dual forgetting—of "things" and of the self.

The past, therefore, operates on the trajectory of creation in at least two interdependent dimensions, as a reminiscence and forgetting. For Gadamer (2011, pp. 169–170), from one perspective or the other, man will always transform the past into an object.

> When we come upon something in a tradition that we understand, it is always an event. When someone takes a word from the tradition, when they speak the work, something will occur to them. This is not the mere understanding of that which is encountered in the past, challenging us and concerning us. To express this, I may have chosen a formula that is far to ambiguous: that our historical understanding is always determined by a historical-effectual consciousness. The meaning of this formulation is that upon removing ourselves from the event and placing ourselves before the event, were turn the past into an object.

With the passing of time, for Gadamer (2011, pp. 159–173), things do not become old because their "now" has already passed, but rather because an event made them old in the meaning that the subject gives (or gave) to the something. This old something, however, may influence the "now" that occurs and thereby become a true reminiscence. Even if Gadamer sees reminiscence as only existing because of the possibility for forgetting, this reminiscence is still fundamental to the temporal dimension of creation. In this sense, both what has-not-yet-been and that which is-no-longer are both capable of intervening as active forces on the trajectory of creation, of the game, or the subject's symbolic action.

To Benjamin (1994, 1997, 2011), everything that can be forgotten, but not because the subject wishes it so, comes back in the subject's manner for action in the world with an operating power equal in strength to data in the present. If the aesthetic configuration of the work, according to Gadamer, leads the subject to forget himself, it can be said, based on Benjamin's statements of forgetting, that the moment of forgetting is exactly where the subject becomes intensely present in the work, speaks profoundly in the work, even if this may not consciously desired or perceived. In being able to lose the self, forget the self, in a work, the subject truly reveals himself.

Nonetheless, those who play together, who act with, even when totally given to this dual dimension of forgetting, know what they play, what they represent (or present, within the meaning of nowness), but most importantly, know that they play or act. To Gadamer (1975),

> Play fulfills its purpose only if the player loses himself in his play. It is not that relation to seriousness, which directs us away from play, but only seriousness in playing makes the play wholly play. One who doesn't take the game seriously is a spoilsport. The mode of being of play does not allow the player to behave towards play as if it were an object. The player knows very well what play is, and that what he is doing is "only a game"; but he does not know what exactly he "knows" in knowing that. (p. 92)

Thus, all games and all representations are a played being, a represented being. "The attraction of a game, the fascination it exerts, consists precisely in the fact that the game tends to master the players" (Gadamer, 1975, p. 95). It is important to consider that all creative efforts, in the sense of a game, produce a reverberation in return, to the person who exerts the effort. This reverberation in return guarantees the constant proliferation of expectations, even in the nowness of the phenomenon.

Thus, the unpredictability produced by the bidirectional nature of affection in artistic creation, considering the creation as effectively complete only when in front of a user, is constructive, therefore, of the work of performing art. Repetition as the maintenance of the same in new contexts, from this same perspective, in turn guarantees the continuation of unpredictability.

From this perspective, the supremacy of selectivity with regard to fate once again becomes clear, given that even when faced with unpredictability, playing will always be to play something: to play this or that, within these or those standards, rules, and structures, with this or those partners and so on.

> Every game presents the man who plays it with a task. He cannot enjoy the freedom of playing himself out except by transforming the aims of his behavior into mere tasks of the game. Thus the child gives itself a task in playing with the ball, and such tasks are playful ones, because the purpose of the game is not really the solution of the task, but the ordering and shaping of the movement of the game itself. (Gadamer, 1975, p. 96)

Just like in the game, in representing (presenting), performing the "task" of representing, the subject identifies himself with what is being presented; this self-representation of what is being represented ends up as the self-representation of the subject that is representing. Nonetheless, this act of self-representation in the game is an end in and of itself, while the act of self-representation in art is simply directed at someone. This someone, therefore, is a part of what is being self-represented.

> Even a theatrical drama remains a game, i.e., it has the structure of a game, which is that of a closed world. But the religious or profane drama, however much it represents a world that is wholly closed within itself, is as if open toward the side of the spectator, in whom it achieves its whole significance. The players play their role as in any game, and thus the play is represented, but the play itself is the whole, comprising players and spectators. (Gadamer, 1975, p. 98)

Therefore, bringing the subjects (at least two—actor and spectator) for the constitution of the self-representation of the work (theater) implies in the recognition that the phenomenon of repetition, as the maintenance of the present, is not only inherent in theatrical phenomenon but is also inherent in those who make its existence possible. Those who act in the same show are always the same and different—they do not coincide; in existing in an uninterrupted trajectory of time, the subject allows for the cyclical temporal dimension of the theatrical phenomenon. The subject, existing in a continual flow of time, is intentionally capable of allowing himself to experience, while proving, the theatrical phenomenon as an experience, and thus suspends the temporal relation in which it is built.

PERSISTENCE AND DURATION—TEMPORAL INCLINATIONS OF AESTHETIC CONFIGURATION

The following will return, based on the approximation of the artistic process to games, as in Gadamer (1975), to the notion of duration and its implications in the experience of time in the performing arts. Duration will be discussed herein in opposition to the lapse of time, introducing, as a complement to duration, the sensation of the eternal that the experience of performing arts may provide the creative subject in their experience of time.

The duration of the theatrical phenomenon, in accordance with Gadamer (1975), faced with repetition, does not become a *lapse of time*, an important characteristic, for example, of the idea of duration of something as in Heidegger (1927/2009). This, if Gadamer speaks of repetition as the maintenance of/in the past, because Heidegger's repetition seems to be focused on the perpetuation of cycles that lead to a finished being-to-death. In short, the movement that repetition represents in Heidegger is in the quest for new possibilities to lead to a concrete end.

In this sense, if repetition, in Gadamer (1975), favors an operation (power of the actor over) the not-yet occurred as an absence in the trajectory of creation, in Heidegger (1927/2009), repetition seeks to make the not-yet into a real fact. While repetition in Gadamer aids in thinking about the "same, but different," the concept according to Heidegger seems to favor the different in the same, that which is capable of projecting a phenomenon into the future. Thus, one is faced with two distinct temporal inclinations:

In Gadamer (1975), expectation, possibility, initially as a future, will flow into the past if not executed, from where they will influence the present; repetition tends to ensure that the influence of this past force is maintained, in always dealing with new possibilities and not reach an end, but rather to create a given dimension of consistency ("eternity"). It is on this consistency that the subject will act, both symbolically and intentionally, in the confrontation of the self as unable to coincide phenomenonologically.

Heidegger (1927/2009) maintains that the future flows into the past, but repetition does not seek to maintain forces that operate on the present from the past but rather the materialization of these forces as events, if they are able to bring new perspectives of the finite to the phenomenon facing the idea of infinite time.

According to Ricoeur (1985, p. 129), based on Heidegger (1927/2009),

> The cardinal function of the concept of repetition is to reestablish the balance that the idea of a handed-down heritage tipped to the side of having-been, to recover the primacy of anticipatory resoluteness at the very heart of what is abolished, over and done with, what is no longer. Repetition thus opens potentialities that went unnoticed, were aborted, or were repressed in the past. It opens up the past again in the direction of coming-towards. By sealing the tie between handing-down and resoluteness, the concept of repetition succeeds at once in preserving the primacy of the future and in making the shift toward having-been.

Even so, when considering theatrical phenomenon from its festive and ritualistic perspective, it is important to understand that if repetition meets the needs of maintaining the phenomenon (temporal circularity), the act, the event in itself (temporal linearity), tends to run out. The subject experiences the single theatrical act as ended in and of itself, with a set beginning and end. Nonetheless, the same subject experiences the continuity of the phenomenon in that which is to come. Thus, "eternity" and the finite are both essential parts of the performing arts. The phenomenon cannot be inclined to the future, given that the *claim to duration* seeks the temporal suspension characteristic of artistic experience. The theatrical event, however, once it has ended, guarantees the creative subject the possibility of acting in analogue with what has just happened. This sensation that it may be repeated returns the subject to the self, making them revisit the symbolically acted universe in the (one-time) duration of the show upon presentation. This revisiting is an intention with the end in desired future states, therefore seeking to reorganize the subject's symbolic actions.

It must be made clear, however, that the trajectory for performance creation does not include the force that operates from the past on the materialization of something from the past in the present for ends in a given future. To the contrary, it is created simultaneously in these two processes, plus the dimension/possibility of forgetting. Thus, instead of administrating supremacy, creation forces the cohabitation of inclinations of temporality, from whence emerges the plurality of the experience that the creator subject makes of time in time.

According to Simão (2009), however,

> in short, understanding of the notion of temporality requires, as one of the most important aspects, the consideration—based on Heidegger's philosophy in *Being and Time*—that the being in the world tends towards the future, meaning, the being existentially looks towards what they can be, seeking and anticipating possibilities for existence. This anticipation, this projection, leads the being to advance ahead of himself, towards the unknown. In Heidegger, the being's advancement ahead of himself is inevitably an advancement towards the finite, towards death, which is the source of the being's constant uneasiness. But the being in the world can also rely on, in their life, and for this reason move forward, a certain power of forgetting that lives in and for this quest, evolving in daily life and concerned for the self, that allows the being to distance himself from this uneasiness and, in the end, from the self.

Therefore, the encounter between the user and the work in their "now," required for the existential fulfillment of the latter, is an encounter between the being-there of Heidegger (1927/2009), focused toward the future with the possibility of suspending this inclination and later taking it back up. Once this process has been installed, the encounter with the art can be conceived as a *leave-the-self* to *return-to-the-self*, as proposed by Gadamer (1975). The game of art is therefore characterized, in the end, as a constant temporal confrontation. And it is in being the being-there created as the being-in-the-world, that the experience, like the game of art, does not close in on itself but rather refers to all experience, as described above.

In a complementary sense, the temporal confrontation only exists because the subject exists, given that time as a phenomenon only reveals itself through a being; time is not inherent in objects. The theatrical performance, from this perspective, in existing in a given temporal relationship, reveals this temporal dynamic. Even so, the relationship between the being and time, as in Kant (1971/2010), is not a relationship of creation; the creator (artist) does not create time in the work; quite to the contrary, the creator builds within time and in presenting the work reveals time, assuming that, based on Kant, time comes before all phenomenon (Ricoeur, 1985, pp. 44–59). The subject who creates, however, in revealing time and acting temporally, is also affected by time:

> We are temporally affected insofar as we act temporally. Being affected and producing constitute one and the same phenomenon. "The understanding does not, therefore, find in inner sense such a combination of the manifold, but produces it, in that it affects the inner sense." Kant was not wrong in calling this self-affecting of the subject by its own acts a paradox. (Ricoeur, 1985, p. 56)

In opposition to Heidegger's (1927/2009) *duration*, the nowness of Gadamer's (1975) duration is not found in doing something once again and differently, therefore illuminating the finite nature of the phenomenon, but rather in doing something differently to offer persistence. From this perspective of *lasting*, Gadamer seems to channel Husserl, who, according to Ricoeur (1985, p. 29), also understands duration in terms of persistence. "That something persists in chance—this is what enduring means. The identity that results from this is therefore no longer a logical identity but precisely that of a temporal totality."

From which, a one-time occurrence of the theatrical phenomenon as a whole, in its finite nature that is a part of persistence, duration, therefore, will always be the same and the other, given that the duration articulates the simultaneousness of the same and the different in each occurrence of the phenomenon. It is important to note that this dynamic, however, does not guarantee, in accordance with Husserl (1984/2012), a break in the linear, irreversible perspective of time. This supposition is only possible under Gadamer (1975), when faced with the possibility of suspending the subject in relation to the temporal experience: the experience in and of itself. In short, even in accordance with Gadamer, time does not cease to be ontically irreversible and linear, but rather the experience of time and affective-cognitive meaning created due to the aesthetic-temporal confrontation allows for the assumption of a cyclical dimension of time, even if time as a phenomenon does not effectively undergo any constituent alteration.

An effective contribution from the idea of duration within Husserl (1984/2012) to the notion of *experience in aesthetic configuration* in accordance with Gadamer (1975) can be found in the attribution of a longitudinal intent to "now;" this "now," under Husserl, is not a one-time instant; in the case of sound, for example, the same "now" is made up of the exact instant of the emission of the sound, the retention of the sound, and the protension of the imminent phase.

In this sense, temporality in the festive theatrical manifestation carries a new dimension: the one-time event of the phenomenon (a specific public presentation within a season of presentations) as a unit of duration. This unit of duration will therefore be the act that allows for the perception of persistence of the theatrical phenomenon. At the same time, this one-time event does not cease to represent a flow of linear, irreversible time in daily life and the operation based on new possibilities to allow for the "best performance possible" of the show as a whole; from this point, the theatrical performance could claim its right to tend toward an end within Heidegger's (1927/2009) conception of repetition. Nonetheless, one must note the difference between the perspective of duration of the one-time occurrence of the performance, if valid, and the duration of the theatrical phenomenon in itself.

In Husserl (1984/2012), however, even if taking this perspective of duration of the "now," there is a certain supremacy of the past, given that the present is, in Husserl's terms, the addition of pasts. The present-past relationship, in these terms, is a set of *dégradés* of retention. For Husserl, in the words of Ricoeur (1985, p. 30),

> But even now it can be asserted that the present and the recent past mutually belong to each other, and that retention is an enlarged present that ensures not only the continuity of time but the progressively attenuated diffusion of the intuitive character of the source-point to all that the present instant retains in itself or under itself. The present is called a source-point (*Quellpunkt*) precisely because what runs off from it "still" belongs to it. Beginning is beginning to continue. The present itself is thus "a continuity, and one constantly expanding, a continuity of past." Each point of the duration is the source-point of a continuity of models of running-off and the accumulation of all these enduring points forms the continuity of the whole process.

Duration as persistence establishes, upon forcing the cohabitation of temporal inclinations toward the future and the past in different dimensions of existence of the same phenomenon, the temporal confrontation, in the experience of aesthetic configuration, in accordance with Gadamer (1975), as inherent in the same process of construction of the work; this confrontation being prior to the confrontation that takes place in the presence of the user-subject.

The wholeness of the present as the *unit of duration* of the one-time theatrical event, with a set beginning and end, creates a distance from the possibility of having only one instant as the present. For Husserl (1984/2012), this differentiation evokes the Augustinian conception of a triple present (Ricoeur, 1985, pp. 23–44). For Saint Augustine, the present time is a distension: nothing begins or ends during the present, things are already as they are; it is the changes and restructuring that are the possibility of duration.

> What is now clear and unmistakable is that neither things past nor things future have any existence, and that it is inaccurate to say, "There are three tenses or times: past, present and future," though it might properly be Said, "There are three tenses or times: the present of past things, the present of present things, and the present of future things." These are three realities in the mind, but nowhere else as far as I can see, for the present of past things is memory, the present of present things is attention, and the present of future things is expectation. If we are allowed to put it that way, I do see three tenses or times, and admit that they are three. (Augustine, 1997, p. 300)

ETERNAL AND INFINITE—PERSPECTIVES OF TEMPORALITY IN THEATRICAL PHENOMENON

The finite dimension of the theatrical event in relation to an infinite time becomes an event in the Heidegger's (1927/2009) concept of repetition and Husserl's (1984/2012) concept of duration. Both were taken as parts of given dimensions of temporality in the theatrical phenomenon, as explained above. In a complementary sense, however, the concepts of experience and repetition according to Gadamer (1975) were associated, in this text, with the possibility of persistence and the constitution of a cyclical time in the theatrical phenomenon, which creates another perspective, in a certain sense, different from the finite-infinite duo.

Assuming Gadamer's (1975) repetition, as one of the facets of the artistic phenomenon, seeks to maintain the "presents," in the suspension of the experience of time by the ordinary subject, one must ask if the time of the experience is a continuous flow, in the sense of occurring within innumerable finite moments with a certain infinity, or if the possibility of repetition of the suspension is proof of a static time and therefore eternal. Initially, thoughts might tend toward the simultaneousness of the infinite and the eternal in processes of creation and artistic enjoyment, given what has been presented, as long as the work is taken as a *configuration*, to use Gadamer's terms, and carries—as a phenomenon, single event, and unit of duration—different dimensions of temporality.

> I call this development, in which human play finds its true perfection in being art, "the transformation into structure." Only through this development does play acquire its ideality, so that it can be intended and understood as play. Only now does it emerge as detached from the representing activity of the players and consist in the pure appearance of what they are playing. As such the play—even the unforeseen elements of improvisation—is fundamentally repeatable and hence permanent. It has the character of a work, of an ergon and not only of energeia. In this sense I call it a structure. (p. 99)

In this sense, the present is not an ephemeral moment but rather the wholeness of time. One must pay close attention, however, to the fact that the suspension of the temporal experience, responsible for the possibility of this wholeness, only exists between two instants that are absolutely rooted in the "natural course" of time, that which tends toward the infinite. From this perspective, eternity is configured as merely a metaphor for the very possibility of an experience of eternity; as if an aesthetic experience could maintain the subject suspended in daily life in the firmness of their existence. In being only metaphorical, the subject returns to ordinary life and upon doing so is faced with the true experience: immediacy and permanence. What remains is time. Nonetheless, the immediate nature of Gadamer's (1975) experience is characterized by the suspension of the temporal experience, and here, maybe, may find the possibility for a metaphor of the eternal.

In the immediacy of the true experience, therefore, past and future would not be "beings" of time but rather qualities of this metaphor for eternity that exists as a whole present; the present becomes an indivisible unit. The metaphor for eternity, in these terms, does not carry a measurement of time and will always be threatened by the finite nature of the event: the being in true aesthetic experience, thus, does not feel obliged to undertake the intellectual activities of comparison and the pragmatic activities of measurement.

> The eternal present does not appear to be a purely positive notion except by reason of its homonymy with the present that passes. To say that it is eternal, we must deny that it is the passive and active transit from the future toward the past. It is still insofar as it is not a present that is "passed through." Eternity is also conceived of negatively, as that which does not include time, as that which is not temporal. In this sense, there is a double negation: I must be able to deny the features of my experience of time in order to perceive this experience as a lack with respect to that which denies it. It is this double and mutual negation whereby eternity is the other time that, more than anything else, intensifies the experience of time. (Ricoeur, 1984, p. 236, n. 35)

Artistic (theatrical) creation, therefore, is only possible when, instead of removing time through the metaphor of eternity, it is elevated to the highest level of experience. The idea of artistic experience suspending daily reality (Gadamer, 1975) is proof of a trace of eternity in the experience of things, but does not create eternity, not were it able to do so. In this sense, if Heidegger's (1927/2009) repetition displaces the focus to the parts in order to satisfy the finite nature of the whole, the immediacy of Gadamer's experiences seeks the permanence of the whole as an indivisible unit.

It is up to the subject to make the effort to make the things in the consistency of this indivisible unit present, without interruption. In a certain sense, the metaphor of eternity requires that things do not elapse in time and that the subject alters (whenever necessary) the nature of their relation with things, without making them into something else or letting them remain the same;

> Nonetheless, when a true account is given of past events, what is brought forth from the memory is not the events themselves, which have passed away, but words formed from images of those events which as they happened and went on their way left some kind of traces in the mind through the medium of the senses. This is the case with my childhood, which no longer exists: it belongs to past time which exists no longer, but when I recall it and tell the story I contemplate the image of it which is still in my memory. (Augustine, 1997, p. 299)

Conceiving the change in the nature of relations with things assumes that words, images, and vestiges are aspects of their reference phenomenon and not other things that remind the subject of the item. Phenomenon (artistic), from this perspective, only exist when words, images and vestiges, expectations, possibilities, meaning that they only exist when symbolically acted on by the subject; otherwise, they are-no-longer or have-not-yet-been and therefore cannot belong to the aesthetic configuration; it is in the present, based on the past and the future as qualities of the metaphor for eternity, that the change in the nature of relations with things operates. Material, thus, undergoes turmoil and changes, and the result already existed in the material.

The possibility of the suspension of the temporal experience, in this case, is not pulled from an external dimension to an internal dimension; it already existed in the ritualistic theatrical phenomenon and is revealed in the temporal confrontation in its different dimension and the closing of the cycle of creation upon encountering the user; once again, the performance reveals the whole of existence of the temporal being as a being-to-death.

Therefore, the flow of time does not matter in the experience of the metaphor of the eternal in experiencing the artistic configuration; 1 minute or 100 minutes do not determine the meaning of the suspension of time; time ceases to exist as a concern, as a measurement. What does matter, however, is the experience of the change in the nature of relationships with things, as the maintenance of the "now" in a unit of duration; a rare experience of not existing in time. Once again, however, the subject may only desire to experience not being a part of time when they know they are a part of time, after recognizing the finite nature of things.

> In eternity nothing passes, for the whole is present, whereas time cannot be present all at once. Can they not see that whatever is past has been pushed out of the way by what was future, and all the future follows on the heels of

> the past, and the whole of both past and future flows forth from who is always present, and is by him created? Who shall take hold of the human heart, to make it stand still and see how eternity, which stands firm, has neither future nor past, but ordains future and past times? (Augustine, 1997, p. 294)

This, however, should not lead to the confusion of the true possibility of divine eternity, as stated by Saint Augustine (1997), with the metaphor for eternity in aesthetic experience. While eternity, in philosophy, comes before time, in experience, the eternal (metaphor) only exists to the extent that it is the other of time. Both Augustine and Gadamer (1975), however, see a need for the simultaneousness of the eternal, in the suspension of the temporal experience. Eternity and its metaphor are experienced in analogy with the constant simultaneousness of facts and things. What sets them apart is the possibility for change. In the metaphor of eternity, consistency requires change in the nature of relationships with things and in a certain sense changes in the nature of things; in Augustine's eternity, consistency means immortality, given that time is what causes things to get old. This statement coincides with the analysis by Ricoeur (1985, pp. 12–22) of the relationship established by Aristotle between movement and time.

> As if carried along by the power of the words themselves, Aristotle admits that we can say that "a thing, then, will be affected by time" and He accepts the saying that "time wastes things away, that all things grow old through time, and that people forget owing to the lapse of time." (Ricoeur, 1985, p. 17)

To Augustine (1997),

> This does not mean that one thing was said, and then, when that was finished, another thing, so that everything could be mentioned in succession; no, all things are uttered simultaneously in one eternal speaking. Were this not so, time and change would come into it, and there would be neither true eternity nor true immortality. (p. 290)

From this perspective, those who experience the metaphor of eternity know that they will return to the ordinary flow of time, thereby proving to be a being-to-death, or a finite being (Heidegger, 1927/2009). It is the confrontation of return that, in returning to the self, as a temporal being, ensures that the subject cannot be the same. Leaving-the-self and returning-to-the-self, in Gadamer's (1975) terms, given the metaphor for the eternal experienced in the game of art, means allowing oneself to transit through and to ensure the coexistence of different dimensions of temporality and of transgressions through their assumed nullification. In short, aesthetic experience, by requiring permanence, pushes the subject to return to himself. This is not remaining in the phenomenon but rather that which may remain from the mutual affective relationship between phenomenon and subject in the subject.

Theatrical phenomenon therefore carries a dual experience of suspension in the relation of time: festive repetition allows for the displacement of the experience of linear time to cyclical time and a metaphor of eternity of the immediacy of experience in the feeling of no-time, both referring to the ordinary linear time of Heidegger's (1927/2009) being-in-the-world, where something may truly remain. Only because of the units of duration that repeat throughout the existence of theatrical phenomenon that one may observe and experience any form of temporal confrontation.

FINAL CONSIDERATIONS

The case of the subject-time relationship based on the performing arts, aside from allowing for considerations of the nature of the work of performing art and the subject creator's modes for action, has become emblematic of the understanding that time, linearly and chronologically irreversible, may be experienced by the subject as something cyclical, reversible, and able to be controlled and operated. The meaning made by the subject of their experience in time is therefore simultaneous to the affective-cognitive meaning that the subject builds regarding time. Furthermore, as explained above, there are certain creative contexts that may lead the subject to assume that there is no time and maybe desire time. Nonetheless, it is important to note that the experiences that the subject has of time directly remind the subject of the obligatory linear organization of time.

Evidently, all experiences that a subject has in time and with regard to time are crucial for the constitution and reorganization of their modes for future action, given the resistances that the subject may face from the world. Furthermore, once again, the performing arts are seen by the subject as constituency in the temporal confrontation; however, just like ordinary life, it is not a primordial (conscious) pretension of the performing arts, and the subject as a whole, to experience the reversal of time, but rather to act on the transformation of relationships and the nature of things in order to satisfy desires and expectations created throughout the course of life; and it is in doing so that the subject reveals his existence in time and thus experiences time as external to the self.

Repetition as a temporal category, in this sense, does not serve the subject in merely altering something in what has been maintained, but before anything else, the subject may recognize something that is maintained which they seek to alter. In doing so, not only will the thing be altered but also the very potential for symbolic action on the world.

Thus, the subject acts symbolically and continuously in an attempt to make things present, without interruption, making the present, even if felt as ephemeral, into an indivisible unit. The subject experiences his actions as units of *duration*, tied to other durations that guarantee coherence in the experience that the subject has of his place in the world—*persistence.*

Persistence, however, even if it reveals time to the subject, does not create time for the subject, but rather gives it meaning. It is from this perspective that we see the performing arts as crucial for the understanding of human life in temporality.

ACKNOWLEDGMENTS

This chapter is part of the doctoral thesis of the first author, guided by the second author; supported by the National Council for Scientific and Technological Development, Brazil (CNPq).

NOTES

1. On the relation between the real work and the original work, see Gadamer (1975), p. 119 forward.
2. For more on temporality and the ritualistic character of theater, see Gadamer (1986).
3. On the historic path of the word experience, its assumption as a concept as of Dilthey, the implication of intentionality for a concept as of Husserl, and its association with the art experience and as an alegory, see Gadamer, 1975, pp. 55–72.
4. Integrating oneself with all of daily life guarantees a different experience than any other adventure. In the latter case, even if the conditions and committements of daily life are suspended, it is clear that there should be no attempt to integrate the two—adventure and ordinary life—given the serious risks in maintaining the latter. (Gadamer, 1975, pp. 60–63).
5. Even though Gadamer does point out the relationship between creation and forgetting in a number of exerpts in his work and approaches the past from Heidegger's point of view, as in the 1965 text, *The Continuity of History and the Instant of Existence*, [QA: Please provide a reference for this and cite a year here.]he does not effectively dedicate any writing to the theme, at least in the works to which we have access. A possible explanation for this question seems to be in the approximation between Gadamer and Walter Benjamin with regard to the interrelation between alegory and creation after the baroque period, passed on the perspective of the past taken on by Heidegger. Nonetheless, given the proposed focus of this text, we will leave this effort to a future initiative. On the relationship between forgetting and creation, see Benjamin (1994, 2011), Gagnebin (1999), and Sampaio (2011).

REFERENCES

Augustine (1997). *The confessions.* Hyde Park, New York, NY: New City.

Benjamin, W. (1994). *Obras escolhidas—I. Magia e técnica, arte e política* (S. P. Rouanet, Trans.). São Paulo: Brasiliense.

Benjamin, W. (1997). *Obras escolhidas – II. Rua de MÃO única* (R. Rodrigues, T. Filho, J. Carlos, & M. Barbosa, Trans.). São Paulo: Brasiliense.

Benjamin, W. (2011). *Origem do drama alemão.* Belo Horizonte, Brazil: Autêntica Editora.

Berger, P. L., & Luckmann, T. (2005). *Modernidade, pluralismo e crise de sentido. A orientação do homem moderno* (E. Orth, Trans.). Petrópolis, Brazil: Vozes.

Bergson, H. (2006). *Duração e simultaneidade: A propósito da teoria de Einstein.* São Paulo, Brazil: Martins Fontes.

Boesch, E. E. (1991). *Symbolic action theory and cultural psychology.* Berlin, Germany/New York, NY: Heidelberg/Springer.

Boesch, E. E. (1993). The sound of the violin. *Reveu Suisse de Psychologie, 52*(2), 70–81.

Deleuze, G. (1999). *Bergsonismo.* (L. B. L. Orlandi, Trans.). São Paulo, Brazil: Ed. 34.

Gadamer, H.-G. (1975). *Truth and method.* London, UK: Sheed & Ward.

Gadamer, H.-G. (1986). The festive character of the theater. In R. Bernasconi (Ed.), *The relevance of the beautiful and other essays* (pp. 57–65). Cambridge, UK: Cambridge University Press.

Gadamer, H.-G. (2011). *Verdade e método II – Complementos e índices.* Petrópolis, Brazil: Vozes.

Gagnebin, J. M. (1999). *História e narração em Walter Benjamin.* São Paulo, Brazil: Perspectiva.

Heidegger, M. (2009). *Ser e tempo.* Petrópolis, Brazil: Vozes. (Original work published 1927)

Husserl, E. (2012). Investigações Lógicas. Investigações para a fenomelogia e a teoria do conhecimento. Rio de Janeiro: Forense universitária. (Original work published 1984)

Kant, I. (2010). Crítica da Razão Pura. Tradução de Manuela Pinto dos Santos e Alexandre Fradique Morujão. 7. ed. Lisboa: Fundação Calouste Gulbenkian. (Original work published 1971)

Kierkegaad, S. (1986). Entweder/Oder. u. hg. v. E. Hirsch, H. Gerdes u. H.M. Junghans 2.Teil, in Gesammelte Werke, Düsseldorf/Köln.

Ricoeur, P. (1984). *Time and narrative. Vol. 1.* Chicago, IL: University of Chicago Press.

Ricoeur, P. (1985). *Time and narrative. Vol. 3.* Chicago, IL: University of Chicago Press.

Sampaio, J. C. C. (2011). *Dramaturgias consensuais: A interação verbal no ato criativo.* Master's thesis, Instituto de Psicologia, Universidade de São Paulo, Brazil.

Simão, L. M. (2007) Why otherness in the research domain of semiotic-cultural constructivism? In L. M. Simão & J. Valsiner (Eds.), *Otherness in question: Labyrinths of the self* (pp. 11–35). Charlotte, NC: Information Age.

Simão, L. M. (2009). *Alterity and temporality in the semiotic-cultural constructivism in psychology.* Research Project (209-2012) granted by The Brazilian Council for the Scientific and Technological Development (CNPq).

Simão, L. M. (2010). *Ensaios dialógicos: Compartilhamento e diferença nas relações eu-outro.* São Paulo, Brazil: Editora Hucitec.

Valsiner, J. (1998). *The guinded mind: A sociogenetic approach to personality.* Cambridge, MA: Harvard University Press.

Valsiner, J. (2007). Human development as migration—Striving toward the unknown. In L. M. Simão & J. Valsiner (Eds.), *Otherness in question: Labyrinths of the self* (pp. 348–378). Charlotte, NC: Information Age.

CHAPTER 17

BLACK GOD, WHITE DEVIL AND BEHIND THE SUN

Destinies in Modern Brazilian Cinema

Renato Tardivo

To Ismail Xavier

The reflection that follows stems from the research group in art psychology developed by João A. Frayze-Pereira and his advisees in the Department of Social Psychology at the Institute of Psychology of the University of São Paulo, Brazil. According to Frayze-Pereira, the interflow between the arts and psychology is not a recent one; the approximation between these areas predates the actual advent of psychology as an autonomous discipline (Frayze-Pereira, 1994). Devised in the 18th century by Baumgarten, aesthetics was based on the idea that beauty amounted to "a kind of sensitive knowledge, misleading and inferior to rational knowledge." A while later, however, with Kant's philosophy, "the subject of the beautiful will turn into the subject of the 'aesthetic experience.'" From this point on, aesthetics distances itself from the "metaphysical domain to near the experimental and psychological domain." Thus, even if psychology had not

Temporality: Culture in the Flow of Human Experience, pp. 413–427

been conceived as a separate discipline, aesthetics had already conversed with the "psychology that was yet to come," since it brings to the fore the subject of the relationship between the work of art and the spectator (Frayze-Pereira, 1994).

This reflection will be based on the relationship between work and spectator. In this regard, it is worth saying that the stance of the researcher, as a reader of works, raises the subject of alterity, hereby understood, following Merleau-Ponty (2009, p. 187), who thinks of the Being as "that which demands creation of us so that we are able to experience it." From this perspective, I will perform a comparative analysis between the films *Black God, White Devil*, a preeminent work in Brazilian modern cinema, directed by Glauber Rocha (1964), and *Behind the Sun* (2001), by Walter Salles, attending to the temporality contained in each of the films and to some historical, aesthetic, and psychological implications.

In spite of belonging to different historical moments and of referring to different times, *Black God, White Devil* is from 1964 and takes place in the 1960s; *Behind the Sun* is from 2001, but takes place in 1910—both films address similar subjects. Although the first focuses on the oppression implied by power relations in the society of the time and the latter addresses the dispute between two families entrapped in tradition, both question the possibilities and limits for the countryman's destiny. Evidently, for historical and aesthetic matters, the primary reference is Glauber Rocha, that is, Walter Salles was certainly aware of *Black God, White Devil* when he made his film. However, while not disregarding this fact, our purpose is not to assess to what extent *Behind the Sun* draws on elements present in *Black God, White Devil* in a vertical relationship, but rather to comparatively assess some aspects of both films, here understood as autonomous works.

The method for the reading and analysis of the works is inspired in "The Moses of Michelangelo" by Sigmund Freud (1914/2007a), since, according to Paul Ricoeur's commentary (as cited in Frayze-Pereira, 2010, p. 79) on Freud's essay, "The interpretation of the work is made in the manner of a dream interpretation, *stemming from detail*" (emphasis added) and is concerned with historical and aesthetic dimensions. It is in this sense that by delineating a field of activity in art psychology, a field to which this reflection belongs, Frayze-Pereira (1994) postulates that "the openness of the social psychologist towards art will mostly depend on his disposition, as an art spectator, to enter this abyssal field ... at the risk of vertigo and the loss of fixed points." Thus, the spectator, "by opening up to the field of the works," will have to deal with "transferential issues" and "consequently, to implicate himself."

The End in the Beginning—Backwards

Black God, White Devil (1964), by Glauber Rocha, is one of the most important films of modern Brazilian cinema. The work, which belongs to the movement that became known as "Cinema Novo," incorporates sophisticated work within the scope of cinematic language to the political motivations of the time. Glauber has coined this way of filming as "aesthetics of hunger," whose "characteristic lies in a style of filmmaking that turns lack of resources into a trigger rather than a barrier, creating a language that attends to the filmmaker's close relationship with social reality and with its own precarious condition within it" (Xavier, 2009, p. 25).[1] However, this characteristic does not imply lack of care toward the work. In this regard, the opening sequence—a traveling shot that scans the backlands from up above—confronts the viewer with signs designed by Lygia Pape and a soundtrack by Villa-Lobos, renowned Brazilian artists. And what is most pressing, as pointed out by Ismail Xavier, one of the leading researchers of Glauber Rocha's work:

> If his better known slogan was "camera in hand, idea in head," the rejection of the *a priori* value of great equipment was always stressed, as well as the defense of an "authorial cinema" as opposed to the immediate consumption object in the cinematic trade routine. The idea is not to go about filming, which results in a lot of nonsense. It is a creative effort that is quite demanding, for aesthetics are really made effective when one has ideas in hand and the cinema in one's head. When separated, they result in onscreen Sociology, which is neither good sociology nor good cinema. It is about hunger, but it is not aesthetics. (Xavier, 2009, p. 25)

There is thus a refinement of aesthetic order: polyphonic language, complex *mis-en-scène*, not to mention the radical editing work and the reflection it evokes. The film is constructed "by the clash between the singer's voice, the camera style, Sebastião and Corisco's prophecy, Antônio das Mortes' speech, the herdsman's gestures, Manuel and Rosa's journey" (Xavier, 2009, pp. 20–21).

After the initial travelling shot, there is a close-up shot of a (still decomposing) animal carcass. Then follows a close-up of Manuel, the herdsman who is the leading character of the plot. A certain parallel between "countryman" and "decomposing animal" is established right away. Manuel mounts his horse and rides through the backlands. We realize, as Xavier (1983) notes, that the shots present "sky" and "land" without making it possible to glimpse a "horizon" that integrates the pairs of opposites. This aspect of *Black God, White Devil*'s black and white photography may be interpreted as a symbol of the countryman's lack of perspective, a captivity that is also observed in a few framings in which Manuel is *imprisoned* between two vertical poles (tree trunks, fence posts, etc.).

The herdsman's oppression is lived to its limit before Colonel Moraes, who blurts out, "The law is on my side." Manuel, who loses all his cattle in the negotiation, gets into an altercation with the colonel and, as a consequence, his gunmen kill the herdsman's mother. Devoid of anything else, "only destiny," Manuel and Rosa, his wife, join Santo Sebastião, whom Manuel had already seen and been dazzled by. Rosa takes on a more skeptical stance, while Manuel believes in the miracle professed by Sebastião: "land, abundance, happiness." "Across" the backlands, there would be an island. With this belief in mind, Santo Sebastião pronounces the film's catchphrase: "And the backlands will become sea, and the sea will become backlands."

Manuel's frailty, present since the beginning through the parallel with the animal's carcass, is now evidenced by his suggestibility to Santo Sebastião's prophecies. The herdsman practically leaves Rosa to join the saint. The woman, more thoughtful, responds to Santo Sebastião's prophecies with the question, "Why should one run away and find disgrace in hope?" Meanwhile, Antônio das Mortes ("the henchmen killer") appears; he receives from a priest a substantial amount of money for Santo Sebastião's life. He also says that Sebastião has a pact with God and with the Devil (literally quoting the film title). Surprisingly, though, Rosa is the one who, before the killer, takes Santo Sebastião's life.

Let us consider this. Manuel, when oppressed by colonel Moraes, kills him. Analogously, Rosa, once disgraced by the excess of hope in Santo Sebastião's prophecies and sacrifices, also kills her oppressor. If the colonel's death triggers the herdsman's mother's death and frees them for the search throughout the backlands, Santo Sebastião's death allows for the appearance of henchman Corisco, "Lampião's devil," "Lampião's spirit," who had been recently murdered and who, in the absence of Santo Sebastião in the second half of the film, will have the task of "fixing the backlands."[2] In this sense, Xavier (1983) alerts us to the near absence of action in the film: there is an accretion of tension, a discharge when it reaches its limit, a new accretion of tension. We will return to this characteristic later.

And "the story goes on in the trails of the backlands," chants the blind singer. From Corisco's, that is, the *cangaço's*, perspective, hunger is solved through violence: "not with the rosary . . . but with the rifle and the dagger." The henchman baptizes Manuel with a new name, Satanás, which marks an ideological inversion in the herdsman's perspective: first, religion (God); now, the war (Devil). They leave, accompanied by the women, Rosa and Dadá, and by gunmen, to "kill the government men," in order to solve the social problems that plague the Brazilian Northeast through violence and terror. As Sebastião, Corisco also says the catchphrase of the sea becoming backlands and vice-versa, and Manuel's (or rather Satanás') look indicates that he recognizes himself in it, that is, he recognizes himself in this prophecy, a sort of synthesis of the film:

> In *Black God, White Devil*, the double movement—criticism and incorporation of religion—defines, on the one hand, the necessary overcoming of the religious phase of rebellious consciousness to that of revolutionary lucidity. On the other hand, the *telos*—the Revolution—is only palpable if we take the mystical praxis of Sebastião and Corisco as *prefigurations* of the "great war without God and the Devil" announced by the metaphysics of Antônio das Mortes and the blind singer. One must assume an order of things in the backlands within which everything is a sign that the promise of salvation will be fulfilled. Glauber Rocha's choice is clear: the mediation through cordel literature[3] organizes the narrative, which speaks the language of prophecy. (Xavier, 2006, p. 139, emphasis in original)

The war that Antônio das Mortes refers to is, as seen above, "the great war without the blindness of God and the Devil"; that which will only be accomplished "after everyone dies." In this sense, the catchphrase, "the backlands will become sea and the sea will become the backlands"

> condensates a basic principle of the film: the recap of peasant revolts is there to teach us that the Revolution is a sure destiny in the backlands, not only because there are flagrant injustices in the present time, but above all because there is a continuity between the tradition of rebellion in the past and future redemption through violence. (Xavier, 2006, p. 119)

In the end, Antônio das Mortes pulls "the knot tightly," so that the group of henchmen has nowhere to escape to. Manuel and Rosa are joined in their desire to have a child; Antônio das Mortes finally arrives: "They've killed Corisco, they've shot Dadá," chants the singer. Manuel and Rosa leave promptly; the woman falls down, Manuel goes on; and the singer: "The land belongs to man, not to God nor the Devil." The traveling shot that shows him running through the backlands is turned into an aerial shot of the sea. That is, the initial panorama is recuperated, but backwards. Waves cut the frame, alluding to the very idea of transformation.

The sea goes to the backlands.

To Forget Is to Create

Behind the Sun (2001), a contemporary film by the Brazilian director Walter Salles, also takes place in the backlands of Brazil, although it is based on the eponymous book by Ismail Kadaré, whose plot unfolds in cold

Albania. Introducing the initial scene, there is the subtitle—"Backlands of Brazil, 1910"—and then a long frontal shot of a boy who, entirely covered by shadows, walks through the backlands while the light evokes dawn. He says his name is Pacu, but since it is a new name, "he hasn't gotten used to it yet." He walks while trying to remember a story: "sometimes I remember ... sometimes I forget." At this moment, there is a cut to the sequence shot of the boy still walking, but viewed from behind; that is, the camera takes on the perspective of what was left behind. This transition is suggestive, since it happens right when the boy refers to "another story," one that he remembers well: "I cannot get it out of my head ... it is my story, my brother's story and the story of a shirt blowing in the wind." The walk is interrupted to give way to a bloodstained shirt blowing in the wind. The trail has a hint of terror. The patriarch translates what the family's—the parents and two children: a young man and the boy from the previous sequence—eyes witness: "The blood is already turning yellow."

Behind the Sun, from Pacu's point of view, tells the story of an endless land dispute between two families: a war ruled by the Law of Talion, which tends to decimate them. The boy in the opening sequence is the youngest of the Breves family, on which the narrative focus concentrates. The shirt whose blood is turning to yellow belongs to the older brother, recently murdered by the rival family. When the blood turns yellow, Tonho, the other brother, will be responsible for "avenging the deceased's blood." Incidentally, the portraits of the dead are lying in a hallway of the shadowy house, as if to haunt them. The film's photography draws attention to the contrast between light that originates from the outside, either from the scorching backlands sun or the heat that emanates from the cauldrons of jaggery[4] (produced by the Breves family), and the sparse light in the house interior, in the family interior—the death that does not cease.

Thus Tonho, whose posture has always denoted an antagonistic sensibility to the brutality of that war (a brutality that abounds in his father, although he has escaped death in spite, or because, of it), plays his designed role and claims his brother's blood. The traveling and reverse shot that shows him chasing the member of the rival family is an emblem of the power with which the archaic battle impacts the arid land that divides both families. Now it is Tonho's days that are counted. As the patriarch of the rival family claims, "Whenever the clock strikes plus one, plus one, plus one ... it will be telling you 'minus one, minus one, minus one.'"

The time that weights upon the backlands in *Behind the Sun* is the time that advances toward the past.

In the region known as "Riacho das Almas," the creek has dried and only the souls are left, as the boy wisely says. Incidentally, this line takes place at an important moment of the plot: the appearance of traveling circus performers—Salustiano and his goddaughter Clara. The artists sympathize with the boy's spontaneity, and the girl presents him with a book. Although he is illiterate, the boy can "read the images" on the book. Actually, the bloodstained shirt and the portraits of the dead on the house's walls are images that he has always had to read—the language that (dis)organizes life. This may be the film's paradox: life is subjected to death, for the living are subjected to the dead.

In "Civilization and its Discontents," Freud (1930/2007c), developing his notions on the death drive, proposes that, in order for civilization (culture, collectivity) to exist, individuals must renounce, on some level, instinctual satisfaction, subjecting it to cultural norms—the law. The unrestricted satisfaction of instinctual drives would bring disorder, chaos, and destruction, since it would also be an expression of Thanatos—the death drive. To Freud, dealing with these (unsatisfied and thus subjected to repression) drives inevitably implies some level of discontent. Hence individuals, if they are not able to be completely happy, avail themselves of culturally permitted resources to deal with this dissatisfaction: sublimation, religion, and such.

It just so happens that, in *Behind the Sun*, strict obedience to the law is what brings death and destruction. The language of a petrified past, which hangs in the air, in low light, in the blood on the shirt, is an emblem of the death drive. In this sense, the fact that up to a certain point in the narrative the boy does not even have a name is not arbitrary; it is an existence that tends to nonexistence. Resuming our reading of Freud, we can suggest that, in Salles' movie, the discontent is taken to its final consequences and destructiveness is culturally permitted, not because the law is violated but precisely because of the opposite. There is a perverted order of things.[5]

Thus, repetition is such that no one ever moves. In this sense, Costa (2002) writes about *Behind the Sun*:

> Cruelty is usually not a shapeless apparition, as in the scares of afternoon movies or in the metaphysical enormities *a la* Lovecraft. It is a capillary poison that invades the routines we call habit. We live in habits and, because we make a habit of life, we become puppets of the repetition compulsion. Life stuck to habit is surely efficient. But it is an efficacy of the mills, in which sugarcane goes in and bagasse comes out. The wheel of habit, created to deal with sameness, jams when facing diversity, it shatters and injures to death those who put it in motion.

The oxen in the mill, who eventually walk on their own; the swingset where the brothers play; and even the logic of revenge that is perpetuated are many of the elements that indicate repetition. Habit is corrosive. In the opposite direction, novelty finds its place in the image of the circus performers. Clara, the girl, introduces new images to the boy—a whole new world. The contact with this new universe makes him confront the logic of repetition and the accumulated and accumulating past. Through images of the bottom of the sea and of the mermaid, which is entangled with Clara herself, the boy discovers/creates a story, which he sometimes remembers, sometimes forgets. In order for a story to be forgotten, it must exist in the first place. To forget is to create.

Meanwhile, Tonho makes contact with the circus performers when, selling tapioca to the grocery store in the city with his father, he too is dazzled by the "mermaid." In the evening, breaking the habit of oppression and aridness, he sneaks out with the boy to view the spectacle. Clara breathes fire; as her name already indicates, she brings light. The boys are dazzled and, in the end, go meet the artists. The exchange of glances between Tonho and Clara denotes a mutual enchantment. In this moment, Salustiano baptizes the boy with the name Pacu. Back home, the father, a living ghost, spies on them. Tonho, for the first time, does not subject to his law. In return, he is beaten with a whip.

Tonho, in search of life, leaves into exile. He joins the walking circus performers and accompanies them to another city. If the boy creates his story through the images on the book, Tonho, who is older and has no time to lose, lives his own book. He and the girl play on a rope (a situation analogous to the one on the swingset at his home), but unlike the earlier circularity, time passes, so much so that he starts to spin the girl when the day is clear and only stops at nightfall. Meanwhile, Clara also lives a kind of captivity, having to work with, and for, her uncle and godfather.

But Tonho eventually returns to his family. When he appears in the land of the Breves, father, mother, and brother, who were working at the mill, stop for a moment; they seem pleased. Tonho's return to the circularity of the mill is, to some extent, a return to death.

Tonho's time is up. The member of the rival family goes after him. That same evening, Clara arrives. She brings rain to the backlands. Tonho and Clara make love sheltered near the family house. She rips the black ribbon—his death certificate—from his arm. The rival, hindered by the rain, gets closer. Later, Pacu, seeing Tonho's ribbon on the floor, is led to believe that his brother has been killed. But Tonho actually sleeps, after making love to the girl. Clara also leaves (as he had previously done himself), but says she is waiting for him. Pacu, thinking that his brother died, takes his place and he ties the black ribbon around his arm. He goes out walking.

The film returns to the beginning. The initial long shot is repeated. The whole story was a flashback: Pacu, who lived it, was able to tell it. The circular logic has been broken. "April,",which means "youth, exuberance," here marks a new, and painful, cycle. The young man of the rival family, whose glasses have been broken and who is now as blind as his grandfather, kills the boy (by mistake). Tonho does not arrive in time. His father, in desperation, demands that Tonho avenge his blood immediately. Tonho, sad but resolute, enters the house and leaves it. But he leaves the gun. His father threatens to kill him, but he is detained by his wife, who tearfully says, "It is over." The couple embraces. Tonho, who had returned to die, does not stay to kill. The trail that links Riacho das Almas to the other villages and the city has a junction. When they would go to the city, they would take the left path. Now, however, Tonho takes the other path. He has tears, not blood, in his eyes.

There is a cut to a frontal shot of Tonho, from bottom to top. The viewer does not know what the character contemplates, but Tonho's look denotes something inaugural. There follows an open take, and sand can be seen around. Tonho continues his walk. This time without interruption, the camera follows him until he gets away. He is on a beach. When he touches the water, there is a close-up of his face and a cut to the final shot. Tonho's body, viewed from behind, divide the waters. Huge waves break in front of him.

The countryman goes to the sea.

Destinies

Let us now proceed to the comparative analysis of both films. From the perspective of cinematic devices, *Black God, White Devil* is more complex. The lack of material resources operated with aesthetic refinement results in a more *difficult* cinema. The interference of editing, the plethora of sound codes, historical references, the notion of participant camera, the language of trance, all of these are elements that point to this direction. *Behind the Sun*, on the other hand, is a slower, contemplative film: "Characters inhabit a universe where 'one speaks with one's mouth shut' and acts with minimal feelings and gestures. The camera illuminates this scarceness at its maximum and makes us see the 'more' that emerges from the 'less'" (Costa, 2002).

In Glauber Rocha's film, ruled by the logic of the prophecy, there is, "generally speaking, what I would now call a naïveté regarding the historical process. There was an idea that the desired, hoped-for transformations were something that could happen with a certain brevity" (Xavier, 2009, p. 72). Within this concept, there is a Marxist[6] definition of ideology—ideas detached from reality that aim to explain it in order to meet specific interests so as to reinforce power relations. As a criticism to the naturalization of such interests, Marx and Engels' historical materialism considers reality from a historical point of view (praxis) and as such, there is "the possibility that different social actors, who are able to perceive the asymmetries existing in the system itself, act and think so as to revise the very conditions of life" (Bosi, 2010, pp. 66–67).

As seen in the beginning, there is practically no action in *Black God, White Devil*; tension is accumulated and discharged, when it reaches a limit; continuity between the rebellious past and future redemption through violence (Xavier, 1983) is an emblem of such. If we return to the Freudian notions (1930/2007c) of "Civilization and its Discontents," we could then say *Black God, White Devil* is made of a succession of catharses, which is observed in the interpretation of "revolution" as the "backland's sure destiny," in the shape of

> [Glauber Rocha's] totalizing impulse when representing moments of rupture where society lives the drama of change or conservation, a "moment of truth" after which one cannot go back to being entirely what one was. In his cinema, this "moment of truth" has its triggering element in violence ... it is through Antônio das Mortes' violence that ruptures are achieved and the Revolution is announced. (Xavier, 2006, pp. 119–120)

Freud (1930/2007c) cites religion as one of the resources that we use to deal with—if not to eliminate—the discontent in civilization. Manuel joins Santo Sebastião in order to break away from the backlands' extremely oppressive environment. But the strict adherence to Sebastião seems to contribute even further to the discharge of the instinctual drive—including its destructive element—than to its displacement or softening. In this sense, it is Corisco who, in the second part of the film, presents the possibility, through the use of violence, to break away from the oppressive draught cycle of the Northeast. Finally, again in a violent context, ruptures are processed through the "great war without God and the Devil," in the figure of Antonio das Mortes.[7]

In *Behind the Sun* though, (trivialized) violence is not liberating; it is imprisoning, and the reflection of its repetition is the means toward destiny. The traveling circus performers bring a breath of fresh air to the backlands. Once again in the company of Freud (1930/2007c), the circus performers hint at the possibility of, through the arts, sublimating the instinctual drive to some extent, softening this environment of discontent. And not only do they diminish the discontent, the contact between the Breves brothers with the circus performers—"outsiders," "foreigners"—is what allows for the inscription of new realities, of a new story.

Thus, violence, in Salles' film, is the target of criticism, not the means toward disruption and change. We may truthfully point out certain parallels between the films in the traveling shots of the chasing scenes that result in death. The meaning acquired by these events in each of the scripts, however, is divergent. One can even consider the meaning of blindness in both films; in *Black God, White Devil*, through the blind singer, blindness is an organizer of the narrative, whose destiny is achieved through violence; in *Behind the Sun*, through the old patriarch and his grandson, who kills the boy by mistake, blindness is also associated to violence; but violence, in Salles' movie, is the target of criticism, not a mean toward a rupture: "Among the blind, a one-eyed man is thought to be crazy by everyone else," says the boy.[8]

The Breves brothers, in Salles' film, vindicate well-being:

> Because we change, we are always choosing and producing different futures. Tradition is only the world's image according to our ancestors' force and talent. To fix it into a skeleton of rules and principles is to subtract life from its creative impulse. The Good in life is always "on the road,"[9] always in passing, always in the transitional area between "not anymore" and "not yet." (Costa, 2002)

Behind the Sun addresses existential issues and has a timeless atmosphere[10] (the very possibility of adapting a story that takes place in Albania to the Brazilian backlands stress this idea). The film is not ruled by the logic of prophecy, but rather by the reframing of experience. It is the childlike naïveté and creativity that trigger the rupture: the petrified past that weighs upon the backlands is reconstructed by the boy. He rewrites history. The outcome is uncertain.

In several moments of *Behind the Sun*, however, the metaphor of water/sea in the backlands also appears: in the rain (and its absence), the mermaid story, the tears, and finally, the sea. But "a blood stain does not wash out," and must therefore be felt and mourned. Hence the water being, as in a baptism, an emblem of birth in Salles' film.

In *Black God, White Devil*, the concern is different. The gaze turns to the future: "The 'backlands that will become sea' is the central metaphor ... because there is a teleology in the film that—in accordance to the prophecy—reiterates a final term that seems to be already given, inevitable, just like destiny, and not as freedom" (Xavier, 2009, p. 21). Thus, the final shot, which dialogues in reverse to the initial one, is another catharsis, perhaps the greatest of them, in the sense that it encompasses the others—an expression, as we have seen, of Glauber Rocha's totalizing impulse.

In Walter Salles' film, there is not a belief in that the desired transformations can happen promptly. There is also a communication between beginning and end, but the future, a still (and always) open rebirth, is only specifically glimpsed after the ultimate shattering—the boy's death. Thus, *Behind the Sun* may be more in agreement with Adornian ideological critique (Adorno, 1983), according to which ideologies do not lose strength if they are only revealed, for reality and ideology run toward each other; that is, ideology, from this point of view, is not only an envelope but the very image of the world. In this case, Walter Salles seems to be asking, "How does one break the fatal cycle of a history that has been naturalized, that has lost its human role, and of a nature that has been artificialized and become ghostly, unrecognizable and strange to the man who inhabits it?" (Matos, 1999, p. 159).

In Rocha, the sea is indeed the sure destiny in the backlands. In his polyphonic cinema, there is no truce in the path toward change. In *Behind the Sun*, the destiny is constructed through a series of transformations within the family and its generational confrontations. The film is therefore constituted through the reflection and reframing of the crossing—between the lack (of the backlands) and the plentifulness (of the sea).

Resignifying the Past From a Psychological Perspective

Black God, White Devil resorts to the language of trance in its teleological project. Glauber Rocha's totalizing impulse certainly seems to be in line with the propositions of Antonin Artaud (1993), especially his theory of the double and of inverted language. From this perspective, the issuer is simultaneously what he issues and the receiver of the message. In Glauber Rocha's film, these are the emblems of this perspective—the embodied lens that explores the visible and the invisible, the emphasis on close-ups and on the geography of the bodies, the distorted images.

Behind the Sun, by its turn, adopts a contemplative aesthetics that invites reflection, due to the resignification of the past. The film features a sense of time that is in line with the phenomenology of Merleau-Ponty (2006). According to the philosopher, time only exists when a subjectivity disrupts the plenitude of being *per se* and delineates a perspective. The cavity in which time is formed is the mark of this subjectivity. Past, future, and present are articulated in a movement of temporalization. Past and future are expressed in the present: the flow between the new and the old is fundamental to subjectivity which, by breaking away from the plenitude of the being *per se*, delineates a perspective. Thus, time flows by and into itself; and nothing could be more coherent with *Behind the Sun*'s structure, especially if we take into account the initial sequence, in which the boy starts to narrate the story which, flowing by and into itself, is resumed at the end of the film.

In *Black God, White Devil*, there is no space to think the past. It implies such an oppressive present that the change of the current reality is the priority, even if through the means of violence. In Salles' film, the discourse of older generations also seems to take this direction: the dead oppress the living, whose obligation is to reclaim their blood. However, as we have seen, the change derives precisely from breaking away from such discourse. Thus, the time that rules Glauber Rocha's polyphonic aesthetics is *a time that does not pass,* since there is no space to resiginify the past, there is only concern about the future. Polyphonic language is, when taken to its limits, paralyzing.

The time that does not pass also appears in *Behind the Sun* in the discourses and habits of the families. It just so happens that Salles' aesthetics, which uses Tonho's and Pacu's points of view as reference points, brings the new through the resignification of the old. In this sense, it is possible to draw an analogy between Merleau-Ponty's temporality and Freud's *après-coup* temporality, marked by a combination of times and their constant resignifications (Freud, 1915/2007b). That is to say, from a psychic point of view, the time can only pass if there is the possibility of making contact with what was lived, even if through substitute memories—the screen-memory. In this regard, it is clear that Pacu's fixation on the mermaid figure in the story that he (re)creates is related to his attraction to Clara, his mermaid. Also, as we have seen, *Behind the Sun* is almost entirely narrated (lived) in *flashback*, after Pacu figured that Tonho had been assassinated (his trauma) and took hold of the black ribbon—taking his brother's place in *après-coup*. But the final trauma is the boy's death, *after* which Tonho finally breaks free and reaches the sea.

NOTES

1. (Xavier, 2009) refers to the book *Encontros—Ismail Xavier,* edited by Adilson Mendes, which compiles interviews given by the critic, professor and researcher of the University of São Paulo, Ismail Xavier. However, to indicate that the quotes are by Xavier, not by Mendes referring to the critic and researcher, which would justify the inclusion of "(Xavier as cited in Mendes, 2009)," we have opted for indicating simply "(Xavier, 2009)" in the text. This situation will reoccur throughout this study.
2. *Henchmen* is our translation of cangaceiros, men of the cangaço movement. *Cangaço* was the name given to the movement in the Brazilian backlands (*sertão*) that began in the late 19th century and lasted until the first half of the 20th century, which sought to do justice to social inequalities through the use of violence. Its better-known leader was Virgulino Ferreira da Silva, aka "Lampião."
3. Traditional literature of the Brazilian Northeast, written in rhyme, often illustrated and sung with the accompaniment of a *viola caipira* (Brazilian 10-string steel acoustic guitar). Readings or declamations are usually enthusiastic.
4. Jaggery (*rapadura*) is a traditional Brazilian candy made of cane sugar.
5. We observe a similar situation in a study of *Lavoura Arcaica,* both the book (1975) and the film (2001), important works in the Brazilian arts scene, in which strict obedience to the law of the father takes discontent to its final consequences and validates the incest between siblings (Tardivo, 2012).

6. In his analysis of Glauber Rocha's film, Ismail Xavier (1983) highlights the influences of Marx and the Bible. Regarding the emphasis in the future and the affirmation of a practically sure destiny, Xavier also brings to the debate Sartre and his dialectics of human freedom.
7. Another emblem of Glauber Rocha's totalizing impulse and of the catharses that take place throughout *Black God, White Devil* is the kiss between Corisco and Rosa toward the end of the film, in a moment when Manuel and Dadá are not featured in the scene.
8. Altering the popular saying which goes: "Among the blind a one-eyed man is king."
9. Coincidentally, Walter Salles would come to direct On the Road (2012), an adaptation of Jack Kerouac's eponymous novel.
10. In a different context and with explicit historical references, Walter Salles (in collaboration with Daniela Thomas) would also question, in the film *Linha de Passe* (2008), the trivialized violence which marks an oppressive present, showing that attention must be given to the path to be taken (Tardivo, in press).

REFERENCES

Adorno, T. (1983). Ideologia. In T. Adorno & M. Korkheimer (Eds.), *Temas Básicos de Sociologia*. São Paulo, Brazil: Cultrix.

Artaud, A. (1993). *O teatro e seu duplo*. São Paulo, Brazil: Martins Fontes.

Bosi, A. (2010). *Ideologia e contraideologia*. São Paulo, Brazil: Companhia das Letras.

Costa, J. F. (2002, April 28). O último dom da vida (*Abril despedaçado*). Retrieved May 5, 2012, from http://jfreirecosta.sites.uol.com.br/artigos/artigos_html/dom.html

Frayze-Pereira, J. A. (2010). A alteridade da arte: Estética e psicologia. *Psicologia–USP. São Paulo, 5*(1/2), 35–60.

Freud, S. (2007a). El Moises de Miguel Angel. Buenos Aires, Argentina: Amorrortu. (Sigmund Freud, *Obras Completas*, vol. 13.) (Original work published 1914)

Freud, S. (2007b). La Transitoriedad. Buenos Aires, Argentina: Amorrortu. (Sigmund Freud, *Obras Completas*, vol. 14.) (Original work published 1915)

Freud, S. (2007c). El malestar en la cultura. Buenos Aires, Argentina: Amorrortu. (Sigmund Freud, *Obras Completas*, vol. 21.) (Original work published 1930)

Matos, O. (1999). *A escola de Frankfurt: Luzes e sombras do iluminismo*. São Paulo, Brazil: Moderna.

Merleau-Ponty, M. (2006). *Fenomenologia da percepção*. São Paulo, Brazil: Martins Fontes.

Merleau-Ponty, M. (2009). *O visível e o invisível*. São Paulo, Brazil: Perspectiva.

Rocha, G. (Director). (1964). *Deus e o diabo na terra do sol* São Paulo, Brazil: Versátil Home Vídeo.

Salles, W. (Director). (2001). *Abril despedaçado*. São Paulo, Brazil: Imagem Filmes.

Tardivo, R. (2012). *Porvir que vem antes de tudo—Literatura e cinema em* Lavoura Arcaica. Cotia, Brazil: Ateliê Editorial/Fapesp.

Tardivo, R. (in press). Invisible children behind the arrow: social humiliation, perversion and confusion of tongues in *Linha de passe*. In W. Salles & D. Thomas (Eds.), *Revista estudos e pesquisas em psicologia* (UERJ).

Xavier, I. (1983). *Sertão mar: Glauber Rocha e a estética da fome*. São Paulo, Brazil: Brasiliense/Embrafilme/MEC.

Xavier, I. (2006). *Cinema Brasileiro moderno*. São Paulo, Brazil: Paz e Terra.

Xavier, I. (2009). *Encontros–Ismail Xavier* (Ed., A. Mendes). Rio de Janeiro, Brazil: Azougue Editorial.

PART IV

TEMPORALITY AND ITS FUTURE CHALLENGES FOR PSYCHOLOGY

CHAPTER 18

TEMPORALITY AND THE CHALLENGE TO GENETIC CULTURAL PSYCHOLOGY

Cor Baerveldt

> I can't go back to yesterday because I was a different person then.
>
> —Lewis Carroll
> (*Alice's Adventures in Wonderland,* 1865/1998, p. 155)

Toward a Radically Genetic Psychology

Time as it relates to psychology and culture refers not just to our human experience of temporality but also to the understanding of psychology as genesis. The so-called *genetic method* in psychology is intrinsically connected with the pioneering work of such figures as James Mark Baldwin, Heinz Werner, Jean Piaget, and Lev Vygotsky, but has received inspiration from a long tradition of genetic, historical, and expressivist thought, through the writing of Giovanni Battista Vico, Johann Gottfried von Herder, Johann Wolfgang von Goethe, Georg Wilhelm Friedrich Hegel, Karl Marx, Friedrich Wilhelm Nietzsche, Martin Heidegger, Henri Bergson, and Ludwig Wittgenstein, to mention only some of the most influential thinkers who

Temporality: Culture in the Flow of Human Experience, pp. 431–461

each in their own way saw time as an immanent structure of life. To take a genetic approach in psychology is to understand "the psyche" or psychological "stuff" inherently in terms of time or temporality. As such, "genetic psychology" is not merely that subdiscipline of psychology that concerns itself with human development. Rather, it is a perspective on the psychological discipline as a whole, which holds that psychological phenomena, because of their inherently historical nature, require understanding in terms of their *genesis* or *becoming*. Both Kant and Goethe already saw a natural connection between the formative principles involved in cultural and artistic expression and the organism as the expression of a nonreducible whole. Yet, within developmental biology, the study of such processes as cell differentiation, morphogenesis, and metamorphosis would soon take a largely mechanistic route, while—perhaps with the notable exception of Cassirer's work on symbolic forms (Cassirer, 1923/1953, 1925/1955, 1929/1957)—this link would be largely forgotten as well within the cultural and human sciences. Yet, while developmental sociology or developmental anthropology would never really come off the ground, the aforementioned pioneers kept the genetic method alive as a small but significant undercurrent in psychology.

The question of genesis is also the question of *roots* and hence we could argue that a genuinely genetic perspective must be a *radical* perspective and perhaps that it poses a radical alternative to the psychological mainstream. Charles Taylor (1985) still considered genetic psychology's main rival to be traditional learning theory or the incremental view of learning. Now, more than a quarter of a century later, it seems that its main rival might rather be the kind of cognitive psychology that conflates the understanding of cognitive skills with the understanding of their *functional design*. Functional design (from Latin *de*, "away" and *signum* "a mark") is inherently *ahistorical*. In order to understand the working of a digital computer, for example, it is irrelevant how, exactly, the computer came about, how its hardware and software were built, or how the computer was historically conceived. The computer's history is not part of a description of its operations in terms of programmable functions. What matters, we could say, is just that *it works*. But precisely herein lies the problem with the computer as a model for understanding genuine cognition. Whether the computer "works" is determined entirely by the human observer, who evaluates its "output" in terms of a human frame of reference. Indeed, the computer becomes important for people *only* because it can be programmed in a way that is "semantically constrained." According to the biologist Humberto Maturana, we tend to understand the operations of a computer *allo-referentially*, that is, by reference to a domain of significance or a normative domain that essentially remains external to the operations of the computer itself. All functional accounts of cognition are allo-referential accounts in that sense and so are semantic accounts that understand the cognitive nature of a system's operations in terms of their *designation* of something external to them (Maturana, in Maturana & Varela, 1980, p. xiii).

Design or History?

A living systems could be understood in allo-referential terms as well, but doing so would obscure the principal feature of its organization that makes it a living system: living systems are systems that produce and maintain themselves and as such their principal organization needs to be understood as *self-referential*. Self-reference must therefore be a key notion for both developmental biology and any genuinely genetic or developmental psychology. In this chapter, I will be presenting an enactive genetic perspective on psychology that is based on the fundamental insight that only self-referential systems can be historical systems. By "historical," I do not simply refer to the idea that a system, organism, or human agent can be explained in terms of some causal sequence that essentially remains external to it, but rather to the idea that the *particular history* of a system is *constitutive* of the system. The history *of* a system cannot be the cause of that system, because it already implies the system whose history it is indeed in a self-referential way. History, as I will argue, implies *constitution* and not cause. Recognizing this constitutive relation is neither to imply some kind of essentialism, as if the system were already there before it engaged in history. Quite the contrary, it is the recognition that there is no system apart from its history, because the system *is* precisely its history. The nonfoundational view that a biological or psychological system is nothing else than its history of self-constitution is what I will call here the *radical genetic perspective*.

Self-referential biological systems are systems that by virtue of their peculiar circular organization enact their own domain of significance. They inhabit what Jakob von Uexküll (1982/1940) has called an *Umwelt* and what I will call here a *medium*. To understand the implications for the psychological study of human conduct, we must recognize that human domains of significance are inherently normative. By *normativity*, I refer to the idea that all genuinely psychological conduct is conduct in light of a sensitivity to what it means "to get it right." The radical claim in this regard is that, as a consequence of their own particular life history, humans acquire dynamical dispositions to see, feel, believe, think, and act in an irreducibly normative way, that is, in light of their developing sensitivity to what is proper or improper, right or wrong, honorable or dishonorable, and so on. Herder (2002/1772) referred to this peculiar normative sensitivity as *Besonnenheit* (reflective awareness)[1] and believed that it lies at the basis of the human capacity for language. What psychology conventionally refers to as perceptual, cognitive, and behavioral *functions* is on our enactive account better conceived of as normative *skills*, *competences*, and *styles*. This applies in particular to what Vygotsky (1987) has called the *higher mental functions*. Functional and semantic accounts of human behavior fail precisely because they conflate the normative or descriptive domain of an external observer with the normativity of the domain in which human agents constitute or realize themselves, their *normative Umwelt*, so to speak. Here too it should be stressed that these acquired skills and competences are inherently historical. Wittgenstein (1958/1974, 1956/1978) in particular has shown that the "possession" of a skill and the particular way in which the skill is acquired are formally indistinguishable: both are temporally extended processes characterized by repeated demonstrations within a realm of normative correction, and the only difference lies in the practical framing that allows the performances of "cultural experts" to acquire normative status relative to those of novices (as captured in Vygotsky's [1978] notion of the *Zone of Proximal Development*). It follows again that the particular historical process by which a normative skill is acquired is *constitutive* of that skill. We cannot define the skill in advance, in functional or semantic terms, or as by design.

Conceptually, one of the primary questions for a genetic cultural psychology lies in the particular way in which we might connect the self-referential or autonomous nature of biological systems to the inherent normativity of human conduct. One way is to conceive of it as a dialectical relation between "inside" and "outside," or between "sense" and "meaning" as Valsiner (1998) has proposed. In this chapter, I argue that this kind of dialectics can ultimately not account for the generativity of "life" in the triple sense of organic processes, personal histories, and historical practices. The inner/outer dialectics, I will argue, cannot account for *genuine temporality* and *historicity*. I propose instead an *expressive/enactivist* account and argue that psychological development is the acquisition of a normative style rather than the constructive internalization of pregiven meaning. I take my starting point in the thinking of Henri Bergson (1859–1941), whose seminal account of time as *duration* provides an explicit alternative to dialectical thinking. Connecting Bergson's insights to enactive thinking through Maturana's work on biological systems and Merleau-Ponty's understanding of life as expression and style will help us to propose the outlines for an enactive cultural psychology that is genuinely genetic.

Henri Bergson's Challenge to Genetic Thinking

Bergson can be listed among those thinkers whose ideas have had a profound influence on genetic thinking, together with such illustrious figures as Baldwin, Werner, Vygotsky, and Piaget. Indeed, Valsiner (1998) recognizes "a clear intellectual closeness between Bergson's ideas and James Mark Baldwin's 'postulates of genetic science'" and points out that Baldwin's developmental ideas were a direct influence on Bergson. Where at the beginning of the 20th century Bergson was arguably one of the most influential philosophers of his time, beginning in the 1930s his influence, like that of Baldwin, would start to dwindle. This is not the place to dwell on the historical circumstances of their shared fate (for an excellent discussion of the historical role of Baldwin's thinking, see Broughton, 1981; for a discussion of Bergson's importance for genetic psychology, see Valsiner, 1994), but one of the most obvious reasons seems to lie in the fact that their work and thinking ran very much against the current of an increasingly mechanized view of particularly the life sciences. Still, genetic thinkers in developmental and sociocultural psychology remember Bergson for his profound insights into the nature of temporality and the *irreversibility of time* and as a forerunner (particularly through his intellectual link with Pierre Janet) of contemporary activity theories (e.g., Valsiner, 1994, 1998, pp. 179–181).

Legacies and lineages can in hindsight be constructed in ways that naturally seem to support a certain intellectual conclusion. Whereas genetic thought in psychology is already struggling against a cognitivist mainstream that is profoundly ahistorical and a-developmental, it now seems to connect naturally with an alternative tradition of dialectical, structuralist, and constructivist thought that offers great possibilities for a psychology that is historical and cultural rather than mechanical and reductionist. Appropriating Bergson to a dialectical or constructivist psychology, however, may very well obscure his most original contributions. Following the foundational work of Baldwin and Werner, genetic thinkers have come to cast psychological development primarily as a process of cognitive "distancing" and separation, and following the pioneering work of Jean Piaget, they have characterized cognitive development chiefly as a process of abstraction and schematization. The focus within the genetic tradition of psychology is overwhelmingly on phenomena like concept formation and the development of abstract thought, which all seem to require that the psychological agent liberate herself from the immediate influence of the environment. It has been far less concerned with the development of embodied skills, personal styles, and participative modes of understanding. In contrast, although Bergson (1889/1910, 1911, 1922) clearly recognized the constructive role of language and concepts in the formation of consciousness, he tended to see this constructive role as one that hampers rather than facilitates genuine development and sought to connect the notion of development not primarily to concept formation but to the boundless expressions of "life." Genuine development, for Bergson, is not a separation, distancing, or even an overcoming of the opposition of subject and object in an ever-higher form of synthesis, but a movement toward increasingly differentiated and authentic participation in the wholeness of life. Whereas Baldwin's genetic logic is ostensibly Hegelian, Bergson's thinking is rather at odds with the mediational thinking that comes out the Hegelian tradition. Indeed, Bergson's philosophy of temporality, development, and *authentic difference* is part of a powerful and elaborate alternative to the Hegelian dialectics with important implications for a genetic cultural psychology.

Temporality as Self-Differentiation

According to Deleuze (1966/1988a), who arguably played a central role in reviving Bergson for current debates (see Lawlor & Moulard, 2012), one of Bergson's most important and radical contributions to Western thinking is the highly original way in which he conceives of *difference*. In a long tradition of Western thinking that culminates in Hegel, difference is understood essentially on a model that sees it as difference *from something else*. Hegel, says Deleuze, reduces difference to negation: an entity or concept "a" is only a in contrast to all that is "not-a," so that a is the result of a *double negation*: "not not-a." For Hegel, life is not an affirmation of what it is, but a negation of that what it is not. The logic of negation forms the basis of Hegel's conception of change or becoming: "being" generates its universal negation, "nothingness" and this opposition is overcome in a higher order concept: "becoming." It forms the basis of Hegel's semiology as well, which in turn has been a formative influence on Saussure's semiology (for an excellent discussion, see Derrida, 1971/1993). The concept of negativity lies at the heart of dialectical thinking, which, according to Deleuze, endorses therefore an *inauthentic,* derivative, or externalist conception of difference. The singular, the unique, and the nonrepeating, on the dialectical account are not accessible in their own right but only through the *mediation* of universals. Bergson's merit, says Deleuze, is to have given us an *authentic* conception of difference, where difference is *originary* and *affirmative* and not derivative or mediated. In Bergson's thinking, "difference is gathered all at once rather than represented on condition of a prior resemblance, identity, analogy, or opposition" (Lawlor, 1998, p. 16).

Bergson's radical genetic alternative to transcendentalist and dialectical thought anticipates in important ways the nonfoundationalism that would decades later come to characterize poststructuralist thinking. In the Western tradition, *becoming* and *development* are typically understood as the becoming and development of some kind of being, whether understood in terms of its form or structure, or in terms of its essence. Being is understood as grounded or foundational. Poststructuralists, on the other hand, tend to deny that there is any pregiven structure, be it that of consciousness or that of a semiotic system like language, that can serve as a foundation or starting point for knowing the world. To put the issue in a somewhat simplified form, we could say that poststructuralists reverse the priority of "structure" and "becoming": rather than speaking of the becoming or development of already given structures, poststructuralists set out to explain the emergence or genesis of structures without assuming or fixing anything in advance with regard to the structure or essence of that which develops.

Foundational thinking fundamentally understands entities, actions and events as "in" time and "in" space, that is, as entities situated within a spatial or temporal order that essentially remains external to them. Time and space, in the tradition of foundational thinking, are merely principles of external ordering, and change is the change of a structure that has already been identified in advance. Because entities, actions, and events are situated this way, they can be compared with each other and deemed "the same" or "different" by an observer. In psychology, for example, we can talk about individual differences by thinking of individuals as discrete entities located in space and time and of difference in terms their quantitative "positions" on universal spatialized dimensions, like "self-esteem," or "need for achievement." Likewise, we can think of change in terms of quantitative shifts on such spatialized dimensions, for instance when we speak of an increase or decrease of self-esteem. In Bergson's terms, difference is seen, then, as mere *differences in degree* (Bergson, 1911, 1922). On a nonfoundational account, however, entities and events are not located *in* space and time, but temporality, one could say, is their very essence—it is what they *are*. Seemingly discrete entities in such a nonfoundational sense are "made of" what Bergson (1910/1889) has called *duration*. Duration is ceaseless change, but not against the background of a homogenizing space and time, but as the expression of a *qualitative multiplicity* that is nonetheless part of a unified whole. Bergson refers to differences in this sense as *differences in kind*.

According to Bergson (1889/1910, p. 121), there are two kinds of multiplicity or two senses of the word "distinguish": one quantitative, external, and inauthentic, the other qualitative, internal, and authentic. Discrete or quantitative multiplicity casts difference in terms of an externalizing, homogenizing "spatialized" time, while qualitative multiplicity casts difference in terms of *duration* or authentic temporality. For Bergson, entities, actions, and events cannot be the same or different in a merely external sense and neither can the same event or action be repeated in time. Representing difference this way would be to substitute inauthentic or quantitative difference for authentic or qualitative difference. Entities are not merely different from other entities; they are *inherently* differential, that is, they differ first of all from themselves. Talking about life, Bergson understands this internal difference in the active sense, as *self-differentiation*. Life, according to Bergson, *affirms* itself as a ceaseless movement of self-differentiation and not merely as the negation of a negation.

Mediation or Expression?

The radical critique of the Hegelian dialectics casts a new light on the seemingly natural alliance between genetic thought and the idea of semiotic mediation. Semiotic thinking is a way of thinking the particular through the general, a way of understanding individual development as the appropriation or internalization of an already established semiotic system. As we have recently put it elsewhere, a mediational cultural psychology tends to explain *sense making* on the basis of sense already made (Baerveldt & Verheggen, 2012, p. 174). As much as it may seem obvious that meaning as a historical semiotic system precedes our own individual ways of sense making, the crucial question remains how semiotic meanings come to make sense in the lives of actual persons. Indeed, this question is at the basis of most genetic approaches to cultural psychology. Whereas mediational thinkers argue that personal sense is a creative modification of impersonal semiotic meanings, which involves some process of constructive internalization (e.g., Valsiner, 1998; Wertsch, 1991), enactive thinkers make no principal distinction between inner and outer in the psychological or ontological realm, but see all meaning as *con*-sensual. We have proposed an *enactive* account that follows Wittgenstein in considering the "inner" (intrapsychological) and the "outer" (interpsychological) as merely belonging to different *language games*, in each of which we can, over time and in coordination with others, acquire a more elaborate and differentiated competence (Baerveldt & Verheggen, 2012). Both the language of internal experiences and the language of external behaviors belong to a consensual normative realm (for a discussion of the differences between Wittgenstein and Vygotsky in that regard, see Berducci, 2004).

It may seem peculiar to dispute that there is a common order of meaning, a consensual understanding of some kind that we may reflectively refer to as "culture." But rather than understanding the particulars of our experience through the mediation of the pregiven, the universal, or the culturally shared, Bergson's (or Deleuze's) attempt at providing an authentic conception of difference helps us to understand meaning generatively, more like an ongoing *attunement* or *rhythmization* of different durations, than a semiotic system that is conceptually shared. On such an account, even the conceptual competence displayed by adult participants to the culture is dynamic, expressive, and performative and therefore inherently temporal in its organization. It may seem peculiar as well to deny that we have an "inner life" that is modified and transformed through interaction with others. However, the enactive perspective I propose does not deny the significance of inner life, but merely the ontological, psychological, or dialectical opposition of inner and outer. I consider inner and outer as modalities of expression rather than ontological substances, pseudolocations or moments in a Hegelian dialectics. Indeed, to conceive of "insideness" and "outsideness" in an expressive rather than a dialectical or psychological way is what lies at the heart of a broad variety of genetic and generative thinking from Spinoza to Herder, Goethe, Dilthey Bergson, Bakhtin, and Merleau-Ponty.

The challenge that Bergson still poses to genetic thinking does not conceal the many problems with his own philosophy. This is not the place to discuss the merits of Bergson's entire way of thinking. Such divergent thinkers as Gaston Bachelard, Georges Canguilhem, Mikhail Bakhtin, and Maurice Merleau-Ponty would each in their own way critique and extend Bergson thinking, and their attempts would result in a broad variety of genetic, generative, or poetic thinking that provides a rich alternative to the mediational thought of Hegel and the genetic thinkers in his footsteps. My aim here, however, is not simply to propose a genetic approach along Bergsonian lines. Instead, I want to point out some possibilities for genetic thought that may have been closed off prematurely by a cultural psychology that privileges mediation over direct participation and connect those possibilities to an alternative account of the genetic legacy. In an overview of recent developments in cultural psychology, Valsiner (2009, p. 10) states,

> In the 1950s psychologists were disputing the immediacy of perception (à la Gibson) in contrast to the constructive nature of the perceptual act (à la Bruner & Postman, 1950; cf. Ansbacher, 1937 for the origins); then 50 years later we find a similar dispute in cultural psychology around the issues of enactivism—focusing on the immediate nature of cultural actions—and mediation—which centers on the distancing from (yet with) the immediate action (Christopher & Bickhard, 2007; Verheggen & Baerveldt, 2007).

If the constructed or mediated versus the "immediate" nature of perception, cognition, and action is indeed the issue, as Valsiner (2009) claims, then it is important that we reconstruct the history of genetic and generative thought in a way that not only adequately reflects what is at stake in the debate, but that also acknowledges the multifaceted and polysemic nature of thinkers that are now routinely presented as champions of mediational thinking. For example, following Wertsch's authoritative account of Bakhtin as a thinker who "outlined a unique and powerful approach to semiotic mediation" (Wertsch, 1991, p. 93), Bakhtin has been appropriated into a dialectical or pragmatist framework that identifies signs with pragmatic tools and *speech genres* with mediational "tool kits." Elsewhere (Cresswell & Baerveldt, 2009, 2011), we have pointed out that this view largely overlooks Bakhtin's own concern with the concrete materiality of embodied practice. Bakhtin was not a narrow pragmatist, but neither was he a dialectic or mediational thinker in the Hegelian sense. It is well known that Bakhtin, in spite of his debt to both Hegel and Marx, was also highly critical of Hegel, characterizing the Hegelian dialectics as a dialogue that has been lost in abstraction (e.g., Bakhtin, 1986, p. 147). Bakhtin's critique of Hegelian abstraction and his own concern with the unique, nonrepeating, and nongeneralizable moments in dialogue have a distinctively Bergsonian ring. It is therefore particularly relevant that in addition to the influence of such figures as Karl Bühler and Ernst Cassirer, Bakhtin's philosophy owes as much to the *Lebensphilosofie* of Johann Wolfgang von Goethe, Henri Bergson, Wilhelm Dilthey, Friedrich Nietzsche, and Georg Simmel. Already in his early work, Bakhtin (1993) was interested in the uniquely constituted lived experience of *Being-as-event*, which, according to him, we can understand only through our own unique participation and not through abstraction or conceptual mediation. If Bakhtin engaged semiotic thinking, it was primarily to liberate our understanding of *semiosis* from the view that life necessarily passes through the mediation by pregiven meanings.

Genetic Thinking and the Return to Life

Bergson belongs to a current of genetic thought that resists sacrificing the particular and the unique to abstract generalization. Part of what makes his work historically and ontologically relevant to the program of a genetic cultural psychology is the unique way in which he was situated in relation to the scientific, philosophical, and psychological developments of his time. The period of Bergson's active career is one in which particularly the biological sciences and their relation to psychology were given a decisive push in the direction of a reductionist mechanical worldview. Indeed, Prigogine and Stengers (1984, pp. 91–92) conclude that in a crucial sense, Bergson had already conceded science to a mechanical worldview that can merely account for physical reality in terms of external "causal" relations between conceptually demarcated objects. Bergson concluded that science is principally incapable of grasping duration, since it reduces time to a spatialized sequence of states linked only by deterministic law. For Bergson, therefore, philosophy provided both an alternative and a complement to science.

The enactive approach I am presenting here has the advantage of being able to draw on developments within the biological sciences that were not yet available to Bergson. In spite of Bergson's continuous effort to remain in constructive debate with the biological and evolutionary thinking of his time, he might in hindsight not have fully realized that particularly the life sciences were already at the dawn of a radical transformation that would in the next decades lead to the scientific and mathematical disciplines of complexity, nonlinear dynamics, and self-referential systems. Ironically, however, those disciplines in themselves hardly address Bergson's original challenge, and the problem of the singularity versus generality of life is more contested than ever. Even the language of complex dynamical systems fails to grasp life in its *duration*, not because it cannot address the radically new, but, as I will argue in a moment, because such a formal language belongs to the descriptive domain of an external observer and not to the domain of operations by which a living system constitutes itself as a unity. It is the latter that Bergson referred to as *duration*, and if we are to conceive of a truly genetic and generative psychology, we will have to develop a methodological approach that avoids reducing *duration* or authentic temporality to the external or *a priori* categories of an observer. As long as the proper epistemic boundaries between those different phenomenal domains are not respected, biology and the other life sciences will necessarily proceed as largely mechanicist or externalist endeavors, and psychology will remain caught up in a view that can address "individual differences" merely in terms of universal dimensions or as ideosyncratic ways of taking up a generic semiotic system. The emancipation of genuine "life" in psychology is a process that has hardly just begun.

Revitalizing Vitalism or the Return to the Organism

Thinkers like Bakhtin, Bachelard, Canguilhem, Merleau-Ponty, and Deleuze, although incomparable and even incompatible in many ways, each carried on some of Bergson's vital impulse, an impulse that can be characterized as an attempt to recover "life" from its ensnarement in artificial concepts and categories. Both Canguilhem and Deleuze have referred to their own work as a new kind of *vitalism.* Unfortunately, vitalism is nowadays merely associated with the refuted idea that living organisms are animated by some kind of nonphysical vital principle. Bergson, however, proposed no such thing. His vitalism was never posed in opposition to the mechanical worldview as such, but rather to both a finalism that sees in history and evolution the realization of an ultimate purpose and to the kind of mechanicist thinking that understands life merely in terms of external causal relations between artificially isolated entities: "That life is a kind of mechanism I cordially agree. But is it the mechanism of parts artificially isolated within the whole of the universe, or is it the mechanism of the real whole?" (Bergson, 1922, p. 32).

Bergson's holistic understanding of life and organism would find a particularly rich articulation in the work of Kurt Goldstein (1939), who saw himself as an heir to Goethe's organismic philosophy of life and who in turn would become a major source of inspiration for such diverse thinkers as Vygotsky (especially with his theory of inner speech), Bakhtin, Voloshinov (partly through long-term mutual influences between Goldstein and Cassirer; see Brandist, 2002), Canguilhem, and Merleau-Ponty. What characterizes the work of each of those thinkers is their particular concern with the relation between the singular and the general in a way that does not simply subsume the former under the latter. Goldstein not only based his seminal account of the organism on individual case studies, but explicitly referred to Goethe's notion of *Anschauung* (intuitive perception) as a way of intuitively grasping particulars in relation to a nonreducible organic whole without making this whole into an abstract generality. The link with Goethe, like that with Bergson, points at a particular tradition in biological thinking that is both holistic and genetic.

Jaan Valsiner has recently proposed that one of cultural psychology's major challenges lies in the need to establish conceptual links with contemporary biological sciences—especially developmental biology (Valsiner, 2012, p. 1096). Recognizing the vital convergences in the genetic thought of such diverse thinkers will help us to reconcile the language of "life" as an organic process with the language of "life" as we live it in our everyday cultural practices. Long before Wundt's attempt to emancipate psychology as a science of mental causation from physiology, the scientific study of human nature had already divorced itself from a concern with questions of "spirit" and morality. The rise of experimental physiology in the 1830s and 1840s marked the advance of the natural sciences and the break with the discourse of moral philosophy. Physiology itself developed from a field that was primarily medically oriented into a discipline that was centrally concerned with physical and chemical processes—the mechanics—of life.

The concept of "life" was contested throughout the 19th century, yet this history is now much obfuscated, as it is typically written by those who have already adopted a reductionist/mechanistic view of science. It is a telling sign, however, that an eminent scientist like Johannes Müller, one of the leading physiologist of the 1830s, still showed a commitment to vitalism, not by opposing mechanical explanation but by describing the mechanics of the circulatory, lymphatic, respiratory, digestive, endocrine, nervous, and sensory systems by reference to the *organism* as an indivisible whole. Where vitalism is now routinely described as the refuted belief that "living organisms are fundamentally different from non-living entities because they contain some non-physical element or are governed by different principles than are inanimate things" (Bechtel & Richardson, 1998), characterizations like Müller's, as well as those of later biologists and physiologists like Hans Driesch (1867–1941), Kurt Goldstein (1878–1965), Viktor von Weizsäcker (1886–1957), and F. J. J. Buytendijk (1887–1974) stand rather in the tradition of holistic and organismic thinking initiated by Goethe and philosophically elaborated by thinkers like Dilthey and Bergson (also see Baerveldt & Verheggen, 2012). With Valsiner, I propose therefore an integrative approach capable of reconciling our understanding of the "mechanics" of life with that of what it means to *live* and *experience* life. Indeed, enactive cultural psychology is an attempt to propose and develop such an integrative genetic approach by linking the temporality of consensual normative practice to the operational integrity or *duration* of the living organism.

Structure, Language, and Duration

The enactive cultural psychology we have proposed elsewhere (e.g., Baerveldt & Verheggen, 1999, 2012) is a *nonrepresentational* and *nonmediational* account of culture and sense making that finds its roots partly in Varela, Thompson, and Rosch's (1991) seminal discussion of embodied cognition, but particularly in Maturana's ideas on "life," language, and sociality (see in particular Maturana, 1970, 1978, 1988; Maturana & Varela, 1980; Maturana & Verden-Zöller, 2008). The crucial concept that lies at the basis of all of Maturana's thinking is that of *structural determinism*. Structural determinism is the idea that although the external circumstances of a natural system may perturb the system, or "select" certain trajectories of change, those circumstances cannot determine the nature of the changes or behaviors of that system. A system's changes and behaviors are determined by the system's own structure. By *structure*, Maturana refers to the particular way in which a natural system is realized: its actual components and the relations between components that constitute it as a composite unity. A system's structure determines the space in which it exists and can be perturbed (Maturana, 1975, p. 316). Maturana distinguishes a system's structure from its *organization*. Organization refers to the set of relations between components of the system that determine its class identity or its systemic integrity. While the structure of a system goes through changes with every interaction the system undergoes, its organization remains constant for as long as the system maintains its integrity. The system's organization specifies the domain of interactions in which a system can engage without losing its defining identity.

The formal distinction between structure and organization allows Maturana to give a formal operational definition of life, or living systems: a living system is a system that maintains or produces its own defining organization by keeping its structural changes within the operational boundaries that are required if it is to maintain its identity. Maturana, together with Varela, coined the term *autopoiesis* (from Greek *autos*, "self"+ *poeisis*, "make, create") to refer to such self-producing systems (see Maturana, 1970; Maturana & Varela, 1980). Like Bergson, Maturana sees life essentially as self-differentiation. Through its own productive operations, a living system produces itself both as an internally differentiated structure and as a unity distinct from its medium. The cleavage between organism and medium is made and maintained "from within," so to speak.

Enactive cultural psychology elaborates Maturana's idea, following from the principle of structural determinism, that the interaction *between* living systems cannot be a mutually *instructive* one, as the changes each system can undergo are determined by its own structure and not by what affects it from the outside. Living systems can merely *trigger*, but not determine structural changes in each other. As a consequence of this mutual "triggering," different operationally closed systems can engage in a history of interlocked conduct or consensually coordinated behaviors, which gives rise to a cooperative domain of interactions or *consensual domain*. Elsewhere (Verheggen & Baerveldt, 2007, 2012), we have argued in detail that a consensual domain cannot be understood as a domain of socially shared cognitions or shared mental content. Rather, a consensual domain is essentially a form of "attunement" or "rhythmization" that may over time and with the participation of multiple cognitive agents become increasingly complex and differentiated. If such a dynamic consensual domain becomes sufficiently differentiated, elements or aspects of the consensual domain may come to be used for the consensual coordination of other aspects. Such a recursive consensual coordination *of* consensual coordinations of action gives rise to a closed generative domain of interactions, like language. Dance, music, and mathematics are other examples of closed generative domains.

Recursive consensual coordinations of actions allow certain actions to acquire normative status relative to other actions and lie at the basis of all cultural practices, including linguistic practices. Indeed, linguistic behaviors give rise to the recursive consensual distinctions that we refer to as semiotic tools or signs. Human speech testifies to the fact that within a domain of recursive consensual coordinations of actions, even the production of vocal sounds may come to coordinate other actions. Operationally speaking, semiotic tools do not explain meaningful or consensually coordinated conduct and neither do they mediate it. Rather, what we recursively recognize as a "sign" is itself a complex form of consensually coordinated behavior. Indeed, whereas Bergson was reluctant to see language as anything else than an artificially imposed system, unable to capture duration, Maturana's account of the genetic origins of language allows us to conceive of living language as a *dynamic recursive "attunement" and "rhythmization" of durations*. While, ironically, Bergson maintains a conception of language as external to life, Maturana allows us to see it as emerging out of the dynamic flow of life itself, as a particular way in which life gets coordinated. For instance, where mutual attunement in consensually coordinated practice may serve to establish routine or repetitive sequences of action, those actions become truly linguistic only when they are in turn coordinated in a way that allows them to be "right" or "wrong." On this account, it is the recursivity of consensual practice (and not a set of mechanical rules in the head, as conventional linguistics would have it) that accounts for the *normativity*, or what linguistic prefer to call the *grammaticality* of language.

Temporality and the Fallacy of Phenomenal Reduction

According to Maturana (Maturana & Varela, 1980, p. 18, the operations of the nervous system at any given moment in time are entirely determined by its own structure. In this sense, the nervous system can be said to function wholly in the *present*, where "present" can be defined simply as the interval required for an interaction to take place.

But if the nervous system inherently operates in the present, how can we understand the normative sameness and repetition in behaviors that are constitutive of all culturally meaningful actions? Moreover, what criteria would be required for a psychologist or social scientist to identify any behavior or set of behaviors as behaviors of a certain kind? Indeed, how do we understand the notions of "sameness" and "repetition" mindful as well of Bergson's earlier admonition that we cannot simply impose the homogenizing categories of an external observer on the behaviors of a living organism and claim to have explained those behaviors?

According to Maturana (in Maturana & Varela, 1980),

> An organism is a unit to the extent that its conduct results in the maintenance of its basic circularity (and hence identity), and two modes of conduct are equivalent if they satisfy the same class of requirements for this maintenance. (p.18)

Consequently, that what an observer may call "the same" behavior may, operationally speaking, be realized by neuronal processes that are quite different with respect to the specific requirements for maintaining the organism's basic circularity. Inversely, through what an observer may call "different" behaviors the organism may satisfy its subordination to the same class of operational requirements. It follows that seen from the perspective of the nervous system, no neuron or cluster of neurons can have a fixed functional role in the realization of behavior. By the same token, however, "sameness" in the domain of external behaviors of the organism is defined entirely with respect to the semantic or functional criteria of an observer and cannot be mapped directly onto the activity of the nervous system. Whereas an observer may identify sameness in behavior in terms of goals, reference, or function, from the perspective of the nervous system itself, "sameness" in behavior is a matter of "sameness" in relative nervous activity, and its relevance is determined solely by the significance it has for the maintenance of the system's circular organization. In other words, the relation between neuronal dynamics and behavior cannot be understood in causal or representational terms. Behavior and nervous activity belong to two nonintersecting phenomenal domains.

Maturana has used the term *phenomenal* or *phenomenic reduction* (Maturana, 1988; Maturana & Varela, 1980) to refer to the logical error of collapsing two phenomenal domains into one: "A scientific explanation does not constitute a phenomenic reduction, but on the contrary it constitutes the validation of a generative relation between otherwise independent nonintersecting phenomenal domains" (Maturana, 1988, p. 45). According to Maturana (1988), each living system can be seen as to act in three nonintersecting phenomenal domains. First, we can recognize the domain of the system's internal processes, organized around the uninterrupted realization of its autopoiesis. From this perspective, the system appears as a composite unity that produces and maintains its own unity and integrity through a dynamic network of internal molecular processes. Second, we can observe the domain in which the system or organism operates as a unity or relational totality. From this perspective, we can see the behavior of the system as a whole, but observing it this way obscures the molecular dynamics that take place inside of it. Phenomena that are typically conceived of as psychological, whether behavioral or mental, belong to the second or relational domain. Finally, we can see the domain of the organism/niche unity, that is, the domain in which the system realizes its relational living in the operational dynamic congruence with its niche. Seen from this perspective, we distinguish the entire history of structural changes in the organism/niche coupling, organized around the realization and conservation of the organism's autopoiesis and adaptation.

From Cause to History

One of the crucial implications that follow from the principles of structural determinism and phenomenal irreducibility is that the relation between phenomenal domains cannot be stated in causal terms. From observations of what happens within the behavioral domain, we cannot deduce anything about what precisely happens inside the system, and neither can we deduce from observations of the system's internal dynamics anything about its behavior as a relational totality. Of course, a cognitive neuroscientist or a behavioral scientist who, as an external observer, beholds both domains simultaneously may propose an explanatory process of a pseudomechanistic nature. Psychology, in its eagerness to establish itself as a natural science, has traditionally catered in such pseudomechanistic explanations. However, whereas the processes that take place within each domain in its own right are structurally determined, the relation between these domains cannot be structural or causal but only correlational in nature. Indeed, deductions, as Maturana points out, reveal structural coherences whereas correlations reveal historical coherences (Maturana, 2012, p. 157). The fundamental implication is that any account that simultaneously comprehends the operational domain and the behavioral domain needs to be historical in nature. Indeed, the very notion of human behavior is historical and hermeneutic and cannot be captured in merely operational terms. As Varela et al. (1991) have argued, the hermeneutic viewpoint entails that "knowledge depends on being in a world that is inseparable from our bodies, our language, and our social history." The hermeneutic view concerns the body as *historical expression*. Whereas accounts of the body as an *organism* are necessarily structural, accounts of the body as an *expression* are necessarily hermeneutic in nature. Since no phenomenal reduction is possible, a psychology that purports to straddle both the natural and the hermeneutic sciences or, perhaps to put it more adequately, a psychology that inserts itself as a "third science" between those disciplines, necessarily needs to be a genetic discipline.

What Can Psychology Learn?

Psychology has historically found itself caught up in a rather fruitless debate on the question of the hereditary versus environmental determination of behavior. The enactive perspective rejects both explanatory theses, as well as the idea that phenotypical variance and change is due to the "interaction" between those two influences. Instead, it sees both the organism's organization and its adaptation to a medium as invariants while conceiving of its history as a flow of structural congruent changes. Although an organism goes through continuous structural change, its defining organization and its adaptation to its medium are necessarily preserved or else it disintegrates. The organism's history of structural changes is what we call its *ontogeny* and this ontogeny is determined neither by its genes nor by its environment but by its entire structure as it enfolds dynamically over time. As Maturana has more recently stated, "the ontogenetic phenotype ... arises in the process of *epigenesis*, i.e., along a path of interactions starting from the initial structure of both system and medium, along the conservation of its living" (Maturana, 2007, p. 21, emphasis added). We could say that ontogenetic development is necessarily a process "from within," even though a developing organism must maintain its adaptation with its medium. What appears to an observer as "external" influences or environmental demands can, operationally speaking, only be a modulation of the organism's entire structure so that structural change can never be interpreted as a straightforward effect of those external influences and demands. Only a proper historical account avoids imposing an external frame of reference in terms of which the conduct of the organism would merely be relevant in relation to external (e.g., functional or semantic) criteria.

The mistake of collapsing different nonintersecting phenomenal domains may be compared with the fallacy of casting "life" in terms of spatialized or inauthentic time. Living systems are structurally determined and operationally closed, and as such, we cannot give an external characterization of the internal operations that constitute them as a unity. Living systems are characterized precisely by their self-demarcation and self-differentiation. The common mistake in the evaluation of both Bergson and Maturana is that the impossibility of giving a fully external description of life in its vital characteristics is seen as an argument for subjectivism or solipsism. Bergson's (1910) *duration* is depicted by his critiques as merely "subjective time," in contrast with the presumably "objective time" of physics or the conventional or practical time of culture. Similarly, operational closure in Maturana's sense is seen by critiques of the enactive approach as an argument for a new kind of monadology (Kreppner, 1999) that would cut off the human organism from the possibility of internalizing the shared semiotic meanings of culture. Both critiques miss entirely the point that duration as well as operational closure are notions aimed at reconnecting living systems to their concrete situation by liberating them from their conceptual entrapment in external categorizations.

Bergson's (1910) crucial insight was that we can understand other durations not through operations of conceptual demarcation, but only through our own duration in what might best be seen as a process rhythmic attunement. Maturana (1988, p. 46), following the radical implications of the idea of structural determination concluded that two dynamic structural systems might engage in a history of interlocked conduct, not by a process of mutual instruction, but only through a process of congruent structural change. If those structures are living systems that mutually coordinate their actions with regard to each other, the result is a dynamic consensual domain, and if the elements of such a consensual domain become themselves coordinated, the result is a closed generative domain of interaction, like language. Whereas Bergson eschews the role of language, Maturana (1978, pp. 30–35, 1988, p. 48; Maturana & Varela, 1980) sees humans as living in language, but he sees language not as an external semiotic system but as a generative, sense-making practice.

Concepts as Recursive Consensual Coordinations

Maturana is a not a psychologist, but his understanding of language as a recursive consensual coordination of actions allows us to take the first steps toward an enactive developmental account of concept formation and the acquisition of semiotic meaning. It follows from the foregoing discussion that psychologically speaking, concept formation cannot be a matter of abstraction or generalization in the purely classificatory sense of that word. Concepts are not merely the result of cognitive generalizations based on stimulus features, but neither are they external mediational means that can be entered into the stream of activity to make such generalizations possible. Operationally speaking, they are *coordinations of coordinations of action* rather than entities that mediate action. Psychologically speaking, they are like *skillful dispositions* more than they are logical or semiotic structures. Learning to express meaning semiotically is to acquire the dynamic dispositions that allow one to remain normatively attuned to others in a closed generative domain of interactions. If concept formation is a form of generalization, it is not the generalization of subsuming particulars under a genus, but the generalization of recursively recognizing one's own behavioral and expressive conduct at different moments in time as "the same." Such recognition of sameness is necessarily normative and facilitated only by the way our conduct is embedded within regular or repetitive community practices. Normative sameness, in other words, is not pregiven, but is *dynamically enacted* within the historical practices of a community.

Genetic Thought Between Mediation and Expression

One of the central ideas of Vygotsky's genetic approach is that while language and thinking are fundamentally different in terms of their genetic origin, their relationship transforms in the course of ontogenetic development. In the Vygotskyan tradition, this transformation is seen as involving a process of *internalization*. Valsiner has stressed that internalization should be conceived of as bidirectional or co-constructive, which implies some kind of *duality* or *dialectics* between the "inner" and the "outer" (Valsiner, 1998, pp. 113–116). Vygotsky was a monist, however, perhaps even more in the sense of Spinoza than in the sense of Marx, and his understanding of the duality of inner and outer should therefore not be confused with the Cartesian dualism of subject and object.

It is not altogether clear how Vygotsky reconciled the influence of Spinoza on his own thinking with the dialectical thought of Marx and particularly Hegel. What is known, however, is that Hegel, who admired Spinoza, faulted him nonetheless for maintaining a static, nontemporal conception of substance: "But for Spinoza substance and its absolute unity has the form of an unmoved unity, that is, a unity which is not self-mediated, a rigidity in which there is no place yet for the concept of the negative unity of the self, of subjectivity" (Hegel, 1832/2010, p. 212). Indeed, for Hegel, Spinoza was not yet truly a dialectical and hence a temporal thinker, because his philosophy lacks the concept of negativity. Vygotsky is commonly considered to be a dialectical thinker, but his link with Spinoza may point at a tension in his work with potential for a different interpretation of the genetic legacy. Indeed, Spinoza was admired by Goethe and Herder, who saw in him as an ally in developing an *expressivist* rather than dialectical form of genetic thought. Strikingly, beginning in the late 1960s, poststructuralist thinkers like Althusser (see Althusser & Balibar, 1970) and Deleuze (1968/1990, 1970/1988b) started appealing to Spinoza in their revisionist account of Marxism, aimed at disembarrassing Marxism of Hegel's influence (also see Balibar, 1998). For poststructuralist thinkers, it was the abstract logic exemplified in Hegelian dialectics that blocked the possibility of true generativity by making it dependent on some form of mediation. We saw already that Deleuze would develop his own expressionist kind of vitalism particularly through a creative reinterpretation of Spinoza, Nietzsche, and Bergson. This kind of expressionism is not simply a return to a romantic notion of naïve immediacy (if there ever were such a notion), but an alternative *nonmediational* form of genetic thinking that fully recognizes the reality of our concrete historical and material conditions.

Perhaps ironically, this anti-Hegelianism found support from another genetic thinker. Piaget already applauded Althusser's revisionist Marxism in its attempt to break away from Hegelianism (Piaget, 1968/1970). Throughout his academic career, Piaget was quite critical of the dialectics of negation and contradiction he observed in both Hegelian and Marxist thinking, which, according to him, failed to recognize that only statements, not actions, can contradict each other (e.g., Piaget, 1980, pp. 225–227). The mind, in its actions, naturally tends toward affirmation rather than negation, according to Piaget. The critique of mediational thinking was already anticipated by Kierkegaard and Nietzsche and would, after Bergson, be further developed by such divergent thinkers as Bakhtin and Merleau-Ponty. Whereas Piaget would remain stuck in an overly mentalist framework, the latter two thinkers, each in his own way, understood duality *expressively*, as belonging to a generative field of creative juxtapositions in which each position is understood in terms of its participative attunement to an entire *style* of being (Baerveldt, 2013). It could be argued that for

Merleau-Ponty, the notion of style became the unifying principle of his entire philosophy, starting from his earlier attempts to account for the unity of perception to what would become a full-fledged *genetic ontology* in his later work. It allowed him to understand the generality of "sense" without the mediation of semiotic signs. It furthermore allowed him to understand the genetic relation of language to our perceptually guided and consensually coordinated actions.

Merleau-Ponty and the Notion of Style

In the *Phenomenology of Perception*, Merleau-Ponty (1962) remarks,

> Everything appears to us through a medium to which it lends its own fundamental quality; this piece of wood is neither a collection of colors and tactile data nor even their total *Gestalt* but something from which there emanates a woody essence; these "sense data" modulate a certain theme or illustrate a certain style which is the wood itself and which creates round this piece of wood and the perception I have of it a horizon of significance. (p. 450)

Whereas Bergson depicted perception and space as part of a nonauthentic grasp of duration, Merleau-Ponty's understanding of perception is one in which things do not just appear as discrete entities in external spatial relations, but as part of an experiential field that is always both unified and provisional. Hence, when Merleau-Ponty turns to language, rather than the relation between language and conception, Merleau-Ponty connects language to the peculiar primacy and unity of perception. For Merleau-Ponty, the unity of perception is not something that can be imposed on it externally or conceptually. Rather, it is *physiognomic* in nature and therefore allusive. The body's hold upon the world is one of acquired *habits*, *sensori-motor intentions*, or *embodied norms*, which allow us to respond in increasingly differentiated ways to the solicitations of the world. Those solicitations present themselves with a certain motor physiognomy, a living sense or significance that directly "demands" from us a proper response, much in the same way as the words and gestures of the other demand a response. The world's solicitations as well as the words and gestures of the other are grasped not by an act of intellectual interpretation but by participating in a general *style* of being.

For Merleau-Ponty (1962), bodily expression and language are never really opposed. On the one hand, Merleau-Ponty speaks of a primary "dialogue" between body and world, an *operative language* that already functions at the level of perception (p. 310). This operative language is instituted by a pact of bodily powers with the perceived world, a dynamic relational network of "norms and deviations," which allows the body to respond to the "sense" of the world's promptings and which already stylizes at the level of even the first perception (Merleau-Ponty, 1964a, p. 54). On the other hand, language never fully breaks its bond with the lived significance of bodily perception and gesture. In living speech, there is no separation between what is expressed and the manner in which it is expressed. Like Vygotsky, Merleau-Ponty's understanding of word meaning in the *Phenomenology of Perception* (1962) is essentially a genetic one. Although there is no clear evidence of a common source in this regard, it is striking that Vygotsky's understanding of the difference between word sense (*smysl*) and word meaning (*znachenie*) was inspired directly by the work of the French philosopher Frédéric Paulhan (1856–1931) (see Vygotsky, 1987, p. 101), while Merleau-Ponty's understanding of the actualization of the idea by the word owes much to the writing of Paulhan's son, the journalist and literary critic Jean Paulhan (1884–1968) (see particularly Merleau-Ponty, 1973a, 1973b). Like Vygotsky, Merleau-Ponty argues that we don't simply take up words as ready-made instruments. Instead, the very nature of word meaning transforms in the developmental process. This transformation, however, is not primarily a matter of decontextualization or distancing but rather of the formation of a certain behavioral style. "What remains to me of the word once learned is its style as constituted by its formation and sound" (Merleau-Ponty, 1962, p. 180).

According to Merleau-Ponty (1962), words always maintain a certain concreteness, and their meaning is never completely conceptual. Once acquired, they have their own physiognomy: they mean without us having to interpret them intellectually, and their meaning lies at the surface just like the meaning of our gestures. Words "carry a top coating of meaning which sticks to them, and which presents the thought as a style" (p. 182). The word doesn't need to be linked to its meaning since, as a phonological gesture, it belongs to the normative field of bodily practice which speaker and listener already have in common: "Its generality, is not that of the idea, but that of a behavioral style 'understood' by my body" (p. 403). Years later, in a work that would remain unfinished, Merleau-Ponty (1968) would say, "Every concept is first a horizontal generality, a generality of style" (p. 234).

Style is, for Merleau-Ponty, neither concept nor mere association. Likewise, the transformation of style in ontogenetic development is not a move from concrete sensori-motor schemas to conceptual structures, but the differentiation of a dynamic complex of bodily dispositions that is always simultaneously personal and normative. In perception as well as in language, "style" serves as the relational network against the background of which particular manifestations acquire their sense and their relative unity. Merleau-Ponty describes style not as a fixed structure but as a *norm*, together with a certain "logic" of transpositions, deviations, and equivalences. In the *Phenomenology of Perception* (1962), he takes up Husserl's idea of an operative (*fungierende*) intentionality, which prior to any act of judging, intending, or reflecting reveals a world that already has a distinctive physiognomy and style in terms of which we are already meaningfully situated and oriented toward the future. "Ahead of what I can see and perceive, there is, it is true, nothing more actually visible, but my world is carried forward by lines of intentionality which trace out in advance at least the style of what is to come" (Merleau-Ponty, 1962, p. 416). Merleau-Ponty recognizes, however, that such a notion of intentionality can no longer be derived from a philosophy of consciousness or a psychology of individual mental life. Indeed, this *intentional arc* is embodied in the sense that the skills and habits we acquire "are reflected back to us in the solicitations of situations correlative with our dispositions to respond to them," to use the words of Hubert Dreyfus (1996). As the world becomes more differentiated for us, we also reach new levels of participation and responsiveness and a new immediate sense of the necessity of certain courses of action. Through the habitual powers of the body, we acquire a particular style of being in the world as a coherent deformation of a style we already receive at birth:

> To exist for an embodied and historical being is to exist in a particular manner: I am a psychological and historical structure and have received with existence a manner of existing, a style. All my actions and thoughts stand in relation to this structure. (Merleau-Ponty, 1962, p. 529)

This is not the place to discuss in detail Merleau-Ponty's encounter with Saussure. Merleau-Ponty was instrumental in reintroducing Saussure to a French audience even before structuralist thinkers like Claude Lévi-Strauss took him up. Unlike the structuralists, however, he remained interested in what he called a "lived equivalent of structure" (Merleau-Ponty, 1964a, p. 119). His reading of Saussure helped him to see that nothing expresses or means in its own right. Yet unlike Saussure (but like Bakhtin), Merleau-Ponty was interested not in language as a superpersonal structure but in the normativity of lived speech practice. This normativity, he realized, cannot be grounded in individual bodies or individual minds but requires a field of divergences that is inherently social and diacritical in nature. Moreover, this normativity is dynamic, as captured in the German word *Normierung*, which is always the *birth* or *formation* of a norm. Merleau-Ponty considers language not primarily as a denotative or designative system but as a closed self-referential system. Language opens up unto a world not in spite, but because of its self-reference. Like other genetic thinkers, Merleau-Ponty was interested in the developmental shift that takes place when the child progresses from prelinguistic babbling to linguistic signification. He argued that what is acquired by the child in the process of learning to talk is not a code for making reference to internal or external states of affairs but rather an active *principle of distinction*. To enter the world of language is to enter the world of human expression, a world in which each action and each expression is *stylized* according to the internal logic of a normative system of self-differentiation. Words are expressive and generative, and their form can therefore not be finalized or conceptually demarcated,` or they cease to be words: "The marvel is that this simple power of differentiating significations which it cannot give to what does not have them, namely, speech, nevertheless seems to contain and carry significations within it" (Merleau-Ponty, 1973a, p. 103). Indeed, as Merleau-Ponty (1973a) puts it elsewhere, "The meaning of expressions in the process of becoming habitual can in principle be, as it were, a lateral or oblique meaning which results from the commerce between the words themselves (or from the available significations)" (p. 46). Such available significations, however, do not simply constitute a system of semiotic means, but are part of a generic style of being:

> Style is what makes all signification possible. Before signs or emblems become for everyone, even the artist, the simple index of already given significations, there must be that fruitful moment when signs have given form to experience or when an operant and latent meaning finds the emblems which should liberate it. (Merleau-Ponty, 1973a, p. 58)

CONCLUSION: TEMPORALITY AS GENESIS

Genetic thinking in psychology is a legacy that owes much to the seminal work of Baldwin, Vygotsky, Werner, and Piaget; but true to the spirit of genetic thinking, this legacy needs to be a living one. Baldwin clearly understood that the success of any genetic method in psychology hinges critically on our ability to articulate the *genetic logic* required to understand both change and emergence in biological, cultural-historical, and psychological development. Bergson challenged us to see life inherently in terms of self-differentiation and authentic temporality and to go beyond the "spatializing" categories of science that only manage to grasp external relations between conceptually demarcated entities. However, once we recognize that the inherent temporality of "life" cannot be fully grasped conceptually or from the "outside," we need not necessarily abandon language and concepts altogether and resort to a Bergsonian kind of intuition. Maturana's ideas of structural determination and the irreducibility of different phenomenal domains points at the possibility of a historical account of the way our natural expressivity is brought into accordance, "in tune," or "in sync" with the consensually enacted normative practices of a community. Such a historical account allows us to see cultural practices as closed generative domains, produced by a recursive consensual coordination of our own activities. Acquiring culture, on such an account, is not to acquire a pregiven norm but to acquire certain dynamical dispositions or normative skills that allow one to remain attuned both to others and to the ever-changing domain itself. Rather than a process of internalization, this involves a history of *congruent structural changes* between an embodied agent and her or his cultural medium.

As I have argued here, Merleau-Ponty's notion of *style* may help provide an ontological basis for such a historical account. Merleau-Ponty understands style as the "coherent deformation" of a norm, but recognizes as well that a "norm" is itself always a *logos* that remains without positive terms, so to speak. For him, each "stage" in the development of either culture or person is the *advent* of an entire style of being, which is neither just contingent on the past nor already anticipated in advance. Style "unifies" in a way that does not require conceptual or propositional agreement. As psychological agents, we always participate in a particular cultural or historical style in our own uniquely stylistic way. The notions of *attunement*, *consensual coordination*, and *style* offer an alternative to conceptions of culture and community based on the psychological similarity of individuals. Culture, on such an alternative conception, is not based on actors being "the same," but on actors coordinating and stylizing their *differences*. For instance, each man can be seen as taking up the generic "norm" of masculinity in a uniquely stylistic way, but the norm itself is a style that can never be fully captured in conceptual or propositional form.

If the generativity of language, music, dance, and other domains of culture is due to their recursively consensual nature, we might reasonably infer that the acquisition of normative competence in any of those domains also follows a recursive logic. Almost all learning situations in culture are structured in a way that allows cultural apprentices to reapply certain principles of distinction (particularly the distinction between "getting it right" and "getting it wrong") to the outcome of their own previous conduct. Such recursive operations allow the cultural learner to come to see particular behavioral outcomes as *necessary* or *normative* rather than merely contingent on the past, and this in turn leads to the coherent deformations we recognize in personal style. The idea that in the course of development our conduct becomes *recursively coordinated* bears resemblance to Baldwin's notion of *persistent imitation*, which involves a move from the mere repetition of other people's behavior to "a method of absorbing what is present in others and of making it over in forms peculiar to one's own temper and valuable to one's own genius" (Baldwin 1906, p. 22; see also Valsiner, 1989, pp. 52–53; Valsiner & van der Veer, 2000 pp. 151–166).

On the account I have presented here, these processes are inherently consensual and expressive and should not too quickly be connected to a process of internalization. Whereas internalization implies a duality of "inner" and "outer," I propose an alternative in the idea of an increasingly differentiated sensitivity and responsivity to the particulars of one's own historically unique situation. Such an account does not necessarily dispose of the notion of "innerness," but would see it as performative and produced in what metaphorically could be described as a "folding back" of expressive life unto itself. In his attempt to wrest the notion of spatiality from the concept of pure exteriority, Merleau-Ponty repeatedly spoke of a primal dimension of "depth," in which inside and outside are recursively entwined or "turning about one another" and not just dialectically opposed (Merleau-Ponty, 1968, p. 268). Addressing Bergson's apprehension, "depth" could be seen as allowing the painter's vision to be more than "a view upon the outside, a merely "physical-optical" relation with the world" (Merleau-Ponty, 1964b, p. 181). Depth can also bee seen as allowing our "inner" lives to be in plain sight, so to speak, rather than hidden in the interior recesses of the mind. I believe that if we are to maintain the Vygotskyan language in this regard we would have to conceive of internalization quite clearly as an expressive and consensually coordinated act. Of course, unique for human development is the increasingly differentiated capacity for "self-correction" and "self-regulation," which on a genuinely genetic account, leads not merely to additive change over time but to generative or qualitative transformations in the very way a human being is disposed to the world. However, the recursive operations that underlie such generative transformations can take place only as dynamic or temporally enfolding *attunements* to our normative practices. In turn, however, it is the logic of *recursion* that truly makes those practices and our participation in them historical.

This finally brings us back to the question of the implications of our understanding of temporality for a genetic cultural psychology. In this chapter, I connected the notion of *temporality* to that of *generativity* and argued that the inherent generativity of action and expression is due to the peculiar recursive nature of human consensual domains. Consensual domains are generated and maintained by the mutual attunements of their participants, not just to each other, but foremost to the normativity or "style" of the consensual domain as a whole. Within human consensual domains, each recursive consensual coordination gives rise to a new closed generative domain. Hence, considered from a semiotic point of view, what may seem as a process of cognitive distancing is viewed from an expressive/enactive point of view the entrance into ever-new, recursively emerging, generative domains of interaction. None of those generative domains is reducible to any other domain, although different domains can be coupled in a *history* of congruent structural changes. Likewise, what may appear as *mediated* action is viewed from our expressive/enactive perspective rather a consensual coordination *of* consensual coordinations of action. For those who have acquired expertise in any of those generative domains, mathematics, conceptual thinking, and geometry are as participative, involved, and "immediate" as are ritual, dance, music, and everyday conversation.

Now we are also able to make some tentative remarks with regard to the implications for the methodology and ethics of psychological research. As Goethe already recognized, in its dynamic and temporal nature, life cannot be understood by merely imposing external conceptual forms on it. Goethe's "delicate empiricism" involved a way of moving from a careful observation of the particulars of living processes, through a process of "exact sensory imagination" to an intuitive grasping of the implicit wholeness at their source. Goethe and Herder were the first to propose the *genetic method* as a way to grasp the *formative* principles and processes at work in life and history. In Goethe's account of the genetic method, what is first presented *successively* in one's empirical experience is subsequently seen *imaginatively* as a continuous transformative process and is finally grasped *simultaneously* in the intuitively perceived idea (see Miller, 2009). Grasping living systems requires a responsive "tuning in" or what John Shotter (2005) has called "withness understanding" as distinguished from the "aboutness understanding" of a disengaged spectator. Each human expression and each human understanding, however, takes place within a closed generative domain of consensually coordinated action. As Merleau-Ponty recognized most aptly, to understand a human expression requires one to dwell in the style of that expression. From here, the meaning of an expression can never be pinned down or finalized. Indeed, such understanding, if genuine, is itself generative, as it participates in the very creative processes that it attempts to articulate.

ACKNOWLEDGMENTS

Research for this chapter was made possible by a grant from the Social Sciences and Humanities Research Council of Canada (SSHRC 410-2009-0891).

NOTE

1. The German *Besinnung* is typically translated as reflection. However, Forster in his translations of Herder's *Philosophical Writings* (Herder 2002) points out that for Herder *Besonnenheit* is a precondition of *Besinnung* and translates the two terms *Besinnung* and *Besonnenheit* as *taking-awareness* and *awareness* respectively. I choose the term *reflective awareness* to capture the idea that this awareness implies normative sensitivity, or, in the words of Taylor (1995), a sensitivity to "getting it right" (see also Baerveldt & Verheggen, 2012).

REFERENCES

Althusser, L., & Balibar, E. (1970). *Reading capital* (B. Brewster, Trans.). London, UK: New Left.

Baerveldt, C. (2013). Culture and dialogue: Positioning, mediation or style? *New Ideas in Psychology, 32,* 66–72.

Baerveldt, C., & Verheggen, T. (1999). Enactivism and the experiential reality of culture: Rethinking the epistemological basis of cultural psychology. *Culture & Psychology, 5,* 183–206.

Baerveldt, C., & Verheggen, T. (2012). Enactivism. In J. Valsiner (Ed.), *Oxford handbook of culture and psychology*. New York, NY: Oxford University Press.

Bakhtin, M. M. (1986). *Speech genres and other late essays*. Austin: University of Texas Press.

Bakhtin, M. M. (1993). *Toward a philosophy of the act* (V. Liapunov, Trans.). Austin: University of Texas Press.

Baldwin, J. M. (1906). *Mental development in the child and the race*. New York, NY: Macmillan.

Balibar, E. (1998). *Spinoza and politics* (P. Snowdon, Trans.) London, UK: Verso.

Bechtel, W., & Richardson, R. C. (1998). Vitalism. In E. Craig (Ed.), *Routledge encyclopedia of philosophy*. London, UK: Routledge.

Berducci, D. (2004). Vygotsky through Wittgenstein: A new perspective on Vygotsky's developmental continuum. *Theory & Psychology, 13,* 329–353.

Bergson, H. (1910). *Time and free will: An essay on the immediate data of consciousness* (F. L. Pogson, Trans.). London, UK: Swan Sonnenschein. (Original work published 1889)

Bergson, H. (1911). *Matter and memory* (N. M. Paul & W. S. Palmer, Trans.) New York, NY: Macmillan.

Bergson, H. (1922). *Creative evolution* (A. Mitchell, Trans.). New York, NY: Palgrave Macmillan.

Brandist, C. (2002). *The Bakhtin circle: Philosophy, culture and politics.* London, UK: Pluto.

Broughton, J. M. (1981). The genetic psychology of James Mark Baldwin. *American Psychologist, 36,* 396–407.

Carroll, L. (1998). *Alice's adventures in Wonderland.* Chicago, IL: Volume One Publishing. (Original work published 1865)

Cassirer, E. (1953). *The philosophy of symbolic forms, Vol. 1: Language* (R. Manheim, Trans.). New Haven, CT: Yale University Press. (Original work published 1923)

Cassirer, E. (1955). *The philosophy of symbolic forms, Vol. 2: Mythical thought* (R. Manheim, Trans.). New Haven, CT: Yale University Press. (Original work published 1925)

Cassirer, E. (1957) *The philosophy of symbolic forms, Vol. 3: The phenomenology of knowledge* (R. Manheim, Trans.). New Haven, CT: Yale University Press. (Original work published 1929)

Cresswell, J., & Baerveldt, C. (2009). Materiality matters: Reinterpreting the dialogical self along Bakhtinian lines. In T. Teo, P. Stenner, A. Rutherford, E.

Park, & C, Baerveldt (Eds.), *Varieties of theoretical psychology: International, philosophical and practical concerns*. Toronto, Canada: Captus.

Cresswell, J., & Baerveldt, C. (2011). Bakhtin's realism and embodiment: Toward a revision of the dialogical self. *Culture & Psychology, 17*, 263–277.

Deleuze, G. (1988a). *Bergsonism* (H. Tomlinson & B. Habberjam, Trans.). New York, NY: Zone. (Original work published 1966)

Deleuze, G. (1988b). *Spinoza: Practical philosophy*. San Francisco, CA: City Lights. (Original work published 1970)

Deleuze, G. (1990). *Expressionism in philosophy: Spinoza* (M. Joughin, Trans.). New York, NY: Zone. (Original work published 1968)

Derrida, J. (1993). Speech and writing according to Hegel. *Man and World, 11*, 107–130. (Original work published 1971)

Dreyfus, H. (1996). The current relevance of Merleau-Ponty's phenomenology of embodiment. *Electronic Journal of Analytic Philosophy, 4*. Retrieved from http://ejap.louisiana.edu/EJAP/1996.spring/dreyfus.1996.spring.html

Goldstein, K. (1939). *The organism: A holistic approach to biology derived from pathological data in man*. New York, NY: American.

Hegel, G. W. F. (2010). *The science of logic* (G. Giovanni, Trans.). Cambridge, UK: Cambridge University Press. (Original work published 1832)

Herder, J. G. (2002). Treatise on the origin of language. In M. N. Forster (Ed.), *Herder: Philosophical writings* (pp. 65–166). Cambridge, UK: Cambridge University Press. (Original work published 1772)

Kreppner, K. (1999). Enactivism and monadology: Where are Baerveldt and Verheggen taking the individual and cultural psychology? *Culture & Psychology, 5*, 207–216.

Lawlor, L. (1998). The end of phenomenology: Expressionism in Deleuze and Merleau-Ponty. *Continental Philosophy Review, 31*, 15–34.

Lawlor, L., & Moulard, V. (2012, Fall). Henri Bergson. In E. N. Zalta (Ed.), *The Stanford encyclopedia of philosophy*. Retrieved from http://plato.stanford.edu/archives/fall2012/entries/bergson/

Maturana, H. R. (1970). *Biology of cognition*. Biological Computer Laboratory Research Report BCL 9.0. Urbana: University of Illinois.

Maturana, H. R. (1975). The neuropsychology of cognition. In P. Garvin (Ed.), *Cognition: A multiple view*. New York, NY: Spartan.

Maturana, H. R. (1978). Biology of language: The epistemology of reality. In G. A. Miller & E. Lenneberg (Eds.), *Psychology and biology of language and thought: Essays in honor of Eric Lenneberg* (pp. 27–63). New York, NY: Academic.

Maturana, H. R. (1988). Reality: The search for objectivity or the quest for a compelling argument. *Irish Journal of Psychology, 9*, 25–82.

Maturana, H. R. (2007). Systemic versus genetic determination. *Constructivist Foundations, 3*, 21–26.

Maturana, H. R. (2012). Reflections on my collaboration with Francisco Varela. *Constructivist Foundations, 7*, 155–164.

Maturana, H. R., & Varela, F. J. (1980). *Autopoiesis and cognition: The realization of the living*. Dordrecht, The Netherlands: Reidel.

Maturana H. R., & Verden-Zöller G. (2008). *The origins of humanness in the biology of love* (Edited by Pille Bunnell). Exeter, UK: Imprint Academic.

Merleau-Ponty, M. (1962). *Phenomenology of perception* (C. Smith, Trans.). London, UK: Routledge. (Original work published 1945)

Merleau-Ponty, M. (1964a). *Signs* (R. C. McCleary, Trans.). Evanston, IL: Northwestern University Press. (Original work published 1960)

Merleau-Ponty, M. (1964b). Eye and mind. In J. M. Edie (Ed.), *The primacy of perception* (pp. 159–190). Evanston, IL: Northwestern University Press.

Merleau-Ponty, M. (1968). *The visible and the invisible* (C. Lefort, Ed.; A. Lingis, Trans.). Evanston, IL: Northwestern University Press.

Merleau-Ponty, M. (1973a). *The prose of the world* (C. Lefort, Ed.; J. O'Neill, Trans.). Evanston, IL: Northwestern University Press.

Merleau-Ponty, M. (1973b). Consciousness and the acquisition of language (H. J. Silverman, Trans.). Evanston, IL: Northwestern University Press. (Original work published 1964)

Miller, G. L. (2009). Appendix. In J. W. von Goethe (Ed.), *The metamorphosis of plants.* Cambridge, MA: MIT Press.

Piaget, J. (1970). *Structuralism.*(C. Maschler, Trans.). New York, NY: Basic. (Original work published 1968)

Piaget, J. (1980). *Les formes élémentaires de la dialectique* [Elementary forms of dialectic]. Paris, France: Editions Gallimard.

Prigogine, I., & Stengers, I. (1984). *Order out of chaos: Man's new dialogue with nature.* New York, NY: Bantam.

Shotter, J. (2005). Goethe and the refiguring of intellectual inquiry: From "aboutness"-thinking to "withness"-thinking in everyday life. *Janus Head, 8*, 132–158.

Taylor, C. (1985). *Human agency and language: Philosophical papers volume 1.* Cambridge, UK: Cambridge University Press.

Taylor, C. (1995). The importance of Herder. In C. Taylor (Ed.), *Philosophical arguments* (pp. 77–99). Cambridge, MA: Harvard University Press.

Uexkuell, J. von (1982). The theory of meaning. *Semiotica, 42*, 25–82. (Original work published 1940)

Valsiner, J. (1989). *Human development and culture: The social nature of personality and its study*. Boston, MA/Toronto, Canada: Lexington.

Valsiner, J. (1994). Irreversibility of time and the construction of historical developmental psychology. *Mind, Culture, and Activity, 1*, 25–42.

Valsiner, J. (1998). The guided mind: A sociogenetic approach to personality. Cambridge, MA: Harvard University Press.

Valsiner, J. (2009). Cultural psychology today: Innovations and oversights. *Culture & Psychology, 15*, 5–39.

Valsiner, J. (2012). Psychology courting culture: Future directions and their implications. In J. Valsiner (Ed.), *The Oxford handbook of culture and psychology* (pp. 1092–1104). Oxford, UK: Oxford University Press.

Valsiner, J., & van der Veer, R. (2000). *The social mind: Construction of the idea*. New York, NY: Cambridge University Press.

Varela, F. J., Thompson, F., & Rosch, E. (1991). *The embodied mind: Cognitive science and human experience*. Cambridge, MA: MIT Press.

Verheggen, T., & Baerveldt, C. (2007). 'We don't share!' Exploring the theoretical ground for social and cultural psychology: The social representation approach versus an enactivist framework. *Culture & Psychology, 13,* 5–27.

Verheggen, T., & Baerveldt, C. (2012). Mixed up perspectives: A reply to Chryssides et al. and Daanen, and their critique of enactive cultural psychology. *Culture & Psychology, 18,* 272–284.

Vygotsky, L. S. (1978). Interaction between learning and development. In M. Cole, V. John-Steiner, S. Scribner, & E. Souberman (Eds.), *Mind in society: The development of higher psychological processes* (pp. 79–91). Cambridge, MA: Harvard University Press.

Vygotsky, L. S. (1987). *The collected works of L. S. Vygotsky: Vol. 1, Problems of general psychology* (R. W. Rieber & A. S. Carton, Eds.; N. Minick, Trans.). New York, NY: Plenum.

Wertsch, J. V. (1991). *Voices of the mind.* New York, NY: Harvester.

Wittgenstein, L. (1974). *Philosophical investigations* (G. E. M. Anscombe, Trans.). Englewood Cliffs, NJ: Prentice Hall. (Original work published 1958)

Wittgenstein L. (1978). *Remarks on the foundations of mathematics* (3rd ed.; G. E. M. Anscombe, Trans.). Oxford, UK: Blackwell. (Original work published 1956)

CHAPTER 19

TEMPORALITY AND GENERALIZATION IN PSYCHOLOGY

Time as Context

Luca Tateo

"How many times must a man walk down, before you call him a man?" I paraphrase the famous Bob Dylan song to introduce the topic of this chapter: time is not a given natural event nor a solipsistic experience; rather it is the socially constructed context of human psychological life. This context is made of both synchronic and diachronic continuities and discontinuities (Tateo & Iannaccone, 2011). It is constructed at different levels, corresponding to the progressive distance from the individual immediate experience. I will draw a model based on the hints from Bronfenbrenner's ecological model (1979, 2004) and Doise's levels of explication (1986), in which time is considered at different levels (microtime, mesotime, esotime, and macrotime). For a long time, psychology put aside the time dimension from its field of interest (Valsiner, 2011), as far as time is a complex problem to deal with. At the turn of the 19th and 20th centuries, the idea developed that the individual psychological experience of time does not necessarily

Temporality: Culture in the Flow of Human Experience, pp. 463–481

correspond to the actual temporal displacement of events. For instance, Benussi (1913) observed the phenomenon of temporal displacement. He provided evidence that "a series of events, which are therefore in sequence on a physical time gradient, does not necessarily correspond to the series of events that inhabit phenomenal time" (Sinico, 1999, p. 75). Also, William James described the psychological experience of time as synthetic apprehension, in which past and future join to produce the experience of

> a certain duration—the specious present—varying in length from a few seconds to probably not more than a minute, and that this duration (with its content perceived as having one part earlier and the other part later) is the original intuition of time. (James, 1886, p. 406)

James also sketched the complementarity of continuity and discontinuity with respect to time in the subject's experience. Such a dialectic is based on the fact that time is both the way we give continuity to the experience and the rationale we use to segment it. "The great world's ingredients, so far as they are beings, seem, like the rope's fibers, to be discontinuous, cross-wise, and to cohere only in the longitudinal direction" (James, 1981, p. 67). Add to this that for James, human beings deal with continuity through discourse and action, which are social activities:

> Everything gets known by *some* knower along with something else; but the knowers may in the end be irreducibly many, and the greatest knower of them all may not yet know the whole of everything, or even know what he does know at one single stroke: he may be liable to forget. (James, 1981, p. 68, emphasis in original)

If one can say, according to Kierkegaard, that we live forward but we understand backwards, temporality can be understood as a complex experience (Dru, 1938). What is the time perspective of such a type of experience? What is the breadth of the concepts accounting for psychological processes over historical, cultural, and ontogenetic time? I will try to argue that temporality experience emerges from a complex semiotic process that involves several layers of time in which individual and culture intertwine. These questions are strictly related to the problem of the universality of psychological processes and the uniqueness of individual experience. If we put into the cauldron also the cultural dimension, here we have the hot soup of the nature/nurture debate and the person construct reality versus reality construct person dilemma (Flanagan, 1981).

The other side of the coin of generalization is temporality. In what sense can psychological concepts be generalized over time? In other words, what is the temporal range of these concepts. Just think, for instance, of the problem of development. From a perspective in which development was limited to the first years of individual life, psychology came to the idea of life-span development, assuming that this concept covers a wider time range and becomes generalizable to the whole ontogenetic development along life course (Elder 1985; O'Rand & Krecker, 1990). Psychological science has developed different theoretical and methodological tools to generalize its concepts with respect to time. Basically, there are two possible directions of generalization over time: retrospective longitudinal studies that look back in time, and repeated measures designs based on correlations between the same variables over time (Menard, 2008), including for instance the deferred posttest design. The assumption is that if a correlation is quite stable over repeated subsequent measures, it is reasonable to infer that the correlation can be generalizable in a longer time span. Is it worth reflecting about the way psychological experiences occur at different times? Do they occur in a similar way? What does "similar" mean? How do psychological experiences imply the idea of temporality? Looking at temporality can help to better understand concepts such as self, development, change, continuity, and discontinuity, and to different possibilities of generalizing this knowledge.

TEMPORALITY AND EXPERIENCE

Walking in the streets of a Danish town, my attention was drawn to the shop window in Figure 19.1. The clothes shop was designed in a sort of old-fashioned wood style, and the name was probably chosen to confirm this communication strategy. But at the same time, the products in the windows were actually trendy clothes and accessories of a very famous stylist's contemporary brand. The outcome of this juxtaposition was an impression of vintage. Though this is a very common marketing trick, it requires something to be effective: a kind of consensual perception of time despite the phenomenological evidence of fake antiquity.

Even if we are perfectly aware that the picture is not constructing a truthful version of time, we accept this game—someone being attracted by the vintage-fashioned style and even buying the expensive clothes—and this kind of consensual acceptance allows many forms of communication in marketing, art, literature, and such. This game is made possible by the consensual and interpretative nature of what Heidegger calls *datability* of world time in the ordinary understanding of time (2010). Both Heidegger and James (1950) agree upon the fact that time enters the public sphere through acts of segmentation, measurement, and semiotization. Thus, socialized time enters the realm of interpretation and meaning, making possible the communication of time and coordination of collective activities (Couzens Hoy, 2009). Besides, this segmentation and semiotization

Figure 19.1. Construction of time by juxtaposition.

generates the common sense knowledge about time as an infinite sequence of "nows," explaining the difference between the immediate synthetic experience of temporality and the intersubjective analytic experience of ordinary time. Nevertheless, the segmentation/semiotization of time allows the relationship between the "prereflective ways in which Dasein copes with its world and the more reflective or conceptually articulated Interpretation that formulates its commitments in words" (Couzens Hoy, 2009, p. 57). Thus, both immediate and semiotized times interact in the construction of temporality and its meaning. "The interpretative expressing of the 'now,' the 'then,' and the 'on the former occasion,' is evidence that these, stemming from temporality, are themselves time" (Heidegger, 2010, p. 375). This relationship—as every semiotic process—is possible only within a societal framework of meaning-making: the world time. I would say here that temporality is a complex semiotic construction that links the individual experience to a societal construction of time. For instance, in the case of the shop window, the "vintage" temporality experience, with its affective and aesthetic dimensions that are well exploited by advertisement (see Figure 19.2), emerges from the semiotization through juxtaposition of elements (the old-fashioned windows and the trendy clothes) belonging to different moments of ordinary time.

Figure 19.2. Advertisement exploiting the juxtaposition of temporal elements.

This juxtaposition is interpretable only if framed by a societal system of symbols, that is, the shared world time that allows us to place the different elements in the right historical time in the same moment we grasp them as a present whole. We are thus in the presence of at least three different interacting layers of time that concur in generating the temporality experience: the individual, the intersubjective, and the societal. I will argue in the second part of the chapter that this multilayered time can be understood as the time context in which every experience takes place. But first it is worth looking at some concepts developed about time and temporality.

IDEAS ABOUT TEMPORALITY

Temporality in psychology is a controversial issue, probably because time is a very complex concept, involving several dimensions leading to the development of different perspectives. Some important reflections about temporality come from Heidegger (2010), Bergson (2001) and James (1950). Their ideas about time and temporality set the theoretical grounds of the relationship between time and psychological processes, introducing for instance the concept of the vanishing present that has been further elaborated by Valsiner (2011), with idea that psychological processes must be always considered in irreversible time. I will try now to summarize the concepts that I grasped and elaborated in the model of time-as-context I will draw in the following sections.

a. Human beings experience time under different forms: the sense of being in a given moment in time, the living experience of the present time, and the sense of time passing by through the observation of change occurring in the world. James (1886) vividly describes the tension in the experience of time: "The same space of time seems shorter as we grow older—that is, the days, the months, and the years do so; whether the hours do so is doubtful, and the minutes and seconds to all appearance remain about the same" (p. 391).
b. Time looks like a flow, or a stream, after it has been parted and marked off by arbitrary and conventional acts related to a process of sense making, probably based on natural phenomena such as circadian rhythms (James, 1886; McGivalry, 1914). Heidegger (2010) makes a distinction between primordial time, world time, and ordinary time. The ordinary understanding of time cannot grasp its irreversible directionality and its finite nature (Couzens Hoy, 2009). Besides, for Heidegger, time is not like a river flowing from past to the present and toward the future. Once we shift from the primordial time to the ordinary understanding of time, we introduce the process of interpretation, leaving room for "authentic and inauthentic interpretations of temporality" (Couzens Hoy, 2009, p. 57). Heidegger's purpose is to understand the nature of time and its relationship with the human experience. By doing so, he calls for authentic interpretations of time with respect to its own primordial nature. But the aim of psychological science is to understand the experience of temporality rather than the nature of time. In this respect, it does not exist as an "inauthentic" experience, and the process of interpretation is related to the semiotization.
c. The relationship between past, present, and future is not ontological but rather semiotic (Valsiner, 2007). The use of signs is the constitutive element in the human experience of time. Semiotizing time is the way to construct an abstraction of past, present, and future as objects of reflection. They become manageable and shareable beyond the immediate experience. Human beings can then play with them, remembering, planning, regretting, reflecting, imagining, and adjusting life trajectories. Through this primary semiotization of time, the temporal stage for every human action is set up. "The now does not exist on

its own but exists only in relationship to the retention of past events and the expectation of future events that constitute the contextualizing perspective in relationship to which it can be noticed" (Russon, 1998, p. 17).

d. Life time is experienced as irreversible, leaving behind what is no longer and facing the uncertainty of the future (Bergson, 2001; Valsiner, 1994, 2011). This element triggers a process of reducing the uncertainty through an as-if anticipation of future events and at the same time determines the way human beings value both the present and the future potential directions of development. Based on the primary semiotic distinction, the way to envisage alternative trajectories is open. Thus, the primary synthetic unit past-present-future suddenly multiplies in a number of as-could-be, as-could-have-been, as-should-be, as-should-have-been, and so on.
e. Time is related to the causality and change to such an extent that time totally loses its ontological status, acquiring a process nature in which, as in the case of Whitehead (1979) philosophy, time, causality, and change are interwoven concepts. "The world is made of 'actual occasions,' each of which arises from potentialities created by a prior actual occasion" (Klose, 2008, p. 2). This relationship between causality and change is like the two sides of a coin, where on one side there is the effigy of Ananke and on the other side the effigy of Tyche.[1] Thus, on one hand, people are subject to irreversible time, that unfolds its events independently by any attempt to regulate it, just as Ananke superintends that any action is followed by its own effect *by necessity*. On the other hand, every step opens a range of bifurcations, affordances, and possibilities; that's where Tyche rules, and orientation toward the future becomes a social and ethical matter. Time becomes then the semiotic battlefield between constraints and affordances, agency and ineluctability. "All our sign-making is that of creating semiotic constraints to guide us in the immediate future" (Valsiner, 2011, p. 146), making us the only living beings that create by themselves their personal Olympus, populated by Anankes and Tyches.

When time starts to be understood as a complex psychological experience, rather than an external reality affecting the perception in virtue of some kind of mirroring homology between environment and psychophysical functioning, the problem arises of what is the relationship between inner and external time. Sometimes psychology excluded time from its theories, leading to the idea of a time-free psyche (Valsiner, 2011). In other words, the psychological phenomena and their laws were considered independent from time, or at least it was not relevant where they came from and where they were going, despite some scholars like Baldwin (1915), Lewin (1935) and Bronfenbrenner (1979) underlining the genetic and changeable nature of these phenomena.

From the ideas abovementioned, I see a set of epistemological principles that can help to better understand how to study temporality in psychology. First, principles a, b, and c show that psychological processes have to do at least with four different levels of analysis, each one implying a different dimension of time and temporality. Time is in fact experienced as an individual psychological phenomenon, then is shared and communicated at an intersubjective level, leading to a coordination that becomes socially constraining (Fuchs, 2005). Finally, time is historicized, becoming a shared heritage that constitutes the context of future experiences (Tateo & Iannaccone, 2011). All these levels are contemporarily present in time experience, even if psychological theories do not take them all into account.

From the principles c, d, and e we can argue that time is asymmetrical in a twofold sense: we assume that action takes place in the present, with a known and unchangeable past and an unpredictable future. But we can also argue that human beings do not submit passively to time's authority. First of all, temporality is experienced as a relationship rather than essence. The past is not something that has always been past, and the future is something that sooner or later will be. The future is then a product of past and *vice versa*. This is the case, for instance, of historical revisionism, in which the past is reconstructed in the arena of semiotic struggle as a function of future projects (Tateo, 2008). Time is also related to causal relationships, as far as in psychological life, causation is established through a temporal relationship in which the future is caused by past events and in return, the expectations and previsions about the future contribute to determine past events (Benussi, 1913). Asymmetry is instead established with respect to the irreversibility of time, to such an extent that the past always grows, while the future decreases (Valsiner, 2011). The present is the third element of this relationship which actualizes the border between past and future. It is not the only ontological reality, rather the border between no-longer and not-yet (Valsiner, 2011). This border zone, which William James (1886) called "specious present" (p. 377), is an ontological trap that sometimes psychology falls into. "If we speak of now (present) for the actuality of the soul, we become fixated at a singular point void of structure which merely marks the spot along the temporal sequence in which the flow of time has been interrupted" (Diriwächter, 2006, p. 164). Indeed, most parts of psychological concepts and methods are tailored to capture something *that is*, or supposed to be, as the present would have an ontological autonomy. Valsiner (2011) argued how specious the present can be understood as the developmental transition in which A is no-longer B and not yet. Thus, the present represents a border rather than a location, and this border is not the *outcome* of experience but the *context* in which experience develops. The relationship between past and future, occurring at the border of the present, has not only a causal relationship but also a developmental character. In fact, from principles d and e, we can deduce that past events set the ground of potential alternative trajectories we use to anticipate the future(s) in its various forms—as it could be, as it won't be, as it should be, and so on (Valsiner, 2011). This process of incrementing knowledge through past experience is the source of development potentialities, generating the present, in which occasions actualize, as the point in which an indistinct future unfolds in its various forms of "as-could be," "as-should-be," and "as-I-wish" (Klose, 2008; Valsiner, 2011). This idea of temporality is hardly handled by psychology, which is somehow exposed to *a pressure to present*. The implicit assumption of pre-test/post-test and repeated measures experimental design is indeed that what matters is something happening here and now, in which past is nothing but the pre-test state of the subject and future is the result of some intervening, experimentally controlled, modifications. It presupposes a shift rather than a transition, which is also reversible, in the sense that once the intervening element is removed, things can return to the initial state. Here, several problem arise: if we think for instance of Milgram's experiments, Piaget's interviews, or Lewin's action-research, it appears evident that, for the individuals involved, the meaning of the situation is not static, for both the participant and the psychologist, because "the whole situation changes with the process" (Lewin, 1935, p. 33). Also in this case, the problem of generalization is strictly related to time. An experimental design will indeed take into account an arbitrary time frame, establishing it as the context in which psychological phenomena occur. What if a larger time span would be included in the theory? Of course it is not just the idea of deferred posttest, which is not theoretically different, rather the simple idea of including in the experimental design of what happened before the subject enters the lab and what will happen after. How would this change hypotheses and affect their generalization?

GENERALIZATION IN PSYCHOLOGY

Generalization is a creative and reflective act of thought through accessing a higher level of abstraction from meaningful events. Scientific method provides a set of conceptual artifacts, constraints, and norms of sharing that make it possible to reach a partial and temporary agreement about the meaning and the extent of such an elaboration. Thus, generalization is not the mere collection of evidences, is not the generalization of results, rather the generalization of theory about psychological laws. Generalization of psychological theories is also a matter of temporality. The specificity of psychological knowledge is to study phenomena characterized by the uniqueness of experience, the irreversibility of time, and the meaning, understood as sense making for the observed and the observer. These characteristics of psychological phenomena create the problem of deriving general knowledge from the singularity (Salvatore & Valsiner, 2010; Toomela, 2009). The experimental practice of repeated measures goes in the opposite direction, as far as it looks for similarities and repetition, leaving aside what is not recurrent, that is, the process of making sense of the unique experience. If the time context is introduced, then generalization must take into account also the temporal breadth of psychological processes. For instance, if we focus on self or personality traits that are considered quite stable psychological constructs, which temporal context should we focus on? Life span, social-historical context, or the limited time we use to apply research methods to collect data? "What the particular Gestalt of a phenomenon entails, is precisely contingent upon where along the time continuum the researcher places his/her 'dissection-tools'" (Diriwächter, 2006, pp. 165–166). This choice will directly affect the generalizability of the construct, both in diachronic and synchronic dimensions. Theory provides us with resources and constraints, guiding the data collection and interpretation in such a way that even "understanding a case is limited to generalized knowledge" (Toomela, 2009, p. 20). But neither stockpiled evidences nor individual case's idiographic understanding stand for generalization. Thus, generalization implies both an abstraction from the phenomenological level and a perspective-taking by the observer, namely, a time perspective. As James (1886) argued, the psychological experience of temporality is fundamentally a synthetic datum, a Gestalt, depending on the time context in which is experienced. Thus, the challenge of elaborating generalizable knowledge from the uniqueness of experience in time is not accomplished by generalizing neither in an extensive synchronic way (e.g., how many people experience the same) nor in a diachronic way (e.g., for how long). If my arguments are correct—that time is not the result of experience but rather its context, and that the experience of temporality is possible only in time as context—then the relationship between temporality and generalization in psychological knowledge can be understood only by introducing a conceptual and multilevel model of temporality in which this time-context idea is embedded.

TIME AS CONTEXT

I will try to gather the concepts discussed so far to define a theoretical model of time as context. In the experience of temporality, there is a process of semiotic co-genesis between time and forms (Herbst, 1976). From the principles discussed above, it appears that the distinction between "before" and "after" and the distinction between "what is" and "what is not"—no longer and not yet—is generated together with the semiotization of the experience, framed in an intersubjective and societal time. Let me use a short biographical account as an example.

Some time ago, I escorted my lady, who had to go on some errands. She was supposed to be busyfor about an hour. We parted in the street and we made a date later in the afternoon in the same place. I decided to wait on a park bench reading some papers. My account is thus comprised within event A (parting) and event B (meeting again) in the objective time span of 60 minutes (Figure 19.3a).

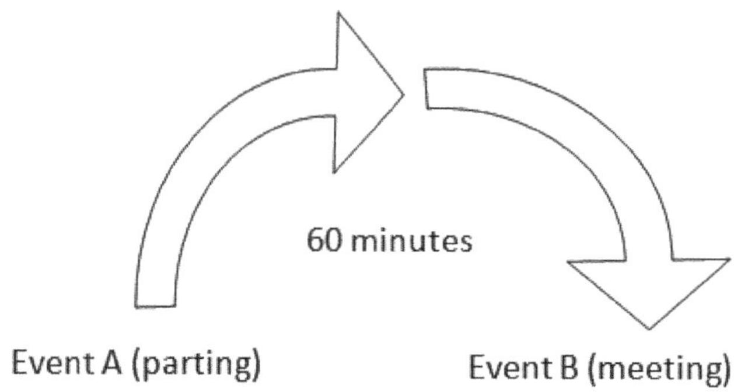

Figure 19.3a: A personal experience of time.

A few minutes after we had said goodbye and parted, I started reading on the bench and felt relief for beginning the new activity. I concentrated on reading, occasionally raising my eyes at people passing by or looking at my watch. After some time, I started to guess that about half an hour had passed and I checked my watch again. I was also surprised at how fast those 30 minutes had elapsed. I said to myself that a sort of border had been overcome, entering a new portion of time: that of expectation for meeting my lady again. Then I started reflecting about the way I had constructed the duration of the waiting and measured time. Thinking back to my acts during the first half of the waiting, I remembered that I (a) had been reading, making hypotheses on the time needed for reading each page; (b) sometimes I looked at my watch; (c) I looked at the shops around and the people passing by; (d) and I looked at the daylight. Through all these elements, I constructed by myself an idea of duration and progressively emerged from my consciousness a *feeling* or an *estimate* of the time. Then a second phase began (see Figure 19.3b). The second 30 minutes seemed longer, with time going slower. I became eager and increased the frequency of the acts aimed at constructing an external reference of the elapsed time. I experienced expectation and haste, and tried to go on reading my papers but with less attention. What happened there? What constructed my experience? What was the relationship between (a) the cognitive process of time estimating; (b) emotional states; (c) and the external elements and artifacts I referred to? What kind of change went on during those 60 minutes?

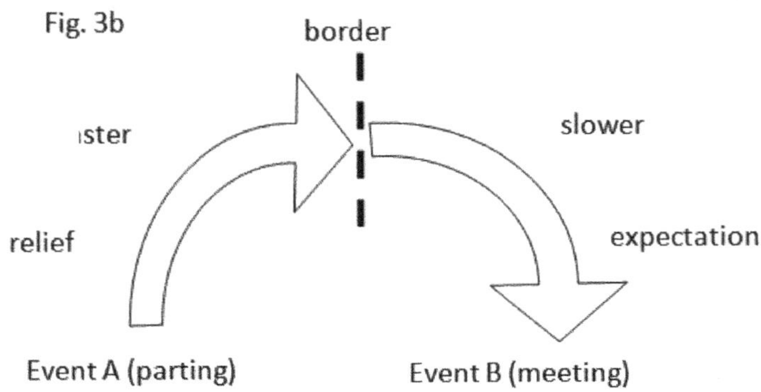

Figure 19.3b: A personal experience of time.

This apparently insignificant life episode can provide an insightful example of how temporality—that for sake of this initial discussion I define as the experience of meaning construction with respect to time—is related to several layers of perception, acting, and intersubjective processes. First of all, in my account, I speak of event A and event B as if these two occurrences in time were actually meaningful points in the history of mankind. They are instead meaningful for me and my lady (perhaps) and become more or less meaningful for the reader of this chapter as well, *but for a different reason*. During the 60 minutes of my very local experience, many other events occurred within the same time frame and space, despite the "I" perceiving them or not. This leads to the first consideration: temporality is characterized by a selective configuration of occurring stimuli—events defined within a context of self-awareness. Selection implies the idea that there is a larger whole, an infinite number of occurrences that I could had been aware of but was not. In other words, my account does not describes "what happened" during a time interval measured by 60 minutes, rather a different phenomenon that has its own features. We are thus in the presence of two different processes, both apperceptable as Gestalt. The first Gestalt is made of all the occurrences that can be grasped in a given set of spatial coordinates, which could be named context, whereas a causal connection is considered meaningful. In this case, the idea of time "is not a line; it is a stream in which events stand in transverse as well as in longitudinal direction *inter se*" (Mc Givalry, 1914, p. 128). Thus, the condition for the construction of this first Gestalt is not the simple contemporaneity, that is, the occurrence within the same time frame, but also any kind of meaningful connection between occurrences, *made possible* by contemporaneity. For instance, what happened in the street while I was sitting on the bench, with people running into each other, trying to avoid each other, interacting, or such is defined, framed, and understood as a Gestalt within the same time context. The second kind of Gestalt, of course, is the experience I was actually experiencing, precipitated in my account of "myself on a bench for 60 minutes." It is somehow related to the first whole but is totally different as well. It would sound trivial to say that it is *contained* in the larger one, because their relationship is more complex. It is a selection of some contemporary occurrences, ignoring much of what was there in the same time frame. Thus, my experience is always a partial negation of something else, is a Gestalt by opposition, and "whenever I speak of it I fail to do justice by underexpressing it, by distorting it, by introducing negation, limitation, rank into seamless, immediate all" (Russon, 1998, p. 14). At the same time, my experience becomes meaningful with respect to the primary distinction I construct, establishing a border that co-generates a "before" and an "after." The question is, What are the criteria for this selection? What is beside the nature of the "border" I experience at a given moment? Is it the boundary between different experiences, or is it a process of change during the same experience? Traditional cognitive theories would simply explain it with the limited amount of cognitive resources that lead to a bias in the selection of infinite stimuli, oversimplifying the perceptive field. This would shift the problem to an ontological issue, that is, reality is complex and the human subject has not the resources to grasp it. Even if this explanation is true, it does not tell me why I experienced that 60 minutes in that particular way. In other words, this type of explanation is about the generalized functioning of experience, while the explanation I am seeking concerns the meaning of that particular experience. Furthermore, it does not explain why, given ontologically limited cognitive resources, in different moments of life, the person enacts a different kind of selection, experiencing different Gestalt. This leads to the paradoxical idea of explaining the uniqueness of experience through an idea of sameness of mental functioning. It is more reasonable to hypothesize that the selection has something to do with the relationship between the two kinds of Gestalt described and to the needs, expectations, and sense-making process involved in one's experience. We can try to theoretically deconstruct this Gestalt for the sake of understanding the multilayered nature of time as a context, but we can't empirically experience it as distinct kinds of temporality.

I argued that time can be considered the context for both the experience of temporality and psychological theories in this experience. Time is also a multilayered and complex concept in which several dimensions intertwine.

> The now does not exist on its own but exists only in relationship to the retention of past events and the expectations of future events that constitute the contextualizing perspective in relationship to which it can be noticed, the background by negation of which it can stand out. Now, in other words is always the now of a larger now. (Russon, 1998, p. 17)

The question is how to include time in a conceptual model that can be used to frame psychological phenomena and to guide theory through a reflective process about its implicit assumptions about time. In the short biographical sketch described above, can be acknowledged different layers of time, creating the time context for the psychological experience. These layers are more or less directly accessible to the individual experience—the historical time or the phylogenetic time are not always accessible and salient to experience—nevertheless they are all present. For instance, there is the subjective experience of temporality, with accelerations and slow downs, with a more or less arbitrary segmentation of time with respect to needs and expectations. There is an intersubjective dimension related to the interaction and consensual construction of temporality. There is also a social dimension of time, represented by the societal practices and artifacts that mediate the intrapersonal and intersubjective experience. Finally, there is a larger temporal context which is made of a historical frame in which experience occurs. The person has a different access to these layers and is differently aware of them. Nevertheless, they all contribute to the definition of the temporal stream and the construction of the relationship between past, present, and future. How can we conceptualize this relationship and how can it become a resource for psychological theory?

I started from two insights—Bronfenbrenner's process-person-context-time model (2004) and Doise's levels of explications model (1986)—to draw an ecological *time-as-context* model (see Figure 19.4).

The aim of this conceptual model is to represent the different layers of time that create the context for the temporality experience. The relevance of time as context is claimed by Valsiner (2007), who argues that all the psychological and intersubjective processes are guided by the fundamental problem of the unknown future, which is confronted by human beings through the powerful semiotic processes as the only way to challenge time's authority. In the model, inspired by Bronfenbrenner (1979) and Bronfenbrenner and Morris (1998) model, time is understood as a multilevel construct, which is experienced by the individual as a concentric system of contextual elements that are progressively more distant from direct experience. There is a *microtime* of individual temporality which is intimately

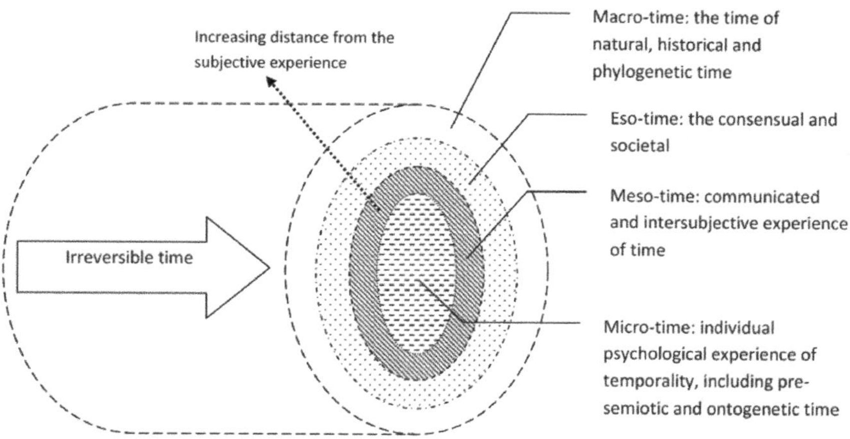

Figure 19.4. S model of time-as-context.

experienced and can follow a different rhythm from other people's one. It requires to be intersubjective *mesotime*, synchronized to make possible the communication and sharing of the individual experience, as in the "vintage shop" example. But this experience is in return possible only through the mediation of intersubjectivity, otherwise, it remains "unsaid" and implicit, not producing any new knowledge in the individual himself. Intersubjectivity is thus the process that triggers the time flow, as far as excessive intersubjective desynchronization is strictly related to pathological states such as depression or schizophrenia (Fuchs, 2005). But intersubjective construction of time as context is inscribed in a wider *esotime*, that is, the societal constructed time, including practices and artifacts related to time measuring and convention (e.g., time devices, social organization, and segmentation of time, scheduling, age and transitions, etc.).

> That time is datable means that it is connected to my practical activities such that different times of day have different significances for me. Time is not simply the bare numbers, but is tied to the coordination of events (for instance, the need to be in class by 1:30 P.M. or at dinner by 8:00 P.M.). (Couzens Hoy, 2009, p. 61)

Even if a person is not directly aware to this level, it is anyway a framework of reference for the sense making of temporal intersubjective experience (e.g., an affective relationship with a person living in a different time zone). Finally, there is a *macrotime*, which represents the larger historical and phylogenetic processes in which human experience is framed. Even if we are not aware of it, the way we experience relationships in *micro-* and *eso-time* are very different from those occurring, for instance, in the 14th century. Nevertheless, the fact that I am experiencing a given kind of relationship with my lady today is both the result of the historical-genetic evolution of couple relationships and the result of the difference between the past and the present models of a couple.

A question arises about what happens at the border between time levels (see Figure 19.5).

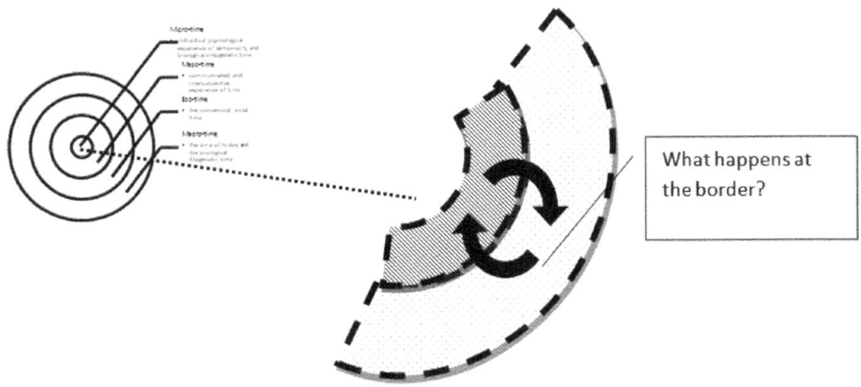

Figure 19.5. What's happening on the border of different time levels.

The different time levels inform each other as far as they are co-generating the temporality experience. For instance, the subjective experience of *microtime* is shaped and shapes in return the intersubjective sharing and synchronization of psychological experience, to the extent that semiotizing the experience of temporality to make it communicable for instance constructs a new experience of temporality for the both the "I" and the "other." In this sense, the *mesotime* co-constitutes the context of temporality experience. Furthermore, the societal *esotime* sets the frame for experiencing the intersubjective experience of temporality, providing constraints and affordances for the development of the psychological experience.

CONCLUSIONS

This model is just a first attempt to conceptualize time as a contextual dimension of psychological processes, avoiding the double temptation of treating it as an entity or a natural phenomenon, or considering it just an epiphenomenon of psychological processes. I started with a short phenomenological account of a temporality experience to show how it occurs in a complex multilayered time frame. I argued that one of the problems of psychology is to establish a relationship between the uniqueness of a temporality experience and the need for generating a generalized knowledge about psychological processes. This cannot be achieved by just putting aside time from theory or using ontological categories such as the limitedness of cognitive resources. Instead, time must become somehow a fundamental part of theory. Time cannot be considered, as well, as a mere constant, as in experimental design repeated measures, because it plays a relevant role in the sense-making process. For this sake, I briefly presented some principles derived from scholars that, especially during the 20th century, reflected the psychological aspects of temporality. From these principles, I argued that time must not be considered the result of experience but rather its context, and temporality can be generalized as a theoretical construct only in this time-context idea. In other words, psychological science must address both Ananke and Tyche when dealing with temporality. It must take into account that time represents the context in which constraints and affordances, agency and ineluctability as constitutive elements of human experience, make sense.

ACKNOWLEDGMENTS

This work has been funded by the Marie Curie IEF-2012 grant "EPICS. Epistemology in psychological science, the heritage of Giambattista Vico and the cultural psychology" at Aalborg University (Denmark).

NOTE

1. Ananke was a primordial ancient Greek goddess of inevitability, the personification of destiny, necessity, and fate, related with the limit and end of life. In ancient Greek literature, Ananke represents necessity and is not usually related with any "moral purpose." Tyche was instead the goddess who opposed evils of the society and punished those unworthy of good with misfortune, and kind people with good luck. Before immortality, when she was a demigod, her name was Aletheia (truth). Unlike Ananke, the action of Tyche is strictly related to human free will and agency.

REFERENCES

Baldwin, J. M. (1915). *Genetic theory of reality*. New York, NY: Putnam.

Benussi, V. (1913). *Psychologie der Zeitauffassung*, Heidelberg, Germany: Winter.

Bergson, H. (2001). *Time and free will: An essay on the immediate data of consciousness*. Mineola, NY: Dover.

Bronfenbrenner, U. (1979). *The ecology of human development: Experiments by nature and design*. Cambridge, MA: Harvard University Press.

Bronfenbrenner, U. (2004). *Making human beings human: Bioecological perspectives on human development*. Thousand Oaks, CA/London, UK/New Delhi, India: Sage.

Bronfenbrenner, U., & Morris, P. A. (1998). The ecology of developmental process. In W. Damon & R. M. Lerner (Eds), *Handbook of child psychology: Vol. 1. Theoretical models of human developmental* (pp. 939–991). New York, NY: John Wiley.

Couzens Hoy, D. (2009). *The time of our lives: A critical history of temporality*. Boston, MA: MIT Press.

Diriwächter, R. (2006). The wandering soul in relation to time. *Culture & Psychology, 12*(2), 161–167.

Doise, W. (1986). *Levels of explanation in social psychology*. New York, NY: Cambridge University Press.

Dru, A. (1938). *The journals of Søren Kierkegaard*. Oxford, UK: Oxford University Press.

Elder, G. H. (Ed.). (1985). *Life course dynamics: Trajectories and transitions, 1968–1980*. Ithaca, NY: Cornell University Press.

Flanagan, O. J. (1981), Psychology, progress, and the problem of reflexivity: A study in the epistemological foundations of psychology. *J. Hist. Behav. Sci., 17,* 375–386. doi:10.1002/1520-6696(198107)17:3<375::AID-JHBS2300170308>3.0.CO;2-U

Fuchs, T. (2005). Implicit and explicit temporality. *Philosophy, Psychiatry and Psychology, 12*(3), 195–198

Heidegger, M. (2010). *Being and time* (D. J. Schmidt, Ed.; J. Stambaugh, Trans.). Albany: State University of New York Press.

Herbst, D. P. G. (1976). *Alternatives to hierarchies*. Leiden, Germany: Martinus Nijhoff.

James, W. (1886). The perception of time. *The Journal of Speculative Philosophy, 20*(4), 374–407.

James, W. (1950). *The principles of psychology* (Vol. 1). Mineola, NY: Dover.

James, W. (1981). *Pragmatism*. Indianapolis, IN: Hackett.

Klose, J. (2008, December 4–6). *Time and temporality from Whitehead to quantum physics*. Paper presented at 2nd International Conference "Beyond Metaphysics? Transcontinental Explorations in Alfred North Whitehead's Late Thought," Claremont Graduate University, Claremont, CA.

Lewin, K. (1935). *A dynamic theory of personality. Selected papers*. New York, NY/London, UK: McGraw-Hill.

McGivalry, E. B. (1914). Time and the experience of time. *The Philosophical Review, 23*(2), 121–145.

Menard, S. (Ed.). (2008). *Handbook of longitudinal research: Design, measurement, and analysis*. San Francisco, CA: Academic.

O'Rand, A. M., & Krecker, M. L. (1990). Concepts of the life cycle: Their history, meanings, and uses in the social sciences. *Annaul Review of Sociology, 16,* 241–262.

Russon, J. (1998). *Reading Hegel's phenomenology*. Bloomington and Indianapolis: Indiana University Press.

Salvatore, S., & Valsiner, J. (2010). Between the general and the unique. Overcoming the nomothetic versus idiographic opposition. *Theory & Psychology, 20*(6), 817–833.

Sinico, M. (1999). Benussi and the history of temporal displacement. *Axiomathes, 10*(1/3), 75–93.

Tateo, L. (2008). The "fascist" discourse in computer mediated communication: The "dual strategy" model of the Italian extreme right. *Psicologia & Sociedade, 20*(2), 277–286.

Tateo, L., & Iannaccone, A. (2011). Social representations, individual and collective mind: A study of Wundt, Cattaneo and Moscovici. *Integrative Psychological and Behavioral Science, 45*(1), 1–13.

Toomela, A. (2009). The methodology of idiographic science: The limits of single-case studies and the role of topology. In S. Salvatore, J. Valsiner, J. T. Simon, & A. Gennaro (Eds.), *Yearbook of idiographic science* (Vol 2, pp. 13–32). Rome, Italy: Firera & Liuzzo.

Valsiner, J. (1994). Irreversibility of time and the construction of historical developmental psychology. *Mind, Culture & Activity, 1*, 25–42.

Valsiner, J. (2007). Human development as migration: Striving toward the unknown. In L. Simão & J. Valsiner (Eds.), *Otherness in question: Labyrinths of the self* (pp. 349–378). Greenwich, CT: Information Age.

Valsiner, J. (2011). Constructing the vanishing present between the future and the past. *Infancia y Aprendizaje, 34*(2), 141–150.

Whitehead, A. N. (1979). *Process and reality*. New York, NY: Free Press.

CHAPTER 20

THE TEMPORALITY OF TRADITION

Some Horizons for Semiotic-Cultural Constructivism in Psychology

Lívia Mathias Simão

I still believe that I am right to claim that we must rigorously deny the possibility that human life can be lived without a future. For, as I see it, it is a distinctively human fact that we must always keep the future open and with it new possibilties.

—Hans-Georg Gadamer (1986/1996b, p. 80)

In this chapter, I would like to propose the relevance and pertinence of invoking the notion of *tradition*—in the hermeneutic philosophical sense espoused by Gadamer—in the understanding of human events regarding temporality, within the theoretical and methodological and ethics perspective of semiotic-cultural constructivism in psychology (Simão, 2005, 2010, 2012, 2014).

This proposal is based on the fact that the hermeneutic notion of *tradition* touches the third among the three interrelated forms of simultaneity in human temporal experience that generate disquieting (Simão, 2007)

Temporality: Culture in the Flow of Human Experience, pp. 483–503

and creative tension (Valsiner, 1998) in I-other-world relationships in accordance with the perspective of semiotic-cultural constructivism in psychology; namely, the simultaneity between

1. temporal continuity and discontinuity in I-other-world relationships,
2. proximity and distancing from oneself and the other,
3. permanence and transformation in forming oneself (*Bildung*) in I-other-culture relationship (Simão, 2012).

Given that these forms are interrelated, I will first touch on the first two forms of simultaneity and then go on to the third, which directly touches *tradition* in the hermeneutic sense.

Temporal Continuity and Discontinuity in I-Other-World Relations

In her study concerning time, Malabou (1996) points to a fact regarding to the origins of the notion of time, which, while quite suggestive, is usually neglected in discussions of its role in I-other-world relationships. According to her, ever since its Indo-European origins, the notion of time has been tied to ideas of *interruption* and *discontinuity*, given that the semantic root *tem* means "to cut." This meaning also appears in the connotation of Greek words such as *tomos* and *atomo*. This same root appears in the vocabulary of neo-Latin languages that connote time, including Portuguese and Italian (*tempo*), Spanish (*tiempo*) and French (*temps*). Nonetheless, Malabou goes on to state that the semantic root *tem* does not merely signify interruption, but a *discontinuity in relation to a whole*: tome as a series of books, atom as the smallest part of a molecule. In this sense, the notion of time embraces, simultaneously and since its very origin, the ideas of *delimiting* and *reunion* (see Malabou, 1996).

With regard to the course of life, this signifies that a person perceives time in separate moments, one following the other, while also perceiving its continuity, forming the entire autobiographical whole of their personal life history. The experience of the simultaneity of temporal continuity and discontinuity in turn reminds us of the experience of temporal flow.

Time as a Flow

When approaching the issue of time, in his *Principles of Psychology*, William James (1890) calls on us to take note that our sense of now is always accompanied by the experience of time as a stream or flow, the same metaphor he employs to address consciousness. Nothing more phenomenologically coherent if we remember that upon experiencing time pass, we are ultimately consciously experiencing the impossibility of stopping time. Time flows independent of us, independent of our will; it is our business to merely cope with this experience with which we occupy ourselves so intensively.

Bergson (1911), in his preface to the French translation of *Pragmatism* by W. James, argues that our understanding of the central propositions of James' work is dependent upon our resistance to the traditional scientific conceptions of *reality* and *truth*.

According to Bergson (1938/2003), the origin and idea that inspired James' pragmatism could be found in that "the truths that are most important for us to know are, for James, the truths that we feel and live before even thinking" (p. 244).

Furthermore, Bergson (1938/2003) goes on to state that,

> from James' point of view, which is that of the pure experience or "radical empiricism," reality ceases to appear as either finite or infinite, but is seen as simply undefined. Reality flows, in which direction we are unable to say, nor do we know if it is the same river that runs everywhere. (p. 241)

"Reality flows; we flow with it; and we call truth any affirmation that, when directing ourselves through this moving reality (*réalité mouvente*), provokes us and places us on firmer ground from which to act" (Bergson, 1938/2003, p. 246).

Time as Duration

As in James, also in Bergson, understanding the relations between *time* and *consciousness* that are most relative to human experience requires one to take a step back from the classic concepts of the science positivist extraction, which were contemporary to them and, for the most part, to us as well. Even so, in Bergson's phylosophy, the past-present-future relationship that is meaningful to human experience in time is not seen as merely a temporal flow, but as *duration*, opening the door, as we will see shortly, for us to touch on change and newness in the time of inner life; in temporality.

In the very first pages of *La Pensée et le Mouvant*, Bergson (1938/2003) argues that the objectives of *accuracy*, *calculation*, and *prediction* of facts lead the positivist scientist to focus his explanations on the mere temporal repetition of discrete events—that which does *not last*, in the sense of going on.

According to Bergson (1938/2003), this is perfectly in keeping with the objectives of classical science, which, interested in the regularity of prediction and control of events, needs to "mask the very duration, whether in movement or in change" (p. 6). For doing so, science operates through the use of intelligence, which apprehends movement as a succession of single positions that a moving object covers in space, indefinitely, and understands transition as an increase in those positions, indefinitely, until the space between them disappears into a miniscule infinity (p. 6).

Given that intelligence "seeks to prepare and illuminate our actions regarding things," and that "our action cannot be *comfortably* exercised, unless we act on fixed points," the efficacy of science regarding its objects is so guaranteed (Bergson, 1938/2003, p. 6, emphasis added).

Once immobilizing duration, classical positivistic science discharges itself of confronting issues that involve changes in the course of events, and the uninterrupted appearance of newness as well; it therefore discharges itself from the principal scope of human creation (see Bergson, 1938/2003, p. 9). "But this duration that science eliminates—that is difficult to conceive and express—we feel it and we live it" (Bergson, 2003/1938).

Intuition is an ability that allows us to gain access to our inner life, in which the experience of time happens as duration—a transformation inherent in living:

> [Intuition] apprehends a succession that is not a juxtaposition, but a growth from within, the interrupted prolonging of the past into a present that stretches out into the future. This is the spirit's direct vision of the spirit. There is nothing else standing in the way; a refracting point through a prism, one side of which is space, and the other language. Instead of states abutting states, which become words juxtaposed to other words, here the indivisible continuity—and for this very reason, substantial—of the flow of inner life. Intuition thereby signifies consciousness; but immediate consciousness, a vision that can hardly make out the object being seen—knowledge that is contact and coincidence. (Bergson, 1938/2003, p. 27)[1]

Consciousness is, then, formed not only by intelligent understanding but also by intuition, which is the only way of grasping time as duration, as it is free from the desire for measuring time by stopping it.

In the subjective scope of duration, perception of time happens therefore in a completely different way than that of an observer who is external to that consciousness. We are then faced with the duality between the time of the world that is perceived by the person as external to oneself, a perception that can be shared intersubjectively by comparing and differentiating (Werner & Kaplan, 1963/1978), thus negotiating between "mine and other's time"; this negotiation is concernd with the use of culturally created artifacts, thereby modulating social relations,[2] and *the time of the inner world*, where the duration (*durée*) is lived subjectively: the *temporality*.

Difference and *simultaneity* between *feeling time as a transformation inherent in the act of living*, through intuition; and *observing changes in the world while time passes by*, through intelligence is then instilled:

> To think intuitively is to think of duration.... Intelligence is usually concerned with things, understood as that which is stable, making change into an accident that will be overlapped by them. For intuition, change is what is essential: as for the thing, given how the intelligence understands it, is an attack on the scope of the future built by our spirit in substitution of the whole. Thought usually represent the new as a new arrangement of pre-existing elements; in thought, nothing is lost, nothing is created. Intuition, tied to duration that is growth, perceives the continuity interrupted by the unforeseeable novelty; it sees, it knows that the spirit takes from itself more than it can afford; that this is spirituality, and that reality, imbibed with spirit, is creation. (Bergson, 1938/2003, pp. 30–31)

Duration thereby implies that any two moments are never the same in consciousness: each time I see the "same thing," I see it differently from all that came before, even though it might be related to those that came before, by similarity, difference, completeness, incompleteness. Duration, therefore, is qualitative multiplicity and memory; creation; it keeps the past not as a copy, but as a difference:

> Duration is essentially memory, consciousness, freedom. It is consciousness and freedom because, first of all, it is memory. Yes, this identity of memory as duration is always presented in two forms by Bergson: "conservation and accumulation of the past in the presen"t; or rather: "the present encompasses in distinctive ways the always growing image of the past, or, better said, testifies, through its continuous change of quality, the load that each of us carries on our shoulders, becoming heavier as far as we become older"; or rather: "memory under two forms: recovering with a pall of remembrances of a background of immediate perception, and contracting a multiplicity of moments" (*L'Energie Spirituelle*, Bergson, 1919, as cited in Deleuze, 1987)

Deleuze (1987) goes on to call our attention to the fact that we should then consider the two forms through which duration is distinguished from a discontinued succession of identical moments that repeat themselves:

> On one hand, "the following moment contains both the preceding moment and the memory that the latter left behind"; while on the other hand, both moments are contracted or condense into each other, as one, nonetheless, does not disappear when the other appears. There are, therefore, two memories, or two aspects of memory that are indissolubly linked: the memory—

> remembrance—and the memory—contraction ... the "present" that lasts is divided at each "instant" in two directions, one pointed to and dilated into the past, the other contracted, contracting into the future. (p. 52, citing Bergson, 1911).[3]

Furthermore, duration concerns a "foggy" perception of time, which is fuzzy, giving it its virtual character and its nature of potential to be. This character, in turn, allows duration, through its relationship with consciousness, to be transformed into memory. This could not happen, however, with intersubjectively objectified time, as it would not possess a virtual character anymore; it could no longer come to be (see Deleuze, 1987).

Ronchi (2011) proposes that the type of coexistence between past and present, in the face of the future, as posited by Bergson in his whole oeuvre, in memory as duration, touches on a hermeneutic synthesis: "In the absence of the unchanging habit of conduct, living becomes interpreting; to live is to face the risks that each interpretation carries with it (p. 130)."

Within the scope of I-other relationships, Bergson (2003) addresses the relationship of sympathy, pointing out that, even though the I and the other both place themselves "in each others' shoes" from the beginning, this placing is then transformed, tinged with different qualities as the relationship goes on.

Thus, far from closing in on ourselves and becoming isolated, Bergson'sperspective of duration (*durée*) places us in the focus of the duration inherent to intra- and intersubjective relations (I-I and I-other), in which permanence and transformation are co-present. Given that intuition is the possible way for accessing the duration, interpersonal relations will occur, through intuition, in the interplay between distancing from our own duration and closing in on the other.

Proximity and Distancing From Oneself and From the Other

Relationships are inherently temporal events. In I-other relationships, there is an intrinsic duality brought by the duality that the very act of being in time encompasses. On one hand, the fact that the I sees the other as an observer outside of the self, with temporal experiences that differ greatly from those of the self, comes into play. For the I, it is the other's time. It is by the social demand to "heed" the other's time that most temporal possibilities and limits are socially and culturally given to the I in the symbolic field of action (for example, a mother calls her child for dinner, because it is late, hoping that the child obeys.).[4] On the other hand, however, there is the self's time to the I (for example, the child who, when called by the mother, feels that playtime has not yet ended.). This interplay can give birth to internal reorganizations within the I to deal with the disquiet (Simão, 2003) resulting from the differences between the *self's time* and the *other's time* (the "10 more minutes, then I'll come in."—which generally seems too long for the mother and too short for the child).

Temporality

The notion of temporality has its origins in the phenomenological analysis of the human experience of time, mainly brought by Edmund Husserl and deeply developed by Maurice Merleau-Ponty. Important to us for the present discussion is that they both highlight that people experience time in at least two simultaneous ways: *we feel ourselves here and now* and *we feel that time passes*.

In Merleau-Ponty, the question of time cannot be seen as isolated from the subject in her perceptual experience of the world. The perceptual qualities of the I and the world perceived by it happen as a totality, being then interdependent and in mutual transformation. One does not precede the other, but rather they are formed in mutually. The only precedence, per se, is that of the selective availability of the I, as a living body, to open up to the experience of the other and the world.

This occurs, in truth, from Merleau-Ponty's wider psycho-philosophical perspective, which touches on the entire relationship between *nature and culture, coming into existence in an intertwined fashion through intersubjective relations*:[5]

> Just as nature penetrates to the center of my personal life and is intertwined with it, behaviors also fall into nature and are deposited on it as the cultural world. I do not have a merely physical world, I do not live merely in the earth, air, and water environment; I am surrounded by highways, plantations, villages, streets, churches, utensils, a bell, a spoon, a pipe. Each of these objects implicitly carries the market of the human action that it serves. (Merleau-Ponty, 1999, p. 465)

According to Merleau-Ponty's (1999) psycho-philosophy, things are never just here in the world, awaiting the subject; nor are they in the subject from the beginning, the subject presuming from perceptual experience to know the object. The subject and things in the world *come to be, reveal themselves, as the subject is experiencing the world*, as time is not a category external to the subject, but a dimension of the subject in its being in the world, *from moment to moment.*

Three relevant aspects for the I-other-world relationships emerge from the above discussion; while these aspects are interdependent, they are broken down herein merely to facilitate understanding:

a. Things comes to their existence *in the I-world relationship at each time,* so that they—the things—are not always there, permanent, visible;
b. The visibility of things by the I demands an involvement of itself in a perceptual organizing work, a synthesizing work, that happens in its experience with things, *from moment to moment;*
c. This demand for organization, *from moment to moment,* in the constant creation of a perceptual whole, also creates the very perception of time as duration as changing movement.

These three aspects, however, correspond to the second of two forms of approaching the relationship between the subject and time, as discussed by Merleau-Ponty (1999).

Since in the first sense, events are understood as a cutout of the space-time totality of the objective world, it is assumed that time flows independent of the observer, in the well-known past-present-future sequence, each one determining the following. To Merleau-Ponty (1999), this case of the Heraclitean metaphor of time flowing like a river is emblematic. According to the author, this metaphor is, in the end, a fallacy, in that it puts events flowing in time as protagonists, independent of the (subjective) experience of the observer, which is nonetheless there, hidden:

> We say that time passes or flows. We talk about the course of time.... If time is similar to a river, it flows from the past towards the present and the future. The present is a consequence of past, and the future a consequence of the present. This famous metaphor is actually quite misleading.... The "events" are cut out, by a finite observer, in the totality of space and time of the objective world. But, if I consider the world in question, there is only one indivisible being that does not change. Change assumes a certain position that I hold and from where I see things pass by; there are no events without someone to originate them, and from where the finite perspective lies as a basis for individuality. Time presupposes a vision of time. Hence, time is not a stream—it is not a flowing substance. This metaphor has survived since the time of Heraclitus because we surreptitiously place a witness in the flowing waters. (p. 511)

In the second manner of conceiving the relation between the subject and time, the latter is born of the perceptual experience of things, in a sense that it cannot be seen as external to the person but rather is constituted and constitutes human subjectivity[6]:

> It is not the past that pushes the present, nor is it the present that pushes the future into being; what is to come is not prepared behind the observer, it is premeditated in front of him, like a storm on the horizon. If the observer, sitting in a boat, flows with the current, one could say that he flows in the current towards his time to come, but that which is to come is made up of the new landscape that awaits in the estuary, and the course of time is no longer the river itself; it is the unfolding of these new landscapes for the moving observer. Therefore, time is not a real processes, an effective succession that I would be limited to register. It is born in *my* relationship with things. Within these things, the future and the past live a certain eternal pre-existence and survival. (Merleau-Ponty, 1999. p. 551)

Thus, the relationship lived in the present, with whatever it may be—a memory, a doll, the sound of a bell, the smell of a perfume—carries in it the past and the future because, as far as temporality being *a dimension in the subject*,[7] the meaning of past and future are already present before any objective creation of time (see Merleau-Ponty, 1999, p. 554).

In the perspective of temporality, time can then be understood as inherent in the process of subjectification of the world and therefore to the very construction of *I-other-world* relationships. Concur to this process our meanings of things that happened to us before the now, affecting us explicitly or implicitly. In such a way, the past never truly ceases to be. The idea of time as a sequence of "befores" and "afters" falls apart, in that it signifies an objectification of the result of time's passage, as if that passage were external and independent of us:

> The past and the future cannot be simple concepts that we created from the abstraction based on our perceptions and memories, they cannot simply be names used to designate the effective series of "psychic facts." Time is thought before thinking the parts of time—it is the temporal relations that make events in time possible. It is therefore necessary, concomitantly, that the subject not be situated therein, to allow him, in intention, being present in the past, just as in the future. We no longer say that time is "data of consciousness," we say, more precisely, that consciousness unfolds or encompasses the time. It is through the ideality of time that it finally breaks itself free of the present. (Merleau-Ponty, 1999, p. 555)[8]

From the notion of "time as data of the consciousness" to that of "time made up by the experience of the I with her things" (temporality) we have, therefore, a shift in the notional character of time: instead of belonging to the objective dimension of the world, temporality is an internal experience, constitutive and constituted by our relations with the world, here included to others.[9]

In its character of "always passing," time has, in Merleau-Ponty (1999), also a character of alterity, as it will always and simultaneously be a negation of the present and a potential for the future. Time is always moving, generating tension between the let-it-happen and the passing, which gives it a character of permanent incompleteness. This then constitutes temporality as a phenomenological dimension on the level of "somewhere" (see p. 556).

Our disquiet (Simão, 2003) in the I-other relationships emerges, in most part, from this incompleteness:

> Once more, my human gaze gives only one face to an object, even if, over the horizon, it aims at all others. It can only be confronted with preceding visions or those of other men via time and language. If I conceive, in resemblance of the mine, the gazes that explore the house from all sides and define the house itself, I will still have only a convergent and undefined series of visions of the object—I cannot get the object in its entirety. In the same sense, although my present condenses in itself the flowed time and the coming time, it only holds them in intention, and, if for example my current consciousness of my past seems to cover exactly that which is gone, the past that I seek to relearn, it is no longer the past in person, it is my past as I see it now—and I may have altered it. In the same sense, in the future I may not recognize the present that I am living now. Thus, this synthesis of horizons is merely a presumptive one; it only operates with certainty and precision in the immediate surroundings of the object. (Merleau-Ponty, 1999, p. 107)

Our temporality as human beings comes exactly from this incompleteness of time. Our real temporal continent encloses more than the actual present; our real present encloses our memories, as well as our expectations, desires, and hopes. Nevertheless, as the past is what is not anymore and the time to come is what is still not actualized, none of them are at hand in our present. For being actualized in our phenomenological world, both require our subjective movement toward them.

Temporality therefore embraces dialogicality (Marková, 2003, 2012) between intersubjective time—implying a consideraton of the position of an observer outside the I—and temporality, which is born of the I's relation with *its* things (Merleau-Ponty, 1999) and cannot be transferred to the other in a complete fashion. It therefore brings *the dual and alter co-presence of the I and other.*

In intersubjective relations, the other inserts itself into the subjective life of the I, requiring a transformation of *temporality as a subjective dimension* in *time objectified by an external observer.* This demand on the I from the other occurs in several different circumstances in cultural sociability in which the present imposes on relations of the "here and now," as well as the past and future emerge as narratives and desires (see, for example, Simão, 2004, 2008).

In interpersonal relations, the I and other assume, in a tacit or explicit manner, certain concomitancy and sharing of

a. the *experience in the present* (what the I and the other do, feel, think in the "here and now" of our co-presence, either concrete, virtual, or imagined);
b. the *future* (what I wish, desire and fear of the other; what I presume the other wishes, desires and fears of me, and what we both wish, desire and fear in and of the relationship);
c. and the *past* (the preconceptions that I and the other bring with regard to ourselves, the other in the relationship, and the world in which we live, in a more general sense).

There, not only the content expressed in one way or the other by the I and the other is at play, but also their positions relative to themselves and the other in the interpersonal relationship (Simão, 2004, 2010).

Thus, I-other relations as intrinsically temporal events can make us otherness to ourselves, in requiring our experiential transition between the dialogicality of "the self's time" and the "other's time," in the dimensions of what is no longer (past), what still is (present), and what has not yet been or will never be (future). They may, for this reason as well, put us into situations of creative tension and reorganization of relationships with the self and the other.

Consequently, they may engender symbolic constructions that allow people to deal with their disquieting feelings (Simão, 2003) brought on by *circumstances of the here and now* (present), the *meaning of which is always tied to something remembered* (past), as well as *something desirable and not guaranteed in the time to come* (future).

In this sense, in her study of time in Greek tragedy, Romilly (2009) shows us one of the forms through which temporality emerges; they are events that suddenly disturb relations of meaning between actions, leaving them disorganized and creating urgency for new organization.

Taking Greek tragedy as paradigmatic of the nature of events regarding temporality, Romilly (2009) points out the fact that the tragic event establishes tensional relationships between the past and the time to come, placing the focus on causes and responsibility, inviting the action protagonists to think on their effects. These relationships express, not necessarily in conscious ways for the actor, a "certain philosophy of time" (see p. 20).

In tragedy, some unique and short-lasting event is installed in present time, demanding that the actor take part on it, in unequivocally ways, not only rationally but mainly emotionally, therefore its strong meaning regarding future consequences. In this sense, time is experiences as an emergent and tensional relationship between a short-lasting present, requiring actions concerning an unexpected event, and a future in which those actions will have decisively and long-lasting repercussions (Romilly, 2009, p. 20). Romilly's (2009) analysis is suggestive for understanding the temporal dimension of I-other-world relations with the focus on *disquieting* that originates in unexpected events, or by events that did not occur as planned by the person, generating tension and demand for symbolic actions that reorganize I-other-world relations. This reorganization that may happen will necessarily be transformative of the self. This also points to the character of the temporal dimension of human subjectivity regarding what is usually not passible of explanation, particularly in experiences felt by the I as abrupt. As far as the past and the future only existing when subjectivity *hits the plenitude of the present*, introducing *something that is not now*, either by memory, wishes, desires, imagination, or fear, *temporality implies*, as we have stated, an *otherness*.

Boesch (1991) points out that this distance separating the present from the future must be traveled subjectively by symbolic action, which fills its gaps step by step, progressively. This progressive, symbolic traveling, which is altered over time in accordance with the present of each moment, "constitutes a necessary condition for the formation of time as an *existential* dimension" (p. 159, emphasis in original).

Experiencing time is therefore more than being conscious of our sequential actions, as it requires that we perceive the difference between *now and then*, anticipating a *duration*, which in turn requires *hoping* and *working* on something that is different from the now. Equally important to Boesch (1991) is the fact that *having to wait* points to a qualitative dimension of time as an *obstacle, a barrier to our wishes and feelings of accomplishment* (see pp. 159–160).

Still according to Boesch (1991), although memories come from the past, they point to the future, which embraces the time passing, where nontransparent processes that cannot be foreseen will happen. As these events do not happen in a causal succession, they, partially or completely, escape from our control. To this extent, the future embraces the realm of the unknown, the amorphous field of the events to come, filled by uncertain menaces and hopes (see pp. 166–177). In such a case, it duplicates our real symbolic field, in a double-handed way: the perception about the present will influence future anticipations and vice versa. Man lives in two worlds: that of the real experiences and that of potential ones. The projections about the future are not arbitrary; besides being controlled by present and past experiences, they are also controlled by social duties. Additionally, these anticipations need to be confirmatory of self-consistency, preferable without menacing the self-image too much.

To remember is also to experience time. We remember events that are relevant in some way to our present situation, giving new meaning to current concerns and guiding our symbolic actions to increase our self-consciousness. Even so, memories can also allow a person to realize their flaws, serving as a basis for change and pointing out new perspectives, leading to self-evaluation and restructuring of the self (see Boesch, 1991, pp. 160–177).

Thus, coming from the past, memories point to the future, to events that cannot be foreseen; events that are thus imagined, desired, presumed, opaque—fuzzy events. In such a way, the future embraces the realm of the unknown, an amorphous field of coming events, filled by threats of uncertainty and hopes of realizations. Thus, the present-future relationship duplicates our real symbolic field in a two-way movement: the perception of the present will influence anticipations of the future, and vice versa (see Boesch, 1991, pp. 160–177).. Hence, Boesch's (1991) statement that mankind lives in two worlds: the world of real experience and that of potential experiences. Furthermore, Boesch states that projections about the future are not arbitrary but rather based on present and past experience and controlled by social obligations. These anticipations must also be confirmations of the self-consciousness, at least to the extent of not posing an excessive threat to self-image (see Boesch, 1991, pp. 166–177). *We are here in the scope of the selectivity of memory*.

Temporality is thereby implied in a dual relationship between intersubjective time and subjective duration. The other comes to us with their ideas, beliefs, feelings, conceptions, values, and purposes, constructed in the realm of *her* temporality. Some of these constructions may be so shared culturally that they do not pose challenges to *our* temporality. They may also represent joint, similar, converging constructions between the I and the other. Nonetheless, some phrase uttered by the other may require that I enter in a distentional movement toward my or the other lived past as in Augustine's philosophy, or toward a future, presumed by me or the other, or both, thereby challenging *my* temporality.

When someone is talking with another person, both trying to capture and create meaning by mutual subjective expressions about a subject—asking, answering, gesturing, silencing—each of them challenges the temporality of the other as they mutually evoke and awaken memories and expectations in each other.

To respond (*in the sense of opening ourselves*) to an meaningful other about "our things" is, in part, to allow ourselves to be challenged by it in our temporality—in our fuzzy perception of time and memories—allowing the temporal meaning of the conversation's subject, together with our relationship with the other, to possibly be transformed as a totality.

The subjective experience of time (temporality) makes us otherness to ourselves, by placing us in constant change, in development, the course of life, thereby creating tensions that may be creative, reconstructive and, other times, destructive of the self.

It is thus that otherness makes itself present in our subjective life through temporality, requiring transformations in our mediational strategies for expressing and communicating memories and hopes, inviting us, providing us incentives, or even pressuring us to re-create ways to deal with time in our processes for relating to the world. In this sense, symbolic actions from the other may be a source of feelings of disquiet (Simão, 2003) that lead us to reorganize our I-other-world relationships through reorganization of past memories and future hopes in the current state of the present.

The other will then be a source of rupture in the ambit of duration and temporality, forcing subjectivity toward intersubjectivity in interpersonal relations, in a constant back and forth movement. In this movement, however, if the subject and the other are not isolated in a time that precedes them or is external to them, neither the world and things are *a priori* of the experience: things are cultural objects in which the presence of the other in the world insinuates: "In the cultural object, I feel, under the veil of anonymity, the close presence of the other" (Merleau-Ponty, 1999, p. 466).

Permanence and Transformation in Forming Oneself *(Bildung)* in I-Other-Culture Relationships

At the beginning of *Truth and Method,* Gadamer touches on Hegel, affirming that "Man is characterized by the rupture with the immediate and natural, which is proper to him, in virtue of the spiritual and rational side of his essence" (1989, p. 41). Thus, human beings are never *a priori of experience*, thereby requiring cultural formation.

I understand *tradition*, in the philosophic hermeneutical sense (Gadamer), as one of the ways for *temporality* manifests itself in human symbolic life. In a cultural creation, in which culture is understood as a field for symbolic human action (Boesch), tradition is an inherent part of this field, allowing for new organizations of the meaning among *present experiences, future expectations and desires*, and *past memories*.

In this respect, and as I have shown before (Simão, 2014), the meaning of tradition laid out by Warnke (1987) also plays here an important role:

> In Gadamer's philosophical hermeneutics, tradition designates the historically pre-given. As socialized human beings we are always already immersed in particular ways of coping with our world. We possess certain forms of practical knowledge, do things in certain ways, and take certain concepts and conceptual relations for granted. These forms of knowing and acting function as deeply rooted pre-agreements, or what Gadamer calls prejudices, that orient our further explorations. (p. 6)

In the same vein, Weiss (2009) warns that "tradition and its given authority are constitutive and immanent elements of comprehension, not something to be protected against or disqualifying" (p. 123). To disqualify or protect against tradition would be to, in the hermeneutic sense, refuse the dialogue because tradition, as "the voice speaking from the past" (Gadamer, 1996a/1975), places our current meanings in the open, making us ask new questions, which will lead to new interpretations of events and our position in relation to them.

Present experience requires the subject to undertake the task of reconstructing the meaning of prior experience, demanding transformation toward a given future, forcing the subject to face temporality in two inextricable meanings. First, given that the contents that are interpellating the person [for this usage of the verb to interpellate, see, please, The Oxford Dictionaires] in the present will only be understood through a transformative process (*Bildung*), which is at the same time transformative of content originating in *tradition*; this is, in summary, a question of dealing with the time to come. Second, it will be impose to the person to deal with its limits, its finitude: the experienced person "knows the limits of all foreseeing and the insecurity of the every plane, ... knows that she is not lord of the time, or the future" (Gadamer, 1975/1996a, p. 433).

Furthermore, in Gadamer, as highlighted by Grodin (2002), understanding implies an agreement in the conversation through the articulation of words, *which are "always mine, but at the same time those of what I strive to understand"*(p. 41, emphasis added). Thus, in Gadamer's hermeneutics, dialogue is a field of tensional relations among past, present, and future, between possible and impossible, but desired and expected.

If, in Gadamer's hermeneutics, on one hand, the absence of consensus validates the authenticity of the other as it confronts the subject with the experience of negation, on the other hand, complete consensus still remains possible and necessary, meaning true understanding. In any event, however, the rupture of expectation of consensus and the constant quest for understanding the challenge of I and the other with a permanent reconstructive task in the dialog, thereby works toward the future.

This process is characterized by the dynamic of the hermeneutic circle, a dialogical construction of meaning in which new meanings emerge through the confrontation between part-whole relations. In this process, the construction of meaning is guided by expectations derived from the context preceding the meeting of that which ones seek to understand. These expectations may, however, be rectified if the object to be understood demands in order to readjust both the expectations and the object in question. Thus, understanding meets in a unit of thought based on the new expectation of meaning.

All this is possible thanks to the anticipation of meaning of the I in interpretive relationship with that which it seeks to understand. If, on one hand, constructions of meaning are guided by expectations derived from the context preceding the meeting with that which is to be understood (*pre-conceptions, according to Gadamer's philosophy*), the expectations will, on the other hand, be changed, readjusted by the demands posed by the object to be understood.

In this process of past-future relations, in the here and now (present) of the reconstructive experience and generation of new expectations, tradition plays a central role: "*the understanding should think of less as an action of subjectivity and more as a self movement towards the happening of tradition,* in which the past and present are in continuous mediation" (Gadamer, 1975 /1996a, p. 360, emphasis added).

Through tradition, the "voice of the past" becomes present to the I through the symbolic actions of the other, creating demands for personal-cultural fitness, that change its horizons (Gadamer, 1996a/1975) and therefore future possibilities; this other could be the oneself, thus the demands of the dialogical selves (Hermans, Kempen, & van Loon, 1992).

Tradition of a people, a group, or a community, in being a collective and continuous reconstruction, is multivoiced and pluritemporal. Therefore, tradition brings an actual demand that people formed in it can manage conflicting perspectives on a given plot; tradition demands a search for intelligibility in I-other dialogues.

The I-other relationship, by its turn, is also an inherently *temporal* event. Herein the twofold dialogical role played by the interlocutors regarding their formation in tradition.

On the one hand, *tradition* brought about by the other's voice in the dialog may allow a momentary reorganization of the lived experience by the I. On the other hand, the aforementioned multivoiced and pluritemporal character of *tradition*, allied with the inherently *noncoincidence* between the I and the other as subjects, makes the other a messenger of meanings to be deciphered by the I.

Herein, the hermeneutic character of the I-other relationship, in which the hermeneutic task of the I is never fully accomplished, in its quest for penetrating the mystery of the unknown, is *the not fully understandable present, the unknown future, the fuzzy and changing past memories*. It is a quest that brings simultaneous feelings of belonging and not belonging to a something; feelings of disquieting by the unknowable as well of being dwelled on by the already known, *being at home* (Boesch, 1991).

In sum, "being cultivated in a tradition," rather than a fixity of the subject's prejudices and passiveness in face of tradition, means continuous changes in its preconceptions (prejudices). Even when "confirming the past," the attitude of the person who is cultivated and formed in her tradition is never of mere distance or naïve freedom concerning what is transmitted to her. On the contrary, as "we are always inside tradition"—therefore not being "insiders above traditions"—nothing that is said by our tradition can be totally strange or alien to us; it can sound as exemplar, boring, not fair, auspicious, and so on. However, it always implies recognition of ourselves not only at the rational level of knowledge, but by transforming us in unnoticed ways, as well as transforming tradition (Gadamer, 1975/1996a, p. 350).

For Gadamer (1996a/1975), the distance in time is what allows the subject to filter preconceptions and express himself in the face of always new demands for meaning in relations with the world, in a process that never ends and is in constant expansion. In this process of distancing for true hermeneutic understanding, tradition plays a central role:

> Only distance in time allows for the true resolution of the critical hermeneutic question, of distinguishing *true* preconceptions under which we *understand* from *false* preconceptions that produce *misunderstandings.* In this sense, a hermeneutic formed consciousness must also be, to a certain point, historic consciousness, becoming aware of the its own preconception that guides itself in understanding, to allow tradition to stand out as a distinct opinion, thereby realizing its right. It is clear what realizing a preconception implies in suspending its validity. This is true because while a preconception is determining our action, we do not even know it, nor do we think on it as a judgment. How, then, can it become visible? It is impossible to put a preconception in front of us while it continues its imperceptible work; it can only be done if the preconception is in some way "stimulated." This stimulus comes precisely from the encounter with tradition, as that which incites understanding must have shown its value in otherness in some form. (p. 369)

In such a way, *tradition*, in philosophical hermeneutic terms, can be understood as one of the ways through which *temporality*, as a personal experience of the time and in time, manifests itself in human life. It is within temporality, as one of the dimensions of human subjectivity, that people build and manifest themselves intra- and intersubjectively, creating their momentary organizations of I-world relations. The manifold meanings that we (re)elaborated with respect to that which has occurred before now, directly and indirectly affecting us in the present, leading us to a certain future, contribute to this process. In this sense, the past never ceases to be, thereby leading to a conception in which time is not a cause-based sequence of "before" and "after," given that this would requires the objectification of the passing of time as externally independent of the subject that experiences it.

Tradition, as temporality, touches simultaneouly on permanence and transformation in cultural formation. This is so because de present is interpellated by the tradition in the symbolism of the voices and meanings constructed on the I-Other-world realtionship, then requiring new meaning. In this sense, it also touches on the two other simultaneities of which we have spoken: on one hand, it guarantees continuity, while on the other it provokes breaks in which Boesch (1991) called the *cycle of knowledge* in individual-culture-individual relations. Finally, as tradition speaks by the voice of the other it requires new articulations that occur by approximation to the self and distancing from the other, and vice versa.

GENERAL CONCLUSIONS

In synthesis, in this chapter, I proposed the relevance and pertinence of invoking the notion of *tradition*—in the hermeneutic philosophical sense espoused by Gadamer—in the understanding of human events regarding temporality, within the theoretical and methodological and ethics perspective of semiotic-cultural constructivism in psychology (Simão, 2005, 2010, 2012, 2014).

For that I have argued that the hermeneutic notion of *tradition* touches the third among the three interrelated forms of simultaneity in human temporal experience that generate disquieting (Simão, 2007) and creative tension (Valsiner, 1998) in I-other-world relationships in accordance with the perspective of semiotic-cultural constructivism in psychology.

Regarding the first simultaneity—between temporal continuity and discontinuity in I-other-world relationships—it was pointed out, from Bergson, the *difference* and *simultaneity* between *feeling time as a transformation inherent to the act of living*, through intuition; and *observing changes in the world while time passes by*, through intelligence. Also, that memory as duration implies anunderstanding of our experience with time as hermeneutic (Ronchi, 2001).

Concerning the second simultaneity—between proximity and distancing from oneself and the other—and given the inherently temporal character of I-other relationships, the differences between the *self's time* and the *other's time* can give birth to internal reorganizations within the I to deal with the disquieting feelings (Simão, 2003) brought in by them. Moreover, the dialogicality (Marková, 2003, 2012) between intersubjective time—which implies considering the position of an observer outside the I—and temporality, which is born of the I's relation with *its* things (Merleau-Ponty, 1999), and cannot be transferred to the other in a complete fashion, brings *the dual and alter co-presence of the I and other.*

Approaching the third simultaneity—between permanence and transformation in forming oneself (*Bildung*) in I-other-culture relationship—it was proposed that the hermeneutic character of the I-other relationship, in what *tradition*, as one of the ways through which *temporality* manifests itself in human life, plays a decisive role.

NOTES

1. In my view, Bergson presents here a cogenetic relationship (Herbst, 1995) between intuition and duration, that will be the theme of a discussion at a later moment.
2. With regard to this matter, see for example, Elias (1997).
3. This aspect of memory dilation toward the past was already found, however, in other terms, in Saint Augustine and will once again be the theme of discussion, although in other terms, by Merleau-Ponty, as we will cover below.
4. In play here—in the I-other relationship—is a tacit conception of time that touches on Merleau-Ponty's discussion of the metaphor of Heraclitus, which we will touch on below, as well as Elias's (1997) conceptions on social time, as mentioned in footnote 3.
5. On the central role of intersubjectivity in Merleau-Ponty's oeuvre, see Bonan (2001).
6. Here we can recognize the manner of conceiving knowledge through experience that permeates the entire work of Merleau-Ponty: the reflexive reversibility of which the discussion of the two hands of a person touching another is famous (see Merleau-Ponty, 1999, p. 137).
7. On the meaning of the notion of dimension in Merleau-Ponty (1999), Bonan (2001) clarifies that "dimension is, in fact, much more than a simple order of reality; from the departure point, it is that what according to which reality is given to me, the perspective that is given, the general frame of the phenomenological process" (p. 17, translation by L. M. Simão).
8. We are here in the ambit of Saint Augustine's (354–430) thesis, in which time is a movement of distension of the mind and cannot therefore be independent thereof, and that Merleau-Ponty (1999) examines in his Phenomenology of Perception.
9. On the central role of otherness in Merleau-Ponty, see Bonan (2001).

REFERENCES

Bergson, H. (1911). Sur le pragmatisme de William James—Verité et realité. In H. Bergson (2003/1938) *La Pensée et le mouvant*. Paris, France:PUF.

Bergson, H. (2003). *La pensée et le mouvant*. Paris, France: PUF. (Original work published 1938)

Boesch, E. E. (1991). *Symbolic action theory and culture psychology*. Heidelberg and Berlin, Germany/New York, NY: Springer-Verlag.

Bonan, R. (2001). *Le probleme de l'intersubjectivite dans la philosophie de Merleau-Ponty—La dimension commune* (Vol. 1). Paris, France: L'Harmattan.

Deleuze, G. (1987). *El bergsonismo*. Madrid, Spain: Ediciones Catedra.

Elias, N. (1997). *Sobre el tiempo*. México City, México: Fondo de Cultura Económica.

Gadamer, H.-G. (1989). *Truth and method*. London, UK: Sheed & Ward. (Original work published 1975)

Gadamer, H.-G. (1996a). *Verdad y método* (Vol. 1). Salamanca, Spain: Ediciones Sígueme. (Original work published 1975)

Gadamer, H-G. (1996b). *The enigma of health—The art of healing in a scientific age*. Stanford, CA: Stanford University Press. (Original work published 1986)

Grodin, J. (2002). Gadamer's basic understanding of understanding. In R. J. Dostal (Ed.), *The Cambridge companion to Gadamer* (pp. 1–12). Cambridge, UK: Cambridge University Press,

Herbst, D. P. (1995). What happens when we make a distinction: An elementary introduction to co-genetic logic. In T. A. Kindermann & J. Valsiner (Eds.), *Development of person-context relations* (pp. 67–79). Hillsdale, NJ: Lawrence Erlbaum.

Hermans, H. J. M., Kempen, H. J. G., & van Loon, R. J. P. (1992). The dialogical self- beyond individualism and rationalism. *American Psychologist, 47*(1), 23–33.

James, W. (1890). *Principles of Psychology*. New York, NY: Holt.

Malabou, C. (1996). *Le temps*. Paris, France: Hatier.

Marková, I. (2003). *Dialogicality and social representations: The dynamics of mind*. Cambridge, UK: Cambridge University Press.

Marková, I. (2012). Challenges to dialogical science. In M. Märtsin, B. Wagoner, E.-L. Aveling, I. Kadianaki, & L. Whittaker (Eds.), *Dialogicality in focus* (pp. 65–76). New York, NY: Nova Science.

Merleau-Ponty, M. (1999). *Fenomenologia da percepção*. São Paulo, Brazil: Martins Fontes.

Romilly, J. de (2009) *Le temps dans la tragédie grecque—Eschyle, Sophocle, Euripide*. Paris, France: Vrin.

Ronchi, R. (2011). *Bergson—Una sintesi*. Milan, Spain: Christian Marinotti.

Simão, L. M. (2003). Beside rupture-disquiet: Beyond the other-alterity. *Culture & Psychology, 9*(4), 449–459.

Simão, L. M. (2004). *Development through self-stories: Using symbolic resources*. Abstracts 18th biennial ISSBD meeting, Ghent, Belgium, p. 189.

Simão, L. M. (2005). Bildung, culture and self: A possible dialogue with Gadamer, Boesch and Valsiner? *Theory & Psychology, 15*, 549–574.

Simão, L. M. (2007). Why otherness in the research domain of semiotic-cultural constructivism? In L. M. Simão & J. Valsiner (EDs.), *Otherness in question: Labyrinths of the self* (pp. 11–35). Charlotte, NC: Information Age.

Simão, L. M. (2008). Ernst E. Boesch's holistic cultural psychology. In R. Diriwächter & J. Valsiner (Eds.), *Striving for the whole: Creating theoretical synthesis*. Somerset, NJ: Transaction.

Simao, L. M. (2010). *Ensaios dialógicos: Compartilhamento e diferença nas relações eu-outro [Dialogical Essays: Sharing and difference in I-other relationships]*. São Paulo, Brazil: HUCITEC.

Simão, L. M. (2012). The other in the self: A triadic unit. In J. Valsiner (Ed.), *The Oxford handbook of culture and psychology* (pp. 403–420). New York, NY: Oxford University Press.

Simão, L. M. (2014). The contemporary perspective of the semiotic cultural constructivism: For a hermeneutical reflexivity in psychology. In G. Marsico, R. Ruggieri, & S. Salvatore (Eds.), *YIS: The yearbook of idiographic science, Volume 6. Reflexivity and change in psychology*. Charlotte, NC: Information Age.

Valsiner, J. (1998). *The guided mind: A sociogenetic approach to personality.* Cambridge, MA: Harvard University Press.

Warnke, G. (2012, December). Solidarity and tradition in Gadamer's hermeneutics. *History and Theory, 51,* 6–22.

Weiss, I. (2009). *Gadamer*. Paris, France: Vrin.

Werner, H., & Kaplan, B. (1978) General nature of developmental changes in the symbolic process. In S.S. Barten & M. B. Franklin (Eds.), *Developmental processes—Heinz Werner's selected writings, Vol. 2, Cognition, language and symbolization* (pp. 487–500. New York, NY: International Univerisities Press. (Original work published 1963)

ABOUT THE AUTHORS

Emily Abbey is associate professor in the Department of Psychology at Ramapo College of New Jersey. Working from an explicitly present-to-future developmental orientation, she is generally curious about the role imagination plays in guiding the transformation of thoughts and feelings over time. Recently, she has published her second edited volume, *Emerging Methods in Psychology* (Transaction) as well as made contributions to the *Oxford Handbook of Cultural Psychology* (Oxford University Press) and the *Cambridge Handbook of Socio-Cultural Psychology* (Cambridge University Press). ADDRESS: Emily Abbey, Ramapo College of New Jersey, Mahwah, NJ, United States.

Cor Baerveldt is a cultural psychologists and associate professor at the University of Alberta, Edmonton, Canada. His work is concerned with the historical, ontological, and epistemological foundations of cultural psychology. He is particularly interested in the embodied and expressive nature of meaning and in the possibility of an enactive cultural psychology with firm roots in the genetic tradition of psychology. To this end, he has developed in his writing the idea of culture as an ongoing consensual coordination of actions rather than an already established symbolic or semiotic system. He is currently involved in a project that focuses on the concept of "life" in cultural psychology, in an attempt to help build conceptual bridges between cultural psychology and the other developmental sciences.

Ana Cecilia Bastos is a professor at the Federal University of Bahia and at Catholic University of Salvador, Brazil, and a researcher for the National Research Council (CNPQ). For two decades, she has been studying the cultural organization of Brazilian families living in poverty, at times of intense social change, focusing parental ideas and childrearing practices. Currently, she is interested in the analysis of meaning-making processes that women from different generations build around childbirth (taken as a marker for understanding developmental transitions) exploring the changes in the meaning of parenthood across the life span. She is the author of *Modos de Partilhar: A Criança e o Cotidiano da família*) (Ways of Sharing: The Child and the Family Everyday Life) and the editor of *Living in Poverty: Developmental Poetics of Cultural Realities* (with Elaine Rabinovich) and *Cultural Dynamics of Women's Lives* (with Jaan Valsiner and Kristiina Uriko).

Nandita Chaudhary teaches at the Lady Irwin College, University of Delhi, India. She has been a Fulbright scholar at the Psychology Department, Clark University, Massachusetts, during the years 1993–1994, and a Senior Fellow of the ICSSR (Indian Council for Social Science Research) from 2012 to 2014. Chaudhary is the author of *Listening to Culture: Constructing Reality From Everyday Life* (2004, Sage) and has co-edited three volumes: *Cultural Realities of Being: Abstract Ideas Within Everyday Lives* (2013, Routledge), *Dynamic Process Methodology in the Social and Developmental Sciences* (2009, Springer), and *Methods for a Cultural Science* (2008, Sage). She is an Associate Editor for the journal *Culture & Psychology* and is on the editorial board of three other journals.

Nelson Ernesto Coelho Jr, PhD, is a professor and researcher in the Institute of Psychology at the University of São Paulo, Brazil. A psychoanalyst, he is currently directing researches in history and philosophy of psychology and psychoanalysis. He is a consultant of research foundations and scientific journals. Among his publications are articles in *Culture & Psychology*, *International Forum of Psychoanalysis*, *Revista Brasileira de Psicanálise*, and *Gaceta de Psiquiatría Universitaria* (Chile). E-mail patnelco@uol.com.br/ ncoelho@usp.br

Carlos Cornejo is full professor at the Pontificia Universidad Católica de Chile. He teaches the history of psychology, theoretical psychology, and psychology of language. His research interests include theoretical and empirical aspects of meaning construction, figurative language, and human interaction. E-mail: cca@uc.cl

Andre M. Cravo completed his undergraduate degree in psychology in 2005 at the University of São Paulo, Brazil, and his PhD at the same university in 2011, with an internship at the University of Oxford. From 2011 to 2012, he was a postdoctoral associate at the Institute of Radiology, FMUSP. He is currently assistant professor in the Center for Mathematics, Computation and Cognition at Universidade Federal do ABC. He has experience in neuroscience and cognition, with emphasis in eletroencefalography and psychophysics.

Basia D. Ellis is a PhD candidate enrolled in the Social and Theoretical Program in the Department of Psychology at the University of Calgary, Canada. Broadly, her research focuses on the production of subjectivity as an embodied, sociocultural, and political process. Whereas her master's project theoretically examined the phenomenological accounts of Maurice Merleau-Ponty and Paul Ricoeur, her doctoral dissertation explores the daily experiences of undocumented migrants in Canada, employing an ethnographic approach. Email: bellis@ucalgary.ca

Maria Francesca Freda is associate professor in clinical psychology at the Department of Humanistic Studies, University of Naples "Federico II." She works on psychosocial intervention in institutions, with particular attention toward educational and health institutions. Many of her studies are about narrations as a methodology for the elaboration of experience's meaning. Her current interests are about study of narration as a device of the psychological clinical intervention, an area of study in which she published the book *Narrazione e Intervento in Psicologia Clinica* [*Narration and Intervention in Clinical Psychology*] (Liguori, Napoli, 2009). E-mail: fmfreda@unina.it

Raffaele De Luca Picione PhD at the Department of Humanistic Studies, Federico II University, Naples, Italy. He is interested in the processes of subjective meaning making in a cultural field and particularly in affective semiotic processes, in reference to the relationship between thought, emotion, and language. Further, he is interested in semiotics, paradigms of complexity, and connections between cultural and clinical psychology. E-mail: raffaele.delucapicione@unina.it

Maria Luisa Martino, PhD at the Department of Humanistic Studies, Federico II University, Naples, Italy. She deals with psychological research intervention in hospital institutions with particular regard to the use and evaluation of different narrative devices as tools to promote well-being. She studies the processing of traumatic experience of illness with particular regard to the construction of meaning-making processes. She published, in collaboration with professor Maria Francesca Freda, the book *Trauma, Scrittura ed Elaborazione in Oncologia Pediatrica. [Trauma, Writing and Processing in Pediatric Oncology]* (Saarbrücken, DE: PAI, 2013). E-mail: marialuisa.martino@unina.it

Anna Gorrese is assistant professor at the University of Salerno, Italy. She received her PhD in methodology of educational research from the University of Salerno. Her major research interests include attachment relationships, emotion regulation, and decision making.

Danilo Silva Guimarães, PhD, is professor at the Institute of Psychology (University of São Paulo, Brazil) where he coordinates a research group on cultural psychology. He has been working with theoretical and methodological issues in the framework of a semiotic-cultural and constructivist psychology. His focus of investigation concerns psychology and Amerindian peoples, psychological approaches to interethnic relationships from a dialogical perspective, the process of symbolic elaborations out of tensional boundaries between cultural identities and alterities. Within this focus, he is exploring related issues on the history and philosophy of psychology. Among his more representative works are *Self and Dialogical Multiplication* (2013), *Scientific Concepts and Public Policies: Semiotic-Cultural Obstacles Concerning Intergroup and Intercultural Relationships* (2012), and Amerindian Anthropology and Cultural Psychology: Crossing Boundaries and Meeting Otherness' Worlds" (2011). E-mail: danilosg@usp.br

Hamilton Haddad attained his BSc in biochemistry and BA in philosophy, both from the University of São Paulo, Brazil. He received his PhD and master's degree in physiology/neuroscience from the same university. He is now a professor at the Institute of Biosciences, University of São Paulo, where he directs the Laboratory for Cognitive Neuroscience and Learning.

Sven Hroar Klempe is associate professor in psychology at the Department of Psychology, NTNU, Trondheim, Norway. His background includes having been full professor in musicology, associate professor in media studies, teacher, and journalist. His research is cross-disciplinary with an emphasis on the history of psychology, culture and psychology, theory of science, communication and music psychology.

Larissa Laskovski is a physiotherapist and a university professor, with specialization in neurofuncional physiotherapy, both from the State University of Londrina, Brazil. In 2012, she concluded her master's dissertation at the University of São Paulo about the processes of disruptions in the courses of the life of patients who have suffered neurological injury and on expectations regarding physiotherapy treatment. Currently, she is doing her doctorate on physiotherapist's professional training, from their choice for the profession to the construction of the physiotherapist-patient relationship.

Giuseppina Marsico is assistant professor of Development and Education Psychology at the Department of Human, Philosophic and Education Sciences (DISUFF), at the University of Salerno, Italy. She has 15 years experience as a researcher, with a proven international research network. Her research track includes studies on developmental risk at school, youth deviance, school-family communication, boundaries, and contexts. She is a member of the editorial board of several international academic journals, such as *Culture & Psychology* (Sage) and *Integrative Psychology & Behavioral Science* (Springer), in addition to editing the book *Crossing Boundaries: Intercontextual Dynamics Between Family and School* (Information Age).

Himmbler Olivares is a doctoral student at the Pontificia Universidad Católica de Chile and the Europa-Universität Viadrina Frankfurt (Oder) in Germany. His research interests are centered on history of psychology, with special emphasis on German holistic psychology and its relationship with the Romantic ideas of the 19th century. His doctoral dissertation focuses on the microgenesis of body movements in communicative interaction and memory. He works at the Language, Interaction and Phenomenology Lab (LiF) in the Pontificia Universidad Católica de Chile. Address: Escuela de Psicología, Vicuña Mackenna 4860, 782-0436 Santiago, Chile. +56 2 2354 5883. E-mail: hgolivar@uc.cl

Vívian Volkmer Pontes, PhD, is clinical psychological and a university professor. Her main research interest concerns issues about the experiences of recurrent ruptures in the life course, such as the experience of women with recurrent pregnancy loss, and the semiotic strategies of dynamic repair of the self. She holds a master's (2009) and a PhD (2013) in psychology, program of postgraduation in psychology at the Federal University of Bahia (UFBA), Brazil. She recently joined the postdoctorate program at the University of São Paulo (USP).

Ruggero Andrisano Ruggieri is assistant professor in dynamic psychology and holds different positions at the University of Salento, including the post of lecturer in organizational psychology and clinical psychology. His research field is particularly the theory of models of mind according to psychological cultural perspectives where the mind conceptualizes a system of sense making. Applications of this theoretical model are addressed in *Passing the Baton of Family Business*, *Idiographic Science*, *Decision Making*,

Methodology of Psychology Intervention, Organization Studies and Community Psychology, Culture Analysis, Mind and Time.

Juliano Sampaio is associated professor at the Federal University of Tocantins (UFT), Brazil, where he teaches undergraduate courses for Arts-Theatre and Philosophy majors; he is also adjunct professor of the Graduate Program in Arts (MA, PhD), at the State University of São Paulo (UNESP), Brazil. He holds his master and PhD at the Institute of Psychology of the University of São Paulo (USP), Brazil, where he is presently a postdoctoral scholar. From 2008, his researches are linked to the research program of the Laboratory of Verbal Interaction and Knowledge Construction of the Institute of Psychology of the University of São Paulo, Brazil. They are conducted the frontiers between the performing arts, psychology and education, with the following topics of interest: Embodiment, Scenic Poetic Composition, Creation Narrative, Knowledge Building and Affective-Cognitive Human Development. He is a member member of the Brazilian Association for Research and Graduate Programs in Performing Arts and the Federation of Art Educators of Brazil. Email: casimiro@usp.br.

Lívia Mathias Simão, PhD, is associate professor at the Institute of Psychology of the University of São Paulo, Brazil, where she coordinates the Laboratory of Verbal Interaction and Knowledge Construction. She is also a sponsored researcher of the Brazilian National Council for Scientific and Technological Development. Her main research interests concern issues embracing the ontological construction of human subjectivity in I-Culture-Other relationships from the perspective of the semiotic-cultural constructivism in psychology, with emphasis in the issues regarding alterity and temporality. Among her more representative works are *Bildung, Culture and Self: A Possible Dialogue With Gadamer, Boesch and Valsiner?* (2005) in *Theory & Psychology*; *Otherness in Question: Labyrinths of the Self*, co-edited with Jaan Valsiner (2007); *The Other in the Self: A Triadic Unit* (2012) in *The Oxford Handbook of Culture and Psychology*; and *Ensaios Dialógicos: Compartilhamento e Diferença nas Relações Eu-Outro* (2010).

Henderikus Stam is professor of psychology at the University of Calgary in Calgary, Canada. His recent interests have focused on contemporary theoretical problems in psychology and the historical foundations of 20th century psychology on which he has published numerous articles and book chapters. He is the founding and current editor of *Theory & Psychology*, a bimonthly journal published by Sage (London). Hank is a Fellow of the American and Canadian Psychological Associations and a founding member and former President of the International Society for Theoretical Psychology; a former President of Division 24 of the APA (Society for

Theoretical and Philosophical Psychology). He has just been elected President of Division 26 of the APA (Society for the History of Psychology), a position he will take up in 2015.

Renato Tardivo, psychoanalyst and writer, lives in São Paulo, Brazil. PhD in social psychology at the University of São Paulo, works with relations among literature, cinema, and psychology. E-mail: rctardivo@uol.com.br Website: www.renatotardivo.com.br

Luca Tateo, PhD, is a Marie Curie Fellow at the Centre for Cultural Psychology, Aalborg University (Denmark). His research areas are the epistemology and history of psychology and the social psychology of education.

Dankert Vedeler is an Associate Professor Emeritus at the Norwegian University of Science and Technology (NTNU), Trondheim, Norway, where he earlier has coordinated an International Master's Course in Human Development. He studied philosophy in France before doing psychology in Sweden. His main interests are the development of infant intentionality and theoretical psychology. His research on infant intentionality is inspired by Merleau-Ponty and Alan Fogel. In theoretical psychology, he explores the contributions to the field of psychology of Gibsonian Ecological psychology, Dynamic Systems Approach, different theories of embodiment, and process oriented approaches to psychology, such as probabilistic epigenesis. In recent years the philosophical implications of theoretical psychology has come more and more into focus. Address: Department of Psychology, NTNU, NO-7491 Trondheim, Norway. E-mail: dankert.vedeler@ntnu.no

Meike Watzlawik is an honorary professor at the International School of Management (Hamburg) and at the Sigmund Freud University (Berlin). She is also still associated with the Technical University of Braunschweig (private lecturer, since 2008), and the University of Osnabrück, where she worked as visiting professor in the Department Development & Culture from 2011 to 2014. From 2008 to 2009, she was a visiting scholar at Clark University, Worcester, Massachusetts, working in the field of theoretical and cultural psychology. Her main research interests are (sexual) identity development and family relations (e.g., transnational relationships).

Printed and bound by CPI Group (UK) Ltd, Croydon, CR0 4YY
15/09/2026
14962371-0002